Handbook of Experimental Pharmacology

Volume 117

Springer
Berlin
Heidelberg
New York
Barcelona
Budapest
Hong Kong
London
Milan
Paris
Tokyo

Diuretics

Contributors

O.S. Better, A. Busch, J. Dørup, H. Endou, R. Greger
W.G. Guder, M. Hosoyamada, M. Hropot, B. Kaissling
T.R. Kleyman, H. Knauf, F. Lang, H.-J. Lang, W. Möhrke
E. Mutschler, T. Netzer, L.G. Palmer, J.B. Puschett
I. Rubinstein, M. Schmolke, F. Ullrich, K.J. Ullrich
H. Velázquez, J. Winaver

Editors

R.F. Greger, H. Knauf and E. Mutschler

Springer

Professor Dr. med. RAINER F. GREGER
Physiologisches Institut
der Albert-Ludwigs-Universität Freiburg
Hermann-Herder-Str. 7
D-79104 Freiburg, Germany

Professor Dr. med. H. KNAUF
St. Bernward Krankenhaus Hildesheim
Medizinische Klinik I
Gastroenterologie-Kardiologie-Nephrologie
Treibestr. 9
D-31134 Hildesheim, Germany

Professor Dr.rer.nat. Dr. med. E. MUTSCHLER
Pharmakologisches Institut für Naturwissenschaftler
Biozentrum Niederursel
Gebäude N 260, Marie-Curie-Str. 9
D-64271 Frankfurt, Germany

With 164 Figures and 32 Tables

ISBN-13:978-3-642-79567-1 e-ISBN-13:978-3-642-79565-7
DOI: 10.1007/978-3-642-79565-7

Library of Congress Cataloging-in-Publication Data. Diuretics/contributors, O.S. Better . . . [et al.]; editors, R.F. Greger, H. Knauf, and E. Mutschler. p. cm. – (Handbook of experimental pharmacology; v. 117) Includes bibliographical references and index. ISBN-13:978-3-642-79567-1 1. Diuretics. I. Better, O.S. II. Greger, Rainer. III. Knauf, H. IV. Mutschler, Ernst. V. Series. [DNLM: 1. Diuretics. W1 HA51L v. 117 1995/QV 160 D6171 1995] QP905.H3 vol. 117 [RM377] 615'.1 s – dc20 [615'.761] DNLM/DLC for Library of Congress 95-3116

Softcover reprint of the hardcover 1st edition 1995

Typesetting: Best-set Typesetter Ltd., Hong Kong

SPIN: 10478734 27/3136/SPS – 5 4 3 2 1 0 – Printed on acid-free paper

Preface

The first edition of this handbook appeared exactly twenty-five years ago. Due to enormous changes in the area of diuretics, the second edition has had to be completely revised. Substantial progress has been made in the functional anatomy of the kidney and in the concepts of how substances and ions are specifically transported across the various nephron segments. No one could have foreseen twenty-five years ago that the late 1980s and the early 1990s have provided us with methodologies to study transport events not only at the single cell level, but even at the level of the single transporter molecule. Many of the transporters for ions and organic substances have been cloned meanwhile by the new methods of molecular biology, and their function can be described more precisely by new transport studies such as the patch-clamp technique. These new insights have also led to a new understanding of how the currently used diuretics act. Just a few months ago, the Na^+Cl^- co-transporter, which is the target of thiazides, the $Na^+2Cl^-K^+$ co-transporter, which is the target of furosemide, and the amiloride sensitive Na^+ channel were cloned. Hence, the targets of diuretics have now been identified at the molecular level.

In addition, during the past twenty-five years extensive studies have been performed on the pharmacokinetics of diuretics. We have learned how changes in liver metabolism and altered renal excretion influence the pharmacology of this class of compounds. Most recent studies have also focused on the transport of diuretics in the proximal nephron and tell us more about the kidney selectivity of these substances and this specific aspect of pharmacokinetics.

Enormous clinical experience, together with our new understanding of the mode of action, has had a major impact on the usage of diuretics. For instance, high dosages of thiazides, which previously were used in the treatment of hypertension, are now regarded as unacceptable and much lower doses have been shown to be equally effective but do not induce comparable side effects.

Intentionally, a major focus of the second editon is on basic mechanisms: functional anatomy, the physiological and biochemical processes involved in kidney function, metabolism, and the transport of diuretics. Furthermore, the specific chapters dealing with the most frequently used groups of diuretics provide a comprehensive update, and the final chapter discusses the clinical use of diuretics as from this year's prospective.

We would like to thank Prof. H. Herken, the editor of the first edition, who helped us with the concept of this second edition, and we are very grateful to all the authors for their expert contributions. Finally, we would also like to thank the publisher for its most competent co-operation.

Freiburg R. Greger
Hildesheim H. Knauf
Frankfurt E. Mutschler
March 1995

List of Contributors

BETTER, O.S., Dr. Rebecca Chutick Crush Syndrome Center and Department of Physiology and Biophysics, Bruce Rappaport Faculty of Medicine, Technion-Israel Institute of Technology, POB 9649, Haifa 31096, Israel

BUSCH, A., Physiologisches Institut, Universität Tübingen, Gmelinstraße 5, D-72076 Tübingen, Germany

DØRUP, J., Department of Cell Biology, Institute of Anatomy, University of Aarhus, DK-8000 Aarhus, Denmark

ENDOU, H., Department of Pharmacology and Toxicology, Kyorin University School of Medicine, 6-20-2 Shinkawa, Mitaka, Tokyo 181, Japan

GREGER, R., Physiologisches Institut der Albert-Ludwigs-Universität, Hermann-Herder-Str. 7, D-79104 Freiburg, Germany

GUDER, W.G., Institut für Klinische Chemie, Städtisches Krankenhaus München-Bogenhausen, Englschalkinger Straße 77, D-81925 München, Germany

HOSOYAMADA, M., Department of Pharmacology and Toxicology, Kyorin University School of Medicine, 6-20-2 Shinkawa, Mitaka, Tokyo 181, Japan

HROPOT, M., Hoechst AG, Herz/Kreislauf Therapeutika, FB Pharmakologie, H 821, D-65926 Frankfurt, Germany

KAISSLING, B., Anatomisches Institut der Universität, Winterthurerstr. 190, CH-8057 Zürich, Switzerland

KLEYMAN, T.R., Department of Medicine, Renal-Electrolyte Division, University of Pennsylvania School of Medicine, 700 Clinical Research Building, 422 Curie Boulevard, Philadelphia, PA 19104, USA

KNAUF, H., St. Bernward Krankenhaus Hildesheim, Medizinische Klinik I, Gastroenterologie-Kardiologie-Nephrologie, Treibestr. 9, D-31134 Hildesheim, Germany

LANG, F., Physiologisches Institut, Universität Tübingen, Gmelinstraße 5, D-72076 Tübingen, Germany

LANG, H.-J., Hoechst AG, Herz/Kreislauf Therapeutika, FB Pharmakologie, H 821, D-65926 Frankfurt, Germany

MÖHRKE, W., Procter & Gamble Pharmaceuticals, Dr. Otto-Röhm-Str. 2-4, D-64331 Weiterstadt, Germany

MUTSCHLER, E., Pharmakologisches Institut für Naturwissenschaftler, Biozentrum Niederursel, Gebäude N260, Marie-Curie-Str. 9, D-60439 Frankfurt, Germany

NETZER, T., E. Merck, Klinische Forschung und Entwicklung Deutschland 1, D-64271 Darmstadt, Germany

PALMER, L.G., Department of Physiology and Biophysics, Cornell University Medical College, 1300 York Avenue, New York, NY 10021, USA

PUSCHETT, J.B., Department of Medicine SL 12, Tulane University Medical Center, 1430 Tulane Avenue, New Orleans, LA 70112-2699, USA

RUBINSTEIN, I., Dr. Rebecca Chutick Crush Syndrome Center and Department of Physiology and Biophysics, Bruce Rappaport Faculty of Medicine, Technion-Israel Institute of Technology, POB 9649, Haifa 31096, Israel

SCHMOLKE, M. Institut für Klinische Chemie, Städtisches Krankenhaus München-Bogenhausen, Englschalkinger Straße 77, D-81925 München, Germany

ULLRICH, F., Du Pont Pharma GmbH, Du Pont Straße 1, D-61352 Bad Homburg, Germany

ULLRICH, K.J., Max-Planck-Institut für Biophysik, Kennedyallee 70, D-60596 Frankfurt, Germany

VELÁZQUEZ, H., Research Office (151), Department of Veterans Affairs Medical Center, 950 Campbell Avenue, West Haven, CT 06516, USA

WINAVER, J., Dr. Rebecca Chutick Crush Syndrome Center and Department of Physiology and Biophysics, Bruce Rappaport Faculty of Medicine, Technion-Israel Institute of Technology, POB 9649, Haifa 31096, Israel

Contents

CHAPTER 1

CHAPTER 3

Renal Energy Metabolism

CHAPTER 1

Functional Anatomy of the Kidney

B. KAISSLING and J. DØRUP

A. Structural Organization of the Kidney

The kidney maintains the homeostasis of body fluids. This is accomplished by a complex process that involves, first, filtration of huge amounts of fluid and solutes from the blood across the wall of specialized capillaries of the glomerulus and, second, transepithelial transport of solutes and water along the tubular system connected to the glomerulus. In the tubular system solutes and water are reabsorbed into the systemic circulation and/or secreted into the tubular fluid. The waste products are excreted in a small fraction of the filtered fluid volume, generally in less than 1%, as the "final urine." In addition, the kidney is an endocrine organ, producing hormones acting at sites outside as well as within the kidney.

I. Microanatomy of the Kidney

1. Nephron

The smallest microanatomical units of the kidney are the nephrons. The nephron consists of (a) the renal corpuscle containing the glomerulus and (b) the tubule that originates from the renal corpuscle. The nephrons are drained by a collecting system that finally delivers the urine into the renal pelvis. Microanatomically the nephron is subdivided into the proximal and distal convoluted parts, connected by a straight portion that is folded into a loop, the loop of Henle. Histologically, the nephron consist of a series of tubular segments, each with structurally and functionally characteristic cell types (Fig. 1).

A single nephron, together with its collecting system, is not sufficient to either concentrate or dilute the primary filtrate. This is the result of the specific juxtaposition of several thousands of nephrons working in parallel, and of their histotopographical relationships to the vascular system. The relationship of tubular segments with each other and with the intrarenal vasculature establishes the characteristic form and architecture of mammalian kidneys.

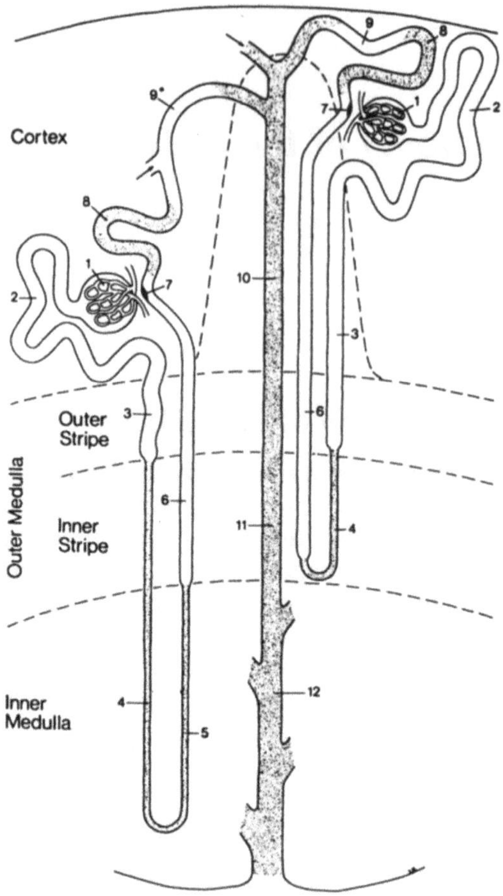

Fig. 1. Schematic representation of the segmentation of the tubular system in a superficial and deep nephron: *1*, glomerulus within the renal corpuscle; *2,3*, convoluted and straight part of the proximal tubule; *4,5*, intermediate tubule; *6*, thick ascending limb (TAL) of Henle's loop; *8*, distal convoluted tubule (DCT); *9*, connecting tubule (CNT); *10*, cortical; *11*, outer medullary; *12*, inner medullary collecting duct. (From KRIZ and BANKIR 1988, with permission)

→

Fig. 2A,B. Organization of the kidney and the renal vasculature. **A** Longitudinal section through a unipapillary kidney (rat); *C*, cortex; *OS*, outer stripe; *IS*, inner stripe; *IM*, inner medulla; *P*, papilla, belonging to the IM; *U*, ureter leaving the kidney at the hilum; the borders between the zones are indicated by *broken lines*; *arrowheads*, arcuate vessels, running along the corticomedullary border; *small arrows*, vascular bundles within the inner stripe; *long arrow*, cavity of the renal pelvis that is embedded in the connective tissue of the renal sinus; *bar* ≈1 mm. **B** Schematic representation of the intrarenal vasculature (not drawn to scale; modified from KRIZ and BANKIR 1988, with permission); abbreviations as in **A**; within the cortex the medullary rays (*MR*) are delineated from the cortical labyrinth (*CL*) by *dashed lines*; *1/1a*, interlobar artery and vein; *2/2a*, arcuate artery and vein; *3/3a*, cortical radial artery and vein; *4*, stellate vein; *5*, afferent arteriole; *6*, efferent arteriole; *7a,b,c*, superficial, midcortical and juxtamedullary glomerulus; *8/8a*, juxtamedullary efferent arteriole, descending vasa recta; *9/9a*, ascending vasa recta (within a vascular bundle and independent from a bundle, respectively); the *dotted lines* represent the capillary plexus, interposed between the arterial and venous vessels; *arrows and arrowheads*, direction of the blood flow (B adapted from KRIZ and BANKIR, with permission)

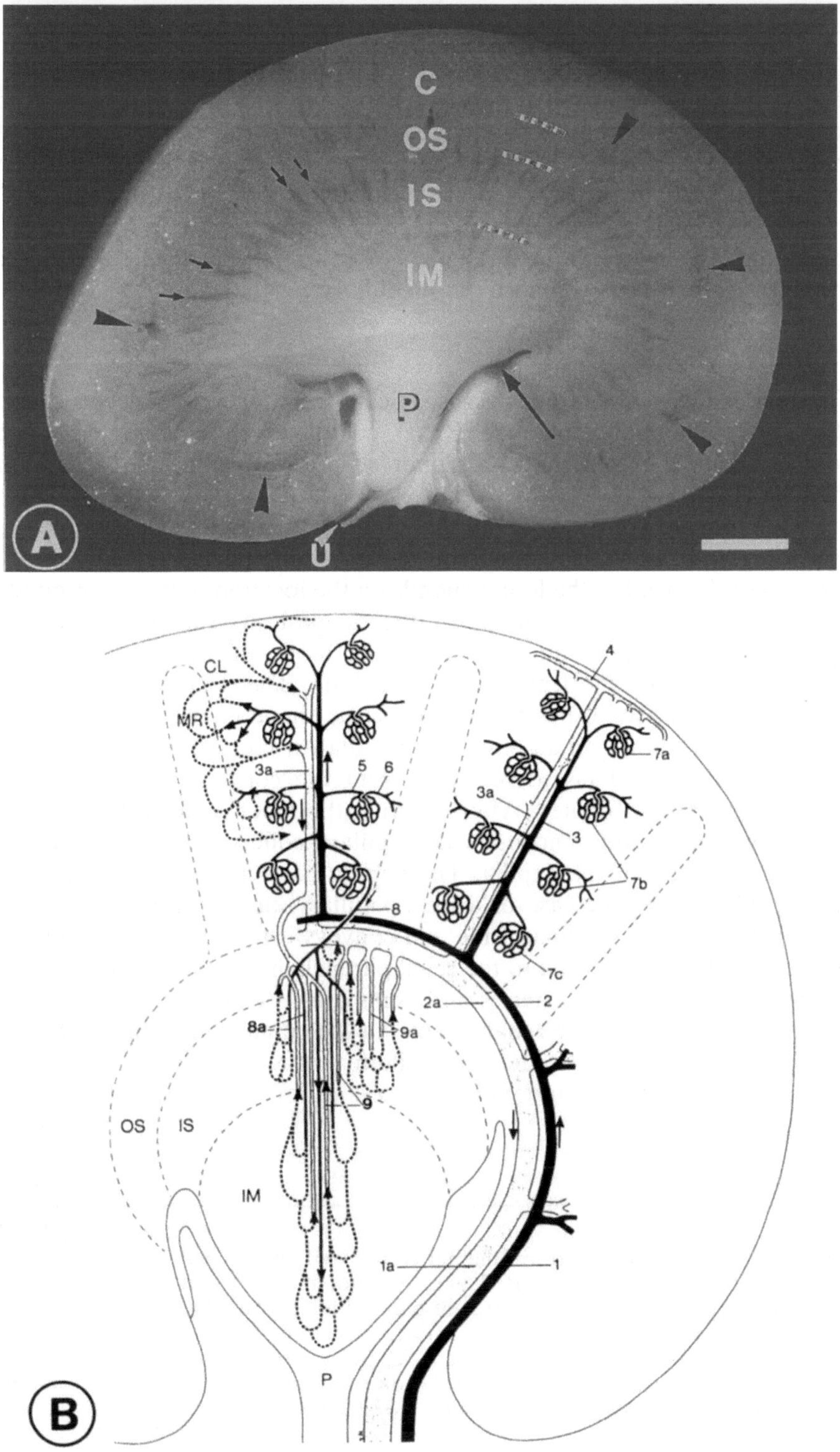
C
OS
IS
IM
P
U
A
CL
MR
3a
5
6
4
7a
3a
3
7b
8
7c
2a
2
8a
9a
9
OS
IS
IM
1a
1
P
B

2. Cortex

The basic architecture of a mammalian kidney is shown in longitudinal section through a unipapillary rodent kidney in Fig. 2A. The renal cortex covers the medulla and is subdivided into the cortical labyrinth and the medullary rays (Fig. 3A). Renal corpuscles and the proximal and distal convoluted parts of the nephron are localized in the cortical labyrinth. The cortical proximal and distal straight parts of the nephron, which are the initial and terminal portions of Henle's loop, together with the cortical part of the collecting ducts compose the medullary rays.

3. Medulla

The medulla displays three macroscopically recognizable concentric layers (Fig. 2A). The outer medulla (OM) comprises the outer stripe (OS), adjacent to the cortex, and the inner stripe (IS), adjacent to the inner medulla (IM).

The tubular constituents of the medulla are the medullary de- and ascending limbs of Henle's loop and the collecting ducts (Fig. 3B–D). The position of the bend of the loop depends on the location of the corresponding glomerulus in the cortex: loops of the deepest juxtamedullary nephrons penetrate the deepest into the medulla, loops of the most superficial nephrons penetrate the least into the medulla. The decrease in number of loops in the IM establishes the characteristic shape of the IM, which in all species tapers from a broad basis to a papilla (or crest in some species).

The collecting ducts, descending from the cortex, traverse the outer medulla as straight tubules. After entering the inner medulla the collecting ducts fuse successively and open as papillary collecting ducts (or ducts of Bellini) at the tip of the papilla. The papilla projects into the cavity of the renal pelvis (or, in the case of a multipapillary kidney, into a calyx of the pelvis) that collects the final urine and conveys it to the ureter. The wall of the renal pelvis is anchored to the renal parenchyma by the connective tissue following the intrarenal arteries and veins and is embedded, together with the larger branches of the renal vessels, in the connective tissue of the renal sinus. The latter opens at the renal hilum through which the renal vessels and the renal pelvis (or ureter) have access to, or leave, the kidney.

Fig. 3A–D. Overview of the histology of the renal zones: **A** Cortex; the *dashed line* delimits a medullary ray from the labyrinth; within the latter are situated the glomeruli (*G*) and the convoluted parts of proximal (*P*) and distal (*D*) tubules; the medullary rays comprise the straight parts of the proximal (*arrow downwards*) of the distal tubules (*TAL*, arrow upwards); *C*, collecting duct. **B** Outer stripe; symbols as in **A**; *asterisks*, ascending vasa recta. **C** Inner stripe. *VB* Vascular bundles. **D** Part of the papilla (inner zone); thin de- and thin ascending loop limbs cannot be distinguished at this magnification. *Bars*, **A,D** ≈200 μm; **B,C** ≈20 μm

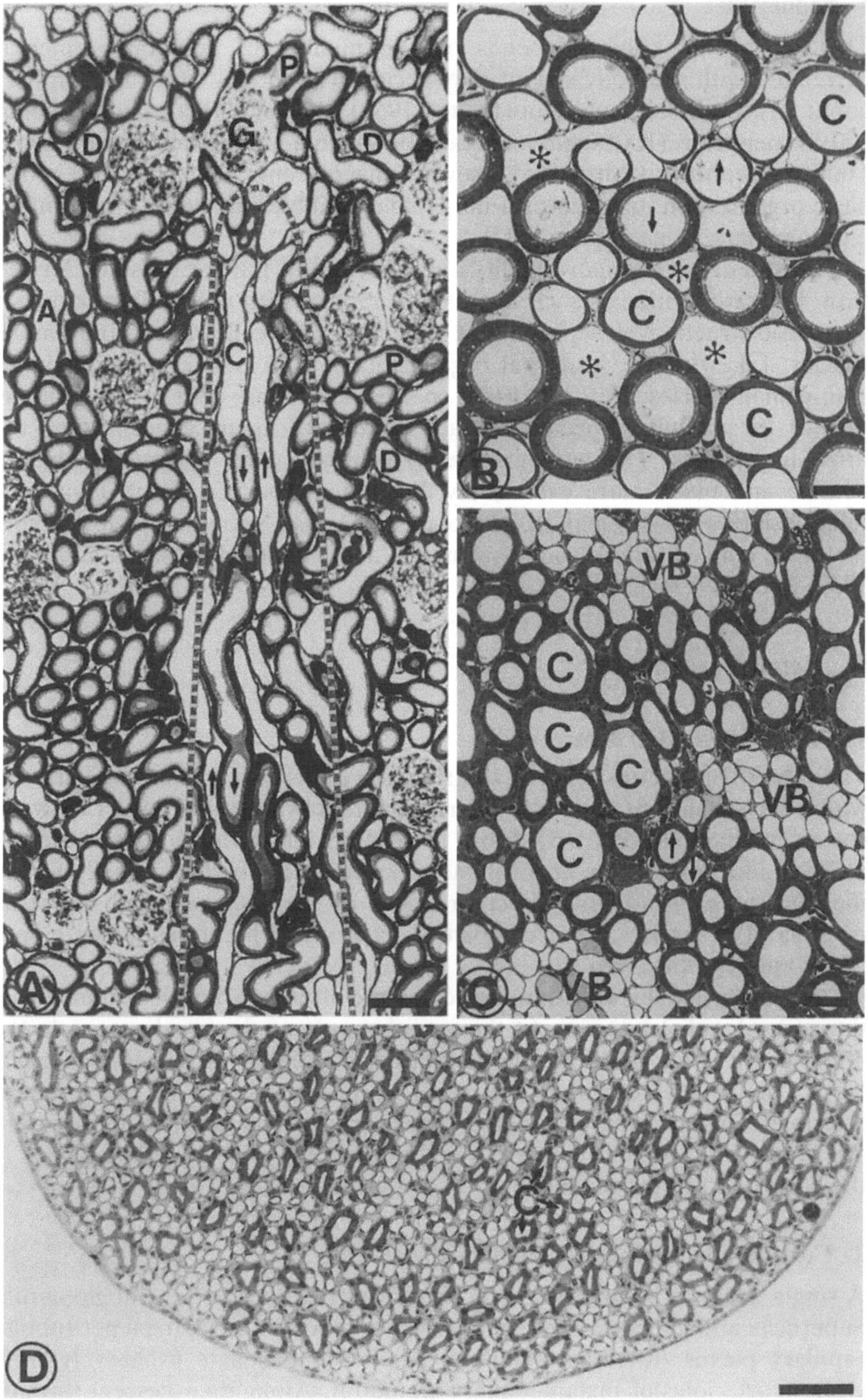
P
G
D
D
A
C
P
D
A
C
C
C
B
VB
C
C
C
VB
C
C
VB
D

4. Kidney Size

The size of the kidney depends on the number of nephrons, which is correlated with the body size of the species. The kidney of a rat comprises about 35 000 (BAINES and ROUFFIGNAC 1969), that of a rabbit kidney about 200 000 nephrons (KAISSLING and KRIZ 1979). In a human kidney with about 1 000 000 nephrons (SMITH 1951), several macroscopic units (reniculi) of the basic organization and size of a unipapillary rabbit kidney are fused into a multipapillary kidney (INKE 1987).

Most data on transport quality and structure of tubular cells are derived from laboratory animals such as rats and rabbits with unipapillary kidneys. The basic structure of nephron and collecting duct cells varies little among species. Likewise the structural organization of the cortex is similar in all mammalian species. Yet, important species differences exist in the architecture of the medulla. Species that are capable of highly concentrating the urine (e.g., mice, *Psammomys obesus*) generally have a larger medullary volume and much more extensively developed, specialized tubulovascular relationships (BANKIR and ROUFFIGNAC 1985) than species with low ability for urine concentration, such as humans and rabbits (KAISSLING and KRIZ 1979; PETER 1927).

II. Renal Vasculature

1. Arteries and Arterioles

The interlobar arteries arise by the successive divisions of the renal arteries within the renal sinus and enter the renal parenchyma at the border between cortex and medulla (Fig. 2B). From there, they follow this border as arcuate arteries and give off the cortical radial arteries that ascend radially within the cortical labyrinth; no direct branches of the arteries enter the renal medulla. The arteries are accompanied by veins with corresponding names (KRIZ and BANKIR 1988).

At successive levels within the cortex the afferent arterioles branch from the cortical radial arteries. At the vascular pole of the renal corpuscle the afferent arterioles split into the glomerular capillaries, arranged in parallel loops, that coalesce within the renal corpuscle usually to one efferent arteriole. This arteriole leaves the renal corpuscle adjacent to the afferent arteriole at the vascular pole and supplies the peritubular capillary plexus in the cortex and in the medulla.

2. Cortical and Medullary Capillary Plexus (Fig. 2B)

A single nephron is surrounded by capillaries perfused by several glomeruli. Superficial and midcortical glomeruli deliver blood to the cortical peritubular capillary plexus, whereas medullary capillaries originate exclusively from efferent arterioles of juxtamedullary glomeruli. Along their descent through the outer stripe these arterioles give off at first a few side branches for the scanty capillary plexus of the outer stripe before they divide into the descending vasa recta. In the inner stripe the descending vasa recta are grouped together with the ascending vasa recta from the inner zone in vascular bundles. The descending vasa recta at the periphery of the bundles spread out at successive levels in the inner stripe and form the dense peritubular interbundle capillary plexus. Only the descending vasa recta in the core of the bundles penetrate the inner zone and deliver blood to the loose peritubular capillary plexus of the inner medulla (KRIZ et al. 1976). Connections between the capillary plexus of the inner medulla with that of the inner stripe are exceedingly rare. In some species, e.g., *Psammomys obesus* (BANKIR et al. 1979), both capillary beds are completely disconnected from each other.

3. Venous Drainage of Cortex and Medulla (Fig. 2B)

The venous drainage of the cortex is effected by the cortical radial veins. It is noteworthy that the blood from capillaries in the medullary ray has to pass through the capillary plexus of the labyrinth before it gains access to the cortical radial veins (KIRZ and KAISSLING 1992). In some species (humans, dogs, cats) the superficial cortex is drained in addition by "stellate veins," which are connected to extrarenal veins in the renal capsule (FOURMAN and MOFFAT 1971).

The venous drainage of the three medullary regions differs. In the inner zone ascending vasa recta ascend between the tubules and join the vascular bundles, at the border between the inner zone and inner stripe at the latest. The blood from the inner zone then traverses the inner stripe within the "highway" of the bundles, in countercurrent arrangement with the descending vessels and without mixing with the blood of the capillary plexus in the inner stripe (BANKIR and ROUFFIGNAC 1985; BANKIR et al. 1989). The close juxtaposition of de- and ascending vasa recta in vascular bundles constitutes an essential feature of the urinary concentrating mechanism (JAMISON and KRIZ 1982). In some species (e.g., *Psammomys obesus*, *Meriones*) with a very high ability for urinary concentration the individual vascular bundles fuse within the inner stripe to a few "giant" bundles and all thin descending limbs of short loops are grouped between the vasa recta (KAISSLING et al. 1975). This countercurrent arrangement of the vessels and tubules seems to enhance considerably the efficiency for recycling of solutes and water (KRIZ 1981, 1983; BANKIR et al. 1989).

Most vessels draining the peritubular interbundle capillary plexus of the inner stripe ascend into the outer stripe without joining the vascular bundles, which break off within the outer stripe and the solute-laden vasa recta from the inner zone and inner stripe, coming into extensive contact with the tubules in that region. This arrangement offers another possibility for recycling of solutes back to the medulla via descending proximal tubular segments. All ascending vasa recta then empty into the arcuate veins or into the basal parts of the cortical radial veins (KIRZ and KAISSLING 1992) (Fig. 2B).

4. Wall Structure of Renal Vessels

The wall structure of arteries corresponds to that in other organs. The arterioles are surrounded by two or one layers of smooth muscle cells. A short distance before the afferent arteriole divides into the glomerular capillaries some of the smooth muscle cells are transformed to granular (epithelioid) renin-containing cells (see Sect. A. V). In efferent arterioles the smooth muscle cell layer disappears gradually before the vessel divides into the capillaries of the cortical peritubular plexus. The endothelium of juxtamedullary efferent arterioles is composed of a strikingly high number of longitudinally arranged cells (KRIZ et al. 1976). In the descending vasa recta the smooth muscle cells are gradually replaced by pericytes, which disappear with the transformation into capillaries.

All peritubular capillaries in the cortex and in the medulla, and the ascending vasa recta have a flat, densely fenestrated endothelium. Similarly, the wall of all intrarenal veins (cortical radial, arcuate) consists of only such a thin endothelium. The close spatial association and countercurrent arrangement of the thin-walled veins to the intrarenal arteries might favor a shunt diffusion of blood gases before blood enters the nutritional capillary plexuses. This may explain at least in part the comparatively high oxygen levels in the venous blood leaving the kidney and the low oxygen tension measured within the cortical regions (SCHUREK et al. 1990).

III. Interstitium, Lymphatics and Renal Nerves

The space between blood vessels and tubules is the "interstitium." All transports from tubules into vessels and vice versa pass through the interstititum. Thus, its composition may influence transport processes. Furthermore, it is involved to a large degree in immunological and inflammatory responses in the kidney. The fractional volume of interstitium is about 5%–7% in the cortex (KRIZ and NAPIWOTZKY 1979; PFALLER 1982; KAISSLING et al. 1993); it is even less within the outer stripe and within vascular bundles, but it increases continuously within the inner medulla towards the

Fig. 4A–C. Cortical renal interstitium: **A** Nerve fibers (*arrows*), extending along an afferent (*a*) and efferent (*e*) glomerular arteriole and between some tubules; *G*, glomerulus. **B** Cortical peritubular fibroblasts (*arrow*), immunostained for ecto-5′-nucleotidase; the brush border of proximal tubules (*P*) also shows a weak specific fluorescence. **C** Periarterial interstitium around a cortical radial artery (*rA*) in juxtaposition with the thin-walled radial vein (*rV*) and with connecting tubules (*CN*); *arrow*, lymphatic within the periarterial connective tissue sheath; *c*, capillaries. *Bars* ≈50 μm

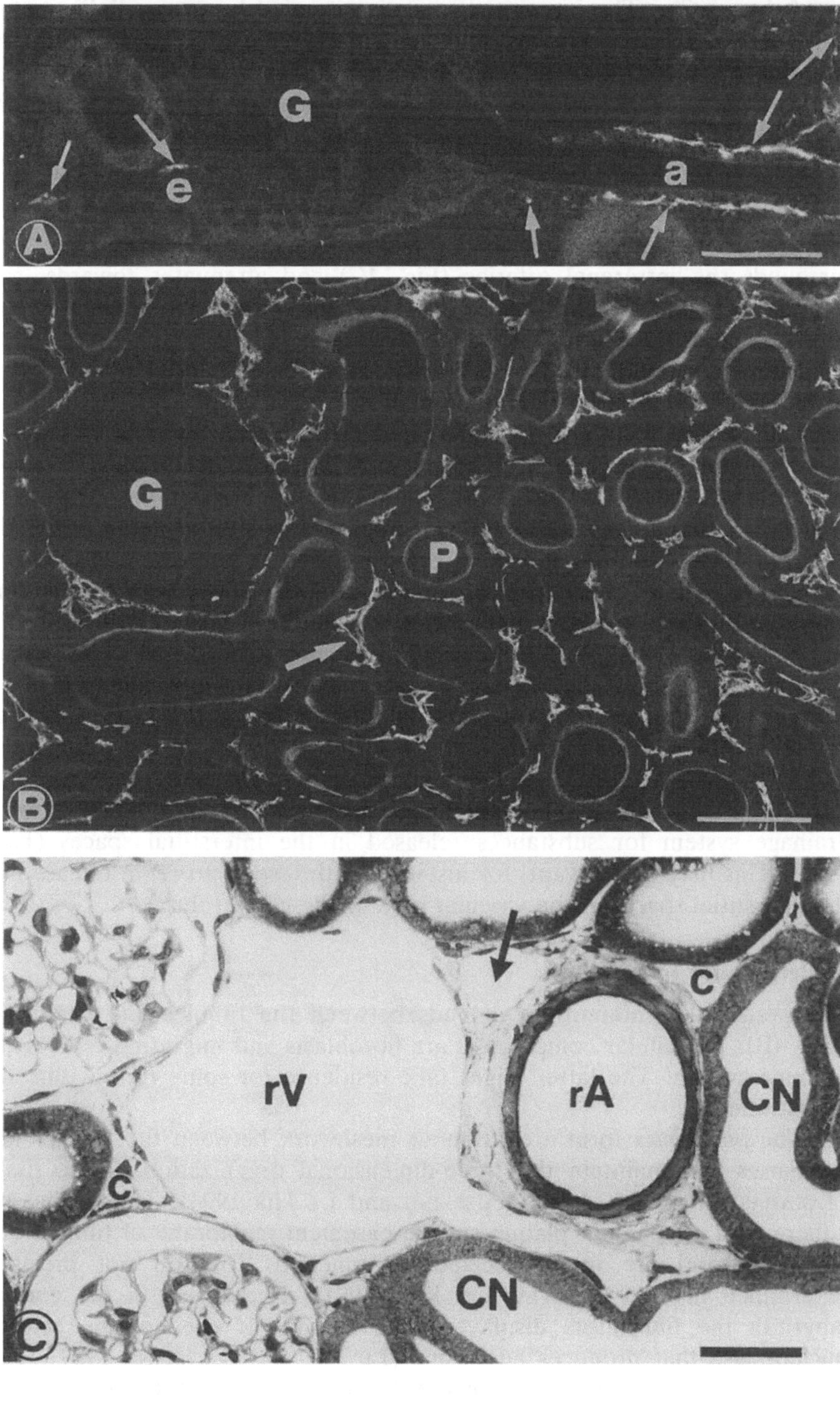
G
e
a
A
G
P
B
rV
rA
CN
c
c
CN
C

tip of the papilla. The interstitial space is occupied by extracellular matrix (proteoglycans, glycoproteins and interstitial fluid), collagenous fibers, microfibrils and different interstitial cell types (Kriz and Kaissling 1992). In the cortex the periarterial and the peritubular interstitium can be distinguished (Fig. 4).

1. Periarterial Interstitium

The periarterial interstitium is a layer of loose connective tissue that surrounds the intrarenal arteries (Fig. 4C) and attenuates towards, and terminates at, the vascular pole of the glomerulus (Kriz 1987). Renal nerves and lymphatics travel with this connective tissue sheath.

Efferent sympathetic nerves usually accompany arterial blood vessels as far as the vessels display smooth muscle cells (Fig. 4A). Pericytes in descending vasa recta are not innervated. The tubular innervation consists of occasional fibers (Fig. 4A) adjacent to perivascular tubules (Barajas 1971; Barajas and Powers 1990; Reinecke and Forssmann 1988). The majority of tubular portions have no direct relationship to nerve terminals. Afferent nerves from the kidney have been shown to be present in the renal capsule as well as in the connective tissue of the pelvic wall that is continuous with the connective tissue sheath around the arcuate arteries (Kriz and Kaissling 1992). Within the renal parenchyma they seem to be sparse.

Lymphatics drain the interstitial spaces. They start approximately in the region of the vascular pole of the glomeruli and leave the kidney through the renal hilum, running within the connective tissue sheath along intra- and extrarenal arteries. No lymphatics ascend from the renal medulla. Periarterial interstitium and lymphatics are regarded as a hilum-directed drainage system for substances released in the interstitial spaces (Kriz 1987). This may be relevant, for instance, with respect to renin release into the interstitial spaces at the vascular pole of the glomerulus.

2. Peritubular Interstitium

The peritubular interstitium extends between the tubules and capillaries (Fig. 4B). Its cellular components are fibroblasts and migrating cells of the immune system. The latter might take residence for some time within the interstitium.

The *fibroblasts* form a continuous meshwork between the tubules and capillaries and maintain the three-dimensional organization of the tissue (Takahashi-Iwanaga 1991; Kaissling and Le Hir 1994). Their processes adhere with attachment plaques to the basement membrane of tubules and capillaries and are interconnected by junctions (of still not precisely determined nature) (Lemley and Kriz 1991). Exclusively in the cortical labyrinth the fibroblasts display a high activity of the enzyme ecto-5′-nucleotidase that produces adenosine (Le Hir and Kaissling 1989, 1993; Kaissling et al. 1993). All known extratubular targets for adenosine are located in the close vicinity of these adenosine-producing cells: smooth muscle cells of renal arterioles, granular renin-producing cells and renal nerves. Recently it has been demonstrated that renal ecto-5′-nucleotidase-positive fibroblasts are the source of renal erythropoietin (BACHMANN et al. 1993; MAXWELL et al. 1993).

The fibroblasts in the inner medulla are called "lipid-laden interstitial cells" because of their content of lipid droplets within the cytoplasm. The number of lipid droplets varies tremendously from species to species, among individuals and from cell to cell (KRIZ and KAISSLING 1992). The profiles of these cells resemble ladder rungs, interposed between the longitudinally running loop limbs and vessels.

Bone marrow-derived "free" cells, in particular dendritic antigen-presenting cells (with high expression of major histocompatibility complex class II antigen) in the peritubular space, reveal a particularly close spatial association with fibroblasts (KAISSLING and LE HIR 1994). In healthy individuals they are rarely found within glomeruli. More or less numerous lymphocytes may also be present within the peritubular interstitium, whereas in healthy individuals the occurrence of classical macrophages is restricted to the periarterial tissue around large arteries and the connective tissue of the pelvic wall and of the renal capsule (KAISSLING and LE HIR 1994).

IV. Renal Corpuscle

1. Organization (Fig. 5A,B)

The renal corpuscles enclose the filtration units of the kidney, the glomeruli. In mice renal corpuscles measure about 100 µm, in humans about 200 µm and in elephants about 300 µm in diameter (PETER 1927; SMITH, 1951). Within a kidney, the diameter of juxtamedullary renal corpuscles (PETER 1927; SMITH 1951; SANDS et al. 1992) and their single glomerular filtration rates (SNGFRs) may exceed those of midcortical and superficial ones by up to 50% (SANDS et al. 1992).

The filtration occurs through the wall of the specialized glomerular capillaries that are interposed between the afferent and the efferent arteriole. This arrangement provides the possibility for effective control of flow rate and filtration pressure in the capillary loops by regulating the diameter of either one or both arterioles. Both arterioles are innervated (GORGAS 1978; BARAJAS and POWERS 1990) and respond to a great variety of vasoactive substances (DWORKIN and BRENNER 1992; SANDS et al. 1992).

Each intraglomerular primary capillary loop gives rise to a glomerular lobule the axis of which is formed by the mesangium. Capillaries and mesangium together form the glomerular tuft. The glomerular tuft is covered by epithelial cells that constitute the visceral epithelium of Bowman's capsule. At the vascular pole occurs the transition of the visceral epithelium into the parietal epithelium of Bowman's capsule consisting of flat squamous epithelial cells. The space between both epithelial layers is the urinary space, which at the urinary pole passes over into the tubular lumen. At the urinary pole the flat parietal epithelial cells are continuous with the epithelium of the proximal tubule.

2. Mesangium (Fig. 5C)

The mesangium consists of mesangial cells, surrounded by the mesangial matrix, and occasionally comprises a few bone marrow-derived cells. The cytoplasmic processes of mesangial cells are attached, either directly or by the interposition of microfilaments, to the glomerular basement membrane (GBM), which covers the mesangium and the capillaries (KRIZ and KAISSLING 1992). Mesangial cells have numerous contractile filaments and are able to ingest exogenous particulate tracers as well as macromolecules such as immune complexes. It has been suggested that the contractility of the mesangial cells maintains the structural integrity of the glomerular tuft against the high pressure inside both the glomerular capillaries and the glomerular mesangium (KRIZ et al. 1990; ELGER et al. 1990).

3. Glomerular Capillaries (Fig. 5D)

The glomerular capillary wall is composed of thin endothelial cells, supported by a thick basement membrane. The latter is covered by epithelial cells, the "podocytes."

A small fraction of the circumference of the endothelial tube of glomerular capillaries is fixed to the mesangium. The peripheral filtration areas of glomerular endothelial cells are extremely attenuated and perforated by densely arranged round to oval pores 50–100 nm in diameter (LARSSON and MAUNSBACH 1980). In adults these pores lack diaphragms (unlike in peritubular capillaries). The luminal membrane of endothelial cells is highly negatively charged due to several polyanionic glycoproteins, including the sialoprotein "podocalyxin," within the cell coat (KERJASCHKI et al. 1984). Podocalyxin is considered to be the major surface polyanion of glomerular

Fig. 5A–F. Renal corpuscle and juxtaglomerular apparatus: *AA*, *EA*, afferent and efferent arterioles; *GC*, granular cells; *MD*, macula densa; *EGM*, *M*, extra- and intraglomerular, respectively, mesangium; *E*, capillary endothelium; *PO*, podocyte; *US*, urinary space; *PE*, parietal layer of Bowman's capsule; *P*, proximal tubule cell. **A** Schematic drawing showing the organization of a renal corpuscle. **B** Section through a rat renal corpuscle in the same orientation as **A**, 1 μm Epon section. **C** Intra- and extraglomerular mesangium in a mouse renal corpuscle, evident by the activity of ecto-5′-nucleotidase; cryostat section. **D** Wall of a glomerular capillary, displaying the components of the filtration barrier; *arrow*, slit membrane; **E,F**, Rat afferent arteriole with granular, renin-containing cells. **E** Immunofluorescence with a-renin. **F** Cross-section through an afferent arteriole; *arrows*, nerve fibers, running along the vessel. *Bars*, **B,C** ≈50 μm; **D** ≈0.1 μm; **E,F** ≈10 μm (A from KRIZ and KAISSLING 1992, with permission)

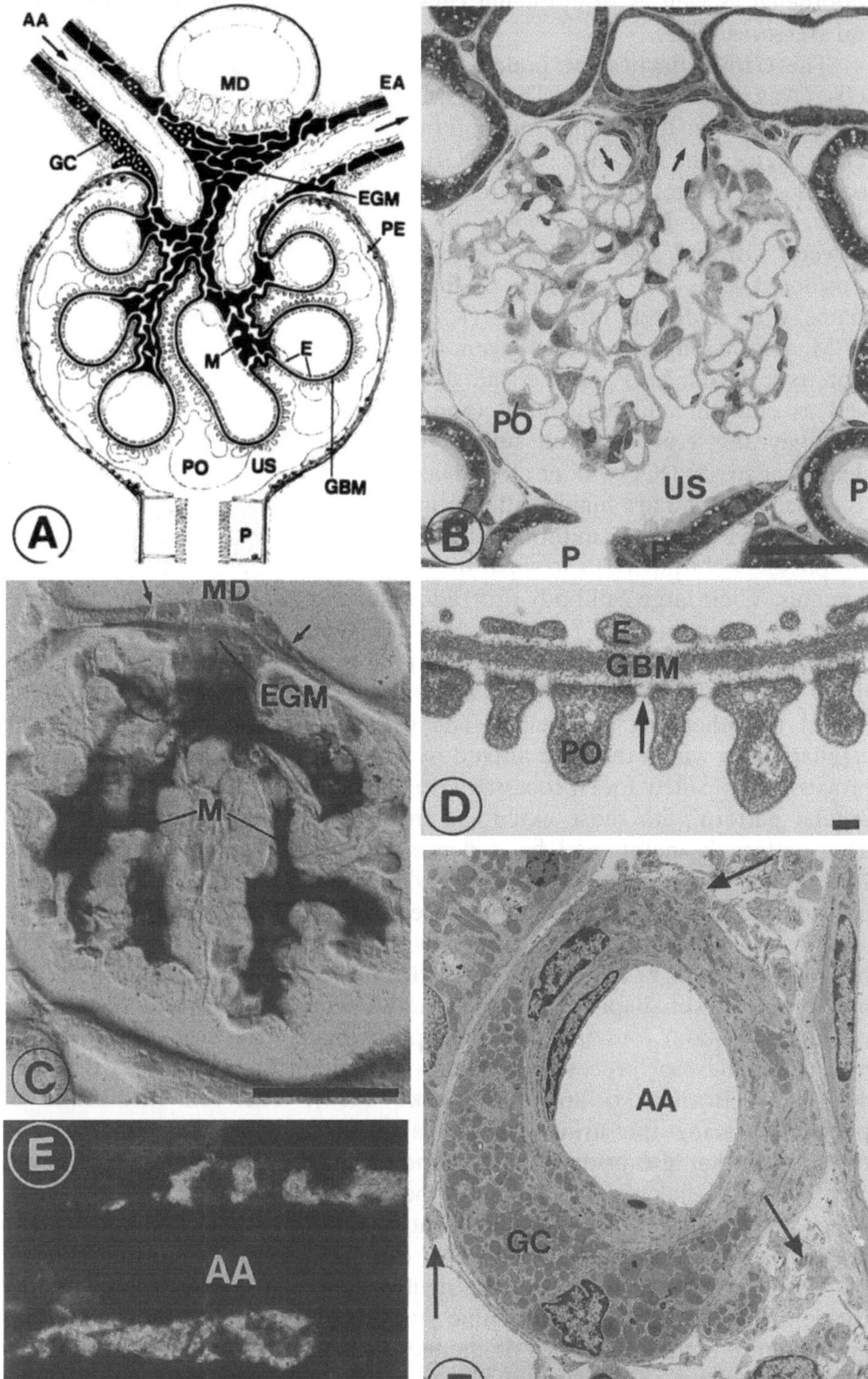
AA
MD
EA
GC
EGM
PE
M
E
GBM
PO
US
P
A
B
C
D
E
F

endothelial as well as of epithelial cells (KERJASCHKI et al. 1984; DWORKIN and BRENNER 1992).

The *GBM* attaches the podocytes to the endothelial and to mesangial cells (ABRAHAMSON 1987). The GBM displays three layers of different electron density: the lamina rara interna, the lamina densa and the lamina rara externa. The thickness of the GBM varies among species and usually increases with age: in humans the width ranges between 240 and 370 nm (JORGENSEN and BENTZON 1968), in rat and other experimental animals between 110 and 190 nm (KRIZ and KAISSLING 1992). The GBM can be considered as a gel, consisting of highly cross-linked meshes of polymer chains with hydrated interstices (KARNOVSKY 1979). Components of the GBM are, among others, collagenous peptides (mainly type 4 collagen, some type 5 collagen) and the sialoglycoproteins laminin and fibronectin as well as entactin and amyloid P (KRIZ and KAISSLING 1992). Polyanionic proteoglycans, e.g., heparan sulfate, confer to the GBM a considerable electronegativity (KANWAR et al. 1984a,b).

The *podocytes* are highly differentiated epithelial cells that, in the adult, seem to have lost their mitotic ability. Unlike other polarized cells they do not possess tight junctions to separate apical from basolateral cell membrane domains. Their large cell body protrudes into the urinary space and contains all major cell organelles (among which the Golgi apparatus is particularly conspicuous). The perikaryon gives off long primary cytoplasmic processes, richly equipped with microtubules and intermediate filaments but containing almost no other cell organelles. The primary cell processes encircle the capillaries, to which they are affixed by numerous interdigitating secondary processes, so-called foot processes (pedicles). The latter interdigitate in a regular pattern with those extending from neighboring primary processes. This pattern is maintained by a dense network of actin filaments at the sole of the processes (ELGER et al. 1993; ERENCKHAHN et al. 1990). The meandering slits between the foot processes are called filtration slits. These are most narrow at their floor, where they range between 25 and 65 nm. The filtration slits are bridged by a 4-nm-thick extracellular structure, the slit membrane or slit diaphragm, which shows regularly arranged rectangular openings of about four times 40 nm, approximately the size of an albumin molecule. The foot processes are anchored within the GBM to a depth of 40–60 nm (RODEWALD and KRANOVSKY 1974). The cell membrane of podocytes facing the urinary space is covered by a thick surface coat (glycocalix) that also bridges the slits and covers the slit diaphragm. The cell coat is rich in sialoproteins, mainly podocalyxin (KERJASCHKI et al. 1984), which are responsible for the high negative surface charge of podocytes and are crucial in maintaining the shape of the foot processes and the width of the filtration slits. In response to a neutralization of this surface charge by cationic substances (e.g., protamine sulfate, poly-L-lysine) the glomerular epithelium undergoes a series of changes including flattening and retraction of foot processes, narrowing of filtration slits and formation of tight junctions between adjacent foot processes (Kurihara et al. 1992; Andrews 1988).

4. Filtration Barrier (Fig. 5D)

The filtration of the primary urine proceeds via (a) the endothelium with large open pores (partly filled by glycocalix), (b) the pericapillary portion of the GBM, and (c) the slits between the podocyte foot processes, bridged by the slit membrane and filled by glycocalix material. All aspects of the capillary wall contribute to the two outstanding characteristics of the barrier: (a) the high permeability for small solutes and water and (b) the very low permeability for plasma proteins and other macromolecules (for refs. see Dworkin and Brenner 1992).

The high permeability for water and small solutes can be explained by the fact that filtration obviously occurs along an extracellular route through the glomerular capillary. The barrier function for macromolecules is selective for size, shape and charge. The size and shape selectivity of the filtration barrier for uncharged macromolecules is established by the dense network of the GBM (Dworkin and Brenner 1992). Such molecules with an effective radius of up to about 2 nm freely pass through the filter. Larger compounds are increasingly more restricted and are totally restricted at effective radii of more than 4 nm.

The charge selectivity is rooted in the dense accumulation of negatively charged molecules throughout the entire depth of the filtration barrier, including the surface coats of endothelial and epithelial cells and all three layers of the GBM. Negatively charged molecules are more effectively restricted than neutral or positively charged molecules of the same size. Albumin molecules with an effective radius of 3.6 nm are restricted by the filter due to their negative charges. Polyanionic macromolecules in the plasma, such as plasma proteins, are repelled by the assemblies of negative charges of the filter, thereby establishing an electronegative shield and minimizing the clogging of the filter. For positively charged molecules the passage across the filter is facilitated. It has been suggested that for positively charged molecules the slit diaphragm exerts the most effective barrier (Dworkin and Brenner 1992).

V. Juxtaglomerular Apparatus

The vascular, glomerular and tubular structures at the vascular pole of the glomerulus (Fig. 5A) constitute a functional unit, the "juxtaglomerular apparatus" (JGA), that controls the glomerular filtration rate of the nephron in response to the salt load present in the tubular fluid at the end of the loop of Henle (Schnermann and Briggs 1992; Briggs and Schnermann 1987). The components of the JGA are (a) the terminal and initial portion of the afferent and efferent arteriole, respectively, (b) the extraglomerular mesangium and (c) the specialized tubular cells of the macular densa, within the epithelium of the end portion of the thick ascending limb.

1. Granular Cells

In the immediate vicinity of the vascular pole the smooth muscle cells of the afferent, occasionally also of the efferent, arteriole produce renin which they store in large specific granules; therefore these transformed smooth muscle cells are called "granular cells" (Fig. 5E,F). Renin synthesis and release are controlled, among others, by renal nerves (KURTZ 1989; HOLMER et al. 1994) that have abundant terminals at the vascular pole (GORGAS 1978). Other factors are adrenal hormones and sodium chloride transport across the macula densa cells (MODENA et al. 1993). In situations that chronically require enhanced renin synthesis additional smooth muscle cells further upstream in the wall of afferent arterioles are recruited and transform into granular cells (see KRIZ and KAISSLING 1992). Renin is released by exocytosis into the surrounding interstitium and may then be distributed retrogradely via lymphatics along arterioles and arteries (KRIZ 1987).

2. Extraglomerular Mesangium

Extraglomerular mesangial cells (EGM cells, formerly termed Goormaghtigh or lacis cells) are surrounded by a conspicuous matrix and form a solid cell complex between the two glomerular arterioles and the macula densa. The EGM is not penetrated by any blood or lymphatic vessels (KRIZ and KAISSLING 1992). Nerves pass on both sides from the afferent to the efferent arteriole but do not enter the cell complex. The EGM cells extend into the stalk of the glomerular tuft (Fig. 5A,C), thus mixing with intraglomerular mesangial cells (SPANIDIS et al. 1982). The EGM cells are coupled to each other as well as to intraglomerular mesangial cells and smooth muscle cells in the glomerular arterioles by extensive gap junctions (GORGAS 1978; TAUGNER et al. 1978; KRIZ and KAISSLING 1992). Because of their central position within the JGA and their constant contact with the macula densa, the EGM cells have repeatedly been considered as a necessary functional link between the macular densa and any possible effector cell within the feedback mechanism of the JGA (GORGAS 1978; TAUGNER et al. 1978; CHRISTENSEN and BOHLE 1978; ELGER et al. 1990).

3. Macula Densa

Shortly before its end the thick ascending limb (TAL) passes between the two arterioles of its original renal corpuscle and contacts the EGM (Fig. 5). At that side of the tubule the epithelium forms a plaque with densely arranged nuclei (Fig. 12); hence, this plaque is called the macula densa (MD). The macula densa may cover in addition to the EGM variable portions of the afferent and/or efferent arterioles (Fig. 12). In contrast to all other tubular portions the base of the MD cells lacks any contact to peritubular capillaries. All solutes transported across this cell plaque will reach at first the extracellular matrix that surrounds the EGM cells and connects all components of the JGA. Although the detailed mechanisms of the feedback control of the GFR via the JGA are still under investigation, it seems to be firmly established that the NaCl transport across the macula densa cells plays a crucial role (BRIGGS and SCHNERMANN 1987; SCHLATTER et al. 1989). The structure of MD cells will be described in Chap. 7, this volume.

B. General Organization of Renal Epithelia and Correlation with Transport

I. Polarity of Epithelia

The single-layered epithelia of the renal tubules are interposed between two extracellular compartments with different physicochemical compositions, the luminal compartment, containing the tubular fluid, and the interstitial compartment (Fig. 6). A prerequisite for vectorial transepithelial transport is the asymmetric distribution of transport proteins between the luminal and basolateral membrane domains, respectively, and the joining of the individual cells to a tight, continuous layer by tight junctions (MATLIN and CAPLAN 1992).

1. Transport Pathways

Transport across the epithelium can proceed either through the cells (transcellular pathway) or between the cells via the tight junctional belt (paracellular pathway). Both transport routes interfere with each other in the lateral intercellular spaces.

Transepithelial movement of solutes via the *transcellular pathway* necessitates the passage of a given solute through at least two cellular membrane domains with different functional qualities. For instance, an Na^+ channel in the luminal membrane allows Na^+ to move "passively" into the cell, following the concentration gradient down. The extrusion of Na^+ across the basolateral membrane has to proceed "actively," against an electrochemical gradient under consumption of energy. The enzyme (Na^++K^+)-ATPase, which promotes the energetically unfavorable transport of (three) sodium ions out of, and (two) potassium ions into, the cell, is localized exclusively in the (baso)lateral membrane (ERNST 1975; KASHGARIAN et al. 1985; KOOB et al. 1988) of renal cells. In renal epithelia most ATP is produced by oxidative phosphorylation. Therefore renal tubules with high active transport rates display large amounts of mitochon-

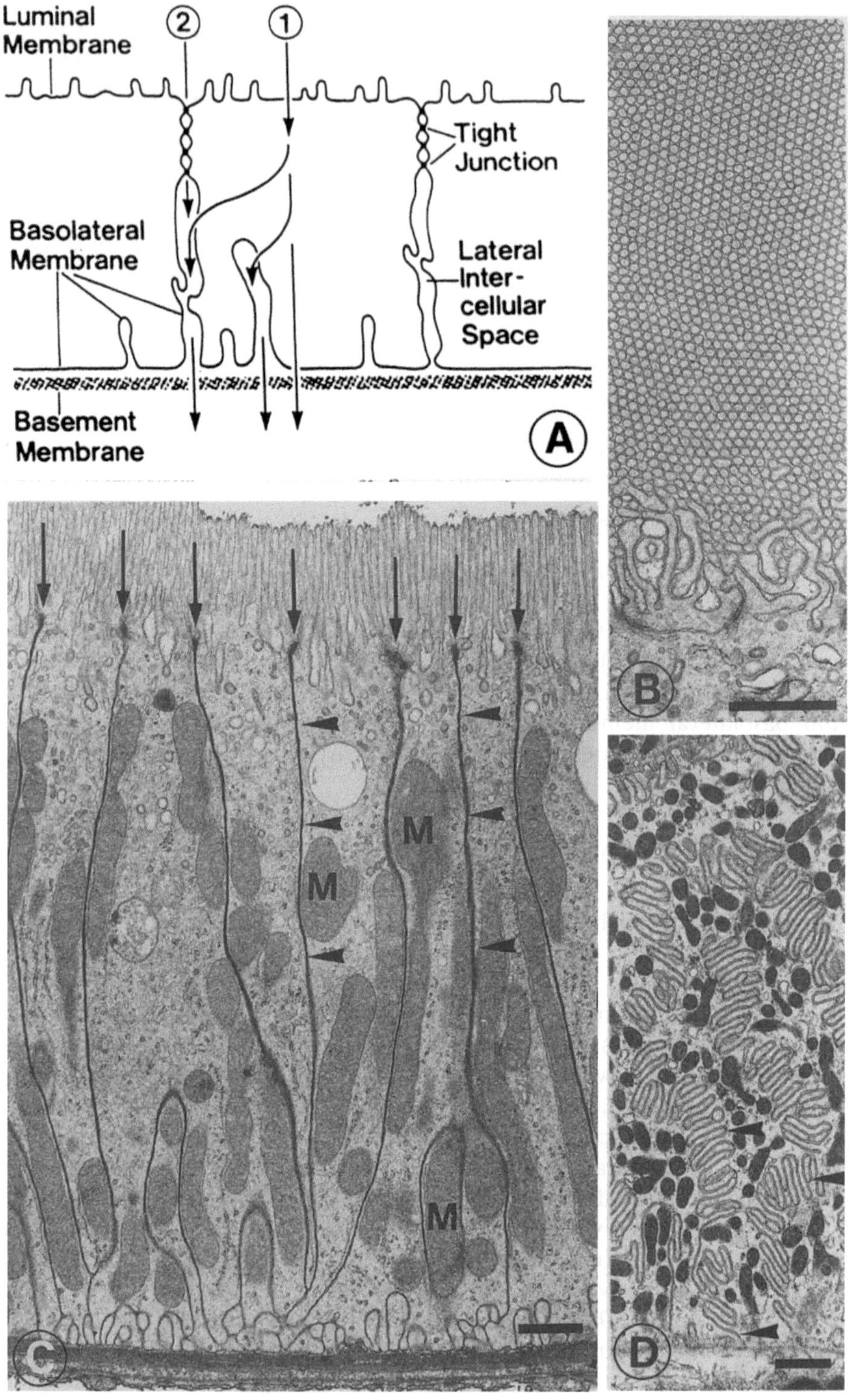
Luminal Membrane
2
1
Tight Junction
Basolateral Membrane
Lateral Inter-cellular Space
Basement Membrane
A
B
M
M
M
C
D

dria. In distal, but not in proximal, cell types glycolysis contributes significantly to energy production (GUDER et al. 1986).

The *paracellular pathway* consists of the tight junction and the lateral intercellular space. Solute movements along the paracellular pathway are passive in nature, but generally are linked directly or indirectly to the transcellular solute movements. The tight junction (or zonula occludens) encircles the cell like a belt. It topographically delimits the luminal from the basolateral cell membrane domains and inhibits the mobility of outer leaflet-lipids from the one domain to the other (MATLIN and CAPLAN 1992). In the tight junctional area the outer leaflets of the neighboring cell membranes are sealed together by so-called strands, apparent by freeze-fracturing of the membranes in the zones of contacts (MATLIN and CAPLAN 1992). The density of particles that compose the strands, and the number and complexity of the strands are the rough structural correlate of the transjunctional resistance. The functional properties of the zonula occludens, such as differences in the permeability for specific cations and anions, are explained by fixed negative charges and probably by the protein composition in the junctional zone. The function of the ZO-1 protein from tight junction membrane fractions seems to be related to the mechanical tightness of the junction (KURIHARA et al. 1992; MATLIN and CAPLAN 1992).

In addition to the tight junction the intermediate junction (zonula adherens) forms a belt which attaches the cells mechanically together. Patches of desmosomes (maculae adherentes) may also be present.

II. Organization of Epithelial Surfaces

According to the different functional requirements the basolateral and apical membrane domains of epithelia are morphologically differently organized.

1. Basolateral Domain

In most renal tubular cells with high transport rates the basolateral cell surface is much greater than the projection area of the cells on the tubular

Fig. 6A–D. Organization of electrolyte-transporting epithelia: **A** Schematic representation of (*1*) the transcellular and (*2*) the paracellular pathway route. **B** Cross-section across a brush border; the individual microvilli are all of similar diameter. **C** Section through the epithelium of a proximal tubule, displaying the apical membrane domain, which is increased by a brush border, formed of microvilli, all of similar length; the basolateral membrane domain (*arrowheads*) borders the intercellular space, filled with an electron-dense tracer substance that stops at the level of the tight junctions (*long arrows*); the basolateral area is increased by lateral plate-like processes, which interdigitate with those of neighboring cells and that enclose large mitochondria (*M*). **D** Basal part of a collecting duct cell with basal membrane infoldings (*arrowheads*) that are *not* adjacent to the paracellular pathway. *Bars* ≈1 μm (A and C adapted from KAISSLING and STANTON 1992, with permission)

basement membrane. Basically, the augmentation of membrane area results from two different modes: formation of lateral cell processes and/or infoldings of the membrane into the cell body.

In some cell types (S1 and S2 of proximal tubule, thick ascending limb cells, distal convoluted tubule cells) almost exclusively the surface of the *lateral* cell membrane is amplified by the formation of large, lamella-like lateral *cell processes* which interdigitate intimately with those of neighboring cells (TISHER and MADSEN 1987; WELLING et al. 1977, 1978; WELLING and WELLING 1975, 1976). The intercellular space confined between the lateral cell membranes of adjacent cell processes is of uniform narrow width (≈20 nm). The lateral intercellular spaces converge towards the tight junctional belt. This type of cellular organization suggests extensive interference between the trans- and paracellular transport routes (Fig. 6C).

In other cell types (collecting duct cells, S3 of proximal tubules, macula densa cells) the lateral cell surfaces are only moderately increased by formation of small lateral folds and microvilli (WELLING et al. 1981, 1983). In such cell types true *membrane infoldings* within the basal cell portion may substantially contribute to the amplification of the basolateral surface (Fig. 6D). The lateral microprojections of the neighboring cells are often connected by numerous small desmosomes. In noninterdigitating epithelia both the intercellular spaces and the spaces between the basally infolded membranes vary considerably in width. Dilated intercellular spaces in such epithelia are generally considered to be the structural evidence for bulk water flow (GANOTE et al. 1968; WOODHALL and TISHER 1973).

2. Apical Domain

The quality and the quantity of transport proteins in the luminal membrane, which allow movement of a given solute into the cell, establish the specific permeability properties of the membrane and, thus, the transport characteristics of individual cells and epithelia. Three modes of structural amplification of the luminal cell membrane are apparent in renal cells:

a) The *brush border* consists of closely arranged microvilli of uniform dimensions that homogeneously cover the luminal cell pole. In renal epithelia it exists exclusively on proximal tubule cells. The brush border membrane displays, in addition to numerous specific transport proteins and channels for inorganic and organic solutes, a great variety of enzymes (including peptidases, phosphatases, glycosidases, nucleotidases) involved in the cleavage of organic compounds into units that can be transported across the membrane (Fig. 6B).

b) In all other renal epithelial cells the luminal surface carries short *microvilli*. Their shape, quantity and distribution is variable (Figs. 7A, 11B).

c) *Microfolds*, occasionally intermingled with microvilli, seem to be associated with rapid modulation of the luminal surface area, occurring with specific membrane recycling processes, for instance, in intercalated cells and collecting duct cells (Fig. 7A,B).

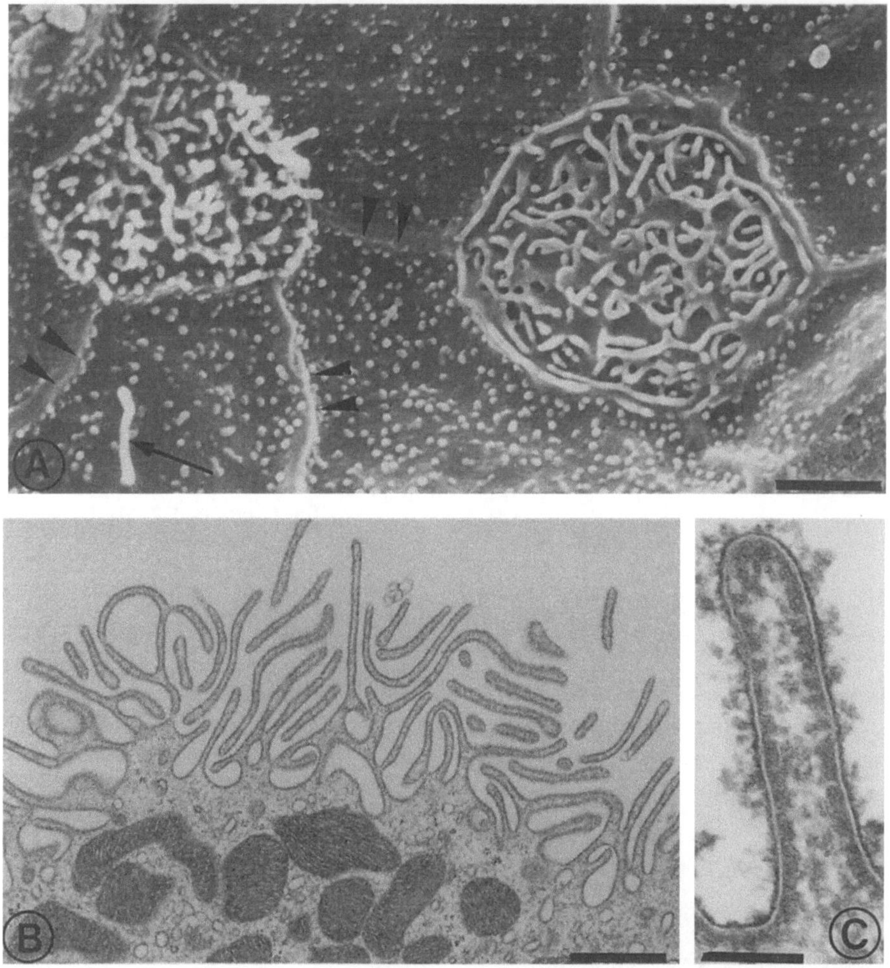

Fig. 7A–C. Luminal surface specializations: **A** Surface of a cortical collecting duct; one intercalated cell with a tuft of microvilli (*left*) and one with complex microfold formation (*right*); the surface of the collecting duct cell is covered with short microvilli; *arrow*, cilium on a collecting duct cell; *arrowheads*, cell borders; **B** Apical cell pole of a type A cell (rat) with extensive microfolds; **C** the cytoplasmic face of the microfold membrane shows a characteristic coat of "studs"; the studs are associated with the electrogenic proton ATPase. *Bars*, **A,B** ≈1 μm; **C** ≈0.1 μm

III. Correlation of Structure with Na^+ Transport Rates

The differences between renal cell types with respect to the area of basolateral membranes are quantitatively correlated with corresponding differences in the activity of $(Na^+ + K^+)$-ATPase (Fig. 8) and the Na^+ transport rate of the given segments (KAISSLING 1982; KAISSLING and LE HIR 1982; LE HIR et al. 1982; PFALLER 1982; KAISSLING and STANTON 1992; STANTON and KAISSLING 1989). For instance, in all the distal segments

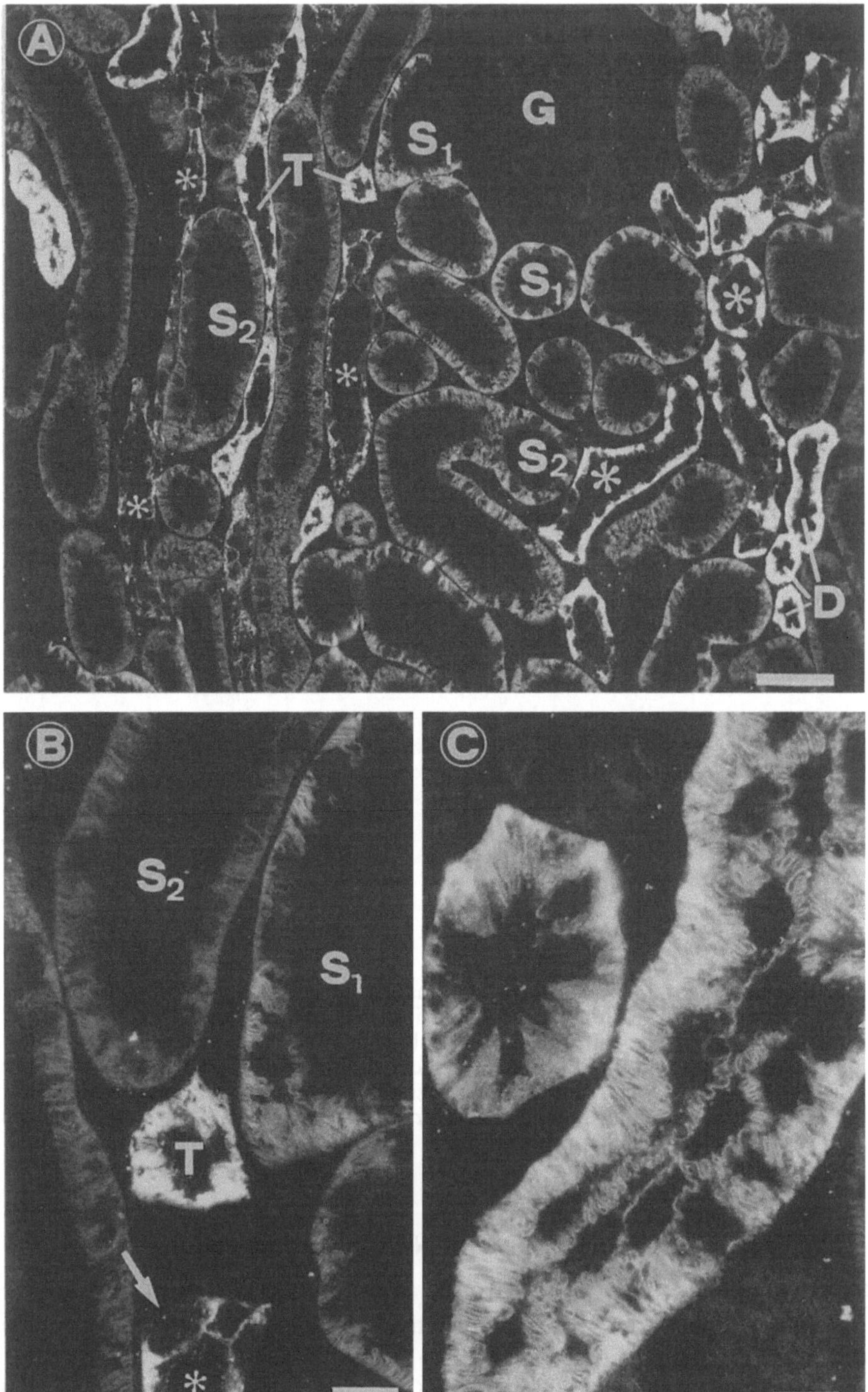
A
G
S1
T
S1
S2
S2
D
B
S2
S1
T
C

membrane area, enzyme activity and Na^+ transport rate are the highest in the distal convoluted tubule (Fig. 8A,C) (KAISSLING 1982; PFALLER 1982). Changes in the basolateral membrane area *along* a given nephron segment generally reflect corresponding axial changes in the sodium transport rate. This has been observed, for instance, in the thick ascending limb, where the basolateral membrane area of the initial portion in the inner stripe is highest and decreases gradually towards its terminal portion in the cortex. Yet, it should be kept in mind that also the enzyme density per membrane area might differ among cell types or along a segment. In distal segments the enzyme density within the basolateral membranes seems to be severalfold higher than in proximal tubules (Fig. 8A). In proximal tubules the enzyme density within the membrane decreases along the tubule, in addition to the decrease in membrane area.

C. Nephron and Collecting Duct Structure

I. Proximal Tubule

The proximal tubule begins at the urinary pole of the renal corpuscle and transforms abruptly to the thin intermediate tubule at the border between the outer and inner stripe of the outer medulla (Fig. 9). The convoluted part of the proximal tubule is located in the cortical labyrinth, the straight part in the medullary rays and in the outer stripe. The volume of proximal tubules has been calculated to amount to approximately 48% of total cortex volume (PFALLER 1982). Within the cortical labyrinth it is about 85% of the total tubular volume (proximal convoluted, distal convoluted, and connecting tubules) (KAISSLING et al. 1993). In the outer stripe the relative volume of proximal tubules is about 54% (PFALLER 1982).

The proximal tubule reabsorbs approximately 70% of the tubular fluid and solutes, and usually 100% of the filtered organic compounds (e.g.,

◄

Fig. 8A–C. Immunostaining of $(Na^+ + K^+)$-ATPase with a monoclonal antibody against the β-subunit of $(Na^+ + K^+)$-ATPase in rat kidney cortex; 1-μm cryostat sections; *G*, glomerulus; *S1, S2*, segments of proximal tubules; *T*, cortical part of thick ascending limb; *D*, distal convoluted tubule; *large asterisk*, connecting tubule: small *asterisk*, cortical collecting duct; exclusively basolateral membranes are stained. **A** $(Na^+ + K^+)$-ATPase in proximal tubules is highest at the urinary pole (*S1*), lower in S2 segments in the labyrinth and lowest in S2 in medullary rays; in all distal segments the staining exceeds that in proximal ones; differences are related to corresponding differences in basolateral cell membrane area. **B** Proximal tubule S1 at the urinary pole and S2 in a medullary ray and in a TAL and cortical collecting duct; note the presence of unstained cells in the collecting system (*arrow*); these are intercalated cells. **C** The highest membrane density is seen in distal convoluted tubules. *Bars*, **A** $\approx 50\,\mu$m; **B,C** $\approx 10\,\mu$m

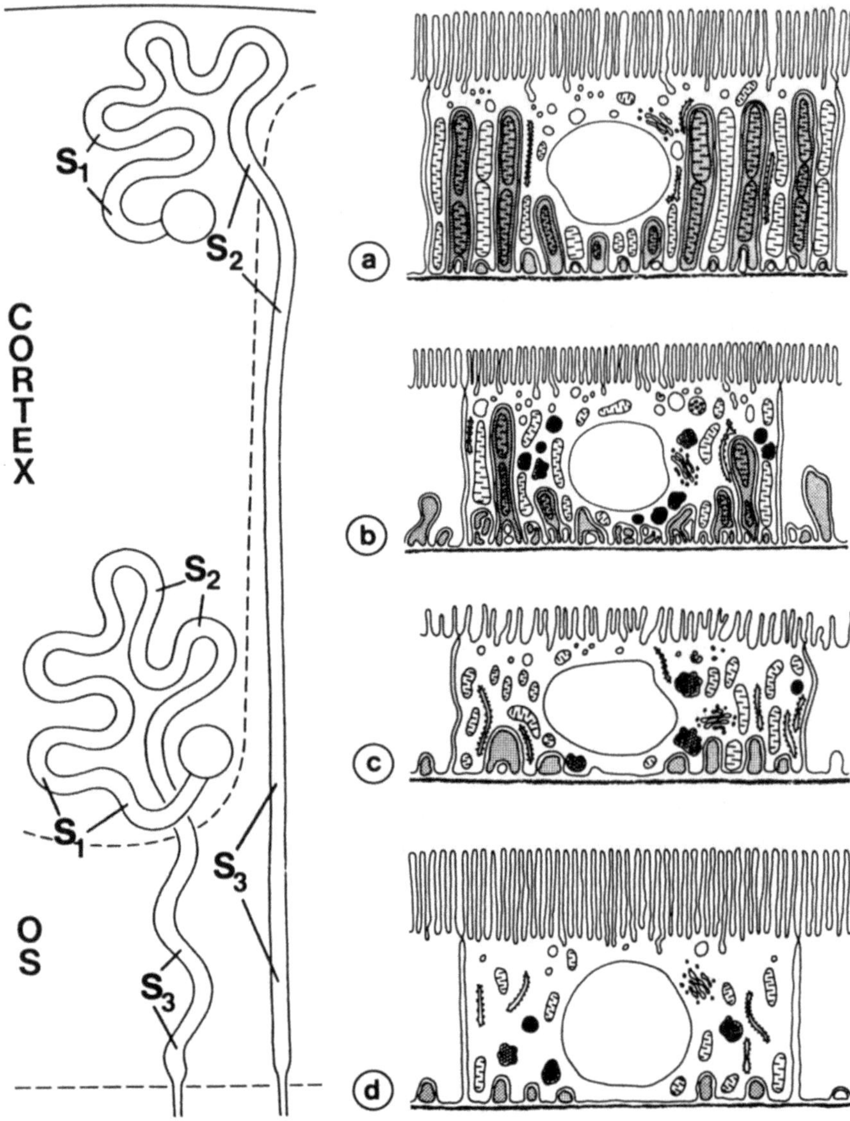

Fig. 9. Survey of segmentation and ultrastructure of the proximal tubule; *dashed line* delimits the cortical labyrinth from the cortical medullary ray and from the outer stripe; the proximal tubule begins with S_1 (*a*) at the urinary pole of the glomerulus, transforming gradually within the cortical labyrinth to S_2 (*b*) and within the lower half of the medullary ray to S_3 (*c*, rabbit; *d*, rat); basolateral cell membrane area decreases from S_1 to S_3. (Modified from KRIZ and KAISSLING 1992, with permission)

glucose, proteins, peptides, amino acids). Accordingly, the luminal membrane area is increased by a brush border (MAUNSBACH and CHRISTENSEN 1992; WELLING and WELLING 1975, 1976; PFALLER 1982), characteristic for proximal tubule cells (Fig. 6B,C), and the transport machinery for electrolytes is extensively developed by formation of large lateral cell proresses, filled with large mitochondria (Fig. 6B) (MAUNSBACH and CHRISTENSEN 1992). The rough and even more the smooth endoplasmic reticulum are well developed. Peroxisomes are often closely associated with smooth endoplasmic reticulum. The structural correlate for endocytotic uptake and digestion of proteins (CHRISTENSEN and NIELSEN 1991) is the so-called vacuolar apparatus, consisting of vacuolar profiles and dense apical tubules in the cytoplasm just below the brush border. The varying amounts of lysosomes in the cells are functionally related to the vacuolar apparatus. Both vacuoles and lysosomes in the cytoplasm have been recognized as acid compartments in the cells (VERLANDER et al. 1989). Protein intake and sex hormones influence the extension of the vacuolar apparatus and lysosomal system (MAUNSBACH and CHRISTENSEN 1992). The cells are electrically coupled by gap junctions between the lateral cell processes. The narrow intercellular space is separated from the luminal compartment by tight junctions of the "leaky" type.

Besides the $(Na^+ + K^+)$-ATPase a number of specific proteins have been identified in the basolateral membrane (MAUNSBACH and CHRISTENSEN 1992). Carbonic anhydrase is abundant not only in the cytoplasm, but also in both membrane domains of the proximal tubule cells (DOBYAN et al. 1982; BROWN et al. 1983; SEKI and FRÖMTER 1992). The brush border membrane houses a great variety of specific transport proteins (e.g., SANDS et al. 1992; BURCKHARDT and KINNE 1992; BIBER et al. 1993; for further refs. see MAUNSBACH and CHRISTENSEN 1992). The constitutive water channel aquaporin CHIP (AQP-CHIP = CHIP 28) amounts to approximately 4% of the total brush border membrane proteins, explaining the high water permeability of the tubule (NIELSEN et al. 1993b; SABOLIC et al. 1992).

The extent of the transport machinery, including the endocytotic apparatus, decreases along the proximal tubule (Fig. 8). On account of these changes the proximal tubule has been subdivided into three segments: S1, S2 and S3. The changes from S1 to S2 occur gradually within the convoluted part. The transition to S3 is situated within the straight part in the medullary ray, and is associated with an abrupt change from S2 cells with interdigitating lateral processes to S3 cells with a simple polygonal outline, numerous small mitochondrial profiles, and the highest density of smooth endoplasmic reticulum and of peroxisomes. In most species the height and density of the brush border decreases towards the end of the proximal tubule, but in rat S3 cells it is the greatest. In addition to axial heterogeneity internephron heterogeneity also exists. From all nephron types juxtamedullary nephrons, which possess the greatest glomeruli with the highest SNGFR, have the best-developed transport machinery.

II. Loop of Henle

1. Organization and Histotopography (Fig. 10)

The loop of Henle comprises a thick descending limb (the straight part of the proximal tubule) and a thick ascending limb (TAL, the straight distal tubule), connected by the thin-walled intermediate tubule. The transition from the proximal to the intermediate tubule occurs for all nephron types at the macroscopically well discernible border between the inner and outer stripes of the outer medulla. The intermediate tubule is differently organized in nephrons with "short" and nephrons with "long" loops. Short loops turn back within the inner stripe (or occasionally in the cortex, "cortical" loops), and their intermediate tubule is present exclusively in the descending portion. Long loops turn back at varying levels within the inner zone, and their intermediate tubule consistently has a thin de- and a thin ascending portion. The transition of the ascending thin intermediate tubule of long loops to the TAL coincides with the sharp demarcation between the inner zone and inner stripe (for refs. see Kaissling and Kriz 1992). The macula densa (MD), situated at the contact side of the TAL with its original renal corpuscle, functionally "closes" the loop.

The ratio of short and long loops varies among species. Most rodent species with high urine concentrating ability (rat, mouse, golden hamster, *Psammomys obesus*, *Meriones*) have more short than long loops (Bankir and Rouffignac 1985). In these species the descending thin limbs of short loops are more or less integrated into the giant vascular bundles of these species within the inner stripe, whereas those of long loops are intermingled with the TALs and the descending collecting ducts in the interbundle compartments (Jamison and Kriz 1982; Kaissling et al. 1975). Some species have only short loops (e.g., mountain beaver). These lack an inner medulla and have a poor ability to concentrate urine (Bankir and Rouffignac 1985; Jamison and Kriz 1982).

2. Intermediate Tubule

A total of four different epithelia are distinguished along the intermediate tubule:

Type 1 epithelium consists of flat simply organized polygonal cells, connected by tight junctions of several anastomosing strands, and by frequent desmosomes (Kaissling and Kriz 1992). It is present in descending thin limbs of short loops exclusively.

Type 2 to 4 epithelium exists in long loops only. Type 2 lines the upper part, type 3 epithelium the lower part of descending thin limbs and type 4 the ascending thin limbs of long loops. Type 2 epithelium has a complex structure with an extensive basolateral surface, with significant (Na^{+}+K^{+})-ATPase activity and relatively abundant mitochondria, suggesting the possibility for active transport processes. Depending on the species, type 2

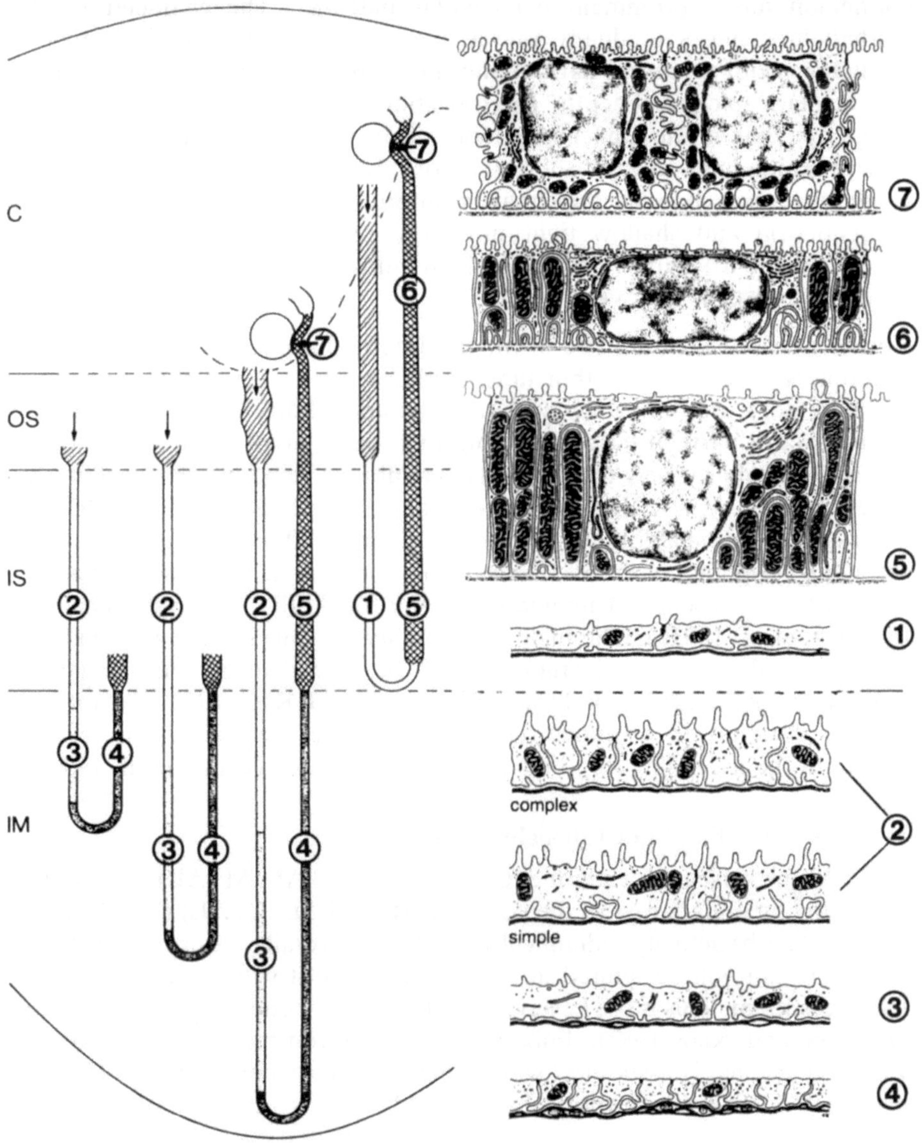

Fig. 10. Survey of organization and ultrastructure of the loop of Henle; the loop of Henle begins with the straight part of the proximal tubule (*arrows*); the different epithelia of the intermediate tubule and their location are indicated by *1–4*; the type 2 cells are of complex structure in, e.g., rat and *Psammomys*, of the simple type in rabbit and guinea pig; *5,6*, epithelium of the medullary and cortical thick ascending limb; *7*, macula densa. (Modified from Kriz and Kaissling 1992, with permission)

epithelium has a prominent paracellular pathway. The character of the epithelium changes gradually to the increasingly more simplified type 3 epithelium with relatively flat noninterdigitating cells and tight junctions of intermediate apicobasal depth. The transitional region is shorter in "shorter" long loops than in the longer long loops. The type 4 epithelium replaces the type 3 epithelium always shortly before the bend. Type 4 epithelium is characterized by flat but heavily interdigitating cells with very few mitochondria and shallow tight junctions, usually consisting of a single junctional strand. Type 4 epithelium is the most uniformly organized among species (KAISSLING and KRIZ 1992).

Correlation to Function. The axial, the internephron, and the interspecies differences in descending thin limb epithelia are in agreement with functional findings, exhibiting a considerable functional diversity of thin nephron segments. All parts of descending thin loop limbs in all species have a high water permeability and their luminal membrane contains the constitutive water channel AQP-CHIP (NIELSEN et al. 1993b; SABOLIC et al. 1992). In contrast, the permeability for Na^+, Cl^- and urea differs among species. The ascending loop limb is almost impermeable for osmotic water flow, in agreement with the absence of immunoreactivity for AQP-CHIP water channels (NIELSEN et al. 1993b). The high permeability for ions in type 4 epithelium, demonstrated in functional studies, might be related to the leaky organization of the paracellular pathway (KAISSLING and KRIZ 1992). In rat and mouse, antidiuretic hormone (ADH) stimulates the adenylate cyclase system in the ascending thin limb epithelium (IMBERT-TEBOUL et al. 1980).

3. Thick Ascending Limb (Straight Distal Tubule)

The TAL traverses the outer medulla (medullary TAL, MTAL) and ascends within the medullary rays of the cortex (cortical TAL, CTAL) (Fig. 11). The TAL reabsorbs actively sodium chloride, but is virtually water-impermeable. Thus, the salt will be diluted in the tubule, while it will be accumulated in the interstitium. This is the key event in the urinary concentrating mechanism (JAMISON and KRIZ 1982). Impairment of salt transport in the TAL, by whatever means, is associated with impaired urinary concentration (BANKIR et al. 1987). Salt reabsorption by the TAL cells is achieved via a specific

Fig. 11A–C. Loop of Henle. **A** Longitudinal section through a cortical medullary ray, with descending (*arrow downwards*) and ascending loop limbs (*arrows upwards*) and a collecting duct (*C*) in close juxtaposition; *asterisks*, intercalated cells. **B** Epithelium of a cortical TAL; the interdigitated cell processes (*open arrows*) contain large mitochondria (*M*); *arrowheads*, tight junctions. **C** Intermediate tubules (*arrow downwards*), TALs (*arrows upwards*), and collecting ducts (*C*) in the interbundle compartment of the inner stripe; part of a vascular bundle is seen in the upper right corner. **A,C** 1-μm Epon section; *Bars*, **A,C** $\approx$50 μm; **B** $\approx$ 0.5 μm

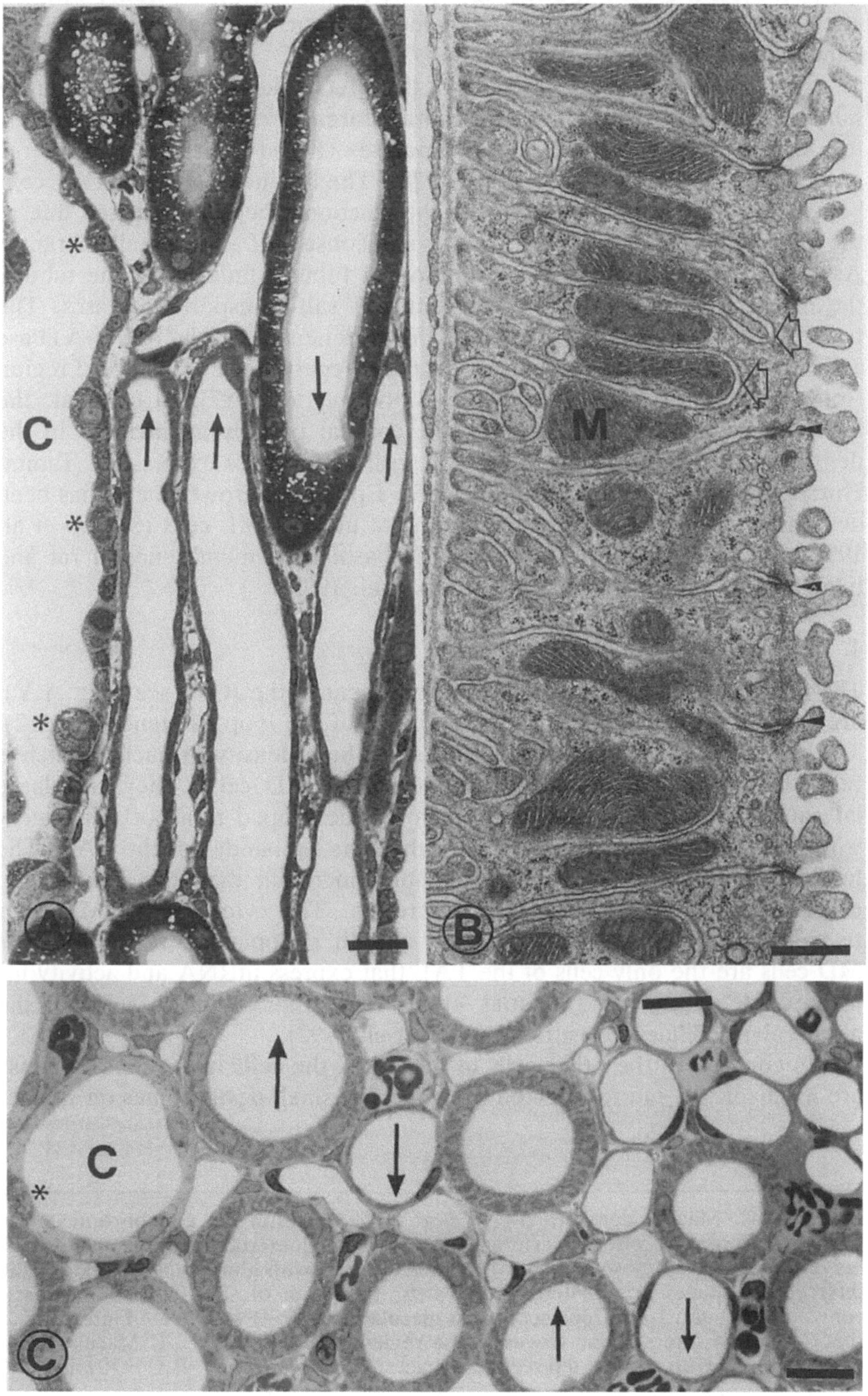
C
M
C

symport system for Na^+, $2Cl^-$, K^+ in the luminal cell membrane, which is effectively and specifically blocked by loop diuretics, such as furosemide and bumetanide (cf. Chap. 7, this volume).

The structure of the epithelium of the TAL reflects its specialization for salt transport. The basolateral membrane area is extensively enlarged by the formation of plate-like lateral processes (Fig. 11B) that enclose large mitochondria (WELLING et al. 1977, 1978). The luminal surface of TAL cells bears short stubby microvilli. The tight junctional belt is elongated due to the interdigitating lateral cell processes and consists of a few parallel strands. Associated with the progressive dilution of tubular fluid along the tubule, the extent of the structural equivalents for salt transport decreases. The reduction of cell height, basolateral membrane area, (Na^++K^+)-ATPase activity and mitochondrial content are the steepest in the transitional region from the inner to the outer stripe (BANKIR et al. 1987). In contrast, the luminal surface augments towards the cortical part by an increase in the density of apical short microvilli. TAL cell membranes express the Tamm-Horsfall protein (BACHMANN et al. 1985). Epidermal growth factor has been demonstrated on the luminal membrane of mouse TAL cells (SALIDO et al. 1991) and receptors for $5\text{-}HT_{1A}$ on the basolateral membranes of rat and human TAL and DCT cells (RAYMOND et al. 1993).

4. Macula Densa

The macula densa cells, the tubular component of the JGA (see Sect. A.V), are specialized cells within the end portion of the loop of Henle (Fig. 12). The cell plaque is prominent because of its high density of nuclei, which is related to the small polygonal outline of the MD cells. They lack large lateral cell processes and have a much less developed transport machinery with less (Na^++K^+)-ATPase activity than the remainder of the TAL. The luminal membrane usually displays more microvilli than the surrounding cells, but lacks the Tamm-Horsfall protein. The cytoplasm of MD cells reveals a particularly well developed smooth endoplasmic reticulum. The MD cells are the only cells of the TAL that express mRNA and activity for nitric oxide synthetase (MUNDEL et al. 1992) that seems to regulate the glomerular capillary pressure (WILCOX et al. 1992).

In contrast to the other cells of the TAL, the cells in the macula densa are mechanically interconnected by numerous small desmosomes on slender

Fig. 12A–C. Macula densa (*between short arrows*) within the end portion of the TAL; *a*, afferent; *e*, efferent arteriole; *EGM*, extraglomerular mesangium; *circles*, nerve fibers. **A** Rat; close contact of the macula cells with glomerular arterioles; the intercellular spaces are dilated; *arrowheads*, transition of the TAL to the distal convoluted tubule; 1-μm Epon section. **B** macula densa (rat) with dilated intercellular spaces; narrow intercellular spaces in the remainder of the TAL. **C** Macula densa (*Psammomys*) with narrow intercellular spaces. *Bars*, **A** ≈50 μm; **B,C** ≈10 μm

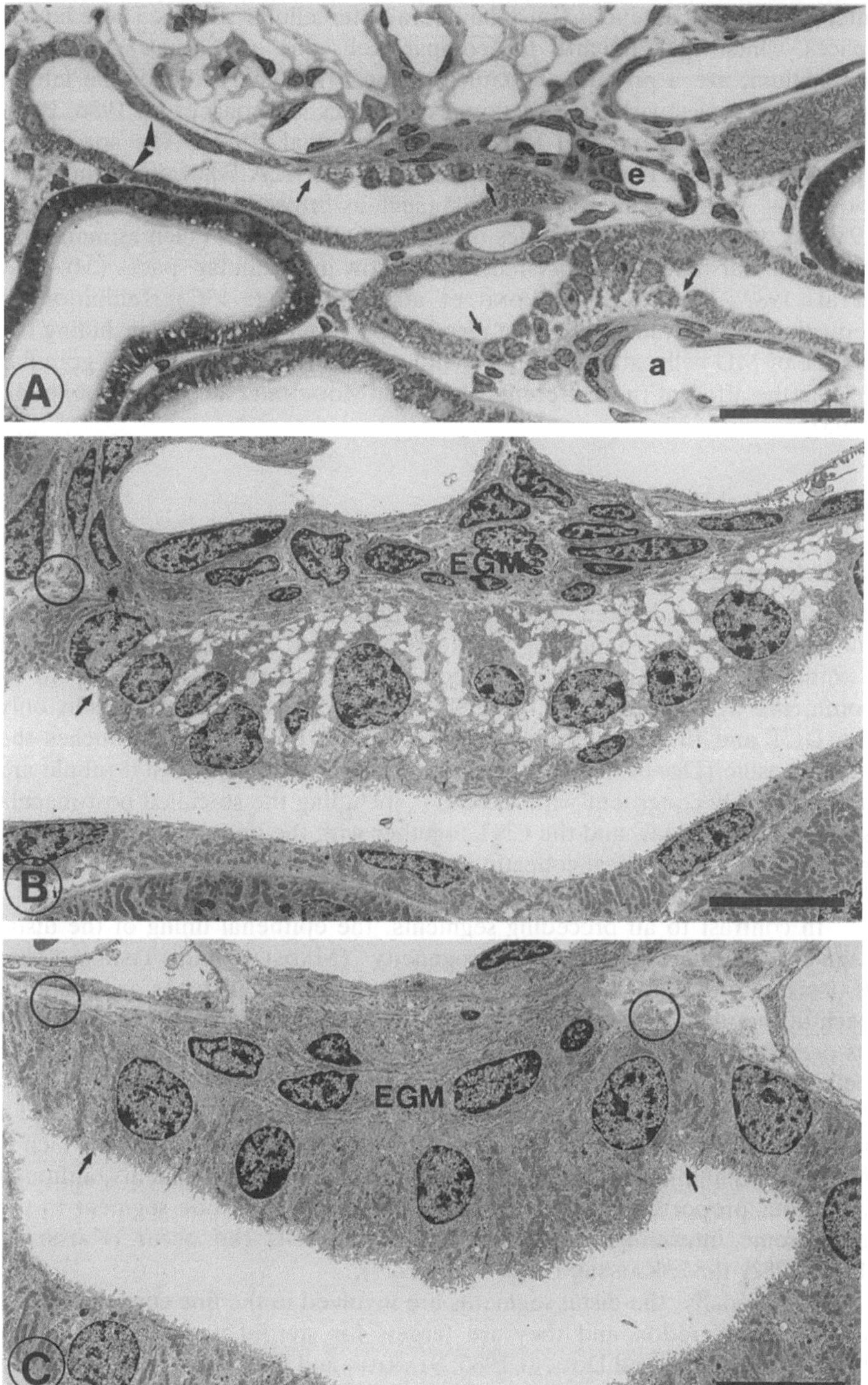
A
e
a
B
EGM
C
EGM

microprojections extending into the lateral intercellular spaces. These lateral spaces, which open towards the compact cell cone of the extraglomerular mesangium, are a prominent feature of the MD. The width of the lateral interspaces varies with the functional conditions (Alcorn et al. 1986; Bell et al. 1985; Kaissling and Kriz 1982; Kirk et al. 1985). The spaces are apparently dilated under control conditions (Fig. 12A,B). Compounds that block Na^+ transport by the MD cells (such as furosemide; Schlatter et al. 1989) as well as high osmolalities or impermeable solutes (such as mannitol) in the tubular fluid are associated with narrow intercellular spaces (Messina et al. 1987; Rasch 1984; Wong et al. 1986) (Fig. 13C). Inhibition or reduction of salt transport by the terminal portion of the TAL, including the plaque of MD cells, stimulates the synthesis and release of renin by granular cells in the afferent (and efferent) arteriole (Modena et al. 1993; Chen et al. 1993).

III. Cortical Distal Segments

1. Structural and Functional Organization

The histologically defined "distal convolution" includes the distal convoluted tubule (DCT), the connecting tubule (CNT), and the initial short portion ("initial collecting tubule") of the cortical collecting duct (CCD) up to its confluence with another CCD (Fig. 14). In superficial nephrons usually only the DCT and the initial part of the CCD, but not the CNT, touches the renal capsule (Dørup 1988). The terms "early" and "late" distal tubule are approximately congruent with the DCT, including the so-called post-macula segment of the TAL, and the CNT together with the initial part of the CCD, respectively. The cortical collecting system comprises the CNT and the CCD (Jones 1985; Kaissling and Kriz 1992).

In contrast to all preceding segments, the epithelial lining of the distal segments displays cellular heterogeneity (Madsen and Tisher 1986; Kaissling and Kriz 1992). Basically, two different cell types are found in each of the segments: one cell type constitutes the majority of the cells, and its occurrence is limited to the given segment. Therefore it is considered to be specific for that segment and is named accordingly: DCT cell, CNT cell or CD cell (formerly principal cell) for the DCT, the CNT or the CD, respectively. The second cell type is the intercalated cell (IC cell), which is interspersed among the segment-specific cells of all three segments, although in various proportions. In the transitional regions from one segment to the next, some intermingling of segment-specific cells can occur (Kaissling 1980, 1982, 1985; Kaissling and Kriz 1979).

Functionally, the distal segments are involved in the fine control of renal electrolyte excretion and they are targets for steroid and several peptide hormones (Morel and Doucet 1986; Stanton and Kaissling 1989; Spielman and Arend 1991; Kaissling and Stanton 1992). The DCT and CNT cells

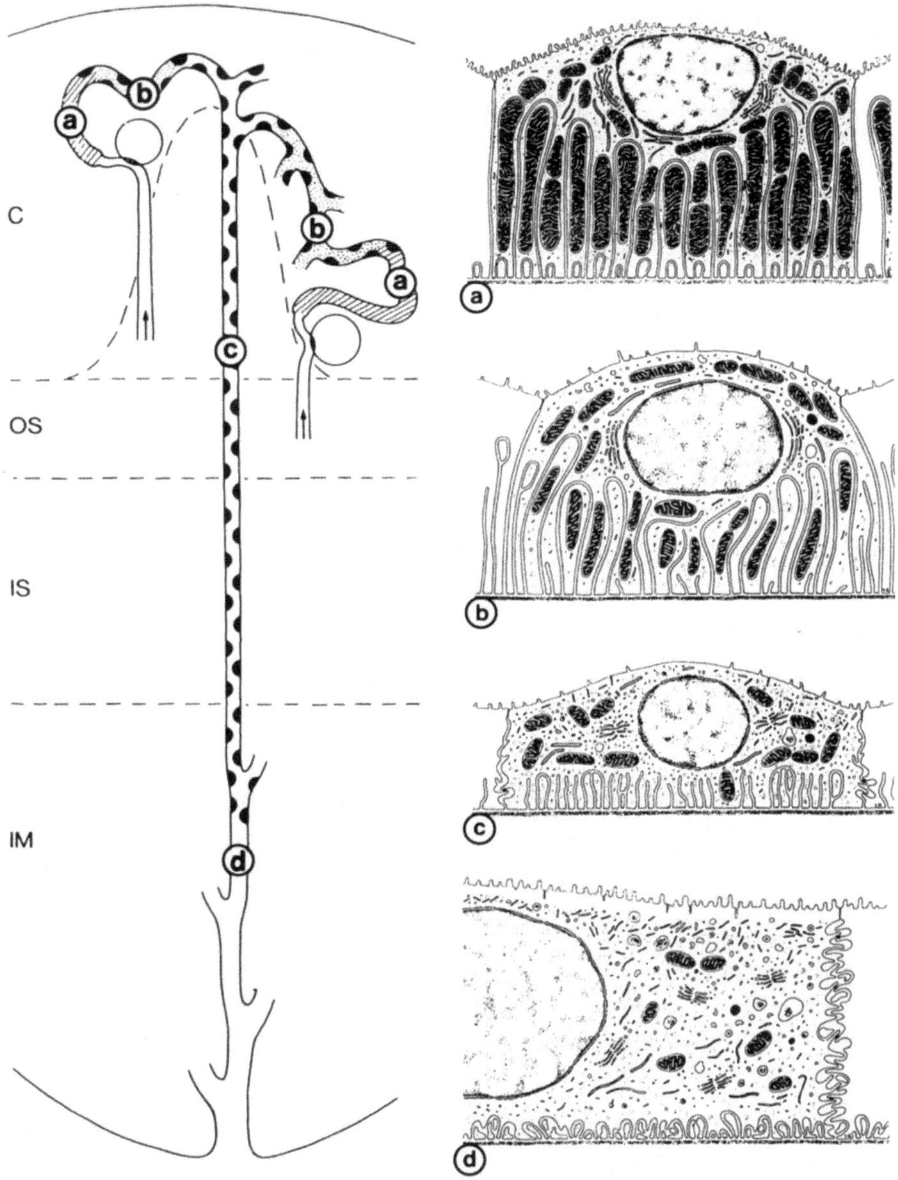

Fig. 13. Survey of distal and collecting duct segments and ultrastructure of segment-specific cell types; *a*, distal convoluted tubule; *b*, connecting tubule; *c*, cortical collecting duct, extending into the outer medulla; *d*, inner medullary collecting duct; the *black half-circles* indicate the distribution of intercalated cells. (Modified from KRIZ and KAISSLING 1992, with permission)

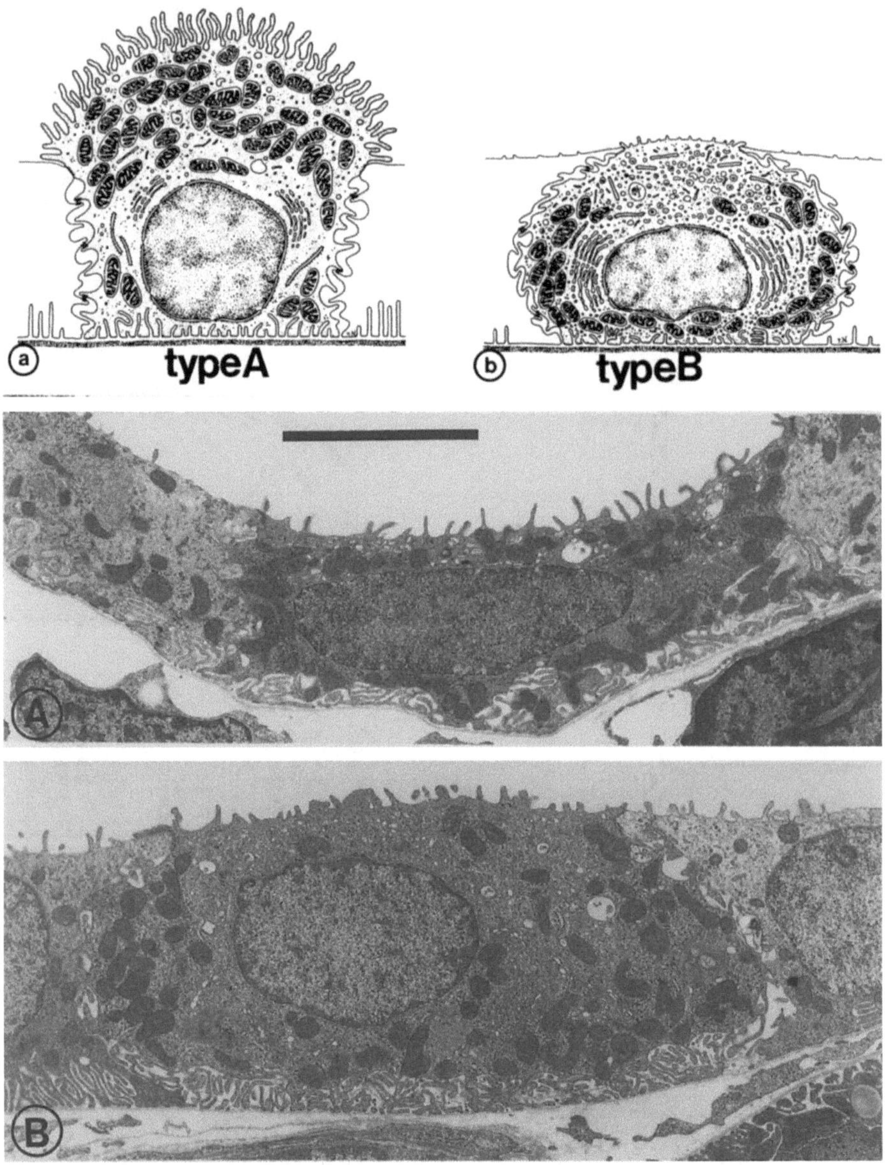

Fig. 14A,B. Intercalated cells; schematic ultrastructure of type A and type B cells. **A,B** IC cells from rat kidney. **A** Type A cell from cortical collecting duct; the luminal surface shows typical microfolds and the apical cytoplasm contains numerous intermediate and some large vesicles. Mitochondria are evenly distributed. **B** Type B cell from cortical collecting duct; small microvilli are present on the luminal surface. The mean diameter of the vesicles and tubulovesicles in the apical cytoplasm is smaller than in A-IC cells; mitochondria are preferentially arranged along the basolateral membranes; Golgi complexes are very promiment and often the *trans* region is oriented towards the base of the cell. *Bar* ≈10 μm (a,b from KRIZ and KAISSLING 1992, with permission)

reabsorb NaCl (STANTON and KAISSLING 1989, KAISSLING and STANTON 1992), possibly also Ca^{2+} (GESEK and FRIEDMAN 1993). Both cell types display high-affinity thiazide-binding sites in their membranes (ELLISON et al. 1993). In contrast to DCT cells the Na^+ reabsorption by CNT and by CD cells in the CD is obligatorily coupled with K^+ secretion. Na^+ and K^+ transports by CNT and CD cells are stimulated by mineralocorticoids, in agreement with the demonstration of receptors for these hormones (FARMAN et al. 1991) and of the 11*b*-OH steroid dehydrogenase in these segments (LOMBÈS 1990; LOMBÈS et al. 1990). CD cells are the only tubular cells that possess receptors for AVP (V_2) and respond to ADH with reabsorption of water (WADE et al. 1992). Receptors for adenosine (SPIELMAN and AREND 1991) and immunoreactivity for atrial natriuretc peptide (ANP) have been localized in collecting ducts (HEALY and FANESTIL 1986; KOSEKI et al. 1986; FIGUEROA et al. 1990), but so far it is still unclear in which cell type and in which membrane domain they are present. The IC cells are involved in the fine control of acid/base metabolism, and reabsorb K^+ (KAISSLING and KRIZ 1992; MADSEN and TISHER 1986; SCHUSTER 1993). Whether transport processes of intercalated cells are directly controlled by mineralocorticoid hormones or by the local environment of the cells or both is still a matter of debate.

2. Distal Convoluted Tubule

The transition from the TAL to the DCT is marked by an abrupt increase in epithelial height (KAISSLING et al. 1977; KAISSLING and KRIZ 1992). The Tamm-Horsfall protein, covering the TAL cells, ceases exactly at the transition (HOYER et al. 1979). In all species the first portion of the segment is lined by DCT cells exclusively. Intercalated cells may be interspersed in later portions of the segment. The organization of DCT cells resembles that of TAL cells but the transport machinery for electrolytes is even more developed than in the TAL (KAISSLING and KRIZ 1992; MADSEN and TISHER 1986). The DCT has the highest surface density of basolateral cell membranes (Fig. 15), the highest (Na^++K^+)-ATPase activity (Fig. 8) and the highest density of mitochondria from all nephron segments (KAISSLING and KRIZ 1992; PFALLER 1982; KAISSLING and LE HIR 1982, 1985). The apical cell pole of DCT cells is densely studded with short microvilli (Fig. 15). The distinct though rather weak labeling of the apical cell membrane for electrogenic proton ATPase (BROWN et al. 1988a) might be functionally related to the important carbonic anhydrase activity of DCT cells (DOBYAN and BULGER 1982; LÖNNERHOLM and WISTRAND 1984), suggesting a proton secretory function. The cytoplasm of DCT cells reveals high levels of calcium-binding proteins and the basolatereal membrane displays, in addition to (Na^++K^+)-ATPase, Ca^{2+}-Mg^{2+}-ATPase (BORKE et al. 1987).

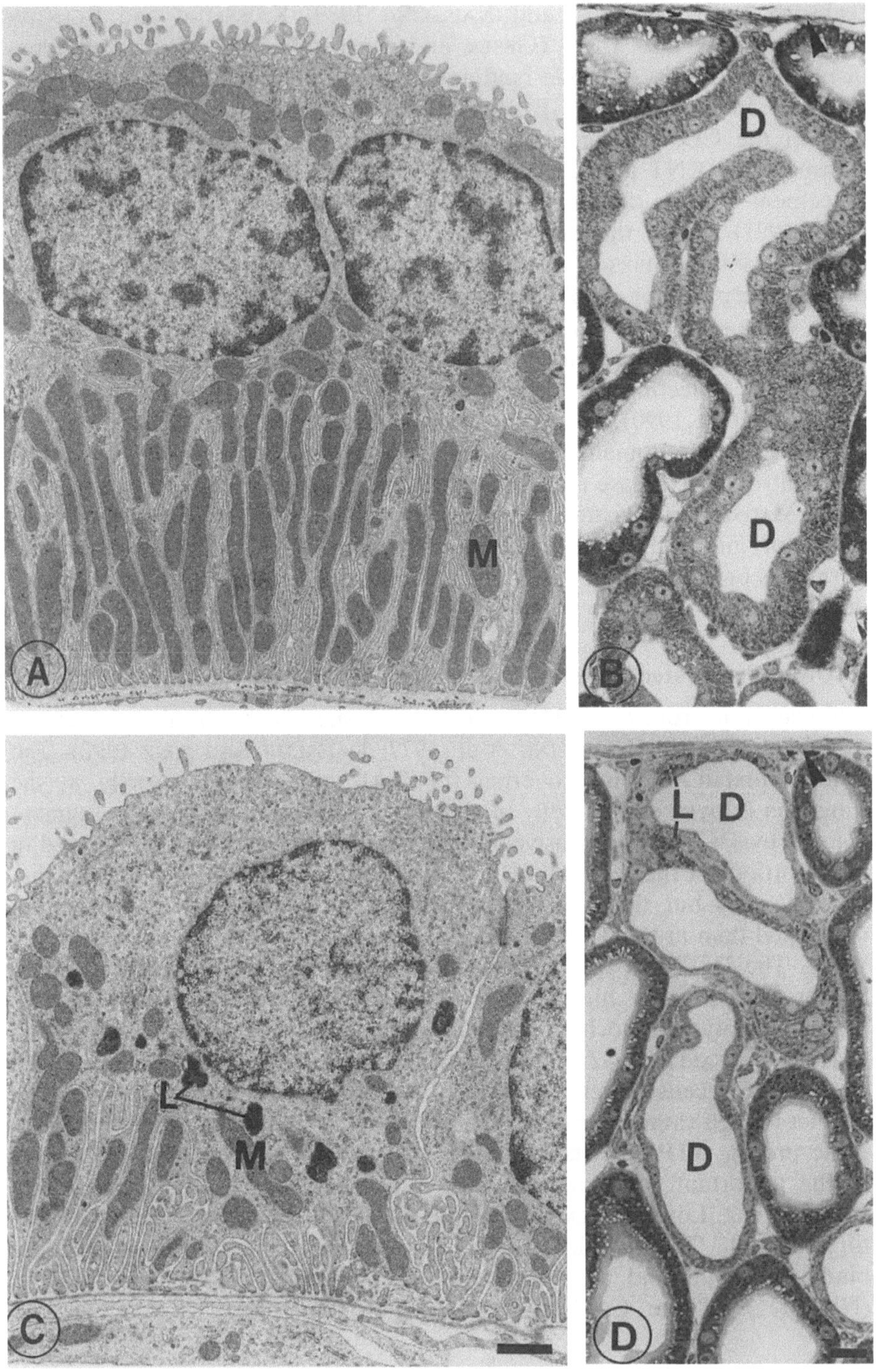
M
A
D
D
B
L
M
C
L
D
D
D

3. Connecting Tubule

a) Organization

The CNT of superficial nephrons connects the DCT with the initial collecting duct. CNTs of each two to six deep and midcortical nephrons fuse to a single tube, the arcade, that ascends along the cortical radial vessels in the cortical labyrinth (KRIZ and KAISSLING 1992; KAISSLING and KRIZ 1979, 1992) and drains the collected tubular fluid to the collecting duct situated in the medullary rays (Fig. 13). This course of the CNTs implies that this nephron segment is more often found adjacent to afferent arterioles and, thus, to renal nerves and to renin-containing granular cells (BARAJAS et al. 1986) than other segments. Generally, the CNT and the immediately neighboring afferent arteriole belong to the same nephron (DORUP et al. 1992).

b) CNT Cells

The epithelium of the CNT is composed of approximately 70%–80% CNT cells and 20%–30% intercalated cells (Figs. 16, 17). In contrast to DCT cells, CNT cells display numerous infolded cell membranes in the basal half, whereas the lateral processes are barely developed. The mitochondrial density is slightly less than in DCT cells. The tight junctions are distinctly deeper than in the preceding DCT and the luminal membrane is rather smooth (KAISSLING and KRIZ 1992). Similarly to in the DCT the intercellular spaces are narrow and the CNT cells also reveal a basolateral Ca^{2+}-Mg^{2+}-ATPase (BORKE et al. 1987) and cytoplasmic vitamin-D-dependent calcium-binding protein (ROTH et al. 1982). The luminal membrane displays small amounts of electrogenic proton ATPase (BROWN et al. 1988a) and in addition nerve growth factor (BARAJAS et al. 1987; SALIDO et al. 1986). CNT cells are the only cell type in the kidney that produce kallikrein (GUDER et al. 1987; VIO and FIGUEROA 1985). Regarding the special histotopographical relationship of the CNT to afferent arterioles, the kallikrein production by CNT cells might be relevant with respect to interactions within the renin-angiotensin system (BARAJAS et al. 1986).

Fig. 15A–D. Distal convoluted tubule (*D*) of rats treated for 6 days with either furosemide (**A,B**) or with hydrochlorothiazide (**C,D**), in both cases with compensation of the salt loss; top, renal capsule; *M*, mitochondria; *L*, lysosomes. **A** Hypertrophied DCT cell, displaying two nuclei and high density of mitochondria and basolateral cells membranes. **B** DCT of the superficial cortex; the epithelium is thicker and the nuclear density is much higher than in untreated animals. **C** DCT cell showing only a few mitochondria and basolateral cell membranes, but unusually numerous lysosomes. **D** DCT in the superficial cortex; the epithelium is thin and reveals numerous lysosomes. **A,C**; same magnification, bar ≈1 μm; **B,D**; same magnification, bar ≈10 μm

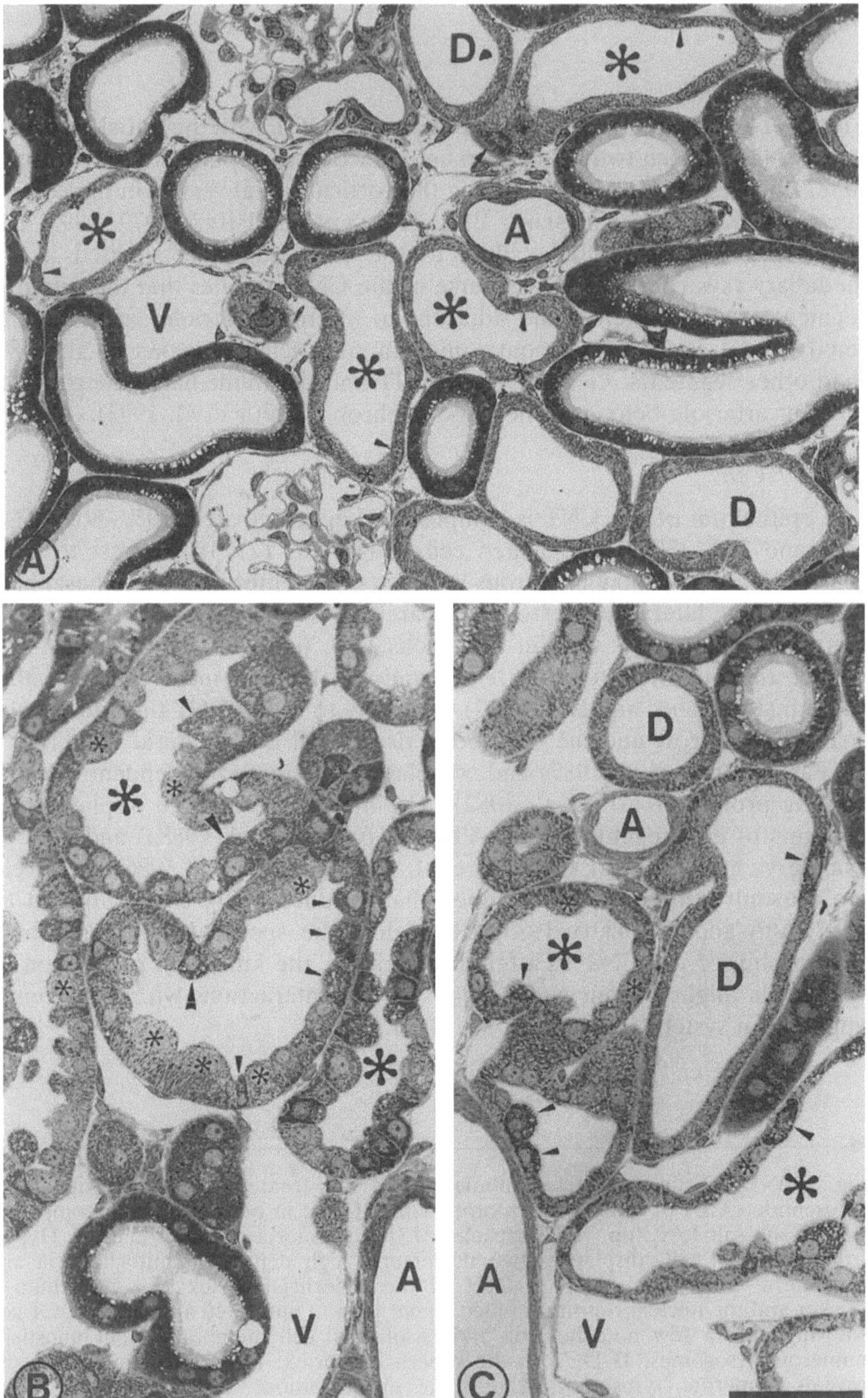
A
B
C
D
A
V
D
A
D
V
A
A
V

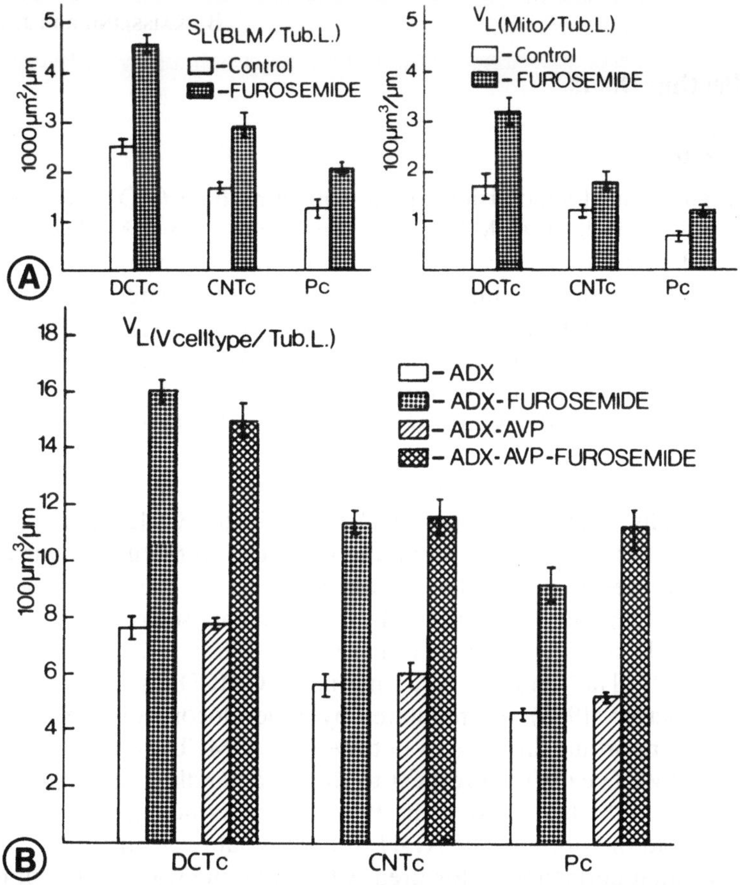

Fig. 17A,B. Morphometric data on distal segment of rats, treated for 6 days with furosemide (compensation of the salt and water loss with the drinking solutions). **A** Surface of basolateral cell membranes (S_L) and of mitochondrial volume (V_L), calculated per unit tubular length in adrenal intact animals in DCT, CNT and collecting duct (*P*) cells; means ± SEM; all differences between untreated and treated animals are significant. **B** Volume (*V*) of segment-specific cell types, calculated per unit tubular length (*L*) in the DCT, CNT and CCD of adrenalectomized, hormone-repleted rats (*ADX*), treated with furosemide (ADX-FUROSEMIDE) or furosemide and arginine-vasopressin (ADX-AVP-FUROSEMIDE) and compensation of the salt and water loss; in the three cell types all data (means ± SEM) are significantly higher in the furosemide-treated rats. (Data from KAISSLING and STANTON 1988a)

Fig. 16A–C. Connecting tubules (*CNT*) in the mid-cortex; the CNTs (*asterisks*) are grouped around the cortical radial vessls (*A*, *V*); *small arrowheads*, intercalated cells; *small asterisks*, CNT cells. **A** In untreated animals the CNT is difficult to discern from DCTs (*D*). **B** After 6 days of furosemide treatment both cell types in the CNT are hypertrophied and the CNT cells can easily be distinguished from intercalated cells; the proximal tubules also seem to be hypertrophied. **C** After 6 days of hydrochlorothiazide treatment, the CNT and intercalated cells, but not the DCT cells, seem to be slightly hypertophied. Same magnification in A–C, bar ≈50 μm

IV. Collecting Ducts

1. Organization

The microanatomical subdivision of the collecting duct (CD) into a cortical (CCD), outer medullary (OMCD) and inner medullary (IMCD) collecting duct, according to their location, does not coincide with the cytological subdivision. The cellular lining of the cortical CD extends into the outer stripe of the outer medulla and that of the OMCD extends through the upper third of the inner medulla. Fusions of collecting ducts occur in the upper half of the cortex and within the inner medulla. Within the outer medulla the collecting ducts are unbranched tubules (KAISSLING and KRIZ 1992).

The CCDs and OMCDs are composed of segment-specific CD cells and intercalated cells. The percentage of CD cells in the CCD of rats is about 60%–65%. It increases slightly along the OMCDs. In the upper third of the IMCD it increases to 90%. The remaining part of the inner medullary collecting duct (papillary CD, duct of Bellini) is composed of inner medullary CD cells (IMCD cells) exclusively (CLAPP et al. 1987).

Functionally, Na^+ and K^+ handling in the OMCDs is similar to that in the CCD (KOEPPEN 1988) and regulated by mineralocorticoids (KASHGARIAN et al. 1987). Important differences between the OMCD and the lower two-thirds of the IMCD exist with respect to the urea handling: as in the OMCD the epithelium in the first third of the IMCD is impermeable to urea, in both the absence and the presence of ADH; the terminal two-thirds of the IMCD has a rather high permeability for urea, which like the water permeability, is further increased by ADH (KNEPPER et al. 1988).

2. Cortical Collecting Duct Cells

CD cells in the cortex are small and have a simple polygonal basal and luminal outline. The basal cell membrane is infolded into the basal cytoplasm and all cell organelles, in particular the few short mitochondrial profiles, are located above the infoldings (Fig. 18A). The intercellular space is sealed from the lumen by a deep tight junctional belt of the "tight" type. CD cells are interconnected by numerous small desmosomes. In the presence of ADH the intercellular spaces between the cells usually appear to be dilated

Fig. 18A–C. Cortical collecting duct cells; *arrowheads*, infolded cell membranes; *L*, *lysosomes*; *circles*, aggrephores; *arrows*, tight junctions. **A** Untreated rat; infoldings of the basal cell membrane in the basal quarter of the cell; the few mitochondria are situated above the infolded membranes. **B** Rabbit, chronically fed with a high sodium, low potassium diet and very low plasma levels of aldosterone; basal infoldings are virtually absent. **C** Rabbit, chronically fed with low sodium, high potassium diet and very high plasma levels of aldosterone; basal infoldings and mitochondrial density are highly increased. B and C, same magnification, bar ≈1 μm (B and C from KAISSLING and LE HIR 1982, with permission)

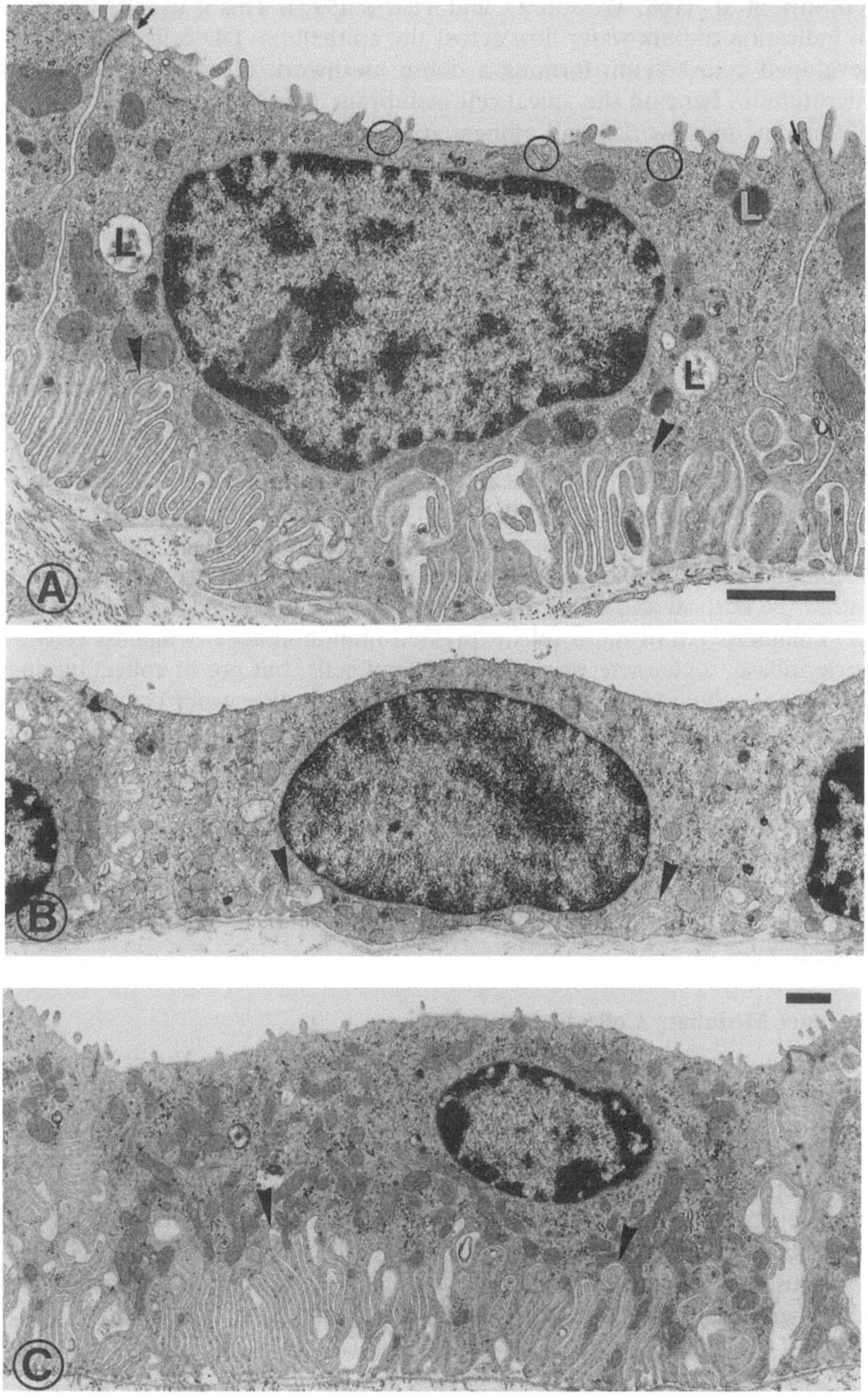
A
L
L
L
B
C

(GANOTE et al. 1968; WOODHALL and TISHER 1973). This is usually taken as an indication of bulk water flow across the epithelium. CD cells have a well-developed cytoskeleton forming a dense meshwork of microfilaments and microtubules beneath the apical cell membrane (KAISSLING and KRIZ 1992). Within this meshwork small elongated tubular profiles are oriented either perpendicularly or at an oblique angle to the luminal membrane (Fig. 18) (KRIZ and KAISSLING 1992; KAISSLING and KRIZ 1992). These vesicles ("aggrephores") carry aggregates of intramembrane particles (KACHADORIAN et al. 1977). In the presence of AVP the aggrephores fuse with the luminal membrane, conveying water permeability to it; after withdrawal of AVP the aggrephores are rapidly retrieved from the membrane (ZALUPS et al. 1985; BROWN 1989). The particles in the aggrephores and in the luminal membrane might well correspond to the proteins of the AVP-regulated water channel aquaporin-CD (AQP-CD = WCH-CD) that have recently been demonstrated (FUSHIMI et al. 1993) in the collecting duct cells. Depolymerization of actin microfilaments inceases water permeability whereas disruption of microtubules has the opposite effect, indicating that micotubules may be needed for movement of aggrephores to the apical membrane whereas micofilaments inhibit this movement (SIMON et al. 1993). A minority of cells within the cortical and outer medullary collecting duct do not express AQP-CD channels. All of these cells expressed immunoreactivity against ecto-5′-nucleotidase, a characteristic of intercalated cells, but not of collecting duct cells proper (NIELSEN et al. 1993a,b). The constitutive water channel AQP-CHIP, which is expressed in the proximal tubule and the descending thin limb of Henle's loop (SABOLIC et al. 1992; HASEGAWA et al. 1993; NIELSEN et al. 1993b), can not be shown in CD cells.

The CD cells undergo gradual though considerable changes from the cortex downstream to the upper third of the inner zone. The surface density of the basolateral cell membrane and mitochondrial volume decrease, whereas the densities of lysosomal elements and of microfilaments and microtubules increase.

3. Inner Medullary Collecting Duct Cells

From the second third of the inner medulla collecting duct the epithelium is composed of a single cell type, the "inner medullary CD cell" (IMCD cell; CLAPP et al. 1987). This cell type is different in structural and functional aspects from the CD cells in the other parts of the CD (KNEPPER et al. 1988; see Sect. C, IV.1). The cell size is considerably greater and the lateral intercellular spaces are conspicuous by their dense assemblies of microvilli and microfolds originating from the lateral cell membranes. The conspicuous, dense cytoskeleton in these cells is thought to counteract and to limit cytoplasmic swelling during water reabsorption.

4. Intercalated Cells

a) General Structural Characteristics

The intercalated cell (IC) type (Fig. 14) is the minority cell type within the epithelium of the DCT, the CNT, the CCD and the OMCD. With scanning electron microscopy IC cells usually show a rounded luminal outline among the polygonal outlines of neighboring cells (Fig. 13). Most IC cells are covered by elaborate microprojections (Figs. 7, 14, 20). The most consistent ultrastructural feature is the numerous vesicles between 80 and 200 nm in diameter (DØRUP 1985a) in the apical cytoplasm (Fig. 14). Many of these carry 10-nm studs on the cytoplasmic face of their membrane (Fig. 20) and similar studs are found on the cytoplasmic face of luminal (Figs. 13, 20) or basolateral membranes (Fig. 20). The studs are ultrastructurally and chemically distinct from clathrin (BROWN and ORCI 1986), which is also frequent. Basolateral membrane infoldings are often well developed but without direct structural association with mitochondria. Mitochondrial profiles are short (BERGERON et al. 1988). Rough endoplasmic reticulum is sparse, but in many IC cells Golgi complexes and smooth endoplasmic reticulum are particularly well developed (KAISSLING and KRIZ 1992).

b) Cytochemistry

Intercalated cells contain little or no $(Na^{+}+K^{+})$-ATPase (RIDDERSTRALE et al. 1988) but large amounts of an electrogenic, vaculolar type (V-type) H^{+}-ATPase (NELSON et al. 1992). The proton pump seems to be associated with the membrane studs mentioned above (BROWN et al. 1987). V-type H^{+}-ATPases are found in all eukaryotic cells and usually serve as the motor for acidification of intracellular organelles such as lysosomes, endocytic vesicles and Golgi-coated vesicles (MORIYAMA and NELSON 1989; NELSON 1992). In IC cells, as in some other cells, however, membrane material can be externalized from vesicles to the surface when a need for extracellular acidification arises (DØRUP 1985b; MADSEN and TISHER 1983, 1984; GLUCK 1992; BROWN et al. 1992). Another proton pump, the P-type, electroneutral $(H^{+}+K^{+})$-ATPase, is also selectively present in IC cells (WINGO et al. 1990; BROWN et al. 1990) and seems to be associated with characteristic rod-shaped particles, seen on the P-face of freeze-fractured membranes of most IC cells. This pump probably mediates active K^{+} reabsorption and in K^{+} depletion its activity is increased (CHEVAL et al. 1991; DOUCET and MARSY 1987) in congruence with the number of intramembrane rod-shaped particles (STETSON et al. 1980). The balance and possible interrelationship between the two pumps is not fully understood. IC cells have a low conductance for Na^{+} and K^{+} but a high Cl^{-} conductance (DIETL et al. 1992). ADH-regulated and constitutive water channels are apparently absent in the membranes of these cells (NIELSEN et al. 1993b). The cytosol of IC cells is rich in carbonic anhydrase (CA) (KIM et al. 1990; HOLTHOFER et al. 1987;

EMMONS et al. 1991) and an $HCO_3^-/Cl-$ anion exchanger is present in the plasma membrane (KOPITO 1990; MADSEN et al. 1992). These collected data on IC cells are consistent with the suggestion that IC cells perform active proton transport coupled to passive bicarbonate transport and possibly also active potassium reabsorption.

c) Subtypes of IC cells

Structural and immunocytochemical differences within the population of IC cells suggest that at least two different subtypes (A and B) exist in the connecting tubule and the cortical collecting duct (Fig. 14). In the outer medullary collecting duct IC cells are homogeneous and resemble A IC cells in the cortex.

Type A cells usually carry luminal microfolds and there appears to be a reverse correlation between the area of vesicular and luminal membrane (DØRUP 1985b). Type A IC cells express the V type proton pump at the apical membrane and the $HCO_3/Cl-$ exchanger band 3 (as the only cell in the kidney) at the basolateral membranes (ALPER et al. 1989; DRENCKHAHN et al. 1985; SCHUSTER et al. 1986; WAGNER et al. 1987). Type A cells are very rich in carbonic anhydrase (KIM et al. 1990) and increase the luminal membrane area in acute metabolic (DØRUP 1985b) and respiratory (VERLANDER et al. 1987) acidosis. Type A cells are thought to have a proton secretory function.

Type B cells usually have short microvilli. The mean diameter of apical vesicles is smaller than in type A cells, and vesicles are absent from a band below the luminal membrane. The cells are particularly rich in smooth endoplasmic reticulum and have large Golgi complexes with the *trans* side pointing towards the base of the cell. Mitochondria are preferentially located along the basolateral membranes. B-IC cells express the V-type proton pump at the basolateral membranes. They contain carbonic anhydrase in smaller amounts than A-IC cells. So far an anion exchanger in membranes of type B cells has not been demonstrated by immunohistochemistry. Available ultrastructural and immunocytochemical data suggest that type B cells are involved in bicarbonate secretion and proton reabsorption.

Between 5% (VERLANDER et al. 1987) and 25% (unpublished observation) of cortical IC cells are structurally intermediary to A and B cells or express structural and immunohistochemical characteristics of both. It is a matter of controversy whether IC cell configurations can change one into the other upon acute or chronic stimulation (SATLIN and SCHWARTZ 1989; SCHWARTZ et al. 1985) or whether the two configurations should be considered as two distinct cell types (HAYASHI et al. 1988).

D. Alterations of Morphology in Electrolyte Disturbances

Alterations in the electrolyte metabolism of the body – whether induced by physiological, pathological or pharmacological means – are associated with alteration of the renal excretion pattern of electrolytes, that – ideally – restore the homeostasis of body fluids. Acute changes in Na^+ transport rates proceed in most renal cell types without recognizable modulation of cell structure and are rapidly reversible; prolonged changes in electrolyte transport rates modify the "transport capacity" of the tubule, i.e., the maximal transport rate per time and tubular length, associated with structural adaptation. These structural and functional changes persist for some time (days to several weeks) after withdrawal of the specific stimulus.

I. Acute Changes in Transport Rates

1. Mechanism

The essential step, leading to an altered transepithelial salt transport rate, is the modulation of Na^+ entry into the cells via sodium channels in the luminal membrane (Stanton and Kaissling 1989). Mediators for changes in the luminal Na^+ conductance are manifold (including many, such as tubular fluid solute-load, peptide and steroidal hormones, growth factors). A rise in cellular sodium load immediately increases by kinetic changes the pumping rate of the $(Na^+ + K^+)$-ATPase in the basolateral membrane (Jorgensen 1986; Blot-Chabaud et al. 1988) and the extrusion of Na^+ out of the cells to the interstitial side. Furthermore, additional Na^+ pumps in the basolateral membrane can be either recruited and/or activated within 1–2 min (Barlet-Bas et al. 1990; Blot-Chabaud et al. 1989; Coutry et al. 1992). This process was found to be reversible within 15 min after ${Na_i}^+$ was lowered and apparently involves neither protein synthesis nor participation of the cytoskeleton.

2. Morphological Changes Associated with Acute Changes in Transport Rates

Most of the acute and rapidly reversible changes in transport rates cannot be detected with today's available morphological methods. A few alterations have been observed in the luminal membrane domain.

Modulation of *expression of transport proteins* in the luminal membrane is apparent within a few hours following an adequate stimulus. This has been observed, for instance, for the expression of a sodium-phosphate transport system in the brush-border membrane of the proximal tubule (Biber et al. 1993; Levi et al., 1994).

In intercalated cells and in CD cells, rapid *modulation of the luminal membrane* area contributes to changes in the luminal conductance for protons and for H_2O, respectively (Brown 1989; Wade et al. 1981). The structural and functional changes are apparent 1–2 min after an acid load in intercalated cells (Gluck et al. 1982; Schwartz et al. 1985; Schuster 1993), and in response to increases in plasma levels of AVP in collecting duct cells (Brown et al. 1988b; Verkman et al. 1988). They are achieved by a specific membrane shuttle system between the luminal membrane and a pool of membrane vesicles in the apical cell pole. Membrane fractions containing the specific transport proteins are inserted and retrieved by exocytotic and endocytotic mechanisms, respectively. These changes proceed without de novo synthesis of specific transport proteins but require a functioning cytoskeleton (Brown et al. 1991; Brown 1989; Simon et al. 1993).

II. Chronic Alteration of Na^+ Transport Rates

1. Mechanism

Under prolonged alterations of transport rates the transport capacity of the epithelium adapts, involving adaptation of the "transport machinery" [surface area of basolateral membranes with correlated $(Na^+ + K^+)$-ATPase activity] and of energy production (mitochondria). These changes are associated with major protein synthesis and profound remodeling of the cell structure that may result in hypertrophy of the epithelium. Often this adaptative "growth" also involves major DNA synthesis and cellular proliferation (Loffing et al., in press; Toback et al. 1993). Eventually, not only the length of the given segment but also the size of the renal zone concerned (e.g., cortex under furosemide treatment; Kaissling et al. 1985) or of the whole kidney (compensatory hypertrophy) may increase. While the initial step for adaptative growth of tubular cells seems to be directly linked to the rise in cellular Na_i^+, knowledge of the detailed signaling events and control mechanisms by growth factors and others involved in functional adaptation is only limited (Hammermann et al. 1993; Wolf 1993).

2. Time Course of Structural Changes

The neosynthesis of $(Na^+ + K^+)$-ATPase units, induced, for instance, by exposure of collecting duct cells to aldosterone, requires at least 1 h (Barlet-Bas et al. 1988; Edelman 1975; Field and Giebisch 1985). Five hours of aldosterone treatment increases K^+ secretion by the initial part of the cortical collecting duct fourfold without recognizable changes in ultrastructure in the segment. After 1 day of aldosterone treatment the basolateral cell membrane area in the collecting duct cells was significantly augmented; a maximum in the membrane area and Na^+ and K^+ transport rates was reached after 14 days (Wade et al. 1990).

The time course of reversibility of such changes, or adaptation to chronically lowered Na^+ uptake by tubular cells, has been less well studied. Measurable reduction of the transport machinery in TAL cells was apparent only after several weeks of treatment that lowered their Na^+ uptake. Three days after adrenalectomy the membrane area in collecting duct cells was found to be significantly smaller than in intact control rats (Stetson et al. 1980; Stanton et al. 1985b; Wade et al. 1990).

III. Adaptation of Proximal Tubule

1. Changes in GFR

Chronically altered glomerular filtration rate (GFR) associated with increased filtered Na^+ load, e.g., in remnant nephrons following reduction of renal mass and treatment with various hormones, diabetes mellitus, and diuretics (SCHERZER et al. 1987; DAVIS et al. 1983; DAVIS and JOHNS 1990; KOECHLIN et al. 1989; ZALUPS et al. 1985), provokes cellular hypertrophy and/or proliferation in the proximal tubule. Accordingly, the greater epithelial volume and transport machinery in the deep proximal tubules than in more superficial ones might well be considered as a "naturally occurring adaptation" to the greater SNGFR of juxtamedullary glomeruli. In fact, in Brattleboro rats with congenital diabetes insipidus (DI rats) nephron heterogeneity with respect to GFR as well as to structure is lacking (TRINH TRANG TAN et al. 1981, 1984; KOECHLIN et al. 1989), but it is restored with chronic AVP treatment, most probably via the effect of the hormone on GFR in deep nephrons (KOECHLIN et al. 1989).

2. Diabetes Mellitus

The hypothesis that a rise in proximal cell sodium causes hypertrophy and increased transport capacity (STANTON and KAISSLING 1989) is supported from data of rats with diabetes mellitus (KUMAR et al. 1988). The increase in filtered glucose seems to increase the rate of Na^+-glucose symport into the proximal tubule cells and thereby increases the intracellular Na^+ concentration (KHADOURI et al. 1987).

3. Reduction of Renal Cell Mass

Pronounced hypertrophy (and hyperplasia) has also been observed in remnant nephrons after reduction of renal cell mass (SALEHMOGHADDAM et al. 1985). It has been postulated that under these conditions the stimulation of the Na^+/H^+ antiport in the luminal membrane might be the requirement to trigger Na^+ uptake and cell proliferation (FINE et al. 1985). Meanwhile this hypothesis has been questioned because experimental studies in cultured cells and after inhibiting the Na^+/H^+ antiport by amiloride in vivo demonstrated that the hypertrophic response also occurred despite low levels of Na^+/H^+ antiport (MACKOVIC-BASIC et al. 1992). It has also been discussed whether the observed increases in Na^+/H^+ exchange might be a consequence rather than the cause of the hypertrophic response.

IV. Adaptation of Thick Ascending Limb of Henle's Loop

The epithelium of the thick ascending limb normally shows a decline in basolateral surface area and (Na^++K^+)-ATPase activity along the TAL.

Several experimental studies suggest that this axial heterogeneity is not a constitutive feature but seems to be correlated with the axial fall in Na^+ delivery and ensuing progressively lower cellular Na^+ uptake along the segment.

1. Variation of ADH Plasma Levels

In rats and mice Na^+ uptake by TAL cells is stimulated by antidiuretic hormone (ADH; arginine vasopressin, AVP; MOREL et al. 1978; MOREL and DOUCET 1992). In rats prolonged endogenously increased levels of ADH by water restriction provoked marked hypertrophy of the beginning of the TAL in the inner stripe (BANKIR et al. 1988). Chronic decreases of endogenous plasma AVP levels, induced by water-rich food, abolished the axial heterogeneity of the TAL. The deep medullary portions became almost as thin as the cortical portions (BANKIR et al. 1988) and the structure of the TAL resembled that observed in TALs of ADH-deficient DI rats with hereditary diabetes insipidus. Treatment of these rats for several weeks with AVP restored the normal axial heterogeneity of the thick ascending limb and the ability for urine concentration (BOUBY et al. 1985).

All described structural changes in the TAL were apparent only after several weeks of treatment. Six days of specific inhibition of cellular Na^+ uptake in TAL cells by blocking the luminal Na^+, $2Cl^-$, K^+ symporter by furosemide (with compensation of the solute loss by adequate drinking solutions) did not significantly reduce the structural equivalents of the transport machinery along the TAL (STANTON and KAISSLING 1988).

2. Variation of Protein Intake

Increased size of the cells in the initial part of the TAL has been observed with chronic high protein diet (BOUBY and BANKIR 1988; TRINH TRANG TAN et al. 1993). Under these conditions the glomerular filtration rate and the salt delivery with the tubular fluid to the TAL are increased. The authors interpreted the structural changes in the TAL as being secondary to the increased salt delivery with the tubular fluid. Structural changes of the TAL epithelium after variation of steroid hormone plasma levels have not been described so far.

V. Adaptation of Distal Segments and Collecting Duct

The structurally defined distal segments differ with respect to luminal conductivities for given solutes, and to distribution of receptors for hormones. Among the known factors that naturally regulate the transepithelial Na^+ and K^+ transport in cortical distal segments (including the CD within the outer stripe of the outer medulla) are several peptide hormones, mineralocorticoids, and K^+ intake and tubular sodium delivery per se (STANTON and KAISSLING 1989; KAISSLING and STANTON 1992). Changes in the acid/base metabolism seem to affect directly proton, bicarbonate and K^+ transport by intercalated cells (see Sect. D.V.4).

1. Role of Tubular Na^+ Load

Data from several lines of experiments in rats and rabbits indicate that the sodium load of the tubular fluid can regulate per se the Na^+ transport rate in distal segments, in particular in the DCT, independently of the hormonal status of the animal (STANTON and KAISSLING 1989).

a) ADH Deficiency

In ADH-deficient DI rats with impaired salt reabsorption by the TAL the Na^+ delivery to the DCT is much higher than in healthy rats, and the DCT cells reveal structural hypertrophy (KOECHLIN et al. 1989), associated with an increase in basolateral membrane area, activity and transport capacity. After restoration of the function of the TAL and lowered Na^+ delivery to the DCT by prolonged treatment of the rats with vasopressin, the size of the DCT cells approached that in healthy rats.

b) Furosemide

Inhibiting salt reabsorption in the TAL by *furosemide* (with compensation of the salt loss by adequate drinking solutions) increases the Na^+ load in the DCT several times. After only 6 days of furosemide treatment of rats (intact adrenal) (KAISSLING et al. 1985; ELLISON et al. 1989) and adrenalectomized rats with hormonal substitution at low physiological levels (KAISSLING and STANTON 1988a,b), the DCT epithelium revealed an almost 100% increase in basolateral membrane area surface (Fig. 15), (Na^++K^+)-ATPase activity and transport capacity. Also in the CNT and CCD a significant, though less pronounced, increase in membrane area was apparent after furosemide treatment (Figs. 16, 17) (STANTON and KAISSLING 1988; KAISSLING and STANTON 1988a; ELLISON et al. 1989).

The increases in the Na^+ transport machinery along the distal segments, occurring as a consequence of increased sodium delivery to the DCT, following inhibition of salt transport in the preceding TAL, might well take part in the described "braking" phenomenon after prolonged treatment with loop diuretics (STANTON and KAISSLING 1989; KAISSLING and STANTON 1992). The structural changes in CNT and CCD cells are consistent with the observed loss of potassium with the urine, occurring with loop diuretics, since in these cell types Na^+ reabsorption is obligatorily coupled with K^+ secretion.

c) Dietary Sodium Intake

In rabbits on a high Na^+, low K^+ intake, associated with very low endogenous plasma levels of mineralocorticoids, from all distal segments exclusively the DCT revealed pronounced hypertrophy (LE HIR et al. 1982; KAISSLING and LE HIR 1982). The segments downstream (CNT, CCD) had even lower basolateral membrane areas than in untreated animals. In rabbits on a low Na^+, high K^+ intake, associated with high endogenous plasma levels of mineralocorticoids, no significant structural effect on the DCT was detected (Fig. 19).

d) Hydrochlorothiazide

Hydrochlorothiazide specifically inhibits the Na^+Cl^- uptake by the epithelial cells of distal tubular segments. After treatment of rats for 1 week with hydrochlorothiazide, DCT cells revealed pronounced autophagocytosis of mitochondria and structural lesions. Structural damage was much less prominent in CNT cells (Figs. 15, 16). Besides degraded cells, cellular regeneration was also observed (KAISSLING and STANTON, unpublished). The distribution pattern of the structural lesions coincided with the distribution pattern of the thiazide-binding site, detected by immunohistochemistry

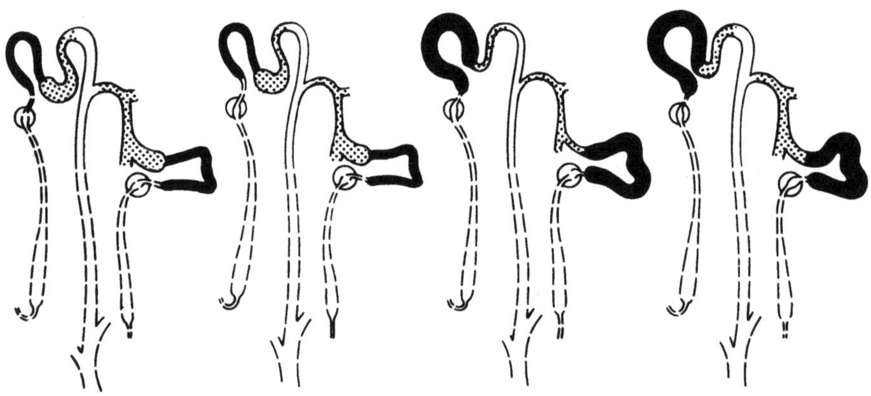

Fig. 19. Schematic presentation of changes of the "transport machinery" along distal segments of rabbits with diets and treatments that influence the endogenous plasma levels of mineralocorticoids; *black*, DCT; *stipled*, CNT; *white*, CCD; *1a*, low sodium, high potassium diet; *1b*, same diet, additional treatment with canrenoate-K^+; plasma levels of mineralocorticoid were lower than in la, but higher than in controls; *2a*, low sodium, high potassium diet associated with very low endogenous plasma levels of mineralocorticoids; *2b*, high sodium diet, potassium intake reduced to half of that of control diet, additional treatment with desoxycorticosterone acetate (DOCA); influence of sodium intake is the most pronounced in the DCT whereas changes in mineralocorticoid plasma levels have little influence on the structural equivalents of the transport machinery in this segment; in contrast, in the CNT and CCD the influence of the hormone is highly evident, even in combination with low potassium intake. (Data from KAISSLING and LE HIR 1982, with permission)

(Ellison et al. 1993) and sites of thiazide-sensitive Na^+Cl^- transport (Ellison et al. 1987).

2. Role of Steroid Hormones

The role of mineralocorticoids in Na^+ reabsorption and K^+ secretion by distal segments has been demonstrated by biochemical (Lombès 1990; Kenouch et al. 1992), functional and morphological studies. While in the first segment beyond Henle's loop, in the DCT, the tubular Na^+ load per se seems to influence function and structure more than steroid hormones, their effect on CNT and cortical CD cells is the most pronounced and evident also in CD cells on the outer stripe of the OM. In the inner stripe and the inner medulla aldosterone seems to be ineffective with respect to stimulation of Na^+ and K^+ transport. Treatment of rabbits on a standard diet with desoxycorticosterone acetate (DOCA) provoked the most significant increase in membrane area in cortical CD cells (Wade et al. 1979). With an additional high Na^+ intake the structural change was even more dramatic (Fig. 19) (Kaissling and Le Hir 1982).

Endogenous high plasma levels of mineralocorticoids, provoked by a low sodium, high potassium or by a high potassium intake alone, stimulated the transport machinery in CNT and CCD cells similarly to that in exogenous hormones (Figs. 18, 19) (Kaissling and Le Hir 1982; Kashgarian et al. 1987; Levine and Jacobson 1986; Le Hir et al. 1982; Stanton et al. 1981, 1985a). Application of the mineralocorticoid antagonist canrenoate-K^+ simultaneously with the low Na^+, high K^+ diet was associated with a smaller basolateral membrane area in CCD cells than in animals on the same diet but without the drug (Kaissling and Le Hir 1982; Le Hir et al. 1982) (Fig. 19).

The described effect of pharmacological doses of glucocorticoids on the structure and function of the collecting duct of rabbits (Wade et al. 1979) is likely to be an indirect one, related to their effect on GFR (Stanton and Kaissling 1989). Changes in the latter and in the ensuing tubular solute load have also been implicated in the hypertrophic response of cortical CD cells to 75% nephrectomy (Zalups et al. 1985).

3. Role of Potassium Intake

a) *High K^+ Intake*

Hyperkaliemia is normally associated with high plasma levels of mineralocorticoids and corresponding structural modulation (Stanton et al. 1981; Kashgarian et al. 1987) (Fig. 17). However, also at fixed aldosterone levels a chronic high K^+ diet increases the transport capacity in cortical cells, probably via an increase in the Na^+ conductance of the luminal membrane of the cells (Stanton et al. 1987).

b) Low K^+ Intake

In rats fed a K^+-deficient diet the most pronounced structural changes were observed within the collecting duct in the inner stripe of the outer medulla (Evan et al. 1980; Hansen et al. 1980; Oliver et al. 1957; Stetson et al. 1980; Toback et al. 1976). They consisted of hypertrophy of both cell types, the collecting duct cells proper and the intercalated cells (Elger et al. 1992). The cellular hypertrophy was even enhanced in rats with 75% nephrectomy (Zalups and Henderson 1992), whereas the cells in the cortical collecting duct were unaffected by this condition (Zalups 1989).

4. Structural Changes in Intercalated Cells

Intercalated cells make a major contribution to the regulation of the acid/base metabolism, including the fine control of K^+ excretion. The ultrastructural response of IC cells to changes in electrolyte and acid-base homeostasis (Fig. 20) has contributed significantly to the understanding of the function of these cells.

a) K^+ Intake

The fundamental study by Stetson and collegues (1980) describing an increase in the surface area of rat medullary IC cells and an increase in the density of rod-shaped intramembranous particles in animals with chronic potassium depletion indicated an involvement of IC cells in collecting duct potassium reabsorption. Since then, evidence for the existence of a P-type $(K^+ + H^+)$-ATPase, analogous to the proton pump in gastric parietal cells, in renal collecting duct IC cells has been presented in a number of studies (Wingo et al. 1990; Brown et al. 1990; Cheval et al. 1991; Stanton 1989; Silver and Frindt 1993; Curran et al. 1992), showing that this pump may be important for both potassium reabsorption and proton secretion by the collecting duct.

Fig. 20A–F. Intercalated cells in midcortical collecting duct from rabbit. **A** Intercalated cell from a rabbit on a low-sodium, high-potassium diet and additional treatment with canrenoate K; the cell resembles a type A intercalated cell and reveals a tremendous density of mitochondria, and a large apical cell pole with elaborate microfold formation. **B** Intercalated cell, resembling a type B intercalated cell, from a rabbit on a high-sodium, reduced potassium diet and additional DOCA treatment; the apical cell pole has a cell organelle-free rim and appears constricted, with a tuft of microvilli. **C** Intercalated cell resembling a type-B intercalated cell, from an untreated control animal; the luminal surface is small, the cell reveals a cell organelle-free rim and the apical cell pole displays numerous small, and a few larger vesicles. **D** Luminal membrane of a type A intercalated cell. **E** Basolateral membrane infolding, often present in type-B intercalated cells. **F** Vesicles in the apical cell pole of an intercalated cell membrane; the membranes, shown in **D–F**, reveal on their cytoplasmic face a coat of specific studs (*arrowheads*) characteristic of intercalated cells; *arrow in* **E**, clathrin-coated pit; *Bars*, **A–C** 1 μm; **D–F**, 0.1 μm

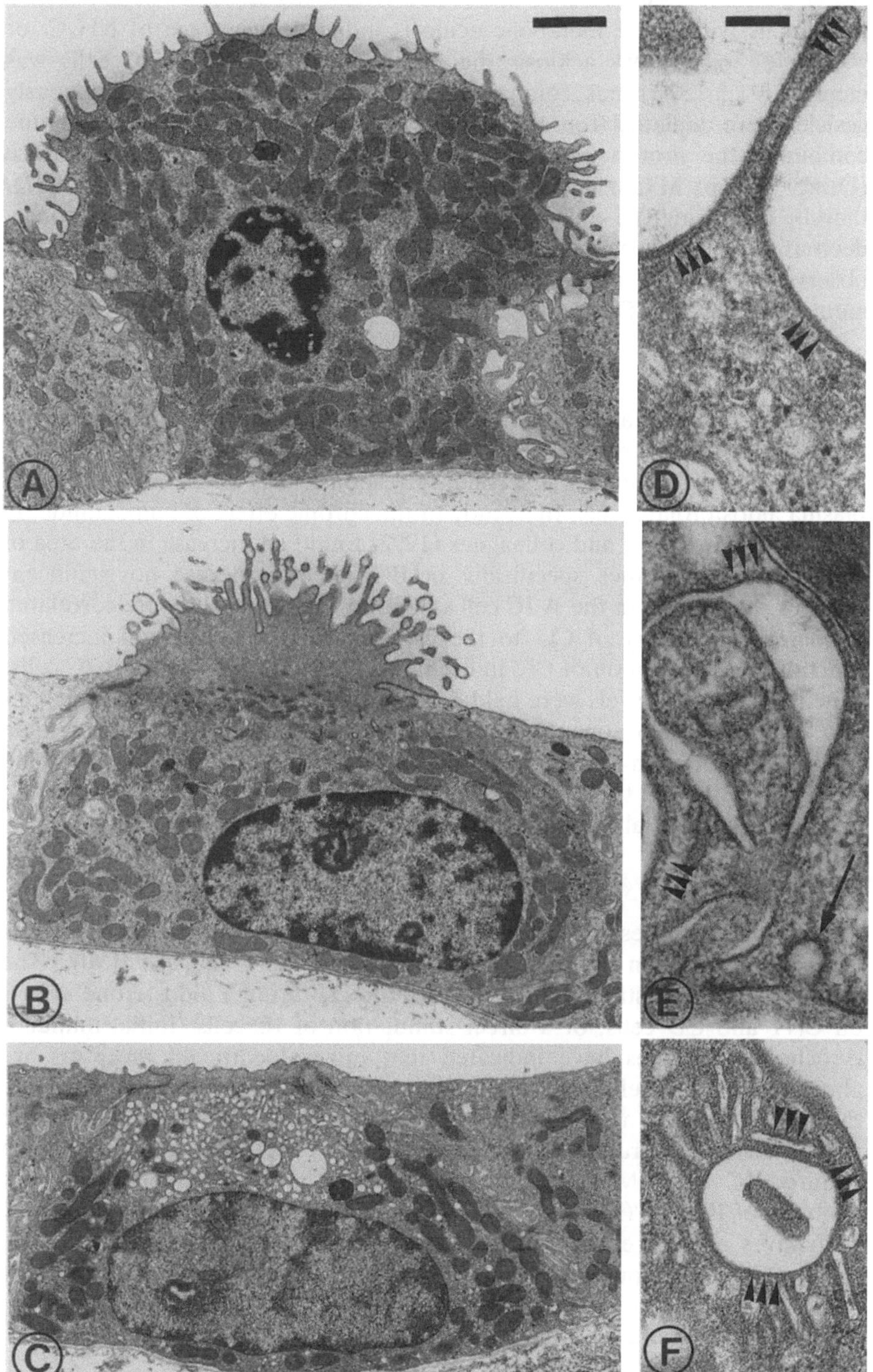
A
B
C
D
E
F

b) Acid/Base Changes

In animals with acute metabolic acidosis induced by gavage of NH_4Cl or with acute respiratory acidosis the luminal surface area of IC cells was increased 1.5–2.4 times compared with control animals. Simultaneously vesicles were depleted from the apical cytoplasm, indicating that membrane containing the proton pump was transferred to the luminal cell surface (DØRUP 1985b; MADSEN and TISHER 1983, 1984; VERLANDER et al. 1987), thereby increasing the capacity for proton secretion. An opposite response, decreasing luminal membrane area and increasing vesicle number, was observed in IC cells from animals with acute metabolic alkalosis after treatment with $KHCO_3$ (DØRUP 1985b).

c) Loop Diuretics

Loop diuretics such as furosemide or bumetamide, through their inhibitory action on TAL Na^+ reabsorption, increase the load of Na^+ and chloride on the distal convolution. As a secondary response they cause an increase in plasma aldosterone levels. In a study of the chronic effect of bumetamide on IC cell structure, KIM and colleagues (1992) found an increase in the area of basolateral membranes specifically in B-IC cells, whereas no significant changes were found in the A IC cell subgroup. These results were correlated to increased delivery of Cl^- to the cortical collecting duct and increased function in reabsorption of Cl^- in exchange of HCO_3^- by cortical B-IC cells. When aldosterone levels were held constant, by adrenalectomy and substitution therapy of hormones and electrolytes, however, chronic furosemide treatment resulted in marked increases in cell area and mitochondrial volume density in all IC cell types in the connecting tubule and the cortical collecting duct (KAISSLING et al., unpublished observations).

d) Aldosterone and Spironolactone

Aldosterone, besides its effects on Na^+ and K^+ homeostasis, stimulates urinary acidification by the collecting duct (STONE et al. 1990). The mechanism of this stimulation may be indirect, through aldosterone effects on CNT and CD cells, or a direct stimulation of IC cells. Initial immunocytochemical studies have indicated that mineralocorticoid receptors are abundant in CNT cells and CCD cells and absent from intercalated cells in the rat (RUNDLE et al. 1989). However, they may be present in a subgroup of IC cells in the rabbit (FARMAN et al. 1991) and, in a recent study in cultured IC cells, MR receptors were found in small amounts in both subtypes of IC cells (NARAY-FEJES-TOTH et al. 1992). Exogenous aldosterone derivatives or high endogenous aldosterone levels (chronic low Na^+, high K^+ diet) have previously been shown to be associated with increased volume density of mitochondria in rabbit collecting duct IC cells (KAISSLING et al. 1981; KAISSLING and LE HIR 1991).

Spironolactone inhibits the effect of aldosterone on the collecting duct through competitive inhibition at the receptor site (Fanestil 1988). In a recent study of chronic spironolactone treatment in rats, we have demonstrated a reduction in cell area and volume density of mitochondria in A-IC cells, whereas no significant structural changes were found in other IC cells (Dørup and Kaissling, in preparation). These results are compatible with inhibition, by spironolactone, of a direct stimulatory effect of aldosterone on A IC cells, but nevertheless the results do not exclude that the structural changes were the result of indirect mechanisms.

Acknowledgments. Most of the author's (B.K.) studies cited in this article were supported by grants from the Swiss National Foundation. We wish to thank Ms. Margrit Müller for photographic assistance.

References

Abrahamson DR (1987) Structure and development of the glomerular capillary wall. Am J Physiol 253:F783–F794

Alcorn D, Anderson WP, Ryan GB (1986) Morphological changes in the renal macula densa during natriuresis and diuresis. Renal Physiol 9:335–347

Alper SL, Natale J, Gluck S, Lodishand HF, Brown D (1989) Subtypes of intercalated cells in rat kidney collecting duct defined by antibodies against erythroid band 3 and renal vacuolar H^+ATPase. Proc Natl Acad Sci USA 80:5429–5433

Andrews PM (1988) Morphological alteration of the glomerular (visceral) epithelium in response to pathological and experimental situations. J Electron Microsc Tech 9:115–144

Bachmann S, Koeppen-Hagemann I, Kriz W (1985) Ultrastructural localization of Tamm-Horsfall glycoprotein (THP) in rat kidney as revealed by protein A-gold immunocytochemistry. Histochemistry 83:531–538

Bachmann S, Le Hir M, Eckardt KU (1993) Colocalization of erythropoietin mRNA and ecto-5′-nucleotidase immunoreactivity in peritubular cells of the rat renal cortex suggests that fibroblasts produce erythropoietin. J Histochem Cytochem 41:335–341

Baines AD, Rouffignac C (1969) Functional heterogeneity of nephrons. II. Filtration rates, intraluminal flow velocities and fractional water reabsorption. Pflugers Arch 308:260–276

Bankir L, Rouffignac Cde (1985) Urinary concentrating ability: insights from comparative anatomy. Am J Physiol 249:R643–R666

Bankir L, Kaissling B, Rouffignac Cde, Kriz W (1979) The vascular organization of the kidney of *Psammomys obesus*. Anat Embryol 155:149–160

Bankir L, Bouby N, Trinh-Trang-Tan MM, Kaissling B (1987) Thick ascending limb – anatomy and function: role in urine concentrating mechanism. Adv Nephrol 16:69–102

Bankir L, Fischer C, Fischer S, Jukkala K, Specht H-C, Kriz W (1988) Adaptation of the rat kidney to altered water intake and urine concentration. Pflugers Arch 412:42–53

Bankir L, Bouby N, Trinh Trang Tan MM (1989) The role of the kidney in the maintenance of water balance. Baillieres Clin Endocrinol Metab 3:249–311

Barajas L (1971) Renin secretions: an anatomical basis for tubular control. Science 172:488–487

Barajas L, Powers K (1990) Monoaminergic innervation of the rat kidney: a quantitative study. Am J Physiol 259:F503–F511

Barajas L, Powers K, Carretero O, Scicli AG, Inagami T (1986) Immunocytochemical localization of renin and kallikrein in the rat renal cortex. Kidney Int 29:965–970

Barajas L, Salido EC, Laborde NP, Fisher DA (1987) Nerve growth factor immunoreactivity in mouse kidney: an immunoelectron microscopic study. J Neurosci Res 18:418–424

Barlet-Bas C, Khadouri S, Marsy S, Doucet A (1988) Sodium-independent in vitro induction of Na-K-ATPase by aldosterone in renal target cells: permissive effect of triiodothyronine. Proc Natl Acad Sci USA 85:1707–1711

Barlet-Bas C, Khadouri C, Marsy S, Doucet A (1990) Enhanced intracellular sodium concentration in kidney cell recruits: a latent pool of Na-K-ATPase whose size is modulated by corticosteroids. J Biol Chem 264:7799–7803

Bell PD, Kirik K, Ribadeneira M, Barfuss D (1985) Direct visualization of the isolated and perfused macula densa. Kidney Int 27:303

Bergeron M, Gaffiero P, Berthelet F, Thiery G (1988) Interrelationship between organelles in kidney cells of adult and developing rat. Pediatr Nephrol 2:100–107

Biber J, Custer M, Werner A, Kaissling B, Murer H (1993) Localization of NaPi-1, a Na/Pi-cotransporter, in rabbit proximal tubules. II. Localization by immunohistochemistry. Pflugers Arch 424:210–215

Blot-Chabaud M, Jaisser F, Gingold M, Bonvalet JP, Farman N (1988) Na^+-K^+-ATPase-dependent sodium flux in cortical collecting tubule. Am J Physiol 255:F605–F613

Blot-Chabaud M, Wanstok F, Bonvalet J-P, Farman N (1989) Cell sodium-induced recruitment of Na^+-K^+-ATPase pumps in rabbit cortical collecting tubules is aldosterone-dependent. J Biol Chem 265:11676–11681

Borke JL, Minami J, Verma A, Penniston JT, Kumar R (1987) Monoclonal antibodies to human erythrozyte membrane Ca^{2+}-Mg^{2+} adenosine triphophatase pump recognize an epitope in the basolateral membrane of human kidney distal tubule cells. J Clin Invest 80:1225–1231

Bouby N, Bankir L (1988) Effect of high protein intake on sodium potassium-dependent adenosine triphosphate activity in the thick ascending limb of Henle's loop in the rat. Clin Sci 74:319–329

Bouby N, Bankir L, Trinh-Trang-Tan MM, Minuth WW, Kriz W (1985) Selective ADH-induced hypertrophy of the medullary thick ascending limb in Brattleboro rats. Kidney Int 28:456–466

Briggs JP, Schnermann J (1987) The tubuloglomerular feedback mechanism: functional and biochemical aspects. Annu Rev Physiol 49:251–273

Brown D (1989) Membrane recycling and epithelial cell function. Am J Physiol 256:F1–F12

Brown D, Orci L (1986) The "coat" of kidney intercalated cell tubulovesicles does not contain clathrin. Am J Physiol 250:605–608

Brown D, Kumpulainen T, Roth J, Orci L (1983) Immunohistochemical localization of carbonic anhydrase in postnatal and adult rat kidney. Am J Physiol 245:F110–F118

Brown D, Gluck S, Hartwig J (1987) Structure of the novel membrane-coating material in proton-secreting epithelial cells and identification as a proton translocating ATPase. J Cell Biol 105:1637–1648

Brown D, Hirsch S, Gluck S (1988a) Localization of a proton-pumping ATPase in rat kidney. J Clin Invest 82:2114–2126

Brown D, Weyer P, Orci L (1988b) Vasopressin stimulates endocytosis in kidney collecting duct principal cells. J Cell Biol 46:336–341

Brown D, Sabolic I, Gluck S (1991) Colchicine-induced redistribution of proton pumps in kidney epithelial cells. Kidney Int 40 [Suppl 33]:S-79–S-81

Brown D, Lui B, Gluck S, Sabolic I (1992) A plasma membrane proton ATPase in specialized cells of rat epididymis. Am J Physiol 263:C913–C916

Brown NL, Madsen KM, Wingo CS, Smolka A, Tisher CC (1990) Translocation of H-K-ATPase to the apical membrane in interalated cells (IC) of rat outer medullary collecting duct (OMCD) during potassium depletion. Kidney Int 37:560 (abstract)

Burckhardt G, Kinne RKH (1992) Transport proteins: cotransporters and countertransporters. In: Seldin DW, Giebisch G (eds) The kidney: physiology and pathophysiology. Raven, New York, pp 537–586

Chen M, Schnermann J, Malvin RL, Killen PD, Briggs JP (1993) Time course of stimulation of renal messenger RNA by furosemide. Hypertension 21:36–41

Cheval L, Barlet-Bas C, Khadouri C, Feraille E, Marsy S, Doucet A (1991) K^+-ATPase-mediated Rb^+ transport in rat collecting tubule: modulation during K^+ deprivation. Am J Physiol 260:F800–F805

Christensen EI, Nielsen S (1991) Structural and functional features of protein handling in the kidney proximal tubule. Semin Nephrol 11:414–439

Christensen JA, Bohle A (1978) The juxtaglomerular apparatus in the normal rat kidney. Virchows Arch [A] 379:143–150

Clapp WL, Madsen KM, Verlander JW, Tisher CC (1987) Intercalated cells of the rat inner medullary collecting duct. Kidney Int 31:1080–1087

Coutry N, Blot Chabaud M, Mateo P, Bonvalet JP, Farman N (1992) Time course of sodium-induced Na(+)-K(+)-ATPase recruitment in rabbit cortical collecting tubule. Am J Physiol 263:C61–C68

Curran KA, Hebert MJ, Cain BD, Wingo CS (1992) Evidence for the presence of a K-dependent acidifying adenosine triphosphatase in the rabbit renal medulla. Kidney Int 42:1093–1098

Davis G, Johns EJ (1990) The effect of angiotensin II and vasopressin on renal haemodynamics. J Med Eng Technol 14:197–200

Davis RG, Madsen KM, Fregly MJ, Tisher CC (1983) Kidney structure in hypothyreoidism. Am J Pathol 113:41–49

Dietl P, Kizer N, Stanton BA (1992) Conductive properities of a rabbit cortical collecting duct cell line: regulation by isoproterenol. Am J Physiol 262: F578–F582

Dobyan DC, Bulger RE (1982) Renal carbonic anhydrase. Am J Physiol 243:F311–F324

Dobyan DC, Magill LS, Friedman PA, Hebert SC, Bulger RE (1982) Carbonic anhydrase histochemistry in rabbit and mouse kidneys. Anat Rec 204:185–197

Dørup J (1985a) Ultrastructure of distal nephron cells in rat renal cortex. J Ultrastruct Res 92:101–118

Dørup J (1985b) Structural adaptation of intercalated cells in rat renal cortex to acute metabolic acidosis and alkalosis. J Ultrastruct Res 92:119–131

Dørup J (1988) Ultrastructure of three dimensionally localized distal nephron segments in superficial cortex of rat kidney. J Ultrastruct Mol Struct Res 99: 169–187

Dørup J, Morsing P, Rasch R (1992) Tubule-tubule and tubule-arteriole contacts in rat kidney distal nephrons. Lab Invest 67:761–769

Doucet A, Marsy S (1987) Characterization of K-ATPase activity in distal nephron: stimulation by potassium depletion. Am J Physiol 253:F418–F423

Drenckhahn D, Schlüter K, Allen DP, Bennett V (1985) Colocalization of band 3 with ankyrin and spectrin at the basal membrane of intercalated cells in the rat kidney. Science 230:1287–1289

Drenckhahn D, Schnittler H, Nobiling R, Kriz W (1990) Ultrastructural organization of contractile proteins in rat glomerular mesangial cells. Am J Pathol 137: 1343–1351

Dworkin LD, Brenner BM (1992) Biophysical basis of glomerular filtration. In: Seldin D, Giebisch G (eds) The kidney: physiology and pathophysiology. Raven, New York, pp 977–1016

Edelman IS (1975) Mechanism of action of steroid hormones. J Steroid Biochem 6:147–159

Elger M, Sakai T, Kriz W (1990) Role of mesangial cell contraction in adaptation of the glomerular tuft to changes in extracellular volume. Pflugers Arch 415: 598–605
Elger M, Bankir L, Kriz W (1992) Morphometric analysis of kidney hypertrophy in rats after chronic potassium depletion. Am J Physiol 262:F656–F667
Elger M, Mundel P, Kriz W (1993) The contractile apparatus of podocytes is arranged to counteract glomerular pressure. Nieren Hochdruckkrankh 22: 474
Ellison DH, Velasquez H, Wright F (1987) Thiazide-sensitive sodium chloride cotransport in early distal tubule. Am J Physiol 253:F546–F554
Ellison DH, Velazquez H, Wright FS (1989) Adaptation of the distal convoluted tubule of the rat. Structural and functional effects of dietary salt intake and chronic diuretic infusion. J Clin Invest 83:113–126
Ellison DH, Biemesderfer D, Morrisey J, Lauring J, Desir GV (1993) immunocytochemical characterization of the high-affinity thiazide diuretic receptor in rabbit renal cortex. Am J Physiol 264:F141–F148
Emmons CL, Matsuzaki K, Stokes UB, Schuster VL (1991) Axial heterogeneity of rabbit cortical collecting duct. Am J Physiol 260:F498–F505
Ernst SA (1975) Transport ATPase cytochemistry: ultrastructural localization of potassium-dependent and potassium-independent phosphatase activities in rat kidney cortex. J Cell Biol 66:586–608
Evan A, Huser J, Bengele HH, Alexander EA (1980) The effect of alterations in dietary potassium on collecting system morphology in the rat. Lab Invest 42: 668–675
Fanestil DD (1988) Mechanism of action of aldosterone blockers. Semin Nephrol 8:249–263
Farman N, Oblin ME, Lombes M, Delahaye F, Westphal HM, Bonvalet JP, Gasc JM (1991) Immunolocalization of gluco- and mineralocorticoid receptors in rabbit kidney. Am J Physiol 260:C226–C233
Field M, Giebisch G (1985) Hormonal control of renal potassium excretion. Kidney Int 27:379–387
Figueroa CD, Lewis HM, MacIver AG, Mackenzie JC, Bhoola KD (1990) Cellular localisation of atrial natriuretic factor in the human kidney. Nephrol Dial Transplant 5:25–31
Fine LG, Badie Dezfooly B, Lowe AG, Hamzeh A, Wells J, Salehmoghaddam S (1985) Stimulation of Na^+/H^+ antiport is an early event in hypertrophy of renal proximal tubular cells. Proc Natl Acad Sci USA 82:1736–1740
Fourman J, Moffat DB (1971) The blood vessels of the kidney. Blackwell Scientific, Oxford
Fushimi K, Uchida S, Hara Y, Hirata Y, Marumo F, Sasaki S (1993) Cloning and expression of apical water channel of rat kidney collecting tubule. Nature 361: 549–552
Ganote CE, Grantham JJ, Moses HL (1968) Ultrastructural studies of vasopressin effect on isolated perfused renal collecting tubules of the rabbit. J Cell Biol 36:355–367
Gesek FA, Friedman PA (1993) Calcitonin stimulates calcium transport in distal convoluted tubule cells. Am J Physiol 264:F744–F751
Gluck S (1992) The osteoclast as a unicellular proton-transporting epithelium. Am J Med Sci 303:134–139
Gluck S, Cannon C, Al-Awqati Q (1982) Exocytosis regulates urinary acidification in turtle bladder by rapid insertion of H^+ pumps into the luminal membrane. Proc Natl Acad Sci USA 79:4327–4331
Gorgas K (1978) Structure and innervation of the juxtaglomerular apparatus of the rat. Adv Anat Embryol Cell Biol 54:5–84
Guder WG, Wagner S, Wirthensohn G (1986) Metabolic fuels along the nephron: pathways and intracellular mechanisms of interaction. Kidney Int 29:41–45

Guder WG, Hallbach J, Fink E, Kaissling B, Wirthensohn G (1987) Kallikrein (kiniogenese) in the mouse nephron: effect of dietary potassium. Biol Chem Hoppe Seyler 368:637–645
Hammermann MR, O'Shea M, Miller SB (1993) Role of growth factors in regulation for renal growth. Annu Rev Physiol 55:305–321
Hansen GP, Tisher CC, Robinson RR (1980) Response of the collecting duct to disturbances of acid-base and potassium balance. Kidney Int 17:326–337
Hasegawa H, Zhang R, Dohrman A, Verkman AS (1993) Tissue-specific expression of mRNA encoding rat kidney water channel CHIP28k by in situ hybridization. Am J Physiol 264:C237–C245
Hayashi M, Schuster VL, Stokes JB (1988) Absence of transepithelial anion exchange by rabbit OMCD: evidence against reversal of cell polarity. Am J Physiol 225:F220–F228
Healy DP, Fanestil DD (1986) Localization of atrial natriuretic peptide binding sites within the rat kidney. Am J Physiol 250:F573–F578
Holmer S, Rinne B, Eckardt KU, Le Hir M, Kaissling B, Riegger G, Kurtz A (1994) Role of renal nerves for the expression of renin in the rat kidney. Am J Physiol 266:F738–F745
Holthofer H, Schulte BA, Pasternack G, Siegel GJ, Spicer SS (1987) Immunocytochemical characterization of carbonic anhydrase-rich cells in the rat kidney collecting duct. Lab Invest 57:150–156
Hoyer JR, Sisson SP, Vernier RL (1979) Tamm-Horsfall glycoprotein ultrastructural immunoperoxidase localization in rat kidney. Lab Invest 41:168–173
Imbert-Teboul M, Chabardes D, Morel F (1980) Vasopressin and catecholamine sites of action along rabbit, mouse and rat nephron. In: Bahlman, J, Brod J (eds) Disturbance of water and electrolyte metabolism. Karger, Basel, pp 41–47 (Contributions to nephrology, vol 21)
Inke G (1987) The protolobar structure of the human kidney. Liss, New York
Jamison RL, Kriz W (1982) Urinary concentrating mechanism. Structure and function. Oxford University Press, Oxford
Jones DB (1985) Scanning electron microscopy of basolateral surfaces or rat renal tubules isolated by sequential digestion. Anat Rec 213:121–130
Jørgensen F, Bentzon MW (1968) The ultrastructure of the normal human glomerulus. Thickness of glomerular basement membrane. Lab Invest 18:42–48
Jørgensen PL (1986) Structure, function and regulation of Na, K-ATPase in the kidney. Kidney Int 29:10–20
Kachadorian WA, Levine SD, Wade JB, DiScala VA, Hays M (1977) Relationship of aggregated intramembraneous particles to water permeability in vasopressin-treated toad urinary bladder. J Clin Invest 59:576–581
Kaissling B (1980) Ultrastructural organization of the transition from the distal nephron to the collecting duct in the desert rodent Psammomys obesus. Cell Tissue Res 212:475–495
Kaissling B (1982) Structural aspects of adaptive changes in renal electrolyte excretion. Am J Physiol 243:F211–F226
Kaissling B (1985) Cellular heterogeneity of the distal nephron and its relation to function. Klin Wochenschr 63:868–876
Kaissling B, Kriz W (1979) Structural analysis of rabbit kidney. Adv Anat Embryol Cell Biol 56:1–123
Kaissling B, Kriz W (1982) Variability of intercellular spaces between macula densa cells: a transmission electron microscopic study in rabbits and rats. Kidney Int 12:S9–S17
Kaissling B, Kriz W (1992) Morphology of the loop of Henle distal tubule, and collecting duct. In: Windhager EE (ed) Handbook of physiology. OUP USA (American Physiological Society), New York, pp 109–167
Kaissling B, Le Hir M (1982) Distal tubular segments of the rabbit kidney after adaptation of altered Na- and K-intake. I. Structural changes. Cell Tissu Res 224:469–492

Kaissling B, Le Hir M (1985) Anpassung distaler Tubulussegmente an Änderung im Elektrolythaushalt. Acta Histochem Suppl XXXI:185–191
Kaissling B, Le Hir M (1991) Aldosterone: influence on distal tubule cell structure. In: Bonvalet JP, Farman N, Lombès M, Rafestin-Oblin ME (eds) Aldosterone: fundamental aspects. Colloque INSERM, Paris, pp 175–185
Kaissling B, Le Hir M (1994) Characterization and distribution of interstitial cell types in the renal cortex of rat. Kidney Int 45:709–720
Kaissling B, Stanton B (1988a) Adaptation of distal tubule and collecting duct to increased sodium delivery. I. Ultrastructure. Am J Physiol 255:F1256–F1268
Kaissling B, Stanton B (1988b) Chronic furosemide treatment alters the ultrastructure of intercalated cells in renal collecting ducts. Clin Res 36:521A
Kaissling B, Stanton BA (1992) Structure-function correlation in electrolyte transporting epithelia. In: Seldin DW, Giebisch G (eds) The kidney: physiology and pathophysiology. Raven, New York, pp 779–801
Kaissling B, Rouffignacs Cde, Barrett JM, Kriz W (1975) The structural organization of the kidney of the desert roden *Psammomy obesus*. Anat Embryol 148: 121–143
Kaissling B, Peter S, Kriz W (1977) The transition of the thick ascending limb of Henle's loop into the distal convoluted tubule in the nephron of the rat kidney. Cell Tissue Res 182:111–118
Kaissling B, Koeppen BM, Le Hir M, Wade JB (1981) Effect of mineralocorticoids on the structure of intercalated cells in renal cortical collecting ducts. Acta Anat 111:72
Kaissling B, Bachmann S, Kriz W (1985) Structural adaptation of the distal convoluted tubule to prolonged furosemide treatment. Am J Physiol 248: F374–F381
Kaissling B, Spiess S, Rinne B, Le Hir M (1993) Effects of anemia on the morphology of the renal cortex of rats. Am J Physiol 264:F608–F617
Kanwar YS, Jakubowski ML, Rosenzweig LJ, Gibbons JT (1984a) De novo cellular synthesis of sulfated proteoglycans of the developing renal glomerulus in vivo. Proc Natl Acad Sci USA 81:7108–7111
Kanwar YS, Veis A, Kimura JH, Jakubowski ML (1984b) Characterization of heparan sulfate-proteoglycan of glomerular basement membranes. Proc Natl Acad Sci USA 81:762–766
Karnovsky M (1979) The structural bases for glomerular filtration. In: Churg J et al. (eds) The kidney disease. Present status. Williams and Williams, Baltimore (IAP monograph 20)
Kashgarian M, Biemesderfer D, Caplan M, Forbush B (1985) Monoclonal antibody to Na,K-ATPase: immunocytochemical localization along nephron segments. Kidney Int 28:899–913
Kashgarian M, Ardito T, Hirsch DT, Hayslett JP (1987) Response of collecting tubule cells to aldosterone and potassium loading. Am J Physiol 253:F8–F14
Kenouch S, Coutry N, Farman N, Bonvalet J-P (1992) Multiple patterns of 11b-hydroxysteroid dehydrogenase catalytic activity along the mammalian nephron. Kidney Int 42:56–60
Kerjaschki D, Sharkey DJ, Farquhar MG (1984) Identification and characterization of podocalyxi – the major sialoprotein of the renal glomerular epithelial cell. J Cell Biol 98:1591–1596
Khadouri C, Bas-Barlet C, Doucet A (1987) Mechanism of increased tubular Na-K-ATPase during streptozotocin induced diabetes. Pflugers Arch 409:296–301
Kim J, Tisher CC, Linser PJ, Madsen KM (1990) Ultrastructural localization of carbonic anhydrase II in subpopulations of intercalated cells of the rat kidney. J Am Soc Nephrol 1:245–256
Kim J, Welch WJ, Cannon JK, Tisher CC, Madsen KM (1992) Immunocytochemical response of type A and type B intercalated cells to increased sodium chloride delivery. Am J Physiol 262:F288–F302

Kirk KL, Bell PD, Barfuss DW, Ribadeneira M (1985) Direct visualization of the isolated and perfused macula densa. Am J Physiol 248:F890–F894
Knepper MA, Sands JM, Nonoguchi H, Star RA, Packer RK (1988) Inner medullary collecting duct. In: Davidson AM (ed) Nephrology. Proceedings of the Xth international congress of nephrology, vol I. Bailliere Tindall, London, pp 317–331
Koechlin N, Elalouf JM, Kaissling B, Roinel N, Rouffignac CDE (1989) A structural study of the rat proximal and distal nephron: effect of peptide and thyroid hormones. Am J Physiol 256:F814–F822
Koeppen BM (1988) Electrophysiology of the outer medullary collecting duct. In: Davidson AM (ed) Nephrology. Proceedings of the Xth international congress of nephrology, vol I. Bailliere Tindall, London, pp 304–316
Koob R, Zimmerman M, Schoner W, Drenckhahn D (1988) Colocalization and coprecipitation of ankyrin and Na^+,K^+-ATPase in kidney epithelial cells. Eur J Cell Biol 45:230–237
Kopito RR (1990) Molecular biology of the anion exchanger gene family. Int Rev Cytol 123:177–199
Koseki C, Hayashi Y, Torikai S, Furuya M, Ohnuma N, Imai M (1986) Localization of binding sites for rat atrial natriuretic polypeptide in rat kidney. Am J Physiol 250:F210–F216
Kriz W (1981) Structural organization of the renal medulla: comparative and functional aspects. Am J Physiol 241:R3–R16
Kriz W (1983) Structural organization of the renal medullary counterflow system. Fed Proc 42:2379–2385
Kriz W (1987) A periarterial pathway for intrarenal distribution of renin. Kidney Int Suppl 20:51–56
Kriz W, Bankir L (1988) A standard nomenclature for structures of the kidney. Kidney Int 32:1–7
Kriz W, Kaissling B (1992) Structural organization of the mammalian kidney. In: Seldin DW, Giebisch G (eds) The kidney. Physiology and pathophysiology. Raven, New York, pp 707–777
Kriz W, Napiwotzky P (1979) Structural and functional aspects of the renal interstitium. Contr Nephrol 16:104–108
Kriz W, Barrett JM, Peter S (1976) The renal vasculature: anatomical-functional aspects. Int Rev Physiol 11:1–21
Kriz W, Elger M, Lemley KV, Sakai T (1990) Mesangial cell-glomerular basement membrane connections counteract glomerular capillary and mesangium expansion. Am J Nephrol 10:4–13
Kumar AM, Gupta RK, Spitzer A (1988) Intracellular sodium in proximal tubules of diabetic rats. Role of glucose. Kidney Int 33:792–797
Kurihara H, Anderson JM, Farquhar MG (1992) Diversity among tight junctions in rat kidney: glomerular slit diaphragms and endothelial junctions express only one isoform of the tight junction protein ZO-1. Proc Natl Acad Sci USA 89:7075–7079
Kurtz A (1989) Cellular control of renin secretion. Rev Physiol Biochem Pharmacol 113:2–40
Larsson L, Maunsbach AB (1980) The ultrastructural development of the glomerular filtration barrier in the rat kidney: a morphometric analysis. J Ultrastruct Res 72:392–406
Le Hir M, Kaissling B (1989) Distribution of 5′-nucleotidase in the renal interstitium of the rat. Cell Tissue Res 258:177–182
Le Hir M, Kaissling B (1993) Distribution and regulation of ecto-5′-nucleotidase in the kidney. Implications for the physiological function of adenosine. Am J Physiol 264:F377–F387
Le Hir M, Kaissling B, Dubach UC (1982) Distal tubular segments in the rabbit kidney after adaptation to altered Na- and K-intake. II. Changes in Na-K-ATPase. Cell Tissue Res 224:493–503

Lemley KV, Kriz W (1991) Anatomy of the renal interstitium. Kidney Int 39: 370–382
Levi M, Lötscher M, Sorribas V, Custer M, Arar M, Kaissling B, Murer H, Biber B (1994) Cellular mechanisms of acute and chronic adaptation of rat renal phosphate transporter to alterations in dietary P_i. Am J Physiol 267:F900–F908
Levine DZ, Jacobson HR (1986) The regulation of renal acid secretion: new observations from studies of distal nephron segments. Kidney Int 29:1099–1109
Loffing J, Le Hir M, Kaissling B (1995) Modulation of salt transport rate affects DNA-synthesis in vivo in rat renal tubules. Kidney Int (in press)
Lombès M (1990) Immunohistochemical localization of renal mineralocorticoid receptor by using an anti-idiotypic antibody that is an internal image of aldosterone. Proc Natl Acad Sci USA 87:1086–1088
Lombès M, Farman N, Oblin ME, Baulieu EE, Bonvalet JP, Erlanger BF, Gasc JM (1990) Immunohistochemical localization of renal mineralocorticoid receptor by using an anti-idiotypic antibody that is an internal image of aldosterone. Proc Natl Acad Sci USA 87:1086–1088
Lönnerholm G, Wistrand PJ (1984) Carbonic anhydrase in the human kidney: a histochemical and immunocytochemical study. Kidney Int 25:886–898
Mackovic-Basic M, Fine LG, Norman JT, Cragoe EJ, Kurtz I (1992) Stimulation of Na^+/H^+ exchange is not required for induction of hypertrophy of renal cells in vitro. J Am Soc Nephrol 3:1124–1130
Madsen KM, Tisher CC (1983) Cellular response to acute respiratory acidosis in rat medullary collecting duct. Am J Physiol 14:F670–F679
Madsen KM, Tisher CC (1984) Response of intercalated cells of rat outer medullary collecting duct to chronic metabolic acidosis. Lab Invest 51:268–276
Madsen KM, Tisher CC (1986) Structural-functional relationships along the distal nephron. Am J Physiol 250:F1–F15
Madsen KM, Kim J, Tisher CC (1992) Intracellular band 3 immunostaining in type A intercalated cells of rabbit kidney. Am J Physiol 262:F1015–F1022
Matlin KS, Caplan MJ (1992) Epithelial cell structure and polarity. In: Giebisch G, Seldin D (eds) The kidney: physiology and pathophysiology. Raven, New York, pp 447–473
Maunsbach AB, Christensen IE (1992) Functional ultrastructure of the proximal tubule. In: Windhager EE (ed) Handbook of physiology: renal physiology. Section 8. Oxford University Press, New York, pp 41–107
Maxwell PH, Osmond MK, Pugh ChW, Heryet A, Nicholls LG, Tan CC, Doe BG, Ferguson DKP, Johnson MH, Ratcliffe PJ (1993) Identification of the renal erythropoietin-producing cells using transgenic mice. Kidney Int 44:1149–1162
Messina A, Alcorn D, Ryan GB (1987) Intercellular spaces between macula densa cells: an ultrastructural study comparing high pressure perfusion fixation with in situ drip-fixation of rat kidney. Cell Tissue Res 250:461–464
Modena B, Holmer S, Eckardt K-U, Schricker K, Riegger G, Kaissling B, Kurtz A (1993) Furosemide stimulates renin expression in the kidneys of salt-supplemented rats. Pflugers Arch 424:403–409
Morel F, Doucet A (1986) Hormonal control of kidney functions at the cell level. Physiol Rev 66:377–468
Morel F, Doucet A (1992) Functional segmentation of the nephron. In: Giebisch G, Seldin DW (eds) The kidney: physiology and pathophysiology. Raven, New York, pp 1049–1086
Morel F, Imbert-Teboul M, Chabardes M, Montegut M, Clique A (1978) Impaired response to vasopressin of adenylate cyclase of the thick ascending limb of Henle loop in Brattleboro rats with diabetes insipidus. Renal Physiol 1:3–10
Moriyama Y, Nelson N (1989) H^+-translocating ATPase in Golgi apparatus. Characterization as vacuolar H^+-ATPase and its subunit structures. J Biol Chem 264:18445–18450

Mundel P, Bachmann S, Bader M, Fischer A, Kummer W, Mayer B, Kriz W (1992) Expression of nitric oxide synthase in kidney macula densa cells. Kidney Int 42:1017–1019

Naray-Fejes-Toth A, Rusvai E, Fejes-Toth G (1992) Distribution of mineralocorticoid receptors (mr) and 11-hydroxysteroid dehydrogenase (11-hsd) in principal and intercalated cells. Proceedings of the 25th annual meeting of the American Society of Nephrology, vol 515 (abstract)

Nelson N (1992) Structural conservation and functional diversity of V-ATPases. J Bioenerg Biomembr 24:407–414

Nelson RD, Guo XL, Masood K, Brown D, Kalkbrenner M, Gluck S (1992) Selectively amplified expression of an isoform of the vacuolar H^+-ATPase 56-kilodalton subunit in renal intercalated cells. Proc Natl Acad Sci USA 89: 3541–3545

Nielsen S, DiGiovanni SR, Christensen EI, Knepper MA, Harris HW (1993a) Cellular and subcellular immunolocalization of vasopressin-regulated water channel in rat kidney. Proc Natl Acad Sci USA 90:11663–11667

Nielsen S, Smith BL, Christensen EI, Knepperm MA, Agre P (1993b) CHIP28 water channels are localized in constitutively water-permeable segments of the nephron. J Cell Biol 120:371–383

Oliver J, MacDowell M, Welt LG, Holliday MA, Hollander W, Winters RW, Williams TF, Segar WE (1957) The renal lesions of electrolyte imbalance. I. The structural alterations in potassium-depleted rats. J Exp Med 106:563–574

Peter K (1927) Untersuchungen über Bau und Entwicklung der Niere. Fischer, Jena, pp 1909

Pfaller W (1982) Structure function correlation in rat kidney. Quantitative correlation of structure and function in the normal and injured rat kidney. Adv Anat Embryol Cell Biol 70:1–106

Rasch R (1984) Changes in macula densa of the juxtaglomerular apparatus in experimental diabetes. Diabetalogia 27:323A–324A

Raymond JR, Kim J, Beach RE, Tisher CC (1993) Immunohistochemical mapping of cellular and subcellular distribution of 5-HT_{1A} receptors in rat and human kidneys. Am J Physiol 264:F9–F19

Reinecke M, Forssmann WG (1988) Neuropeptide (neuropeptide Y, neurotensin, vasoactive intestinal polypeptide, substance P, calcitonin gene-related peptide, somatostatin) immunohistochemistry and ultrastructure of renal nerves. Histochemistry 89:1–9

Ridderstrale Y, Kashgarian M, Koeppen BM, Giebisch G, Stetson DL, Ardito T, Stanton BA (1988) Morphological heterogeneity of the rabbit collecting duct. Kidney Int 34:655–670

Rodewald R, Karnovsky MJ (1974) Porous substructure of the glomerular slit diaphragm in the rat and mouse. J Cell Biol 60:423–433

Roth J, Brown D, Norman AW, Orci L (1982) Localization of the vitamin D-dependent calcium-binding protein in the mammalian kidney. Am J Physiol 243:F243–F252

Rundle SE, Smith AI, Stockman D, Funder JW (1989) Immunocytochemical demonstration of mineralocorticoid receptors in rat and human kidney. J Steroid Biochem 33:1235–1242

Sabolic I, Valenti G, Verbavatz JM, Van Hoek AN, Verkman AS, Ausiello Ad, Brown D (1992) Localization of the CHIP28 water channel in rat kidney. Am J Physiol 263:C1225–C1233

Salehmoghaddam S, Bradley T, Mikhail N, Badie Dezfooly B, Nord EP, Trizna W, Kheyfets R, Fine LG (1985) Hypertrophy of basolateral Na-K pump activity in the proximal tubule of the remnant kidney. Lab Invest 53:443–452

Salido EC, Barajas L, Lechago J, Laborde NP, Fisher DA (1986) Immunocytochemical localization of nerve growth factor in mouse kidney. J Histochem Cytochem 34:1155–1160

Salido EC, Lakshmanan J, Fisher DA, Shapiro LJ, Barajas L (1991) Expression of epidermal growth factor in the rat kidney. An immunocytochemical and in situ hybridization study. Histochemistry 96:65–72
Sands JM, Kokko JP, Lacobson HR (1992) Intrarenal heterogeneity: vascular and tubular. In: Giebisch G, Seldin DW (eds) The kidney: physiology and pathophysiology. Raven, New York, pp 1087–1155
Satlin LM, Schwartz GJ (1989) Cellular remodeling of HCO_3^--secreting cells in rabbit renal collecting duct in response to an acidic environment. J Cell Biol 109:1279–1289
Scherzer P, Wald H, Popovitzer MM (1987) Enhanced glomerular filtration and Na^+-K^+-ATPase with furosemide administration. Am J Physiol 252:F910–F915
Schlatter E, Salomonsson M, Pesson AEG, Greger R (1989) Macula densa cells sense luminal NaCl concentration via the furosemide sensitive NaCl-K cation-sporter. Pflugers Arch 414:266–290
Schnermann J, Briggs JP (1992) Function of the juxtaglomerular apparatus: control of glomerular hemodynamics and renin secretion. In: Seldin DW, Giebisch G (eds) The kidney: physiology and pathophysiology. Ravan, New York, pp 1249–1290
Schurek HJ, Jost U, Baumgartl H, Bertram H, Heckmann U (1990) Evidence for a preglomerular oxygen diffusion shunt in rat renal cortex. Am J Physiol 259: F910–F915
Schuster VL (1993) Function and regulation of collecting duct intercalated cells. Annu Rev Physiol 55:267–288
Schuster VL, Bonsib SM, Jennings ML (1986) Two types of collecting duct mitochondria-rich (intercalated) cells: lectin and band 3 cytochemistry. Am J Physiol 251:C347–C355
Schwartz GJ, Barasch J, Al-Awqati Q (1985) Plasticity of functional epithelial polarity. Nature 318:368–371
Seki G, Frömter E (1992) Acetazolamide inhibition of basolateral Cl^-/HCO_3^- exchange in rabbit renal proximal S3 segment. Pflugers Arch 422:55–59
Silver RB, Frindt G (1993) Functional identification of H-K-ATPase in intercalated cells of cortical collecting tubule. Am J Physiol 264:F259–F266
Simon K, Cao Y, Franki N, Hays R (1993) Vasopressin depolymerizes apical F-actin in rat inner medullary collecting duct. Am J Physiol 265:C757–C762
Smith HW (1951) The kidney. Structure and function in health and disease. Oxford University Press, New York
Spanidis A, Wunsch H, Kaissling B, Kriz W (1982) Three-dimensional shape of a Goormaghtigh cell and its contact with a granular cell in the rabbit kidney. Anat Embryol 165:239–252
Spielman WS, Arend LJ (1991) Adenosine receptors and signalling in the kidney. Hypertension 17:117–130
Stanton B, Kaissling B (1988) Adaptation of distal tubule and collecting duct to increased sodium delivery. II. Na^+ and K^+ transport. Am J Physiol 255: F1269–F1275
Stanton B, Kaissling B (1989) Regulation of renal ion transport and cell growth by sodium. Am J Physiol 257:F1–F10
Stanton B, Biemesderfer D, Wade JB, Giebisch G (1981) Structural and functional study of the rat distal nephron: effects of potassium adaptation and depletion. Kidney Int 19:36–48
Stanton B, Giebisch G, Klein-Robbenhaar G, Wade JB, De Fronzo RA (1985a) Effects of adrenalectomy and chronic adrenal corticosteroid replacement on potassium transport in rat kidney. J Clin Invest 75:1317–1326
Stanton B, Janzen A, Klein-Robbenhaar G, De Fronzo RA, Giebisch G, Wade JB (1985b) Ultrastructure of rat initial collecting tubule. Effect of adrenal corticosteroid treatment. J Clin Invest 75:1327–1334

Stanton B, Pan L, Deetjen H, Guckian V, Giebisch G (1987) Independent effects of aldosterone and potassium on induction of potassium adaptation in rat kidney. J Clin Invest 79:198–206
Stanton BA (1989) Renal potassium transport: morphological and functional adaptations. Am J Physiol 257:R989–R997
Stetson DL, Wade JB, Giebisch G (1980) Morphologic alterations in the rat medullary collecting duct following potassium depletion. Kidney Int 17:45–56
Stone DK, Crider BP, Xie XS (1990) Aldosterone and urinary acidification. Semin Nephrol 10:375–379
Takahashi-Iwanaga H (1991) The three-dimensional cytoarchitecture of the interstitial tissue in the rat kidney. Cell Tissue Res 264(2):269
Taugner R, Schiller A, Kaissling B, Kriz W (1978) Gap junctional coupling between JGA and the glomerular tuft. Cell Tissue Res 186:279–285
Tisher CC, Madsen KM (1987) Anatomy of the kidney. In: Brenner BM, Rector FC (eds) The kidney, 3rd edn. vol 1. Saundes, Philadelphia, pp 3–60
Toback FG, Ordonez NG, Bortz SL, Spargo BH (1976) Zonal changes in renal structure and phospholipid metabolism in potassium-deficient rats. Lab Invest 34:115–124
Toback FG, Kartha S, Walsh-Reitz MM (1993) Regeneration of kidney tubular epithelial cells. Clin Invest 71:871–873
Trinh Trang Tan MM, Diaz M, Grunfeld JP, Bankir L (1981) ADH-dependent nephron heterogeneity in rats with hereditary hypothalamic diabetes insipidus. Am J Physiol 240:F372–F380
Trinh Trang Tan MM, Bouby N, Doute M, Bankir L (1984) Effect of long- and short-term antidiuretic hormone availability on internephron heterogeneity in the adult rat. Am J Physiol 246:F879–F888
Trinh Trang Tan MM, Antras J, Levillain O, Bankir L (1993) Adaptation of the medullary thick ascending limb to dietary protein intake. Exp Nephrol 1:158
Verkman AS, Lencer WI, Brown D, Ausiello DA (1988) Endosomes from kidney collecting tubule cells contain the vasopressin-sensitive water channel. Nature 333:268–269
Verlander JW, Madsen KM, Tisher CC (1987) Effect of acute respiratory acidosis on two populations of intercalated cells in rat cortical collecting duct. Am J Physiol 252:F1142–F1156
Verlander JW, Madsen KM, Larsson L, Cannon JK, Tisher CC (1989) Immunocytochemical localization of intracellular acidic compartments: rat proximal nephron. Am J Physiol 257:F454–F462
Vio CP, Figueroa CD (1985) Subcellular localization of renal kallikrein by ultrastructural immunocytochemistry. Kidney Int 28:36–42
Wade JB, O'Neil RG, Pryor JL, Boulpaep EL (1979) Modulation of cell membrane area in renal collecting tubules by corticosteroid hormones. J Cell Biol 81: 439–445
Wade JB, Stetson DL, Lewis SA (1981) ADH action: evidence for a membrane shuttle mechanism. Ann NY Acad Sci 372:106–117
Wade JB, Stanton B, Field MJ, Kashgarian M, Giebisch G (1990) Morphological and physiological responses to aldosterone: time course and sodium dependence. Am J Physiol 259:F88–F94
Wade JB, Stanton BA, Brown D (1992) Structural correlates of transport in distal tubule and collecting duct segments. In: Windhager EE (ed) Handbook of physiology: renal. Oxford University Press, New York, pp 1–10
Wagner S, Vogel R, Lietzke R, Koob R, Drenckhahn D (1987) Immunocytochemical characterization of a band 3-like anion exchanger in collecting duct of human kidney. Am J Physiol 253:213–221
Welling LW, Welling DJ (1975) Surface areas of brush border and lateral cell walls in the rabbit proximal nephron. Kidney Int 8:343–348

Welling LW, Welling DJ (1976) Shape of epithelial cells and intercellular channels in the rabbit proximal nephron. Kidney Int 9:385–394
Welling LW, Welling DJ, Hill JJ (1977) The shape of epithelial cells in rabbit thick ascending limb of Henle. Kidney Int 10:603
Welling LW, Welling DJ, Hill JJ (1978) Shape of cells and intercellular channels in rabbit thick ascending limb of Henle. Kidney Int 13:144–151
Welling LW, Evan AP, Welling DJ (1981) Shape of cells and extracellular channels in rabbit cortical collecting ducts. Kidney Int 20:221–222
Welling LW, Evan AP, Welling DJ, Gattone VH (1983) Morphometric comparison of rabbit cortical connecting tubules and collecting ducts. Kidney Int 23:358–367
Wilcox CS, Welch WJ, Murad F, Gross SS, Taylor G, Levi R, Schmidt HH (1992) Nitric oxide synthase in macula densa regulates glomerular capillary pressure. Proc Natl Acad Sci USA 89:11993–11997
Wingo CS, Madsen KM, Smolka A, Tisher CC (1990) H-K-ATPase immunoreactivity in cortical and outer medullary collecting duct. Kidney Int 38:985–990
Wolf G (1993) Regulating factors of renal tubular hypertrophy. Clin Invest 71: 867–870
Wong T, Morgan TO, Alcorn D, Ryan GB (1986) Effect of sodium intake and sodium delivery to the macula densa on renal renin content and juxtaglomerular apparatus morphology. Clin Exp Pharmacol Physiol 13:267–270
Woodhall PB, Tisher CC (1973) Response of the distale tubule and cortical collecting duct to vasopressing in rat. J Clin Invest 52:3095–3108
Zalups RK (1989) Effect of dietary K^+ and 75% nephrectomy on the morphology of principal cells in CCDs. Am J Physiol 256:F387–F396
Zalups RK, Stanton B, Wade JB, Giebisch G (1985) Structural adaptation in initial collecting tubule following reduction in renal mass. Kidney Int 27:636–642
Zalups RK, Henderson DA (1992) Cellular morphology in outer medullary collecting duct: effect of 75% nephrectomy and K^+ depletion. Am J Physiol 263: F1119–F1127

CHAPTER 2

Basic Concepts of Renal Physiology

F. LANG and A. BUSCH

A. Introduction

The most obvious function of the kidney is the excretion of xenobiotics, excessive electrolytes and trace elements, and unnecessary and/or harmless metabolic products such as uric acid, urea and ammonia. Through its excretory function, the kidney plays a pivotal role in the regulation of volume and ion composition of the body fluids. For instance, it participates in the regulation of K^+, Na^+, Cl^-, HCO_3^-, Ca^{2+}, Mg^{2+} and HPO_4^{2-} content of the body and thus influences intracellular fluid, extracellular fluid, blood pressure, acid-base balance, mineral metabolism, etc. Accordingly, the kidney is the target of various hormones. Furthermore, the kidney releases or activates hormones itself, such as erythropoietin, calcitriol [1,25-$(OH)_2D_3$], angiotensin, prostaglandins and kinins. The kidney also carries out several important metabolic functions, such as gluconeogenesis, degradation of fatty acids and inactivation of hormones.

In order to effectively serve its function as an excretory organ, the kidney must control large volumes of extracellular fluid at a given time. Indeed, renal blood flow by far exceeds the nutritional requirements of this small organ. Some substances are completely extracted from the blood passing through the kidney and are completely excreted due to epithelial secretion (see Sect. C.II). About 20% of the plasma passing through the kidneys is usually filtered in the glomerula. Substances which have passed through the glomerular filter will eventually be excreted unless they are reabsorbed by the tubular epithelium. Thus, the large renal blood flow and filtration with subsequent epithelial transport allows for efficient renal regulation of body volume and composition.

This chapter will focus on the mechanisms and regulation involved in renal perfusion and filtration, mechanisms of renal epithelial transport processes, and the regulation of renal electrolyte excretion. The renal metabolic functions are dealt with in Chap. 3 of this volume.

B. Renal Blood Flow and Glomerular Filtration

Renal blood flow of both kidneys is in the range of 1.2 l/min, which amounts to approximately 20% of cardiac output. Given the small mass of the kidney

(300 g), its specific perfusion (organ blood flow/organ mass) is higher than that of any other organ.

I. Pressures and Resistances in Renal Vascular Segments

Blood flowing through the kidney must pass through two resistance vessels (afferent and efferent arterioles) and two capillary systems (glomerular and postglomerular capillaries). As shown in Fig. 1, little pressure loss (and thus little resistance) occurs in the renal artery, the interlobar and the arcuate arteries.

A substantial loss of pressure occurs, however, in the interlobular arteries (HÄBERLE 1988; ULFENDAHL and WOLGAST 1992). The highest resistance is usually found in the afferent arterioles, where the pressure usually decreases to less than half the mean arterial pressure, i.e., approximately 35 mmHg. The resistance in the afferent arteriole is under the influence of a variety of factors (see Sect. B.VI, Table 1). The afferent arteriole feeds a great number of glomerular capillary loops, which are arranged in parallel and are very short, thus creating very little resistance. Accordingly, there is very little pressure drop along the glomerular capillaries. These are drained by the efferent arteriole, the second high-resistance vessel in the renal vascular system. The relatively high resistance of the efferent arteriole maintains the pressure in the glomerular capillaries, a prerequisite for efficient glomerular filtration (see below). The remaining vascular segments, the postglomerular capillaries and interlobular and arcuate veins, add little resistance and here the pressure drop is minimal. The efferent arterioles of the juxtamedullary nephrons feed the vasa recta,

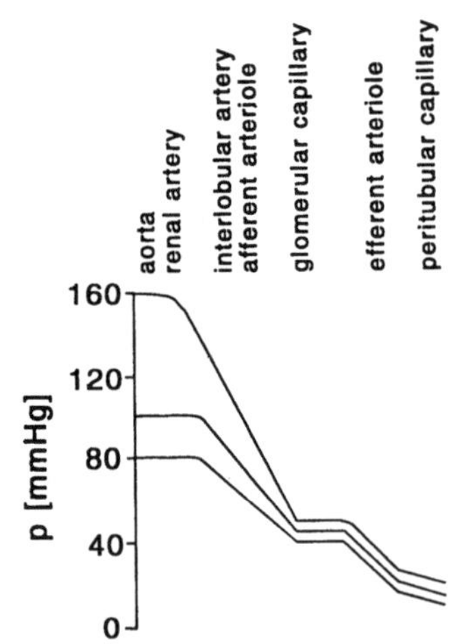

Fig. 1. Pressure in renal vascular segments. Means and upper and lower limits of autoregulatory range

Table 1. Factors modifying renal vascular resistance (according to DWORKIN and BRENNER 1992; HÄBERLE 1988; LANG 1982; MENÉ and DUNN 1992)

Decrease in vascular resistance	Increase in vascular resistance
Hormones and mediators	
Atrial natriuretic factors	Dopamine
Acetylcholine	Norepinephrine
Histamine	Epinephrine
Bradykinin	Angiotensin
Prostaglandins E_1, E_2	Antidiuretic hormone
Glucocorticoids	Leukotrienes C_4, D_4
Glucagon	Thromboxane
Secretin	Adenosine
Calcitonin	Neuropeptide Y
Nitroxide	
Substance P	
Drugs	
Furosemide	Chlorothiazide
Mannitol	Indomethacin
Ethacrynic acid	
Saralasin	
Isoproterenol	
Nitroprusside	
Others	
Dietary proteins	Massive nerve stimulation
Salt intake	Massive calcium infusions
Hyperosmotic infusions	Blood pressure increase
Increased ureteral pressure	Cold
Decreased blood pressure	Heavy work
Blood loss	
Increased hematocrit	
Acute anemia	
Uncomplicated pregnancy	

with long loops of vessels perfusing the renal medulla. Usually, the vasa recta do not create a high vascular resistance despite their length, since many vasa recta are arranged in parallel. However, blood flow in the vasa recta is highly susceptible to impairment by increased viscosity. It has been suggested (MASON 1986) that postischemic acute renal failure is due to stasis in the vasa recta with subsequent impairment of the energy supply to the thick ascending limb cells.

II. Intrarenal Blood Flow Distribution

Large differences in blood supply occur between the renal cortex and renal medulla (PALLONE et al. 1990; SANDS et al. 1992). Virtually all blood flowing through the kidney passes through the glomerula in the renal cortex. Efferent arterioles of the superficial glomerula feed the postglomerular capillaries of the cortex, which supply the proximal and distal convoluted tubules. The efferent arterioles of the juxtamedullary glomerula feed the vasa recta, which supply the renal medulla. The renal medulla, which contributes about 30% to renal mass, receives less than 10% of renal blood supply. The arrangement of the vasa recta in the form of long loops impedes the delivery of O_2 and nutrients to, and the removal of CO_2 and metabolic products from, the kidney medulla, further compromising the supply of the renal medullary cells (see Sect. C.IV).

Factors modifying renal blood flow may have different effects on cortical and medullary blood flow. Preferential medullary vasodilation is elicited by a variety of substances including prostaglandins, acetylcholine and bradykinin.

III. Permeability-Selectivity of the Glomerular Filter

An intact glomerular filter does not allow the passage of molecules larger than 4 nm in diameter, corresponding to a molecular weight of 50 KDa (see also Chap. 1, this volume). Besides the size, the charge of the molecules determines glomerular permeability (see Fig. 2): The passage of negatively charged molecules is impeded by negative charges at the glomerular filter. Since most plasma proteins are negatively charged, their filtration is restricted. During glomerulonephritis, the negative charges at the glomerular

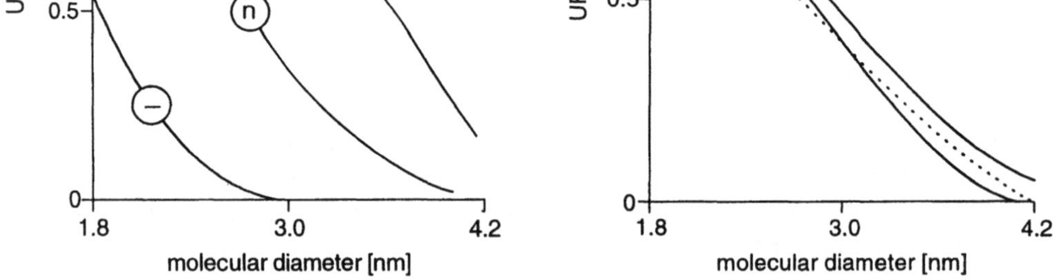

Fig. 2. Permeability-selectivity of glomerular filter. Negative surface charges at the intact glomerular filter (*left*) impede the passage of negatively charged proteins (−), while neutral (*n*) or positively charged proteins (+) permeate more easily. In glomerulonephritis (*right*) the negative surface charges are lost and proteins are filtered irrespective of their charge (modified from BRENNER et al. 1978). The concentration of the respective proteins in ultrafiltrate is plotted versus the effective molecular radius of the proteins

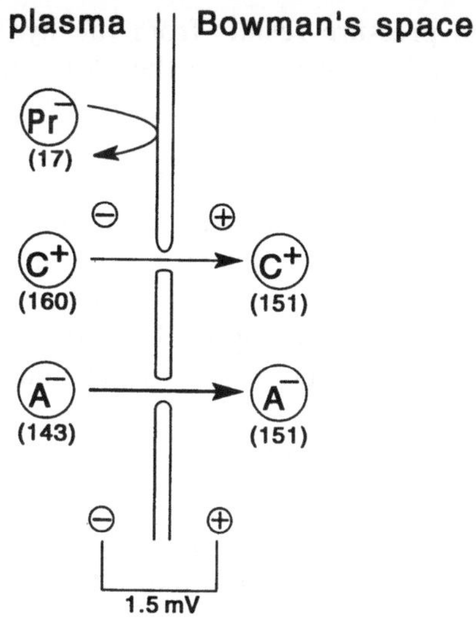

Fig. 3. Generation of the Gibbs-Donnan potential and its influence on the filtration of permeable cations (C^+) and anions (A^-). The negatively charged proteins (Pr^-) cannot permeate the filter. They create a negative charge surplus and thus a potential difference with negative charge on the blood side (so-called Gibbs-Donnan potential). The potential impedes the filtration of cations and facilitates the filtration of anions (respective concentrations in plasma and ultrafiltrate are *in parentheses*)

filter are neutralized and the reduced permeability-selectivity of the glomerular filter allows the filtration of plasma proteins, which usually escape glomerular filtration (Fig. 2). The reflection of the negatively charged plasma proteins by an intact glomerular filter creates a surplus of negative charge on the plasma side, i.e., a plasma-negative potential difference across the glomerular filter of some 2 mV (referred to as the Gibbs-Donnan potential). This potential difference favors the filtration of cations and impairs the filtration of anions. As a result, the concentration of cations is higher and the concentration of anions lower in the ultrafiltrate than in plasma. For monovalent ions the difference is approximately 10% (see Fig. 3).

IV. Determinants of Renal Glomerular Filtration Rate

The volume filtered per unit time (J_v) depends on the area (A) and the hydraulic conductivity (L_p) of the glomerular filter and on the effective filtration pressure (P_{eff}) (Dworkin and Brenner 1992):

$$J_v = L_p \cdot A \cdot P_{eff}$$

The hydraulic conductivity and the area of the glomerular filter cannot be determined separately and are thus combined in a filtration-coefficient (K_f):

$$K_f = L_p \cdot A$$

The effective filtration pressure results from the hydrostatic (Δp) and the oncotic ($\Delta\pi$) pressure difference between the glomerular capillary (p_C, π_C) and Bowman's space (p_B, π_B):

$$P_{eff} = \Delta p - \Delta\pi = p_C - p_B - (\pi_C - \pi_B)$$

The oncotic pressure is created in large part by the non-filtered plasma proteins. These proteins are concentrated by the filtration of plasma water and π_C increases accordingly (see Fig. 4). As a result, the effective filtration pressure declines and may even approach zero towards the end of the glomerular capillaries (filtration equilibrium). The increase in $\Delta\pi$ thus limits the filtration rate.

Theoretically, the glomerular filtration rate may be influenced by any of the above factors (see Fig. 5):

1. A decrease in the filtration coefficient (K_f), i.e., of the hydraulic conductivity and/or the area of the glomerular filter, decreases the glomerular filtration rate, unless the filtration equilibrium is approached by the end of the glomerular capillary. The effect of a reduced K_f is blunted by the decrease in glomerular filtration rate and the reduced increase in oncotic pressure. An increase in the filtration coefficient enhances the filtration rate as long as the filtration equilibrium is not approached.
2. An increased resistance of the efferent arteriole enhances glomerular hydrostatic pressure, thus favoring an increased filtration rate. However, due to the increase in the resistance renal blood flow decreases. Thus, at any given filtration rate, a greater portion of renal blood flow is filtered and the increase in oncotic pressure enhanced. The increase in oncotic pressure then limits glomerular filtration. As a result, an increase in the resistance of the efferent arteriole may increase or decrease the glomerular filtration rate depending on prior renal blood flow. Similarly,

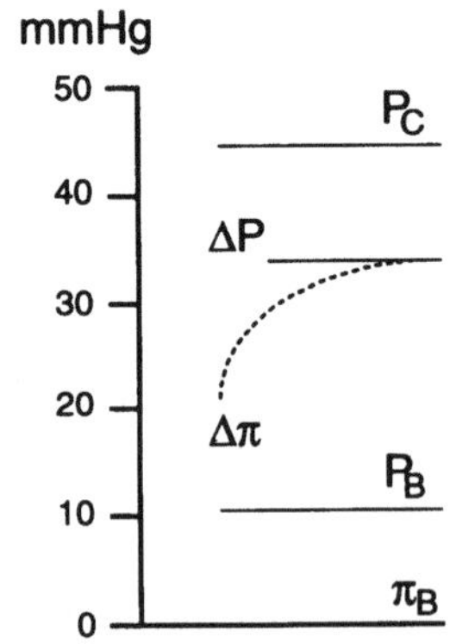

Fig. 4. Driving forces for glomerular filtration. P_C and P_B are hydrostatic and π_C and π_B oncotic pressures in capillaries and Bowman's capsule, respectively

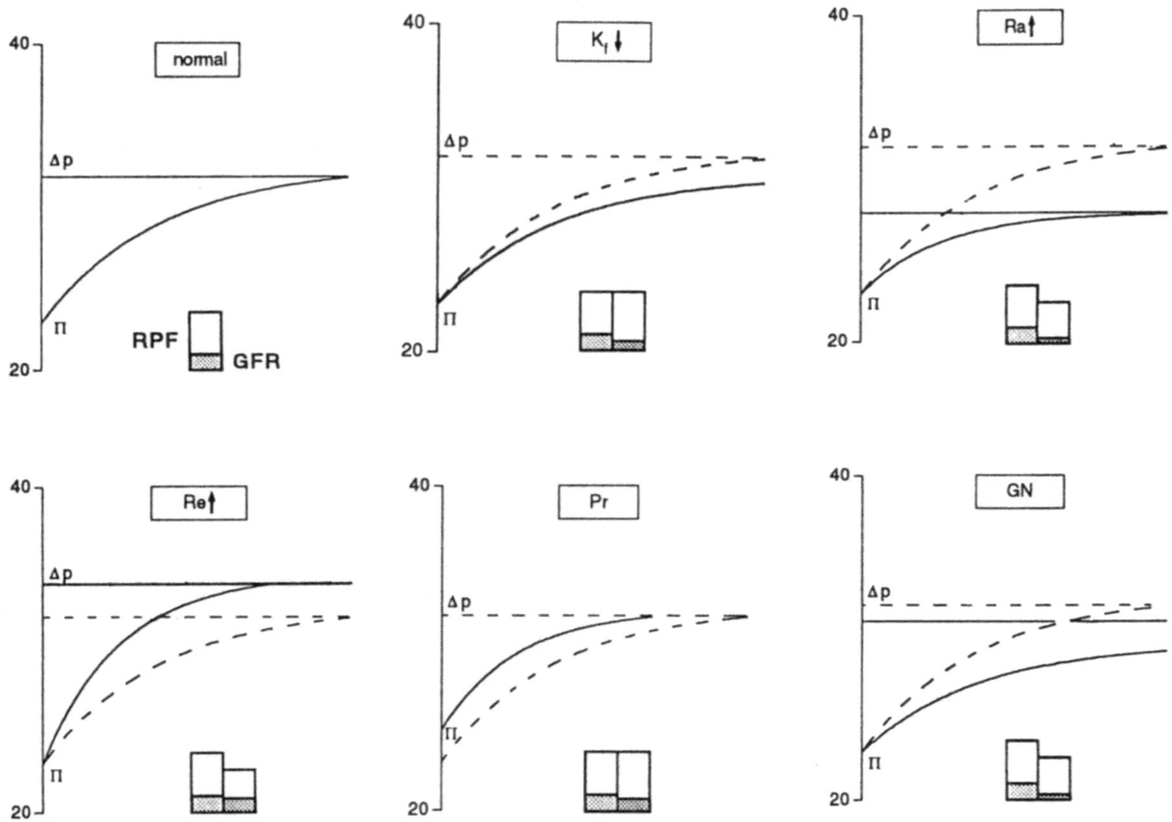

Fig. 5. Influence of various factors on hydrostatic (Δp) and oncotic (π) pressure gradient across the glomerular filter, on renal plasma flow (RPF, *complete column*) and glomerular filtration rate (GFR, *shaded part of column*). The respective values are shown in comparison to the normal situation (*dashed lines and left-hand columns*): *1*, reduced ultrafiltration coefficient (K_f), i.e., reduced hydraulic conductivity or reduced filtering area; *2*, increased vascular resistance of the afferent arteriole (R_a); *3*, increased vascular resistance of the efferent arteriole (R_e); *4*, increased plasma protein concentration (P_r); *5*, glomerulonephritis (*GN*), i.e., combination of increased resistance of afferent arteriole and reduced hydraulic conductivity of the glomerular filter)

a dilation of the vas efferens may decrease or increase glomerular filtration rate.

3. An increase in the resistance of the afferent arteriole decreases glomerular hydrostatic pressure and renal blood flow. The latter leads to a more rapid increase in oncotic pressure at any given glomerular filtration rate. Thus, both effects combine to decrease glomerular filtration rate.
4. An increase in the concentration of plasma proteins enhances the oncotic pressure and thus decreases the glomerular filtration rate. Conversely, a decrease in plasma protein concentration favors an increase in glomerular filtration rate.

During glomerulonephritis the glomerular filtration rate is compromised mainly by the decreased renal blood flow and by the decreased filtration coefficient.

V. Autoregulation of Renal Blood Flow and Glomerular Filtration Rate

With changing systemic blood pressure, renal blood flow and glomerular filtration rate are maintained constant by autoregulation (HÄBERLE 1988). As shown in Fig. 6, autoregulation is almost perfect at mean arterial pressures between 80 mmHg and 180 mmHg. The constancy of renal blood flow is achieved by vasoconstriction with increasing mean arterial pressure and by vasodilation with decreasing mean arterial pressure. Following sudden

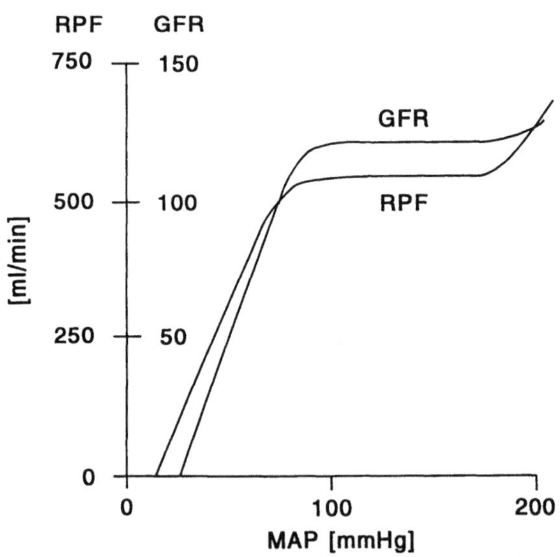

Fig. 6. Autoregulation of renal blood flow (*RPF*) and glomerular filtration rate (*GFR*). RPF and GFR are plotted versus mean arterial pressure (*MAP*)

changes in mean arterial pressure, the kidney takes a few seconds to adjust the vascular resistance. To carry out autoregulation of simultaneous renal blood flow and glomerular filtration rate, the vascular tone must be modified prior to the glomerulus since alterations in vascular tone in the efferent arteriole influence renal blood flow and glomerular filtration rate differently.

The mechanism underlying autoregulation has been a matter of longstanding debate. Several mechanisms probably affect the resistance of different segments of the interlobular arteries and afferent arterioles to perform autoregulation. Three mechanisms are of particular interest (HÄBERLE 1988; LANG 1982):

1. As in some other tissues, renal vascular smooth muscles are activated by stretch due to increased intramural pressure. The subsequent vasoconstriction increases resistance at increased perfusion pressure, thus achieving myogenic autoregulation (HÄBERLE 1988).
2. An increase in glomerular filtration rate enhances the tubular NaCl load. However, the NaCl reabsorption in the proximal tubule and Henle's loop cannot catch up with the increased load, resulting in increased NaCl delivery at the macula densa. This triggers tubuloglomerular feedback, which decreases the glomerular filtration rate (BRIGGS and SCHNERMANN 1986; SCHNERMANN and BRIGGS 1992) (Fig. 7). The mechanisms underlying the tubuloglomerular feedback are still a matter of debate.
3. A decreased medullary blood flow stimulates the release of vasodilating prostaglandins (MENE and DUNN 1992).

None of the three factors fully accounts for the observed autoregulation. Rather, all three mechanisms interact to guarantee constant perfusion of the kidney during variations in mean arterial pressure.

VI. Hormonal Regulation of Renal Blood Flow and Glomerular Filtration

A great variety of factors influence renal blood flow and glomerular filtration (Table 1).

In the past few years, the influence of glomerular filtration rate by protein-rich diet has attracted the interest of nephrologists, since it has been demonstrated that this diet gradually destroys renal glomerula by inducing chronic hyperfiltration (BRENNER et al. 1982). The mechanism underlying protein-induced hyperfiltration is not entirely clear, but it may involve glucagon-induced release of a hyperfiltrative substance from the liver (LANG et al. 1992).

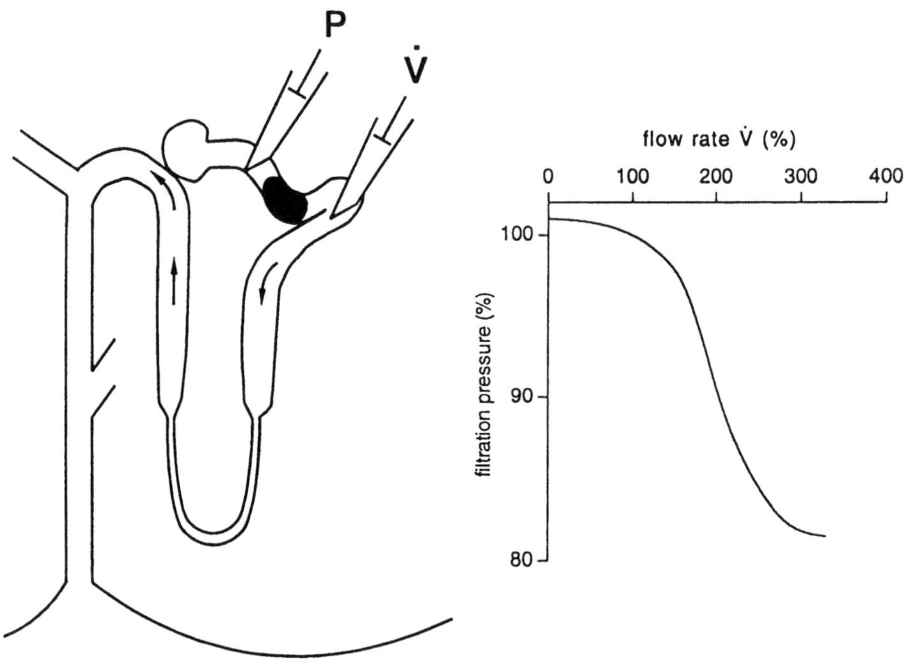

Fig. 7. Tubuloglomerular feedback: relationship between filtration pressure (*P*) and end proximal flow rate (*V*). *P* is determined through a capillary inserted into the proximal tubule, the end proximal flow rate through a second capillary in the same nephron. *P* decreases with increasing *V*. If GFR exceeds the reabsorptive capacity of the proximal tubule and Henle's loop, the enhanced distal tubular NaCl delivery triggers the tubuloglomerular feedback and reduces GFR (according to SCHNERMANN and BRIGGS 1992)

C. Renal Tubular Transport

I. Driving Forces for Epithelial Transport

Transport processes across the epithelia may occur through tight junctions bypassing the cells (paracellular pathway) or through the cells (i.e., through the apical and the basolateral cell membranes). The transport across the tight junctions is always passive. Transport across the cell membranes may be passive, active or secondary active.

1. Water Transport

The passive transport of water is driven by a hydrostatic (Δp) and an osmotic ($\Delta\pi$) pressure gradient (DAWSON 1992; MICHEL 1992; WHITTEMBURY and REUSS 1992). The volume of water transported per unit time (J_v) is dependent on the hydraulic conductivity (L_p) of the epithelial wall:

$$J_v = L_p \cdot (\Delta\pi - \Delta p) \quad (1)$$

The osmotic pressure gradient depends on the concentration difference (Δc_i) and the reflection coefficient (σ_i) of each solute on either side of the membrane:

$$\Delta\pi = R \cdot T \cdot \Sigma(\sigma_i \cdot \Delta c_i) \quad (2)$$

where R and T represent the gas constant and the temperature in o_K. The reflection coefficient σ_i may vary from 0 to 1, depending on the permeability of substance i. Thus, highly permeable substances do not contribute to $\Delta\pi$.

2. Solvent Drag

Conversely, the stream of transported water may drive transport of solutes (solvent drag). The amount (mol) of solute i transported (J_i^s) by solvent drag is:

$$J_i^s = (1 - \sigma_i) \cdot c_i \cdot J_v \quad (3)$$

where σ_i represents the reflection coefficient of the substance and c_i the effective concentration at the membrane, which is usually estimated from the arithmetic mean of the concentrations on either side of the membrane.

3. Diffusion

Another passive transport mode for solutes is diffusion, which is driven by an electrochemical gradient, i.e., a gradient consisting of the voltage and the chemical gradient, for the solute i (KRISTENSEN and USSING 1992; PALMER and SACKIN 1992). Uncharged solutes are driven by chemical gradient, i.e., a concentration difference (Δc_i) across the membrane. The amount (mol) of uncharged solute i transported (J_i^d) by diffusion can be estimated using the following equation:

$$J_i^d = P \cdot \Delta c_i \quad (4)$$

where P represents the permeability of the membrane (or epithelium).

Charged molecules are driven by both a chemical (concentration difference) and an electrical (potential difference) driving force. These two driving forces may oppose each other, thus creating an electrochemical equilibrium. The potential difference across the membrane ($\Delta E_{equ.}$) required to achieve such an equilibrium amounts to:

$$\Delta E_{equ.} = -(R \cdot T/z_i \cdot F)\, ln(c_{i,1}/c_{i,2}) \quad (5)$$

where z_i represents the charge of the substance i, F the Faraday constant, R the gas constant, T the temperature in °K and $c_{i,1}$ and $c_{i,2}$ the concentrations of substance i on either side of the cell membrane. At 37°C, $R \cdot T/F$ amounts to 30.5 mV. If the concentration difference is not too large ($0.5 < c_{i,1}/c_{i,2} < 2$):

$$ln(c_{i,1}/c_{i,2}) \approx \Delta c_i/\bar{c}_i \quad (6)$$

where Δc_i represents the concentration difference across the cell membrane ($c_{i,1} - c_{i,2}$) and $\bar{c}_i$ the mean concentration at the membrane [$\approx (c_{i,1} + c_{i,2})/2$]. At moderate concentration differences across the cell membrane ($0.5 < c_{i,1}/c_{i,2} < 2$) the following equation can thus be used for calculation of transport (J_i^d):

$$J_i^d = -P(\Delta c_i + (z \cdot F)/(R \cdot T) \cdot \Delta E \cdot c_i) \quad (7)$$

P is not necessarily constant but may depend on the potential difference across the membrane, thus complicating the relation. Diffusion may or may not involve specific carriers or ion channels at the cell membrane. If carriers are involved, the term "facilitated diffusion" may be applied.

A special case of diffusion is the nonionic diffusion of weak acids or bases. It applies if the uncharged form of an acid or base is lipid soluble and can traverse the membrane, whereas the charged form cannot. In this event transport is driven by the concentration difference of the uncharged substance only, whereas the concentration of the charged species does not influence transport. Usually a pH gradient is established across the cell membrane and across the epithelium. Accordingly, the dissociation of the acid or base is different in tubule fluid and peritubular space. At acid luminal pH, the dissociation is reduced in the lumen and the portion of uncharged, diffusible acid is larger. Thus, acid luminal pH favors reabsorption of weak acids. Conversely acid luminal pH impedes the reabsorption of weak bases.

4. Active Transport

Active transport allows the transport of a given substance against a chemical and/or electrochemical gradient. It requires the expenditure of energy at the transporting site. Primary active transport depends on chemical energy in the form of ATP. Several ATPases [(Na^++K^+)-ATPase, H^+-ATPase, (H^++K^+)-ATPase, Ca^{2+}-ATPase] operate in the kidney, utilizing ATP breakdown in order to transport the respective ions (DEWEER 1992; GULLANS and MANDEL 1992). Due to the large free energy of ATP breakdown, the primary active transport processes are usually able to transport against steep electrochemical gradients.

Secondary active transport does not utilize ATP for the transport of a given substance, but exploits the electrochemical gradient of another substance (DEWEER 1992; BURCKHARDT and KINNE 1992). For the transport by secondary active mechanisms, the same considerations are valid as for diffusion, but instead of the electrochemical gradient of a single substance the lumped electrochemical gradient for all co- or countertransported substances must be considered. Table 2 lists the driving forces for a few electrolyte transporters operating in the kidney. A variety of substrates for Na^+-coupled transport are listed in Table 3.

Table 2. Driving forces in proximal tubules and thick ascending limbs[a]

	a_i (mmol/l)	a_e (mmol/l)	(EZF)/(RT)	E_o (mV)	E_m-E_a (mV)
Na^+	12	115	2.3	+60	−130
K^+	80	4	−3.0	−80	+10
Cl^-	15	90	−1.8	−50	−20
HCO_3^-	10	20	0.7	−20	−50
Ca^{2+}	0.0001	0.8	9.0	−120	−190
Na^+/H^+ exchange			3.0	–	–
HCO_3^--Na^+ cotransport (3:1)			4.3	−58	−12
Na^+/Ca^{2+} exchange (3:1)			−2.2	−60	−10
NaCl-KCl cotransport[b]			2.8	–	–
KCl symport[b]			1.2	–	–
Cl^-/HCO_3^- exchange			1.1	–	–

[a] Approximate values for ion activities in intracellular (a_i) and extracellular (a_e) fluid, equilibrium potential (E_o) and the difference of E_o and cell membrane potential ($E_m = -70\,\mathrm{mV}$).
[b] At luminal or peritubular ion composition as above, which is not usually the case in vivo.

II. Saturability of Epithelial Transport Processes

Specific transport processes are in principle saturable, irrespective of the energy driving the transport. Saturability may not be apparent, because the affinity is too low for saturation to be achieved at reasonable concentrations. For many substances, however, saturation is a critical feature of renal transport.

1. Michaelis-Menten Kinetics

The most simple relation for saturable transport is that of Michaelis-Menten kinetics, which describes the transport per unit time of a given substance (substrate) i (J_i) as a function of the concentration of this substance (c_i), the maximal transport rate ($J_{i,max}$) and the concentration at which transport rate is half-maximal ($K_{i,1/2}$):

$$J_i = c_i \cdot J_{i,max}/(K_{i,1/2} + c_i)$$

At very low substrate concentrations ($c_i \ll K_{i,1/2}$), the transport rate increases in almost linear proportion to substrate concentration, i.e., the transport system behaves like a nonsaturable transport process. At very high substrate concentrations ($c_i \gg K_{i,1/2}$), the transport rate approaches its maximal value ($J_i \approx J_{i,max}$). $K_{i,1/2}$ is a measure of the affinity of the transport process (see Fig. 8). In the kidney, the transport of a great variety of substances including glucose, amino acids, HCO_3^-, and HPO_4^{2-} follows simple Michaelis-Menten kinetics.

Table 3. Substrates for Na^+-coupled transport in proximal renal tubule (according to BURCKHARDT and KINNE 1992; LANG 1988; SILBERNAGL 1990)

Electrolytes
H^+ (Na^+/H^+ exchange)
Ca^{2+} (Na^+/Ca^{2+} exchange)
HPO_4^{2-}
SO_4^{2-}
Cl^-
Carbohydrates
D-Glucose
α-Methylglucose
D-Galactose
3-*o*-Methyl-D-glucose
Amino acids
L-Phenylalanine
L-Histidine
L-Aminoisobutyrate
L-Aminobicyclo-(2,2,1)-heptane-2-carboxylic acid
L-Lysine
L-Arginine
L-Ornithine
L-Aspartate
L-Proline
Glycine
Organic acids
Lactate
Pyruvate
Nicotinic acid
Picolinic acid
Pyrazinoic acid
Acetacetate
β-Hydroxybutyrate
Succinate
Malate
Oxalacetate
Fumarate
β-Ketoglutarate
Citrate
Biliary acids

2. Transport Kinetics of Whole Kidney

For a variety of substances the saturability of transport determines renal excretion, and it is frequently desirable to determine a given transport system from whole kidney clearnance studies (SCHUSTER and SELDIN 1992). For the evaluation of transport kinetics from renal excretion, the plasma concentration of the respective substance is enhanced by infusion and the urinary excretion plotted versus the plasma concentration. The filtered load (M_f) of the substance can be estimated from the glomerular filtration rate (GFR) and the plasma concentration (P) of the filterable moiety of the

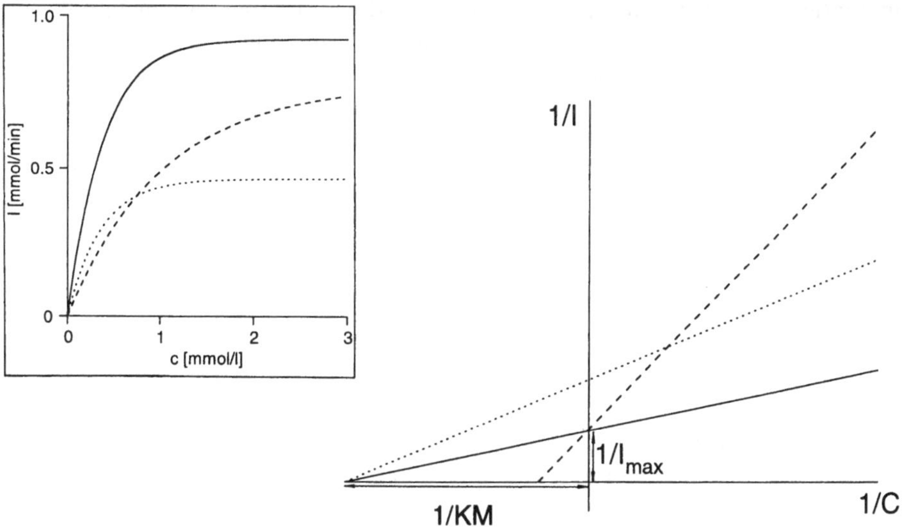

Fig. 8. Transport kinetics: transport rate (I) as a function of substrate concentration (c) plotted directly (*left, upper*) or as a Lineweaver-Burk plot (*right, lower*): *solid line*, I_{max} = 1 mmol/min, K_M = 0.2 mmol/l; *interrupted line*, reduced maximal transport rate (I_{max} = 0.5 mmol/min, K_M = 0.2 mmol/l); *dotted line*, reduced affinity (I_{max} = 1 mmol/min, K_M = 1.0 mmol/l)

substance (M_f = GFR · P) and the transport rate (M_t) can be estimated from the difference in excreted amount of the substance (M_e) and filtered load ($M_t = M_f - M_e$). If net reabsorption prevails, M_e is positive. Following net secretion, M_e is negative. The GFR is not directly amenable to determination, but can be estimated from the excreted amount of a substance which is not transported, such as inulin or creatinine ($M_e = M_f$ = GFR · P).

If the substance is reabsorbed by a high-affinity transport system, virtually no substance is excreted at low plasma concentrations (Fig. 9). This is a result of the filtered load being reabsorbed almost completely as long as the maximal transport rate (TM) is not approached. Very low luminal substrate concentrations are required for maximal activity of a high-affinity transporter, which is thus able to clear the luminal fluid from the substrate. If the filtered load exceeds the maximal transport rate, the excess filtered substance is excreted quantitatively ($M_e = M_f$ − TM). The transition between complete reabsorption and excretion of filtered substance exceeding maximal transport rate is sharp (little splay) and allows the determination of a threshold, i.e., the plasma concentration at this transition.

If the substance is reabsorbed by a low-affinity transport system, it is excreted even at filtered loads far below maximal transport rate. This occurs because at low luminal substrate concentrations the carrier does not approach maximal transport rate and the carrier is not able to clear the luminal fluid from the substrate. With increasing filtered load, the carrier gradually increases transport rate and the excretion does not increase in parallel to excess filtration. Thus, there is not sharp transition between complete reabsorption and excretion (Fig. 9).

If net secretion of prevails, the excreted amount exceeds the filtered amount of substrate. If the substrate is secreted at high affinity (e.g., *para*-aminohippurate, PAH), the renal blood is completely cleared from the substrate as long as the maximal transport rate is not approached. In that case, renal plasma flow (RPF) limits the excretion of the substance (M_e = RPF · P).

The evaluation of transport systems from urinary excretion will lead to very crude results and should be interpreted with several caveats in mind:

1. The plasma concentration of any given substance is obviously not identical to the actual concentration at the carrier, but is modified along the nephron by reabsorption of the substrate and of volume. Thus, the apparent affinity depends not only on the carrier itself but also on the volume reabsorption in the respective and preceding nephron segments.
2. Most substances are transported by several carriers with different kinetic properties. The excretion is the combined effect of these carriers. Frequently, the affinity of the carriers increases towards more distal nephron segments. These segments determine the apparent overall affinity in whole kidney studies.
3. The transport of a given substance may not be identical in all nephrons. Differences have, for instance, been shown between transport in superficial and juxtamedullary nephrons. Different TM/GFR ratios in different nephron segments mimic decreased affinity, since substances are excreted before TM is approached.
4. The increase in plasma or tubular concentration of a substance may indirectly influence its reabsorption, without saturation of carriers. For instance, the infusion of phosphate decreases plasma calcium concentration and this leads to the release of PTH, which inhibits renal phosphate reabsorption. The infusion of calcium leads to increased luminal calcium concentrations. Calcium then decreases the permeability of the tight junctions and thus impairs its own paracellular reabsorption. This phenomenon has nothing to do with saturation of a carrier but leads to similar results.

Despite these reservations, the evaluation of transport from measurements of urinary excretion may yield valuable information about the properties of renal transport systems and may allow the identification of defective affinity or maximal transport rate in tubular transport defects.

III. Segmental Organization of Renal Epithelial Transport

Glomerular filtration imposes a load for tubular reabsorption exceeding 100 ml/min in healthy man. Usually, approximately 99% of filtered water

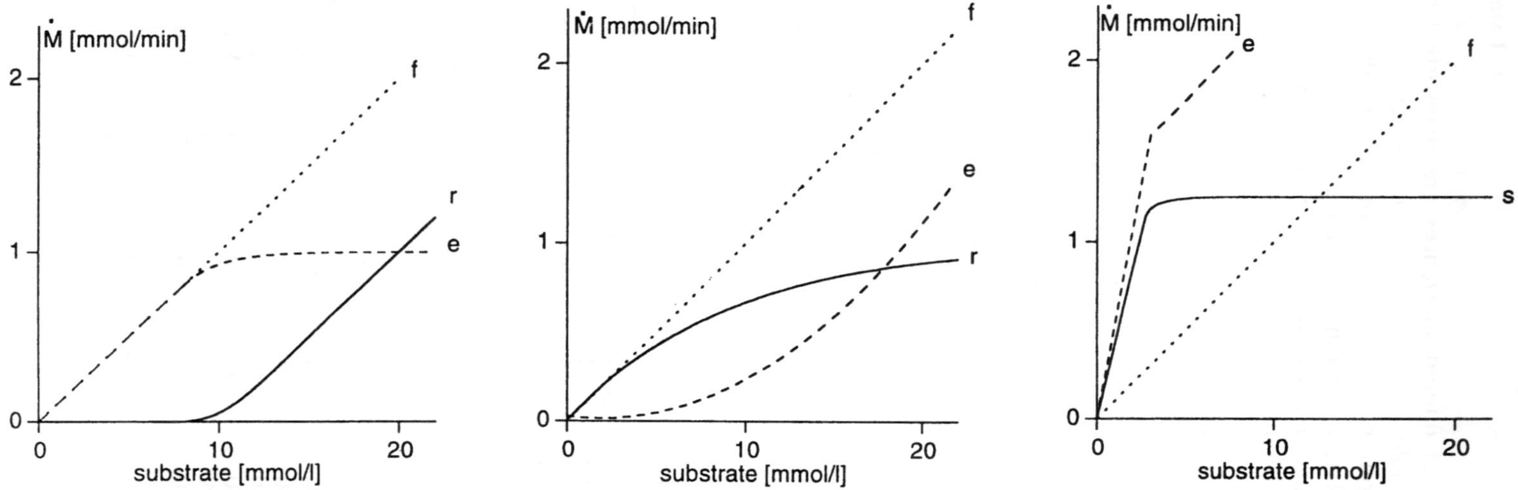

Fig. 9. Filtered (*f*), reabsorbed (*r*), secreted (*s*) and excreted (*e*) substance (*M*). *Left*, high-affinity reabsorption; *middle*, low-affinity reabsorption; *right*, secretion

and more than 90% of filtered substances are reabsorbed by subsequent nephron segments (Table 4). Furthermore, substances are added to luminal fluid by tubular secretion. Tubular transport eventually creates the final urine, which is completely different from filtrate.

The two goals, mass reabsorption and precise tuning of final urine, are achieved by the different nephron segments to varying degrees. In the following section, the transport processes in the proximal tubule, loop of Henle, distal tubule and collecting duct will be discussed. Obviously, none of the nephron segments is homogeneous but varies depending on the exact site within the nephron (more proximal or more distal) and the location of the nephron (superficial and juxtamedullary). For the sake of clarity, these differences will be omitted in this discussion.

Some of the conductive and permeability properties of the various nephron segments are summarized in Table 5. Approximately 45% of filtered substance is Na^+. Thus, reabsorption of Na^+ is quantitatively the most important task of the tubular epithelium in all the segments.

1. Proximal Tubule

The proximal tubule reabsorbs the bulk of filtered water and solutes, such as two-thirds of filtered water and NaCl, 95% of filtered bicarbonate and

Table 4. Segmental reabsorption of filtered substances

Substances	Resorption or secretion in			Urine output	Participating transport mechanisms
	Proximal tubule	Henle's loop	Distal nephron		
Water	60	20	19	1	Along the osmotic gradient
Creatinine	0	0	0	100	–
Na^+	60	34	6	0, 5	Active, diffusion, solvent drag
Cl^-	55	38	6	1	Diffusion, solvent drag, active
K^+	60	25	−5	20	Active, diffusion, solvent drag
HCO_3^-	90	0	10	0.1	Active
Ca^{2+}	60	30	9	1	Active, diffusion
HPO_4^{2-}	70	10	0	20	Active
Mg^{2+}	30	60	0	10	Active, diffusion
D-Glucose	99	1	0	0	Active
Glycine, histidine	90	5	0	5	Active
others,					Active
Amino acids	99	0	0	1	Active (?), diffusion,
Urea	50	−60	60	50	solvent drag
					Active, diffusion
Uric acid	60	30	0	10	Active, diffusion
Oxalate	−20	−10	0	130	

Table 5. General properties of different tubule segments in mammalian kidney[a]

Mammalian	Rte[a] ($\Omega\,m^2$)	Ra, Rb, Rs[a]	Pf[a] (μm/s)
Proximal tubule	≈10	Ra > Rb ≫ Rs	1000–7000
Thin descending limb loop of Henle	700		2500
Thin ascending limb loop of Henle	<10		≈20
Thick ascending limb loop of Henle	10–35	Ra ≈ Rb ≈ Rs	<10
Distal convoluted tubule	10–400		1000 (300)[b]
Connecting tubule	30		10
Cortical collecting duct	100–900	Ra ≈ Rb ≈ Rs	200–1000 (10)[b]
Medullary collecting duct			
× Outer stripe	500	Ra > Rb, Rs	200 (10)[b]
× Inner stripe	500	Ra ≫ Rb, Rs	
Papillary collecting duct	1000–2000		60–600 (40)[b]

[a] Rte, Ra, Rb, and Rs are transepithelial, luminal cell membrane, basolateral cell membrane, and shunt resistance. Pf is the water permeability. Transepithelial and cellular membrane resistances are one order of magnitude higher in amphibian than in mammalian tubules, but the relationships of the various resistances are similar to those in the mammalian kidney.
[b] In the absence of ADH.

almost 100% of filtered D-glucose and amino acids. Furthermore, the proximal tubule secretes several organic acids and bases as well as NH_4^+.

Figure 10 lists the major transport systems in the proximal tubule: Na^+ entry across the luminal cell membrane proceeds through Na^+/H^+ exchange (Boron 1992; Boron and Boulpaep 1989; Murer et al. 1976) and Na^+-coupled transport for D-glucose (Frömter et al. 1988; Hediger et al. 1987; Kinne 1976; Koepsell et al. 1990; Sacktor 1989; Samarzija et al. 1982; Schafer and Williams 1985), amino acids (Hoshi 1990; Hoshi et al. 1976; Samarzija and Frömter 1982a,b,c; Samarzija et al. 1982; Silbernagl 1988; Zelikovic and Chesney 1989), organic acids (Murer and Burckhardt 1983; Ullrich et al. 1982a,b,c), sulfate (Murer and Burckhardt 1983; Turner 1984; Ullrich et al. 1982a) and phosphate (Biber 1989; Biber and Murer 1993). The transport systems are driven by the steep electrochemical gradient for Na^+ across the apical cell membrane (see Table 1). In the straight proximal tubule, Na^+ may enter the cell in part through Na^+ channels (Gögelein and Greger 1986). The potential difference across the apical cell membrane is in part maintained by K^+ channels in the apical cell membrane (Lang and Rehwald 1992), which allow the movement of K^+ from cell to lumen. In addition to the Na^+-dependent transport systems, several exchangers operate in the apical cell membrane of the proximal tubule, allowing exchange of Cl^- with formate (Berry and Rector 1989; Karniski and Aronson 1985; Schild et al. 1986, 1990), of organic anions or hydroxyl ions with organic anions (Burckhardt and Ullrich 1989; Kahn 1989), and of organic cations with H^+ (Ross and Holohan 1983; Wright 1985). Further apical transport systems include an

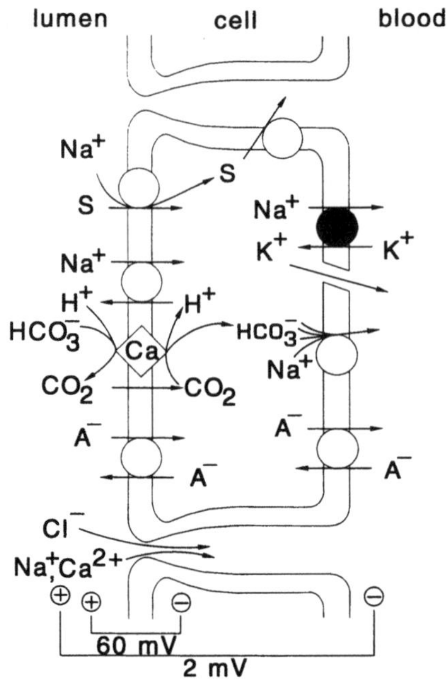

Fig. 10. Synopsis of transport systems in the proximal tubule. *Open circles*, carriers; *closed circles*, $(Na^+ + K^+)$-ATPase; *interrupted cell membrane*, ion channel; *Ca^{2+}*, carbonic anhydrase; *S*, substrate for Na^+-coupled transport, see Table 3; *A^-*, anions

H^+ ATPase (BROWN et al. 1988; KINNE-SAFFRAN et al. 1982; TURRINI et al. 1989), electrogenic organic anion transport (MARTINEZ et al. 1990; WERNER et al. 1990), and H^+-peptide cotransport (SILBERNAGL et al. 1987; TIRUPPATHI et al. 1990a,b).

At the basolateral cell membrane, Na^+ is extruded in exchange for K^+ by the $(Na^+ + K^+)$-ATPase (DOUCET 1988; JOERGENSEN 1980; KATZ 1982). K^+ leaves the cell again via K^+ channels (LANG and REHWALD 1992) and KCl symport (EVELOFF and WARNOCK 1987; SASAKI et al. 1988). The secretion of H^+ across the apical cell membrane leads to cellular formation of HCO_3^-, which exits the cell in large part via an $Na^+(HCO_3^-)_n$ cotransporter (ALPERN 1985; BIAGI and SOHTELL 1986; BORON and BOULPAEP 1983a,b; DIETL et al. 1987; FRÖMTER et al. 1988; WANG et al. 1987; YOSHITOMI et al. 1985). Some HCO_3^- leaves the cell via Na^+-dependent and Na^+-independent Cl^-/HCO_3^- exchange (ALPERN and CHAMBERS 1987; BORON and BOULPAEP 1983b; FRÖMTER et al. 1988; GUGGINO and GUGGINO 1989; KONDO and FRÖMTER 1990; KURTZ 1989). Anion exchangers allow in addition the exchange of organic acids (BURCKHARDT and ULLRICH 1989; ULLRICH et al. 1987a,b,c, 1988, 1989a,b). Further transport systems described include OH^- transport (BURCKHARDT and FRÖMTER 1987), electrogenic transport of organic cations, Na^+-coupled cellular uptake of organic acids (SILBERNAGL 1988) and amino acids (BURCKHARDT and ULLRICH 1989), Na^+/Ca^{2+} exchange (DOMINGUEZ et al. 1989; FRINDT et al. 1988; GAMJ et al. 1979; JAYAKUMAR et al. 1984; TALOR and ARRUDA 1986; YANG et al. 1988) and Na^+/H^+ exchange (BORON and BOULPAEP 1983b; GEIBEL et al. 1989a,b; VÖLKL and LANG 1988).

The transport mechanisms in the proximal tubule are designed to allow the mass reabsorption with relatively little energy expenditure. Thus, only approximately one-third of proximal tubular Na^+ reabsorption is transcellular and thus active. Most of this Na^+ reabsorption is coupled to the transport of other substances (FRÖMTER et al. 1988). The major portion of Na^+ entering the cells serves to extrude H^+ via the Na^+/H^+ exchanger. The extrusion of H^+ generates intracellular HCO_3^-, which then drags part of the Na^+ via the Na^+ $(HCO_3^-)_n$ cotransporter (n is predominantly 3). Thus, a minor part of the Na^+ entering the cell need not be extruded by the $(Na^+ + K^+)$-ATPase, requiring no energy in the form of ATP.

The cellular reabsorption of Na^+ and solutes creates an osmotic gradient, which drives water reabsorption. In the bulk of this reabsorbed water, solutes including Na^+ are reabsorbed passively by solvent drag, which contributes approximately one-third to proximal tubular Na^+ reabsorption (FRÖMTER et al. 1973).

The reabsorption of water leads to an increase in luminal Cl^- concentration; Cl^- leaves the lumen according to its chemical gradient, resulting in a lumen-negative transepithelial potential difference in the second portion of the proximal tubule. This potential difference is only approximately +2 mV, but, due to the high permeability of the proximal tubular epithelium to Na^+, it drives another one-third of the proximal tubular Na^+ reabsorption (FRÖMTER et al. 1973).

The organization of the proximal tubule thus allows the reabsorption of the major Na^+ load without direct expenditure of energy. The price for this advantage is that the tight junctions of the proximal tubule must be very leaky and thus do not allow the establishment of large transepithelial gradients.

2. Loop of Henle

The main task of the next segment, the loop of Henle, is the creation of a hyperosmotic renal medulla, a prerequisite for the concentrating ability of the kidney, as will be detailed below (Sect. C.IV). The loop of Henle consists of three completely different nephron segments. The transport systems of the first, the straight proximal tubule, are qualitatively similar to those in the proximal convoluted tubule (see above).

Transport in the second portion, the thin limb of Henle's loop, is mainly passive (IMAI et al. 1987). The transport systems identified include Cl^- channels in the apical and basolateral cell membrane (YOSHITOMI et al. 1988), K^+ channels, KCl cotransport, $(Na^+ + K^+)$-ATPase and HCO_3^-

transport at the basolateral cell membrane (LOPES et al. 1988). It is not clear at present how these transport processes participate in the integrated transport across this nephron segment.

The third segment of the loop of Henle is referred to as the thick ascending limb of Henle's loop (TAL). As illustrated in Fig. 11, Na^+ entry into the TAL cells is achieved by Na^+, $2Cl^-$, K^+ cotransport (GREGER 1981, 1985; GREGER and SCHLATTER 1983; GREGER et al. 1983, 1984a,b; HEBERT and ANDREOLI 1984a,b; HEBERT et al. 1981, 1984; KINNE et al. 1986; MOLONY et al. 1989; OBERLEITHNER et al. 1982a,b; 1983a,b,c). The steep electrochemical gradient for Na^+ is thus exploited in order to drive K^+ and Cl^- transport across this cell membrane. Na^+ is extruded in exchange for K^+ by the (Na^++K^+)-ATPase at the basolateral cell membrane (DOUCET 1988; JOERGENSEN 1980; KATZ 1982). K^+ taken up by the Na^+, $2Cl^-$, K^+ cotransport largely recycles via K^+ channels back into the tubule lumen (BLEICH et al. 1990; GIEBISCH et al. 1990; HUNTER and GIEBISCH 1987, 1988; HUNTER et al. 1986; KAWAHARA et al. 1990). The K^+ taken up by the (Na^++K^+)-ATPase leaves the cell via K^+ channels (YOSHITOMI et al. 1987) and KCl symport in the basolateral cell membrane (GREGER and SCHLATTER 1983; GUGGINO 1986). Cl^- accumulated by the Na^+, $2Cl^-$, K^+ cotransport leaves the cell via basolateral Cl^- channels and KCl symport (GREGER 1985; GREGER et al. 1990; OBERLEITHNER et al. 1983b; PAULAIS and TEULON 1990).

The apical recirculation of K^+ on the one hand and the basolateral exit of Cl^-, on the other, creates a lumen-positive potential difference across the epithelium. This potential difference drives Na^+, Ca^{2+} and Mg^{2+} through the tight junctions, which are easily permeable for cations (DIETL and

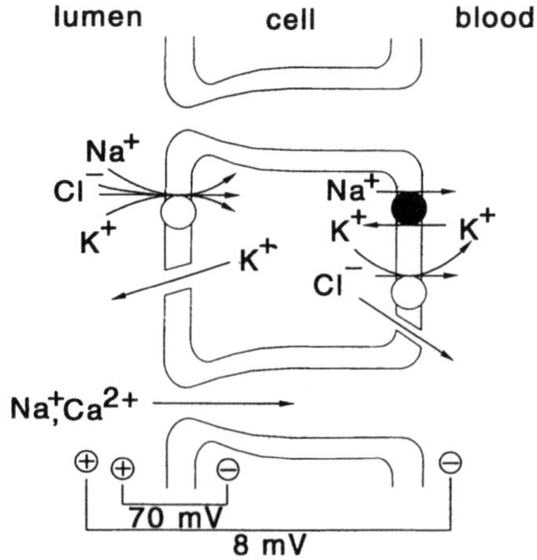

Fig. 11. Synopsis of transport systems in the thick ascending limb of Henle's loop. *Open circles*, carriers; *closed circle*, (Na^++K^+)-ATPase; *interrupted cell membrane*, ion channels

OBERLEITHNER 1987; GREGER 1985; HEBERT and ANDREOLI 1986; SCHWAB and OBERLEITHNER 1988). Thus, a portion of thick ascending limb Na^+ reabsorption is passive.

In the mouse and rat cortical thick ascending limbs, parallel operation of Na^+/H^+ exchange and Cl^-/HCO_3^- exchange has been demonstrated in the apical cell membrane (FRIEDMAN and ANDREOLI 1982; GOOD et al. 1984; GOOD 1985). A similar mechanism has not been found in rabbit thick ascending limb, however (GREGER 1985; HEBERT and ANDREOLI 1984b).

The Na^+, $2Cl^-$, K^+ cotransport accepts NH_4^+ instead of K^+. The resulting reabsorption of NH_4^+ leads to accumulation of NH_4^+ in the renal medulla (GARVIN et al. 1988; KNEPPER et al. 1989). A most important property of the thick ascending limb is its water impermeability, a prerequisite for the diluting and concentrating capacity of the kidney (see Sect. C.IV). The transport systems in the macula densa are qualitatively but not quantitatively similar to those of the thick ascending limb (BELL et al. 1989; SCHLATTER et al. 1990b).

3. Distal Tubule and Collecting Duct

The distal nephron comprising the distal convoluted tubule and collecting duct carries out the fine tuning of urinary composition. This nephron segment has only a very limited transport capacity, but is able to create large transepithelial gradients. Thus, the tight junctions of the distal epithelium are poorly permeable and all transport has to pass the cells. Accordingly, transport in the distal nephron is "costly."

The distal nephron consists of several nonhomogeneous nephron segments and, in any given segment, many functionally diverse cells may operate in parallel (BIEMESDERFER et al. 1989; HOLTHÖFER et al. 1987; MADSEN et al. 1988; MADSEN and TISHER 1986; RIDDERSTRALE et al. 1988). Furthermore, the properties of the cells apparently depend on the functional state of the animal, as would be expected for the nephron segment, which determines the composition of the urine. The predominant ion transport systems in the four most important cell types (the distal tubule cells, the principal cells and the intercalated cells type A and B) will be described (see Figs. 12, 13).

The distal tubule cell is thought to reabsorb Na^+ via an electroneutral, hydrochlorothiazide-sensitive Na^+Cl^- cotransport system (ELLISON et al. 1987; GREGER and VELÁZQUEZ 1987; STOKES 1989; TERADA and KNEPPER 1990; VELÁZQUEZ et al. 1984; VELÁZQUEZ and WRIGHT 1986) or the parallel operation of an Na^+/H^+ exchanger and Cl^-/HCO_3^- exchange (STANTON 1988). Cl^- exits the cell mainly via KCl cotransport at the basolateral and possibly the luminal cell membrane (ELLISON et al. 1987; GREGER and VELÁZQUEZ 1987; VELÁZQUEZ et al. 1987). The cell membrane potential is maintained by basolateral K^+ channels (GREGER and VELÁZQUEZ 1987; MEROT et al. 1989). Distal tubule cells are probably able to reabsorb Ca^{2+}

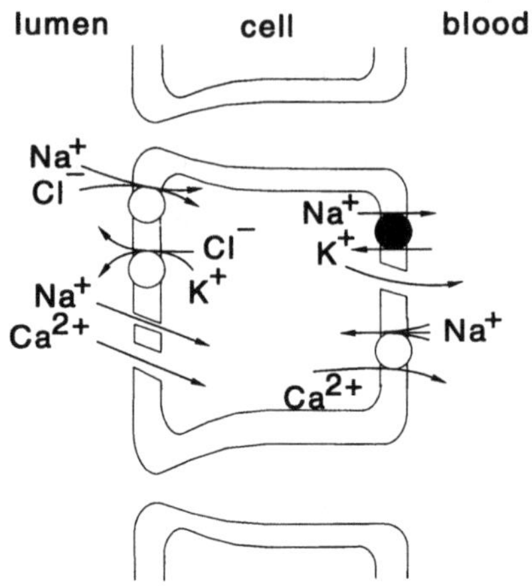

Fig. 12. Synopsis of transport systems in distal tubule cell. *Open circles*, carriers; *closed circle*, ($Na^+ + K^+$)-ATPase; *interrupted cell membrane*, ion channels

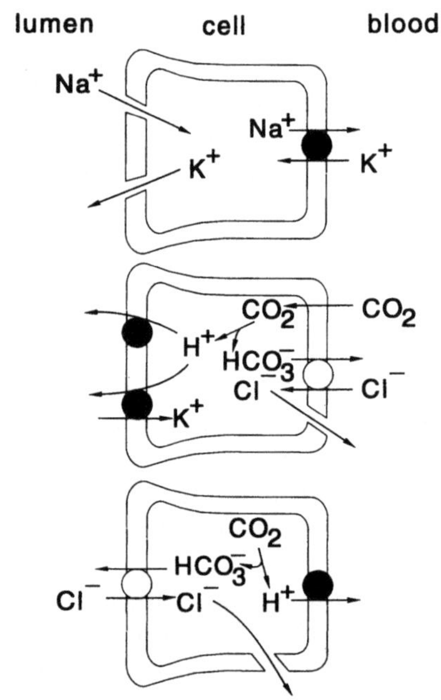

Fig. 13. Synopsis of transport systems in collecting duct. *Top*, principal cell; *middle*, intercalated cell, type A; *bottom*, intercalated cell, type B (*open circles*, carriers; *closed circles*, ATPases; *interrupted cell membrane*, ion channels)

via luminal Ca^{2+} channels and basolateral Na^+/Ca^{2+} exchange (WINDHAGER et al. 1986; cf. Chap. 8, this volume).

The principal cell is characterized by amiloride-sensitive luminal Na^+ channels (PALMER 1990; PALMER and FRINDT 1986, 1988, cf. Chap. 10, this volume), basolateral *($Na^+ + K^+$)-ATPase* (DOUCET 1988; KATZ 1982) and *K^+ channels* on both cell membranes (for review see LANG and REHWALD 1992; SCHLATTER 1989). In principal cells of the rabbit (GROSS et al. 1988; SANSOM et al. 1984, 1989, 1990) but not the rat (SCHLATTER et al. 1990a; SCHLATTER and SCHAFER 1987) Cl^- channels are found in the basolateral cell membrane. Their role in transepithelial transport is not clearly established. The intercalated cells secrete either H^+ (type A) or HCO_3^- (type B):

In type A intercalated cells H^+ transport involves either H^+ ATPase (AL-AWQATI 1986; BROWN et al. 1988; GLUCK and CALDWELL 1988; SCHUSTER 1990; STEINMETZ 1986) or ($H^+ + K^+$)-ATPase in the apical cell membrane (DOUCET and MARSY 1987; GARG and NARANG 1988, 1989, 1990; WINGO 1989). Expression of the ($H^+ + K^+$)-ATPase is probably stimulated by K^+ depletion (DOUCET and MARSY 1987; GARG and NARANG 1989, 1990; STANTON 1989). HCO_3^- formed in the cell is extruded across the basolateral cell membrane via a Cl^-/HCO_3^- exchanger (ALPER et al. 1989; KOEPPEN 1989). Cl^- thus accumulated exits the cell through Cl^- channels (DIETL et al. 1991; KOEPPEN 1987, 1989; MUTO et al. 1987).

HCO_3^- secretion in the type B intercalated cell is carried out by a luminal Cl^-/HCO_3^- exchanger and a basolateral Cl^- channel and H^+-ATPase (ALPER et al. 1989; BROWN et al. 1988; DIETL et al. 1991; MATSUZAKI et al. 1989; MUTO et al. 1990; SCHUSTER and STOKES 1987; SCHWARTZ et al. 1985; STAR et al. 1985). Furthermore, a luminal anion channel allows HCO_3^- secretion and/or Cl^- reabsorption (LIGHT et al. 1990; SCHUSTER 1990). One portion of intercalated cells apparently exhibits K^+ channels, the function of which has not yet been defined (BECK et al. 1988; KOEPPEN 1987; MUTO et al. 1990; SCHLATTER and SCHAFER 1988).

In the inner medullary collecting duct, Na^+ reabsorption is achieved by furosemide- and thiazide-sensitive electroneutral cotransport (ROCHA and KUDO 1990a,b; WILSON et al. 1983a,b). Furthermore, in this segment HCO_3^- and K^+ channels (IMAI and YOSHITOMI 1990) and Cl^- secretion via basolateral Na^+, $2Cl^-$, K^+ cotransport and luminal Cl^- channels (GRUPP et al. 1989; ROCHA and KUDO 1990a,b; SANDS et al. 1986) have been described. At the papillary epithelium, ($Na^+ + K^+$)-ATPase and Na^+, $2Cl^-$, K^+ co-transport have been identified (SANDS et al. 1986).

IV. Urinary Concentrating Mechanism

Depending on the demands of the body, the kidney is able to produce a diluted (50 mosmol/l) or highly concentrated (1600 mosmol/l) urine. Urinary concentration and dilution are accomplished by the loop of Henle (ROY et al. 1992). As described above (Sect. C.II.2) the thick ascending limb reabsorbs NaCl without water. Thus, transepithelial transport decreases the osmolarity of luminal fluid and increases the osmolarity of the interstitium. Due to enhanced interstitial osmolarity, water in excess of solutes is extracted from the descending loop of Henle, leading to an increase in luminal osmolarity. This occurs as far as the bend of Henle's loop, where the osmolarity can reach more than 5 times the isotonicity. While the fluid returns towards the cortex, more solute than water leaves the ascending limb of Henle's loop and the luminal fluid eventually becomes hypotonic. The arrangement of the tubule in a loop allows the creation of high osmolarity at the tip without the establishment of steep osmotic gradiens at any point of the loop.

The antidiuretic hormone renders (via cAMP) the distal tubule and collecting duct water-permeable (KIRK and SCHAFER 1992). As the diluted tubule fluid from the end of the thick ascending limb enters the distal tubule, water is reabsorbed according to the osmotic gradient. The tubule fluid then passes with the collecting duct the highly concentrated medulla and further reabsorption of water occurs until the luminal fluid approaches the osmolarity of he medullar interstitium. In water diuresis the release of ADH is suppressed, water reabsorption in the distal nephron prevented and hypotonic urine excreted.

Besides NaCl, urea participates in the urinary concentrating mechanism: The thick ascending limb of Henle's loop, distal convoluted tubule and cortical collecting duct are poorly permeable to urea. The water reabsorption in the distal tubule and collecting duct thus enhances the luminal urea concentration, thereby creating the driving force for urea reabsorption in the inner medullary collecting duct, which, in the presence of ADH, is highly permeable to urea. The epithelium of the descending limb of Henle's loop is easily permeable to water, but weakly permeable to NaCl. The high interstitial urea concentration exracts water from this nephron segment, thereby increasing the luminal NaCl concentration beyond the interstitial NaCl concentration. The thin ascending limb, on the other hand, is easily permeable to NaCl but poorly permeable to water. During passage of the thin ascending limb, NaCl leaves the tubule lumen following its concentration gradient and leaves a relatively hypotonic luminal fluid behind. Thus, due to the effect of urea, the thin ascending limb can do the same as the thick ascending limb without local expenditure of energy. Primary active transport would not be possible in the deep renal medulla due to the poor energy supply (see below).

The hypertonicity of kidney medulla would be rapidly washed away if the tissue were perfused by normal blood vessels. However, the vasa recta supplying the renal medulla are arranged in loops as are the tubules. Thus, the blood entering the medulla takes up solutes and approaches the same osmolarity as the interstitium. During return of blood from the kidney medulla the blood passes interstitial fluid with decreasing osmolarity. Following their gradient, solutes diffuse out of the ascending vasa recta, which upon arrival at the renal cortex carry virtually isotonic blood. As a result, the countercurrent system effectively prevents loss of solutes from the kidney medulla. However, the countercurrent system similarly prevents the clearance of waste products, such as CO_2 and lactate, and it impedes the entry of glucose and O_2. Oxygen-loaded erythrocytes travelling downstream towards the medulla come into close contact with the oxygen-depleted erythrocytes returning from the medulla, which take over the O_2 from the descending erythrocytes. Thus, the countercurrent system leads to medullary depletion of anything which is consumed in the medulla and leads to medullary excess of anything which is produced. As a consequence, the thin limbs of Henle's loop cannot perform substantial primary active transport and the renal medullary tissue is highly susceptible to ischemia.

C. Regulation of Renal Water and Electrolyte Excretion

I. Water

The hydraulic conductivity (cf. water permeabilities in Table 5) of the proximal renal tubule is extremely high. Accordingly, water reabsorption in the proximal tubule parallels reabsorption of osmotically active substances and the osmolarity of end proximal tubule fluid is close to isotonicity.

As discussed above, no water reabsorption occurs in the thick ascending limb of Henle's loop (cf. Table 5), a prerequisite for the concentrating ability of the kidney. Any factor influencing NaCl reabsorption in the thick ascending limb of Henle's loop will thus impair the concentrating ability of the kidney. The most important factors are hypokalemia, hypercalcemia and loop diuretics. Hypokalemia is probably effective through a decreased luminal K^+ concentration, since the availability of K^+ limits the activity of the Na^+, $2Cl^-$, K^+ cotransporter (GREGER 1985). Ca^{2+} has been shown to block tight junctions of the thick ascending limb (DISTEFANO et al. 1988), which impairs paracellular reabsorption of Na^+, Ca^{2+} and Mg^{2+}. Loop diuretics commonly inhibit the Na^+, $2Cl^-$, K^+ cotransporter, as detailed in Chap. 7 of this volume. As can be deduced from above (Sect. C.IV), further important factors impeding concentrating ability are urea depletion (protein-deficient diet) and dilation of vasa recta (e.g., due to inflammation of renal medulla).

Water reabsorption in the distal nephron (Table 5), i.e., in the distal convoluted tubule and the collecting duct, depends on the presence of antidiuretic hormone (ADH). This hormone stimulates the insertion of water channels into the luminal cell membrane via a cAMP-mediated mechanism (KIRK and SCHAFER 1992). Since water permeability of the basolateral membrane is large, water can then easily cross the epithelial layer following the osmotic gradient into the hypertonic medulla. ADH release is stimulated both by a decrease in intracellular volume and by a decrease in the filling of the cardiac atria (ROBERTSON 1992).

In the absence of ADH the luminal cell membrane is virtually water impermeable (cf. Table 8) and the dilute tubule fluid leaving the thick ascending limb (see above) is further diluted during the passage through the distal nephron.

II. Na^+

A great variety of hormones and mediators influence renal Na^+ excretion by interaction with tubular transport or by modification of glomerular filtration rate and thus filtered load of Na^+ (see Table 6). For most of the hormones listed in Table 2 the influence on renal Na^+ excretion is only one of many effects of the hormones. A detailed description of their effects has been given elsewhere (LANG 1988) but would be beyond the scope of this review. Here, only the most important hormones regulating renal Na^+ excretion will be considered.

Depletion of NaCl leads to a reduction in extracellular fluid, which through capillary exchange (MICHEL 1992) decrease blood volume. This in turn decreases heart filling, stroke volume and thus cardiac output. The tendency of blood pressure to decline is sensed by pressure receptors in the carotid artery, causing a stimulation of sympathetic nerve tone. In the following the perfusion of the kidney will be compromised and maintained only by autoregulation. The compromised perfusion of the kidney activates renin release (LARAGH 1992). Renin catalyzes the formation of angiotensin I from angiotensinogen. Angiotensin I is broken down by the converting enzyme to form the strongly vasoconstrictory angiotensin II (HALL and BRANDS 1992). At physiological concentrations, angiotensin II enhances the proximal-tubular reabsorption of Na^+ and stimulates the release of aldosterone from the adrenal gland. Aldosterone is responsible for augmenting the distal tubular reabsorption of Na^+. In the distal nephron aldosterone activates and stimulates de novo synthesis and incorporation of Na^+ channels, K^+ channels and $(Na^+ + K^+)$-ATPase. In addition, it stimulates the synthesis of mitochondria to ensure sufficient energy supply (ROSSIER and PALMER 1992). Besides its influence on aldosterone release, angiotensin II stimulates the release of ADH. Thus, water and NaCl are retained in the body and a further reduction of extracellular fluid is prevented.

NaCl excess not only depresses the release of renin, angiotensin and aldosterone, but also stimulates the release of natriuretic hormones. The most important of these natriuretic hormones are the atrial natriuretic peptides or factors (ANFs), which enhance NaCl excretion (COGAN 1990; SONNENBERG 1985; ZEIDEL 1990). Furthermore, a nonatrial natriuretic factor with ouabain-like action may contribute to the natriuresis during NaCl excess (BLAUSTEIN 1993).

Table 6. Factors modifying renal Na^+ handling (BLAUSTEIN 1993; GONZALES-CAMPOY and KNOX 1992; BALLERMANN and ZEIDEL 1992; LANG 1988; KUROKAWA et al. 1992; ROSSIER and PALMER 1992)

Hormone	Effect
Mineralocorticoids	+ DL: Na^+/H^+ exchanger + DT, CD: Na^+ channel, (Na^++K^+)-ATPase, energy supply distal nephron
Glucocorticoids	+ GFR, PT: Na^+/H^+ exchanger, TAL, DT, CD: (Na^++K^+)-ATPase – PT: Na^+, HPO_4^{2-} cotransport
Progesterone	Antimineralocorticoid action
Thyroid hormones	+ GFR, PT: K^+ channels, (Na^++K^+)-ATPase, Na^+, HPO_4^{2-} cotransport
ADH cotransport	+ TAL: Cl^- channels, Na^+, K^+, $2Cl^-$ + DT, CD: Na^+ channels
ANF	+ GFR – PT: Na^+, HPO_4^{2-} cotransport – CD: Na^+ reabsorption
Strophanthin	– (Na^++K^+)-ATPase
PTH transport	– PT: Na^+, HPO_4^{2-} cotransport, HCO_3^- + PT: Na^+/Ca^{2+} exchanger,
Calcitonin	– PT: Na^+, HPO_4^{2-} cotransport + TAL, DT: reabsorption of NaCl, Ca^{2+}, Mg^{2+}
Growth hormone	+ PT Na^+-coupled transport processes
Insulin	+ PT: Na^+, HPO_4^{2-} cotransport DT: Na^+ reabsorption, K^+ secretion – PT: reabsorption of Na^+ and Ca^{2+}
Glucagon	+ TAL, DT: reabsorption of Na^+, Ca^{2+}, Mg^{2+}
Angiotensin (pmol/l) (nmol/l)	+ PT: Na^+ reabsorption – PT: Na^+ reabsorption
PGE_2	– TAL, CD: Na^+ reabsorption
Bradykinin	– CD: Na^+ reabsorption
Catecholamines (a_2) (β)	+ PT: Na^+ reabsorption – CD: Na^+ reabsorption + TAL, CNT, CD: NaCl reabsorption
Dopamine	+ GFR – PT: Na^+, HPO_4^{2-} reabsorption
Substance P	– PT: Na^+ reabsorption
Histamine	+ GFR, – Na^+ reabsorption (?)

GFR, glomerular filtration rate; PT, proximal tubule; TAL, thick ascending limb of Henle's loop; DL, diluting segment (corresponds to TAL); DT, distal tubule; CNT, connecting tubule; CD, collecting duct; +, stimulation; –, inhibition.

III. Bicarbonate and Hydrogen Ions

At a urinary pH of as low as 4.5 the free H^+ concentration in urine is only 30 µmol/l. Thus, even the ability of the distal nephron to create high transepithelial H^+ gradients does not allow the excretion of the daily acid load of approximately 100 mmol H^+. The ability of the kidney to eliminate H^+ thus depends on the excretion of buffers. Two buffer systems are of major importance: the NH_3/NH_4^+ system and the $HPO_4^{2-}/H_2PO_4^-$ system, contributing approximately 60% and 30%, respectively, to renal acid excretion (GENNARI and MADDOX 1992; HAMM and SIMON 1987). A minor portion of H^+ excretion is in the form of organic acids such as uric acid, which, given a pK of 5.8, is at a pH of 7.4 and 5.5 97% and 70% dissociated, respectively. Thus, at a urine pH of 5.5, two-thirds of the excreted uric acid have bound H^+ during tubular passage.

NH_3 is a weak base with a pK of 9, i.e., at the blood pH approximately 98% binds H^+ to form NH_4^+. Accordingly, virtually all NH_3 will be excreted in the form of NH_4^+. In order to eliminate acid as NH_4^+, the kidney must produce NH_4^+ (BROSNAN et al. 1988; HALPERIN et al. 1992), as detailed in Chap. 3 of this volume. NH_3 is formed in and released from the proximal tubule cell into the lumen, where it readily reacts to form NH_4^+ (Fig. 7). Proximal tubule secretion of NH_4^+ is fostered by uptake of NH_4^+ across the basolateral cell membrane via ion channels (VÖLKL and LANG 1991). In Henle's loop, NH_4^+ undergoes countercurrent exchange leading to high medullar NH_4^+ concentrations (KNEPPER et al. 1989), as detailed above (Sect. C.IV, Fig. 14).

The pKs for the dissociation of phosphates

$$PO_4^{3-} \leftrightarrow HPO_4^{2-} \leftrightarrow H_2PO_4^- \leftrightarrow H_3PO_4$$

are 2.0, 6.8 and 12.3, respectively. In blood, approximately 80% of phosphate is present in the form of HPO_4^{2-}, approximately 20% as $H_2PO_4^{2-}$ and less than 1% as PO_4^{3-} or H_3PO_4. With an acidic urine of pH 5.8 approximately 90% of the phosphate is in the form of $H_2PO_4^{-1}$. Thus, for each millimole phosphate excreted into a urine of pH 5.8, 0.7 mmol H^+ has been eliminated bound to phosphate. Further acidification of urine does not markedly increase H^+ excretion in the form of phosphate. A further reduction of pH to 4.8, for example, increases H^+ excretion in the form of phosphate only by some 10%. In that range of urinary pH, H^+ excretion in the form of phosphate is mainly a function of phosphate excretion. With more alkaline urine, on the other hand, it is the urinary pH which strongly influences H^+ excretion in the form of phosphate (Fig. 15): At a urinary pH of 7.4, for example, no H^+ is eliminated in the form of phosphate irrespective of phosphate excretion. For the acid base balance of blood it must be kept in mind where the phosphate excreted comes from. If it is released from bone, where it prevails in the highly dissociated form, phosphate has already bound H^+ prior to excretion and the net effect for the acid base balance is enhanced.

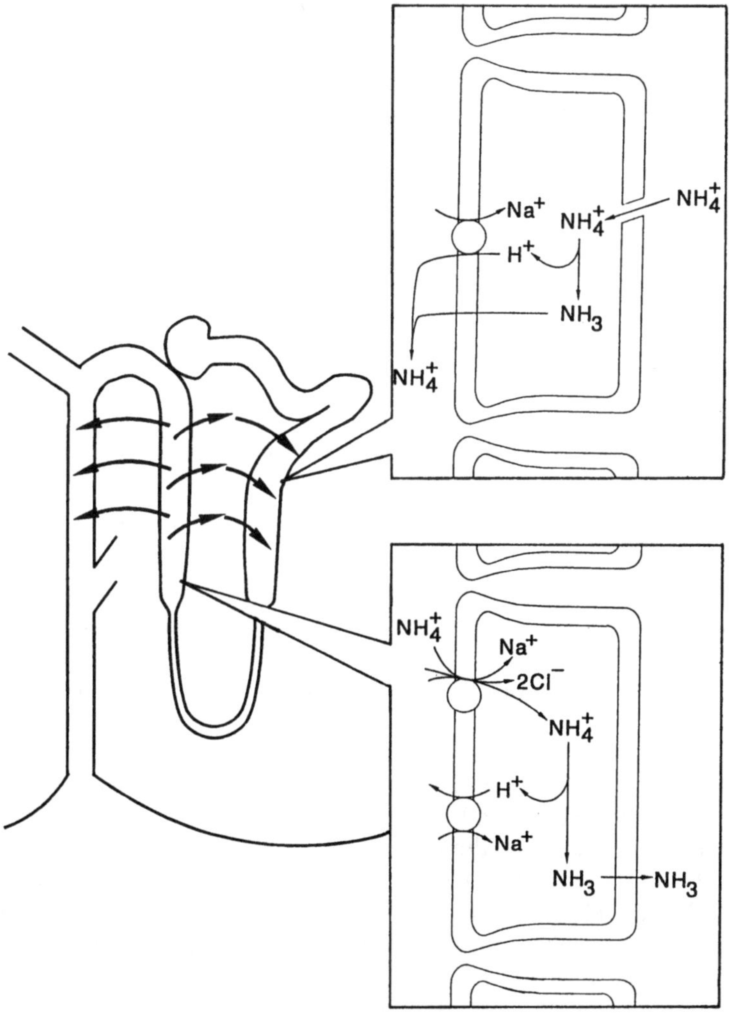

Fig. 14. Recirculation of NH_4^+ in renal medulla. In the thick ascending limb of Henle's loop, NH_4^+ is transported into the cell by the Na^+, $2Cl^-$, K^+ cotransporter and leaves the cell across the basolateral cell membrane as NH_3. NH_3 is thus reabsorbed. In the straight portion of the proximal tubule, HN_4^+ is taken up across the basolateral cell membrane via ion channels and leaves the cell as NH_3 mainly into the lumen, where it is trapped as NH_4^+. Accordingly, NH_3 is secreted. The countercurrent system thus builds up high concentrations of NH_3/NH_4^+ in the renal medulla, allowing for efficient secretion across the collecting duct

Regulation of renal acid elimination may thus be achieved by altered urinary pH and by altered buffer excretion. In the case of acidosis, NH_4^+ formation by the kidney is enhanced (see Chap. 3, this volume) and demineralization of bone facilitated. Accordingly, buffer excretion is stimulated. At the same time, renal tubular bicarbonate reabsorption and

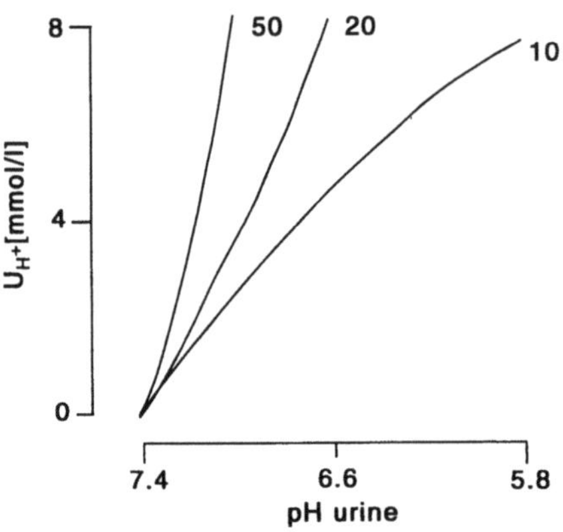

Fig. 15. Significance of urinary phosphate concentration in millimoles/e and of urine pH for the amount of H^+ in urine (U_H^+), which is excreted in the form of phosphate

acid secretion are stimulated, leading to more acid urine (GENNARI and MADDOX 1992).

Renal acid excretion may be modified by altered extracellular K^+ concentration. During hyperkalemia the proximal tubular cell membrane is depolarized and the driving force for the $Na^+(HCO_3^-)_n$ cotransporter reduced. The resulting intracellular alkalosis turns off the Na^+/H^+ exchanger, which is highly sensitive to intracellular pH. As a result, hyperkalemia inhibits proximal tubular HCO_3^- reabsorption, thus favoring the development of acidosis (LANG and REHWALD 1992). Conversely, hypokalemia favors the development of alkalosis.

In metabolic alkalosis, the kidney should excrete HCO_3^-, which requires reduced HCO_3^- reabsorption in the proximal tubule, where approximately 95% of filtered HCO_3^- is reabsorbed. In the proximal tubule, however, decreased HCO_3^- reabsorption is paralleled by decreased Na^+ reabsorption, which is in large part linked to HCO_3^- reabsorption (see above). Thus, if fluid depletion coincides with metabolic alkalosis (e.g., after severe vomiting of acid gastric fluid), the proximal tubule is forced to avidly reabsorb Na^+ (and HCO_3^-) and the kidney is unable to compensate for the alkalosis with bicarbonaturia (referred to as volume depletion alkalosis). Furthermore, during volume depletion aldosterone is released (see Sect. C.III.1), which stimulates distal tubular H^+ secretion (GENNARI and MADDOX 1992).

IV. K^+

K^+ transport in the distal nephron is critical for renal K^+ excretion (WRIGHT and GIEBISCH 1992). Accordingly, all factors enhancing distal tubular K^+ secretion will lead to kaliuresis. The most important factors and their mode of action are listed in Fig. 16. According to the alterations in chemical driving force, K^+ secretion is reduced with increasing luminal K^+ concentra-

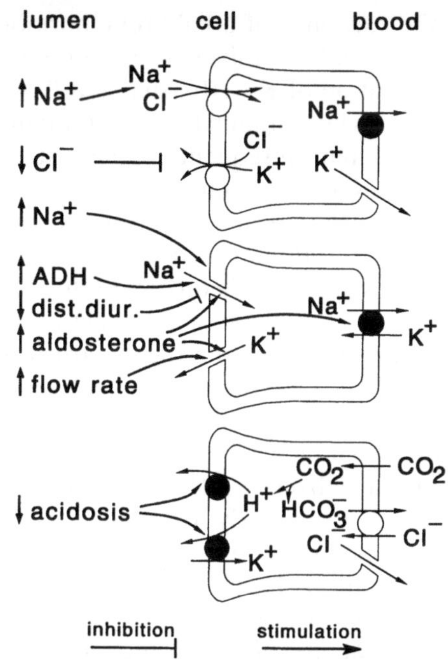

Fig. 16. Factors enhancing (↑) or decreasing (↑) distal-tubular K^+ secretion and their mechanism of action in the distal nephron. The factors are listed on the luminal cell side even though some (e.g., aldosterone) are effective from the peritubular side

tion and enhanced with decreasing luminal K^+ concentration. In addition, a decrease in cellular K^+ concentration may (slightly) reduce the chemical gradient. More importantly, cellular K^+ depletion may lead to cell shrinkage, which in turn closes K^+ channels (Ritter et al. 1991). This again decreases K^+ secretion. Enhanced distal luminal flow rate increases the gradient and stimulates K^+ secretion by "diluting" luminal K^+ (Wright and Giebisch 1992).

The entry of Na^+ across the luminal cell membrane depolarizes the principal cell and thus enhances the electrical driving force for K^+ secretion. Furthermore, Na^+ entering the cell must be extruded by the $(Na^+ + K^+)$-ATPase (Wright and Giebisch 1992), which accumulates K^+ within the cell, thus fostering K^+ secretion. Enhanced delivery of Na^+ to the distal tubule and/or stimulated Na^+ channel activity thus augments the secretion of K^+ (Wright and Giebisch 1992).

The Na^+ channel activity and the $(Na^+ + K^+)$-ATPase activity are stimulated by aldosterone (see above), subsequently strongly influencing K^+ secretion. Volume depletion thus favors the enhanced secretion of K^+ via aldosterone. On the other hand, aldosterone release is stimulated by K^+ excess and blunted by K^+ depletion. By K^+ depletion, aldosterone release and distal tubular K^+ secretion are thus reduced (Wright and Giebisch

1992). Furthermore, the existence of a kaliuretic hormone has been postulated, which should be released from liver or the central nervous system and may enhance renal elimination of K^+ (ROSSIER and PALMER 1992).

Besides aldosterone, ADH stimulates distal tubule K^+ secretion by activation of Na^+ channels and subsequent depolarization of the luminal cell membrane (SCHLATTER 1989). K^+ secretion is further stiumlated by decreased luminal Cl^- concentration and inhibited by increased luminal Cl^- concentration, probably due to the respective alterations in chemical driving force for the luminal KCl symport (WRIGHT and GIEBISCH 1992).

H^+ secretion via the H^+-ATPase of intercalated cells carries positive charge into the tubule lumen, decreasing the electrical driving force for K^+ secretion. Thus, enhanced H^+ secretion reduces K^+ secretion and vice versa (WRIGHT and GIEBISCH 1992). Furthermore, during acidosis proximal HCO_3^- reabsorption is enhanced, paralleled by stimulated proximal Na^+ reabsorption. The decreased distal delivery of Na^+ contributes to the decrease in distal tubular K^+ secretion. Conversely, decreased proximal pubular HCO_3^- and Na^+ reabsorption favor distal tubular K^+ loss in alkalosis.

Ca^{2+} similarly affects distal tubular K^+ secretion by influencing Na^+ reabsorption at more proximal nephron segments. As described above, Ca^{2+} decreases the permeability of tight junctions in the proximal tubule and loop of Henle and by impeding paracellular Na^+ reabsorption in those segments enhances distal tubular Na^+ load. On the other hand, Ca^{2+} more directly inhibits distal-tubular K^+ secretion. The two effects partially cancel each other out.

V. Mg^{2+}

The various factors modifying renal tubular transport and excretion of Mg^{2+} are as follows:

Stimulating
- Extracellular volume depletion
- Mg^{2+} depletion
- Hypocalcemia
- PTH (cAMP)
- Glucagon (cAMP)
- Calcitonin (cAMP)
- ADH (cAMP)

Inhibiting
- Extracellular volume expansion
- Hypermagnesemia
- Hypercalcemia
- Loop diuretics (furosemide, ethacrynic acid)
- Osmotic diuresis (mannitol)
- Phosphate depletion

Acute metabolic acidosis
Carbohydrates
Protein intake
Alcohol ingestion

Most importantly, impairment of Mg^{2+} reabsorption in the thick ascending limb of Henle's loop leads to profound Mg^{2+} loss. The reason for this loss is that no Mg^{2+} reabsorption occurs beyond this nephron site, resulting in the inability to compensate for the decreased Mg^{2+} reabsorption in that nephron segment (ALFREY 1992).

VI. Calcium and Phosphate

A great variety of factors modify the renal reabsorption and excretion of Ca^{2+} (COSTANZO and WINDHAGER 1992; STEWART 1992; LANG 1980) and phosphate (BERNDT and KNOX 1992; LANG 1980; MURER and BIBER 1992). These factors are summarized in Table 7.

Both mineral ions are reabsorbed in a seemingly saturable fashion. Proximal tubular phosphate reabsorption is indeed a saturable process, with a high affinity ($K_M \approx 0.2$ mmol/l) at the straight proximal tubule (LANG 1980). Accordingly, phosphate excretion is governed by a well-defined threshold with very little leeway. Ca^{2+} transport is not a saturable process but Ca^{2+} impedes its own paracellular reabsorption by closing the tight junctions (DISTEFANO et al. 1988). Furthermore, in intact animals increased plasma Ca^{2+} concentration inhibits the release of parathyroid hormone (PTH), which stimulates Ca^{2+} reabsorption.

The most important hormone in the regulation of mineral metabolism is PTH, which is released upon hypocalcemia. PTH releases bone minerals, most importantly Ca^{2+}, HPO_4^{2-} and CO_3^{2-}. In the kidney it stimulates distal tubular Ca^{2+} reabsorption (MUFF et al. 1992) but inhibits the proximal tubular reabsorption of HPO_4^{2-} and HCO_3^- to avoid hyperphosphatemia or alkalosis, respectively (Fig. 17). In parallel, PTH stimulates renal

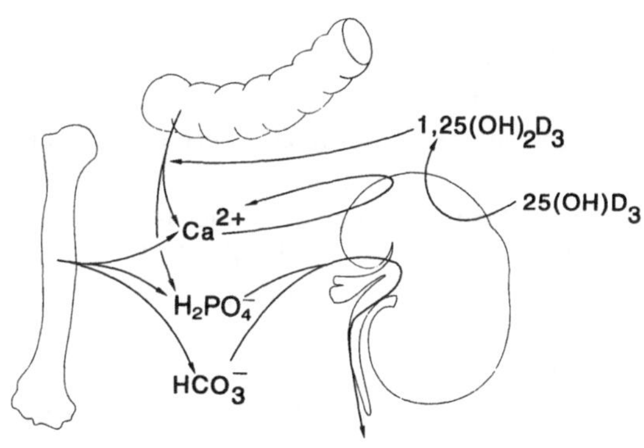

Fig. 17. The major actions of PTH

1,25-hydroxylase, which activates 25(OH)-cholecalciferol to 1,25(OH)$_2$-cholecalciferol (1,25-(OH)$_2$D$_3$ or calcitriol) (BROWN et al. 1992). This renal hormone stimulates intestinal Ca^{2+} and phosphate absorption and thus sets the stage for remineralization of bone (BROWN et al. 1992). Excess PTH paradoxically leads to calciuria despite enhanced stimulation of distal tubular Ca^{2+} reabsorption by PTH. The extrarenal actions of PTH lead to hypercalcemia, which accounts for this calciuria.

Additional hormones which stimulate an increase in renal Ca^{2+} excretion include growth hormone, thyroid hormones, mineralocorticoids, glucocorticoids and insulin. Calcitonin has been described as having both calciuric and anticalciuric effects (Table 7).

Hormones which stimulate an increase in renal phosphate excretion include calcitonin, chronic administration of calcitriol (acting via phosphate excess) and glucocorticoids. Thyroid hormones and growth hormone decrease renal phosphate excretion by stimulating reabsorption (Table 7).

The two mineral ions Ca^{2+} and phosphate mutually influence each other's renal excretion. Hyperphosphatemia leads to anticalciuria, and phosphate depletion produces calciuria. Acute hypocalcemia stimulates and acute hypercalcemia decreases phosphate excretion by the kidney. These effects are in part due to altered PTH release. On the other hand, for unknown reasons, a Ca^{2+}-rich diet leads to phosphaturia and a Ca^{2+}-depleted diet favors antiphosphaturia.

Mg^{2+} infusion leads to calciuria and antiphosphaturia, an effect attributed in part to the inhibition of PTH release and in part to the inhibition of paracellular Ca^{2+} reabsorption (DISTEFANO et al. 1988). Mg^{2+} depletion favors phosphaturia without significantly affecting Ca^{2+} excretion.

Excretion of both Ca^{2+} and phosphate correlates highly with Na^+ excretion. Expansion of the extracellular fluid volume leads to calciuria and phosphaturia and volume depletion to anticalciuria and antiphosphaturia. Regarding Ca^{2+} excretion, the effect is explained in large part by transport of Ca^{2+} and Na^+ via the paracellular shunt, which depends on the driving force and thus indirectly on transcellular Na^+ reabsorption. Regarding phosphate excretion, it appears that inhibition of proximal tubular Na^+ reabsorption during volume expansion includes sodium-phosphate cotransport.

Metabolic acidosis enhances, and metabolic alkalosis somewhat decreases, renal Ca^{2+} excretion. The decreased availability of filtered HCO_3^- during metabolic acidosis decreases proximal tubular $Na^+HCO_3^-$ reabsorption, which compromises proximal tubular fluid and paracellular Ca^{2+} reabsorption. Conversely, the increased filtered HCO_3^- during metabolic alkalosis stimulates proximal tubular fluid and Ca^{2+} reabsorption. Metabolic alkalosis reduces the maximal phosphate transport rate and metabolic acidosis increases, in intact parathyroid glands, tubular phosphate reabsorption. Several mechanisms probably contribue to this effect. Due to enhanced binding to proteins and HCO_3^-, metabolic alkalosis treduces Ca^{2+} in blood and thus stimulates PTH release. Moreover, pH modifies the dissociation of

Table 7. Factors modifying renal phosphate and calcium (Ca^{2+}) reabsorption

Stimulating Ca^{2+} reabsorption
PTH
D-Hormone (1,25-dihydroxycholecalciferol)
Thiazides
Alkalosis
Inhibiting Ca^{2+} reabsorption
Calcitonin
Growth hormone
Thyroid hormones
Adrenal steroids (chronic)
Insulin
Glucose
Acidosis
Extracellular volume expansion
Acetazolamide
Furosemide
Mannitol
Stimulating phosphate reabsorption
Phosphate depletion
Insulin
D-hormone (acute)
Thyroid hormones
Growth hormone
Magnesium
Lithium
Metabolic acidosis
Inhibiting phosphate reabsorption
Phosphate excess
Ca^{2+} excess (chronic)
PTH
Calcitonin
D-hormone (chronic)
Hypocalcemia (acute)
Dopamine
Mg^{2+} depletion
Respiratory acidosis
Metabolic alkalosis
Glucose
Colchicine
Extracellular volume expansion
Thiazides
Furosemide
Ethacrynic acid
Mannitol
Acetazolamide

phosphate and thus the availability of the transported species. Furthermore, pH may directly affect the carrier and modify its regulation by cAMP (LANG 1980).

Other factors influencing Ca^{2+} and phosphate excretion are glucose (glucose infusions may lead to calciuria and phosphaturia) and most importantly diuretics. Their effects will be dealt with separate chapters of this book.

References

Al-Awqati Q (1986) Proton-translocating ATPases. Annu Rev Cell Biol 2:179–199

Alfrey AC (1992) Disorders of magnesium metabolism. In: Seldin DW, Giebisch G (eds) The kidney, 2nd edn. Raven, New York, pp 2357–2374

Alper SL, Natale J, Gluck S, Lodish HF, Brown D (1989) Subtypes of intercalated cells in rat kidney collecting duct defined by antibodies against erythroid band 3 and renal vacuolar H^+-ATPase. Proc Natl Acad Sci USA 86:5429–5433

Alpern RJ (1985) Mechanism of basolateral membrane $H^+/OH^-/HCO_3^-$ transport in the rat proximal convoluted tubule. A sodium-coupled electrogenic process. J Gen Physiol 86:613–636

Alpern RJ, Chambers M (1987) Basolateral membrane Cl/HCO_3 exchange in the rat proximal convoluted tubule. Na-dependent and -independent modes. J Gen Physiol 89:581–598

Beck F-X, Dörge A, Giebisch G, Thurau K (1988) Renal excretion of rubidium and potassium: an electron microprobe and clearance study. Kidney Int 34:455–462

Ballermann BJ, Zeidel ML (1992) Atrial natriuretic hormone. In: Seldin DW, Giebisch G (eds) The kidney. Physiology and pathophysiology, 2nd end. Raven, New York, pp 1843–1884

Bell PD, Lapointe J-Y, Cardinal J (1989) Direct measurement of basolateral membrane potentials from cells of the macula densa. Am J Physiol 257 (Renal Fluid Electrolyte Physiol 26):F463–F468

Berndt TJ, Knox FG (1992) Renal regulation of phosphate excretion. In: Seldin DW, Giebisch G (eds) The kidney. Physiology and pathophysiology, 2nd edn. Raven, New York, pp 2511–2532

Berry CA, Rector FC Jr (1989) Electroneutral NaCl absorption in the proximal tubule: mechanisms of apical Na-coupled transport. Kidney Int 36:403–411

Biagi BA, Sohtell M (1986) Electrophysiology of basolateral bicarbonate transport in the rabbit proximal tubule. Am J Physiol 250 (Renal Fluid Electrolyte Physiol 19):F267–F272

Biber J (1989) Cellular aspects of proximal tubular phosphate reabsorption. Kidney Int 36:360–369

Biber J, Murer H (1993) Towards a molecular view of renal proximal tubular reabsorption of phosphate. Renal Physiol Biochem 16:37–47

Biemesderfer D, Stanton B, Wade JB, Kashgarian M, Giebisch G (1989) Ultrastructure of Amphiuma distal nephron: evidence for cellular heterogeneity. Am J Physiol 256 (Cell Physiol 25):C849–C857

Blaustein MP (1993) Physiological effects of endogenous ouabain: control of intracellular Ca^{2+} stores and cell responsiveness. Am J Physiol 264:C1367–C1387

Bleich M, Schlatter E, Greger R (1990) The luminal K^+ channel of the thick ascending limb of Henle's loop. Pflugers Arch 415:449–460

Boron WF (1992) Control of intracellular pH. In: Seldin DW, Giebisch G (eds) The kidney. Physiology and pathophysiology, 2nd edn. Raven, New York, pp 219–264

Boron WF, Boulpaep EL (1983a) Intracellular pH regulation in the renal proximal tubule of the salamander. Na-H exchange. J Gen Physiol 81:29–52

Boron WF, Boulpaep EL (1983b) Intracellular pH regulation in the renal proximal tubule of the salamander. Basolateral HCO_3^- transport. J Gen Physiol 81:53–94

Boron WF, Boulpaep EL (1989) The electrogenic Na/HCO_3 cotransporter. Kidney Int 36:392–402

Brenner BM, Meyer TW, Hostetter TH (1982) Dietary protein intake and the progressive nature of kidney disease: the role of hemodynamically mediated glomerular injury in the pathogenesis of progressive glomerular sclerosis in aging, renal ablation, and intrinsic renal disease. N Engl J Med 307:652–659

Brenner BM, Hostetter TH, Humes HD (1978) Molecular basis of proteinuria of glomerular origin. N Engl J Med 298:826–833

Briggs JP, Schnermann J (1986) Macular densa control of renin secretion and glomerular vascular tone: evidence for common cellular mechanisms. Renal Physiol 9:193–203

Brosnan JT, Vinay P, Gougoux A, Halperin ML (1988) Renal ammonium production and its implications for acid-base balance. In: Häussinger D (ed) pH homeostasis. Mechanisms and control. Academic, New York, pp 281–304

Brown D, Hirsch S, Gluck S (1988) Localization of proton-pumping ATPase in rat kidney. J Clin Invest 82:2114–2126

Brown AJ, Dusso AS, Slatopolsky E (1992) Vitamin D. In: Seldin DW, Giebisch G (eds) The kidney. Physiology and pathophysiology, 2nd edn. Raven, New York, pp 1505–1552

Burckhardt BC, Frömter E (1987) Evidence for OH^-/H^+ permeation across the peritubular cell membrane of rat renal proximal tubule in HCO_3^--free solutions. Pflugers Arch 409:132–137

Burckhardt G, Kinne RKH (1992) Transport proteins: cotransporters and countertransporters. In: Seldin DW, Giebisch G (eds) The kidney. Physiology and pathophysiology, 2nd edn. Raven, New York, pp 537–586

Burckhardt G, Ullrich KJ (1989) Organic anion transport across the contraluminal membrane – dependence on sodium. Kidney Int 36:370–377

Cogan MG (190) Renal effects of atrial natriuretic factor. Annu Rev Physiol 52:699–708

Costanzo LS, Windhager EE (1992) Renal regulation of calcium balance. In: Seldin DW, Giebisch G (eds) The kidney. Physiology and pathophysiology, 2nd edn. Raven, New York, pp 2375–2394

Dawson DC (1992) Water transport: principles and perspectives. In: Seldin DW, Giebisch G (eds) The kidney. Physiology and pathophysiology, 2nd edn. Raven, New York, pp 301–316

DeWeer P (1992) Cellular sodium-potassium transport. In: Seldin DQ, Giebisch G (eds) The kidney. Physiology and pathophysiology, 2nd edn. Raven, New York, pp 93–112

Dietl P, Oberleithner H (1987) Ca^{2+} transport in diluting segment of frog kidney. Pflugers Arch 410:63–68

Dietl P, Wang W, Oberleithner H (1987) Fused cells of frog proximal tubule. I. Basic membrane properties. J Membr Biol 100:43–51

Dietl P, Schwiebert E, Stanton B (1991) Cellular mechanisms of chloride transport in the cortical collecting duct. Kidney Int 40 [Suppl 33]:S125–S130

DiStefano A, Wittner M, Gebler B, Greger R (1988) Increased Ca^{2+} or Mg^{2+} concentration reduces relative tight-junction permeability to Na^+ in the cortical thick ascending limb of Henle's loop of rabbit kidney. Renal Physiol Biochem 11:70–79

Dominguez JH, Rothrock JK, Macias WL, Price J (1989) Na^+ electrochemical gradient and Na^+-Ca^{2+} exchange in rat proximal tubule. Am J Physiol 257 (Renal Fluid Electrolyte Physiol 26):F531–F538

Doucet A (1988) Function and control of Na-K-ATPase in single nephron segments of the mammalian kidney. Kidney Int 34:749–760

Doucet A, Marsy S (1987) Characterization of K-ATPase activity in distal nephron: stimulation by potassium depletion. Am J Physiol 253 (Renal Fluid Electrolyte Physiol 22):F418–F423

Dworkin LD, Brenner BM (1992) Biophysical basis of glomerular filtration. In: Seldin DW, Giebisch G (eds) The kidney. Physiology and pathophysiology, 2nd edn. Raven, New York, pp 979–1016

Ellison DH, Velazquez H, Wright FS (1987) Mechanisms of sodium, potassium and chloride transport by the renal distal tubule. Miner Electrolyte Metab 13:422–432

Eveloff J, Warnock DG (1987) K-Cl transport systems in rabbit renal basolateral membrane vesicles. Am J Physiol 252 (Renal Fluid Electrolyte Physiol 21): F883–F889

Friedman PA, Andreoli TE (1982) CO_2-stimulated NaCl absorption in the mouse renal cortical thick ascending limb of Henle. Evidence for synchronous Na^+/H^+ and Cl^-/HCO_3^- exchange in apical plasma membranes. J Gen Physiol 80:683–711

Frindt G, Lee CO, Yang JM, Windhager EE (1988) Potential role of cytoplasmic calcium ions in the regulation of sodium transport in renal tubules. Miner Electrolyte Metab 14:40–47

Frömter E (1984) Viewing the kidney through microelectrodes. Am J Physiol 247 (Renal Fluid Electrolyte Physiol 16):F695–F705

Frömter E, Rumrich G, Ullrich KJ (1973) Phenomenologic description of Na^+, Cl^- and HCO_3^- absorption from proximal tubules of the rat kidney. Pflugers Arch 343:189–220

Frömter E, Burckhardt BC, Kondo Y (1988) Mechanisms of basolateral base transport in the renal proximal tubule. Ciba Found Symp 139:106–121

Garg LC, Narang N (1988) Ouabain-insensitive K-adenosine triphosphatase in distal nephron segments of the rabbit. J Clin Invest 81:1204–1208

Garg LC, Narang N (1989) Suppression of ouabain-insensitive K-ATPase activity in rabbit nephron segments during chronic hyperkalemia. Renal Physiol Biochem 12:295–301

Garg LC, Narang N (1990) Effects of low-potassium diet on N-ethylmaleimide-sensitive ATPase in the distal nephron segments. Renal Physiol Biochem 13: 129–136

Garvin JL, Burg MB, Knepper MA (1988) Active NH_4^+ absorption by the thick ascending limb. Am J Physiol 255 (Renal Fluid Electrolyte Physiol 24):F57–F65

Geibel J, Giebisch G, Boron WF (1989a) Effects of acetate on luminal acidification processes in the S3 segment of the rabbit proximal tubule. Am J Physiol 257 (Renal Fluid Electrolyte Physiol 26):F586–F594

Geibel J, Giebisch G, Boron WF (1989b) Basolateral sodium-coupled acid-base transport mechanisms of the rabbit proximal tuble. Am J Physiol 257 (Renal Fluid Electrolyte Physiol 26):F790–F797

Gennari FJ, Maddox DA (1992) Renal regulation of acid-base homeostatsis: integrated resposne. In: Seldin DW, Giebisch G (eds) The kidney. Physiology and pathophysiology, 2nd edn. Raven, New York, pp 2695–2732

Giebisch G, Hunter M, Kawahara K (1990) Apical potassium channels in Amphiuma diluting segment: effect of barium. J Physiol (Lond) 420:313–323

Gluck S, Caldwell J (1988) Proton-translocating ATPase from bovine kidney medulla: partial purification and reconstitution. Am J Physiol 254 (Renal Fluid Electrolyte Physiol 23):F71–F79

Gmaj P, Murer H, Kinne R (1979) Calcium ion transport across plasma membranes isolated from rat kidney cortex. Biochem J 178:549–557

Gögelein H, Greger R (1986) Na^+ selective channels in the apical membrane of rabbit late proximal tubules (pars recta). Pflugers Arch 406:198–203

Gonzalez-Campoy JM, Knox FG (1992) Integrated responses of the kidney to alterations in extracellular fluid volume. In: Seldin DW, Giebisch G (eds) The kidney. Physiology and pathophysiology, 2nd edn. Raven, New York, pp 2041–2098

Good DW (1985) Sodium-dependent bicarbonate absorption by cortical thick ascending limb of rat kidney. Am J Physiol 248 (Renal Fluid Electrolyte Physiol 17):F821–F829

Good DW, Knepper MA, Burg MB (1984) Ammonia and bicarbonate transport by thick ascending limb of rat kidney. Am J Physiol 247 (Renal Fluid Electrolyte Physiol 16):F35–F44
Greger R (1981) Chloride reabsorption in the rabbit cortical thick ascending limb of the loop of Henle: a sodium dependent process. Pflugers Arch 390:38–43
Dreger R (1985) Ion transport mechamisms in thick ascending limb of Henle's loop of mammalian nephron. Physiol Rev 65:760–797
Greger R, Schlatter E (1981) Presence of luminal K^+, a prerequisite for active NaCl transport in the cortical thick ascending limb of Henle's loop of rabbit kidney. Pflugers Arch 392:92–94
Greger R, Schlatter E (1983) Properties of the basolateral membrane of the cortical thick ascending limb of Henle's loop of rabbit kidney. A model for secondary active chloride transport. Pflugers Arch 396:325–334
Greger R, Velazquez H (1987) The cortical thick ascending limb and early distal convoluted tubule in the urinary concentrating mechanism. Kidney Int 31:590–596
Greger R, Schlatter E, Lang F (1983) Evidence for electroneutral sodium chloride cotransport in the cortical thick ascending limb of Henle's loop of rabbit kidney. Pflugers Arch 396:308–314
Greger R, Weidtke C, Schlatter E, Wittner M, Gebler B (1984a) Potassium activity in cells of isolated perfused cortical thick ascending limbs of rabbit kidney. Pflugers Arch 401:52–57
Greger R, Wittner M, Schlatter E, Di Stefano A (1984b) Na^+2Cl^- K^+-cotransport in the thick ascending limb of Henle's loop and mechanism of action of loop diuretics. In: Hoshi T (ed) Coupled transport in nephron. Mechanisms and pathophysiology. Miura Medical Research Foundation, Tokyo, pp 96–118
Greger R, Bleich M, Schlatter E (1990) Ion channels in the thick ascending limb of Henle's loop. Renal Physiol Biochem 13:37–50
Gross P, Minuth WW, Ketteler M, Frömter E (1988) Ionic conductances of cultured principal cell epithelium of renal collecting duct. Pflugers Arch 412:434–441
Grupp C, Pavenstädt-Grupp I, Grunewald RW, Bevan C, Stokes JB III, Kinne RKH (1989) A Na-K-Cl cotransporter in isolated rat papillary collecting duct cells. Kidney Int 36:201–209
Guggino WB (1986) Functional heterogeneity in the early distal tubule of the Amphiuma kidney: evidence for two modes of Cl^- and K^+ transport across the basolateral cell membrane. Am J Physiol 250 (Renal Fluid Electrolyte Physiol 19):F430–F440
Guggino WB, Guggino SE (1989) Renal anion transport. Kidney Int 36:385–391
Gullans SR, Mandel LJ (1992) Coupling of energy to transport in proximal and distal nephron. In: Seldin DW, Giebisch G (eds) The kidney, 2nd edn. Raven, New York, pp 1291–1338
Häberle DA (1988) Hemodynamic interactions between intrinsic blood flow control mechanisms in the rat kidney. Renal Physiol Biochem 11:289–315
Hall JE, Brands MW (1992) The renin-angiotensin-aldosterone systems: renal mechanisms and circulatory homeostasis. In: Seldin DW, Giebisch G (eds) The kidney. Physiology and pathophysiology, 2nd edn. Raven, New York, pp 1455–1504
Halperin ML, Kamel KS, Ethier JH, Stinebaugh BJ, Jungas RL (1992) Biochemistry and physiology of ammonium excretion. In: Seldin DW, Giebisch G (eds) The kidney. Physiology and pathophysiology, 2nd edn. Raven, New York, pp 2645–2680
Hamm LL, Simon EE (1987) Roles and mechanisms of urinary buffer excretion. Am J Physiol 253:F595–F605
Hebert SC, Andreoli TE (1984a) Effects of antidiuretic hormone on cellular conductive pathways in mouse medullary thick ascending limbs of Henle. II. Deter-

minants of the ADH-mediated increases in transpithelial voltage and in net Cl^- absorption. J Membr Biol 80:221–233
Hebert SC, Andreoli TE (1984b) Control of NaCl transport in the thick ascending limb. Am J Physiol 246 (Renal Fluid Electrolyte Physiol 15):F745–F756
Hebert SC, Andreoli TE (1986) Ionic conductance pathways in the mouse medullary thick ascending limb of Henle. J Gen Physiol 87:567–590
Hebert SC, Culpepper RM Andreoli TE (1981) NaCl transport in mouse medullary thick ascending limbs. Am J Physiol 241 (Renal Fluid Electrolyte Physiol 10): F412–F442
Hebert SC, Friedman PA, Andreoli TE (1984) Effects of antidiuretic hormone on cellular conductive pathways in mouse medullary thick ascending limbs of Henle. ADH increases transcellular conductance pathways. J Membre Biol 80:201–219
Hediger MA, Coady MJ, Ikeda TS, Wright EM (1987) Expression cloning and cDNA sequencing of the Na^+/glucose co-transporter. Nature 330:379–381
Holthöfer H, Schulte BA, Pasternack G, Siegel GJ, and Spicer SS (1987) Three distinct cell populations in rat kidney collecting duct. Am J Physiol 253 (Cell Physiol 22):C323–C328
Hoshi T (1990) Electrophysiology of Triturus nephron: cable properties and electrogenic transport systems. Kidney Int 37:157–170
Hoshi T, Sudo K, Suzuki Y (1976) Characteristics of changes in the intracellular potential associated with transport of neutral, dibasic and acidic amino acids in triturus proximal tubule. Biochm Biophys Acta 448:492–504
Hunter M, Giebsch G (1987) Multi-barrelled K channels in renal tubules. Nature 327:522–524
Hunter M, Giebisch G (1988) Calcium-activated K-channels of Amphiuma early distal tubule: inhibition by ATP. Pflugers Arch 412:331–333
Hunter M, Kawahara K, Giebisch G (1986) Potassium channels along the nephron. Fed Proc 45:2723–2726
Imai M, Yoshitomi K (1990) Electrophysiological study of inner medullary collecting duct of hamsters. Pflugers Arch 416:180–188
Imai M, Taniguchi J, Tabei K (1987) Function of thin loops of Henle. Kidney Int 31:565–579
Jayakumar A, Cheng L, Liang CT, Sacktor B (1984) Sodium gradient-dependent calcium uptake in renal basolateral membrane vesicles: effect of parathyroid hormone. J Biol Chem 259:10827–10833
Joergensen PL (1980) Sodium and potassium ion pump in kidney tubules. Physiol Rev 60:864–917
Kahn AM (1989) Indirect coupling between sodium and urate transport in the proximal tubule. Kidney Int 36:378–384
Karniski LP, Aronson PS (1985) Chloride/formate exchange with formic acid recycling: a mechanism of active chloride transport across epithelial membranes. Proc Natl Acad Sci USA 82:6362–6365
Katz AI (1982) Renal Na-K-ATPase: its role in tubular sodium and potassium transport. Am J Physiol 242 (Renal Fluid Electrolyte Physiol 11):F207–F219
Kawahara K, Hunter M Giebisch G (1990) Calcium-activated potassium channels in the luminal membrane of Amphiuma diluting segment: voltage-dependent block by intracellular Na^+ upon depolarisation. Pflugers Arch 416:422–427
Kinne R (1976) Properties of the glucose transport system in the renal brush border membrane. Curr Top Membr Transp 8:209–267
Kinne R, Kinne-Saffran E, Schölermann B, Schütz H (1986) The anion specifity of the sodium-potassium-chloride cotransporter in rabbit kidney outer medulla: studies on medullary plasma membranes. Pflugers Arch 407 [Suppl 2]:S168–S173
Kinne-Saffran E, Beauwens R, Kinne R (1982) An ATP-driven proton pump in brush-border membranes from rat renal cortex. J Membr Biol 64:67–76

Kirk KL, Schafer JA (1992) Water transport and osmoregulation by antidiuretic hormone in terminal nephron segments. In: Seldin DW, Giebisch G (eds) The kidney. Physiology and pathophysiology, 2nd edn. Raven, New York, pp 1693–1726

Knepper MA, Packer R, Good DW (1989) Ammonium transport in the kidney. Physiol Rev 69:179–249

Koeppen BM (1987) Electrophysiological identification of principal and intercalated cells in the rabbit outer medullary collecting duct. Pflugers Arch 409:138–141

Koeppen BM (1989) Electrophysiology of collecting duct H^+ secretion: effect of inhibitors. Am J Physiol 256 (Renal Fluid Electrolyte Physiol 25):F79–F84

Koepsell H, Fritzsch G, Korn K, Madrala A (1990) Two substrate sites in the renal Na^+-D-glucose cotransporter studied by model analysis of phlorizin binding and D-glucose transport measurements. J Membr Biol 114:113–132

Kondo Y, Frömter E (1990) Evidence of chloride/bicarbonate exchange mediating bicarbonate efflux from S3 segments of rabbit renal proximal tubule. Pflugers Arch 415:726–733

Kristensen P, Ussing H (1992) Epithelial organization. In: Seldin DW, Giebisch G (eds) The kidney. Physiology and pathophysiology, 2nd edn. Raven, New York, pp 265–286

Kurokawa K, Fukagawa M, Hagashi M, Saruta T (1992) Renel receptors and cellular mechanisms of hormone action in the kidney. In: Seldin DW, Giebisch G (eds) The kidney. Physiology and pathophysiology, 2nd edn. Raven, New York, pp 1339–1372

Kurtz I (1989) Basolateral membrane Na^+/H^+ antiport, Na^+/base cotransport, and Na^+ independent Cl^-/base exchange in the rabbit S3 proximal tubule. J Clin Invest 83:616–622

Lang F (1980) Renal handling of calcium and phosphate. Klin Wochonschr 58:985–1003

Lang F (1982) Nierendurchblutung. In: Losse H, Renner E (eds) Klinische Nephrologie. Thieme, Stuttgart, p 43

Lang F (1988) NaCl transport in the kidney. In: Greger R (ed) Advances in comparative and environmental physiology. Springer, Berlin Heidelberg New York, pp 153–188

Lang F, rehwald W (1992) Potassium channels in renal epithelial transport regulation. Am J Physiol 72:1–32

Lang F, Häussinger D, Tschernko E, Capasso G, De Santo NG (1992) Proteins, the liver and the kidney – hepatic regulation of renal function. Nephron 61:1–4

Laragh JH (1992) The renin system and the renal regulation of blood pressure. In: Seldin DW, Giebisch G (eds) The kidney. Physiology and pathophysiology, 2nd edn. Raven, New York, pp 1411–1454

Light DB, Schwiebert EM, Fejes-Toth G, Naray-Fejes-Toth A, Karlson KH, McCann FV, Stanton BA (1990) Chloride channels in the apical membrane of cortical collecting duct cells. Am J Physiol 258 (Renal Fluid Electrolyte Physiol 27):F273–F280

Lopes AG, Amzel LM, Markakis D, Guggino WB (1988) Cell volume regulation by the thin descending limb of Henle's loop. Proc Natl Acad Sci USA 85:2873–2877

Madsen KM, Clapp WL, Verlander JW (1988) Structure and function of the inner medullary collecting duct. Kidney Int 34:441–454

Madsen KM, Tisher CC (1988) Structural-functional relationships along the distal nephron. Am J Physiol 250 (Renal Fluid Electrolyte Physiol 19):F1–F15

Martinez F, Manganel M, Montrose-Rafiza-Deh C, Werner D, Roch-Ramel F (1990) Transport of urate and p-aminohippurate in rabbit renal brush-border membranes. Am J Physiol 258 (Renal Fluid Electrolyte Physiol 27):F1145–F1153

Mason J (1986) The pathophysiology of ischemic acute renal failure. Renal Physiol 9:129–147

Matsuzaki K, Stokes JB, Schuster VL (1989) Stimulation of Cl^- self exchange by intracellular HCO_3^- in rabbit cortical collecting duct. Am J Physiol 257 (Cell Physiol 26):C94–C101
Mene P, Dunn MJ (1992) Vascular, glomerular, and tubular effects of angiotensin II, kinins, and prostaglandins. In: Seldin DW, Giebisch G (eds) The kidney. Physiology and pathophysiology, 2nd edn. Raven, New York, pp 1205–1248
Merot J, Bidet M, Gachot B, Le Maout S, Koechlin N, Tauc M, Poujeol P (1989) Electrical properties of rabbit early distal convoluted tubule in primary culture. Am J Physiol 257 (Renal Fluid Electrolyte Physiol 26):F288–F299
Michel CC (1992) Capillary exchange. In: Seldin DW, Giebisch G (eds) The kidney. Physiology and pathophysiology, 2nd edn. Raven, New York, pp 61–92
Molony DA, Reeves WB, Andreoli TE (1989) $Na^+:K^+:2Cl^-$ cotransport and the thick ascending limb. Kidney Int 36:418–426
Muff R, Fischer JA, Biber J, Murer H (1992) Parathyroid hormone receptors in control of proximal tubule function. Annu Rev Physiol 54:67–80
Murer H, Biber J (1992) Renal tubular phosphate transport: cellular mechanisms. In: Seldin DW, Giebisch G (eds) The kidney. Physiology and pathophysiology, 2nd edn. Raven, New York, pp 2481–2510
Murer H, Hopfer U, Kinne R (1976) Sodium/proton antiport in brush-border-membrane vesicles isolated from rat small intestine and kidney. Biochem J 154:597–604
Murer H, Burckhardt G (1983) Membrane transport of anions across epithelia of mammalian small intestine and kidney proximal tubule. Rev Physiol Biochem Pharmacol 96:1–15
Muto S, Giebisch G, Sansom S (1987) Effects of adrenalectomy on CCD: evidence for differential response of two cell types. Am J Physiol 253 (Renal Fluid Electrolyte Physiol 22):F742–F752
Muto S, Yasoshima K, Yoshitomi K, Imai M, Asano Y (1990) Electrophysiological identification of alpha and beta-inter-calated cells and their distribution along the rabbit distal nephron segments. J Clin Invest 86:1829–1839
Oberleithner H, Giebisch G, Lang F, Wang W (1982a) Cellular mechanism of the furosemide sensitive transport system in the kidney. Klin Wochenschr 60:1173–1179
Oberleithner H, Lang F, Wang W, Giebisch G (1982b) Effects of inhibition of chloride transport in intracellular sodium activity in distal amphibian nephron. Pflugers Arch 394:55–60
Oberleithner H, Greger R, Neuman S, Lang F, Giebisch G, Deetjen P (1983a) Omission of luminal potassium reduces cellular chloride in early distal tubule of amphibian kidney. Pflugers Arch 398:18–22
Oberleithner H, Guggino W, Giebisch G (1983b) The effect of furosemide on luminal sodium, chloride and potassium transport in the early distal tubule of Amphiuma kidney. Effects of potassium adaptation. Pflugers Arch 396:27–33
Oberleithner H, Lang F, Greger R, Wang W, Giebisch G (1983c) Effect of luminal potassium on cellular sodium activity in the early distal tubule of Amphiuma kidney. Pflugers Arch 396:34–40
Oberleithner H, Ritter M, Lang F, Guggino W (1983d) Antracene-9-carboxylic acid inhibits renal chloride reabsorption. Pflugers Arch 398:172–174
Pallone TL, Robertson CR, Jamison RL (1990) Renal medullary microcirculation. Physiol Rev 3:885–920
Palmer LG (1990) Epithelial Na channels: the nature of the conducting pore. Renal Physiol Biochem 13:51–58
Palmer LG, Frindt G (1986) Amiloride-sensitive Na channels from the apical membrane of the rat cortical collecting tubule. Proc Natl Acad Sci USA 83:2767–2770
Palmer LG, Frndt G (1988) Conductance and gating of epithelial Na channels from rat cortical collecting tubule. Effects of luminal Na and Li. J Gen Physiol 92:121–138

Palmer LG, Sackin H (1992) Electrophysiological analysis of transepithelial transport. In: Seldin DW, Giebisch G (eds) The kidney. Physiology and pathophysiology, 2nd edn. Raven, New York, pp 361–406
Paulais M, Teulon J (1990) cAMP-activated chloride channel in the basolateral membrane of the thick ascending limb of the mouse kidney. J Membr Biol 113:253–260
Ridderstrale Y, Kashgarian M, Koeppen B, Giebisch G, Stetson D, Ardito T, Stanton B (1988) Morphological heterogeneity of the rabbit collecting duct. Kidney Int 34:655–670
Ritter M, Paulmichl M, Lang F (1991) Further characterization of volume regulatory decrease in cultured renal epitheloid (MDCK-)cells. Pflugers Arch 418:35–39
Robertson GL (1992) Regulation of vasopression secretion. In: Seldin DW, Giebisch G (eds) The kidney. Physiology and pathophysiology, 2nd edn. Raven, New York, pp 1595–1614
Rocha AS, Kudo LH (1990a) Factors governing sodium and chloride transport across the inner medullary collecting duct. Kidney Int 38:654–667
Rocha AS, Kudo LH (1990b) Atrial peptide and cGMP effects on NaCl transport in inner medullary collecting duct. Am J Physiol 259 (Renal Fluid Electrolyte Physiol 28):F258–F268
Ross CR, Holohan PD (1983) Transport of organic anions and cations in isolated renal plasma membranes. Annu Rev Pharmacol Toxicol 23:65–85
Rossier BC, Palmer LG (1992) Mechanisms of aldosterone action on sodium and potassium transport. In: Seldin DW, Giebisch G (eds) The kidney. Physiology and pathophysiology, 2nd edn. Raven, New York, pp 1373–1410
Roy DR, Layton HE, Jamison RL (1992) Countercurrent mechanism and its regulation. In: Seldin DW, Giebisch G (eds) The kidney, 2nd edn. Raven, New York, pp 1649–1692
Sacktor B (1989) Sodium-coupled hexose transport. Kidney Int 36:342–350
Samarzija I, Frömter E (1982a) Electrophysiological analysis of rat renal sugar and amino acid transport. III. Neutral amino acids. Pflugers Arch 393:199–209
Samarzija I, Frömter E (1982b) Electrophysiological analysis of rat renal sugar and amino acid transport. IV. Basic amino acids. Pflugers Arch 393:210–214
Samarzija I, Frömter E (1982c) Electrophysiological analysis of rat renal sugar and amino acid transport. V. Acidic amino acids. Pflugers Arch 393:215–221
Samarzija I, Hinton BT, Frömter E (1982) Electrophysiological analysis of rat renal sugar and amino acid transport. II. Dependence on various transport parameters and inhibitors. Pflugers Arch 393:190–197
Sands JM, Knepper MA, Spring KR (1986) Na-K-Cl cotransport in apical membrane of rabbit renal papillary surface epithelium. Am J Physiol 251 (Renal Fluid Electrolyte Physiol 20):F475–F484
Sands JM, Kokko JP, Jacobson HR (1992) Intrarenal heterogeneity: vascular and tubular. In: Seldin DW, Giebisch G (eds) The kidney. Physiology and pathophysiology, 2nd edn. Raven, New York, pp 1098–1156
Sansom SC, Weinman EJ, O'Nell RG (1984) Microelectrode assessment of chloride-conductive properties of cortical collecting duct. Am J Physiol 247 (Renal Fluid Electrolyte Physiol 16):F291–F302
Sansom SC, Agulian S, Muto S, Illig V, Giebisch G (1989) K activity of CCD principal cells from normal and DOCA-treated rabbits. Am J Physiol 256 (Renal Fluid Electrolyte Physiol 25):F136–F142
Sansom SC, LA B-Q, Carosi SL (1990) Double-barreled chloride channels of collecting duct basolateral membrane. Am J Physiol 259 (Renal Fluid Electrolyte Physiol 28):F46–F52
Sasaki S, Ishibashi K, Yoshiyama N, Shiigai T (1988) KCl co-transport across the basolateral membrane of rabbit renal proximal straight tubules. J Clin Invest 81:194–199
Schafer JA, Williams JC Jr (1985) Transport of metabolic substrates by the proximal nephron. Annu Rev Physiol 47:103–125

Schild L, Giebisch G, Karniski L, Aronson PS (1986) Chloride transport in the mammalian proximal tubule. Pflugers Arch 407 [Suppl 2]:S156–S159
Schild L, Aronson PS, Giebisch G (1990) Effects of apical membrane Cl^--formate exchange on cell volume in rabbit proximal tubule. Am J Physiol 258 (Renal Fluid Electrolyte Physiol 27):F530–F536
Schlatter E (1989) Antidiuretic hormone regulation of electrolyte transport in the distal nephron. Renal Physiol Biochem 12:65–84
Schlatter E, Schafer JA (1987) Electrophysiological studies in principal cells of rat cortical collecting tubules. ADH increases the apical membrane Na^+-conductance. Pflugers Arch 409:81–92
Schlatter E, Schafer JA (1988) Electrophysiological studies in intercalated cells of rat cortical collecting tubules (CCT) (Abstract). Pflugers Arch 411:R101
Schlatter E, Greger R, Schafer JA (1990a) Principal cells of cortical collecting ducts of the rat are not a route of transepithelial Cl transport. Pflugers Arch 417: 317–323
Schlatter E, Salomonsson M, Persson AEG, Greger R (1990b) Macula densa cells reabsorb NaCl via furosemide sensitive Na^+-K^+-$2Cl^{2-}$ contransport. In: Puschett JB, Greenberg A (eds) Diuretics III: chemistry, pharmacology, and clinical applications. Elsevier, New York, pp 756–758
Schnermann J, Briggs JP (1992) Function of the juxtaglomerular apparatus: control of glomerular hemodynamics and renin secretion. In: Seldin DW, Giebisch G (eds) The kidney. Physiology and pathophysiology, 2nd edn. Raven, New York, pp 1249–1290
Schuster VL (1990) Organization of collecting duct intercalated cells. Kidney Int 38:668–672
Schuster VL, Seldin DW (1992) Renal clearance. In: Seldin DW, Giebisch G (eds) The kidney, 2nd edn. Raven, New York, pp 943–978
Schuster VL, Stokes JB (1987) Chloride transport by the cortical and outer medullary collecting duct. Am J Physiol 253 (Renal Fluid Electrolyte Physiol 22):F203–F212
Schwab A, Oberleithner H (1988) Trans- and paracellular K^+ transport in diluting segment of frog kidney. Pflugers Arch 411:268–272
Schwartz GJ, Barasch J, Al-Awqati Q (1985) Plasticity of functional epithelial Polarity. Nature 318:368–371
Silbernagl S (1988) The renal handling of amino acids and oligopeptides. Physiol Rev 68:911–1007
Silbernagl S, Ganapathy V, Leibach FH (1987) H^+ gradient-driven dipeptide reabsorption in proximal tubule of rat kidney. Studies in vivo and in vitro. Am J Physiol 253 (Renal Fluid Electrolyte Physiol 22):F448–F457
Sonnenberg H (1985) Atrial natriuretic factor – a new hormone affecting kidney function. Klin Wochenschr 63:886–890
Stanton BA (1988) Electroneutral NaCl transport by distal tubule: evidence for Na^+/H^+-Cl^-/HCO_3^- exchange. Am J Physiol 254 (Renal Fluid Electrolyte Physiol 23):F80–F86
Stanton BA (1989) Renal potassium transport: morphological and functional adaptations. Am J Physiol 257 (Regul Integrative Comp Physiol 26):R989–R997
Star RA, Burg MB, Knepper MA (1985) Bicarbonate secretion and chloride absorption by rabbit cortical collecting ducts. Role of chloride/bicarbonate exchange. J Clin Invest 76:1123–1130
Steinmetz PR (1986) Cellular organization of urinary acidification. Am J Physiol 251 (Renal Fluid Electrolyte Physiol 20):F173–F187
Stewart AF (1992) Hypercalcemic and hypocalcemic states. In: Seldin DW, Giebisch G (eds) The kidney. Physiology and pathophysiology, 2nd edn. Raven, New York, pp 2431–2460
Stokes JB (1989) Electroneutral NaCl transport in the distal tubule. Kidney Int 36:427–433

Stokes JB (1990) Sodium and potassium transport by the collecting duct. Kidney Int 38:679–686
Talor Z, Arruda JAL (1986) Na-Ca exchange in renal tubular basolateral membranes. Miner Electrolyte Metab 12:239–245
Terada Y, Knepper MA (1990) Thiazide-sensitive NaCl absorption in rat cortical collecting duct. Am J Physiol 259 (Renal Fluid Electrolyte Physiol 28): F519–F528
Tiruppathi C, Ganapathy V, Leibach FH (1990a) Evidence for tripeptide-proton symport in renal brushborder membrane vesicles. J Biol Chem 265:2048–2053
Tiruppathi C, Kulanthaivel P, Ganapathy V, Leibach FH (1990b) Evidence for tripeptide/H^+ co-transport in rabbit renal brush-border membrane vesicles. Biochem J 268:27–33
Turner RJ (1984) Sodium-dependent sulfate transport in renal outer cortical brush border membrane vesicles. Am J Physiol 247 (Renal Fluid Electrolyte Physiol 16):F793–F798
Turrini F, Sabolic I, Zimolo Z, Moewes B, Burckhardt G (1989) Relation of ATPases in rat renal brush border membranes to ATP-driven H^+ secretion. J Membr Biol 107:1–12
Ulfendahl HR, Wolgast M (1992) Renal circulation and lymphatics. In: Seldin DW, Giebisch G (eds) The kidney. Physiology and pathophysiology, 2nd edn. Raven, New York, pp 1017–1048
Ullrich KJ, Rumrich G, Klöss S (1980) Active sulfate reabsorption in the proximal convolution of the rat kidney: specificity, Na^+ and HCO_3^- dependence. Pflugers Arch 383:159–163
Ullrich KJ, Rumrich G, Klöss S (1982a) Active sulfate monocarboxylic acids in the proximal tubule of the rat kidney. I. Transport kinetics of D-lactate, Na^+-dependence, pH-dependence and effect of inhibitors. Pflugers Arch 395:212–219
Ullrich KJ, Rumrich G, Klöss S (1982b) Reabsorption of monocarboxylic acids in the proximal tubule of the rat kidney. II. Specificity for aliphatic compounds. Pflugers Arch 395:220–226
Ullrich KJ, Rumrich G, Klöss S, Fasold H (1982c) Reabsorption of monocarboxylic acids in the the proximal tubule of the rat kidney. III. Specificity for aromatic compounds. Pflugers Arch 395:227–231
Ullrich KJ, Rumrich G, Fritzsch G, Klöss S (1987a) Contraluminal para-aminohippurate (PAH) transport in the proximal tubule of the rat kidney. II. Specificity: aliphatic dicarboxylic acids. Pflugers Arch 408:38–45
Ullrich KJ, Rumrich G, Fritzsch G, Klöss S (1987b) Contraluminal para-aminohippurate (PAH) transport in the proximal tubule of the rat kidney. I. Kinetics, influence of cations, anions, and capillary preperfusion. Pflugers Arch 409: 229–235
Ullrich KJ, Rumrich G, Klöss S (1987c) Contraluminal para-aminohippurate (PAH) transport in the proximal tubule of the rat kidney. III. Specificity: monocarboxylic acids. Pflugers Arch 409:547–554
Ullrich KJ, Rumrich G, Klöss S (1988) Contraluminal para-aminohippurate (PAH) transport in the proximal tubule of the rat kidney. IV. Specificity: mono- and polysubstituted benzene analogs. Pflugers Arch 413:134–146
Ullrich KJ, Rumrich G, Klöss S (1989a) Contraluminal para-aminohippurate (PAH) transport in the proximal tubule of the rat kidney. V. Interaction with sulfamoyl- and phenoxy diuretics, and with β-lactam antibiotics. Kidney Int 36:78–88
Ullrich KJ, Rumrich G, Wieland T, Dekant W (1989b) Contraluminal para-aminohippurate (PAH) transport in the proximal tubule of the rat kidney. VI. Specificity: amino acids, their N-methyl-, N-acetyl- and N-benzoylderivatives; glutathione- and cysteine conjugates, di- and oligopeptides. Pflugers Arch 415: 342–350
Velázquez H, Wright FS (1986) Effects of diuretic drugs on Na, Cl, and K transport by rat renal distal tubule. Am J Physiol 250 (Renal Fluid Electrolyte Physiol 19):F1013–F1023

Velázquez H, Good DW, Wright FS (1984) Mutual dependence of sodium and chloride absorption by renal distal tubule. Am J Physiol 247 (Renal Fluid Electrolyte Physiol 16):F904–F911
Velázquez H, Ellison DH, Wright FS (1987) Chloride-dependent potassium secretion in early and late renal distal tubules. Am J Physiol 253 (Renal Fluid Electrolyte Physiol 22):F555–F562
Völkl H, Lang F (1988) Effect of amiloride on cell volume regulation in renal straight proximal tubules. Biochim Biophys Acta 946:5–10
Völkl H, Lang F (1991) Electrophysiology of ammonia transport in renal straight proximal tubules. Kidney Int 40:1082–1089
Wang W, Dietl P, Oberleithner H (1987) Evidence for Na^+ dependent rheogenic HCO_3^- transport in fused cells of frog distal tubules. Pflugers Arch 408:291–299
Werner D, Martinez F, Roch-Ramel F (1990) Urate and p-aminohippurate transport in the brush border membrane of the pig kidney. J Pharmacol Exp Ther 252:792–799
Whittembury G, Reuss L (1992) Mechanisms of coupling of solute and solvent transport in epithelia. In: Seldin DW, Giebisch G (eds) The kidney. Physiology and pathophysiology, 2nd edn. Raven, New York, pp 317–360
Wilson DR, Honrath U, Sonnenberg H (1983a) Furosemide action on collecting ducts: effect of prostaglandin synthesis inhibition. Am J Physiol 244 (Renal Fluid Electrolyte Physiol 13):F666–F673
Wilson DR, Honrath U, Sonnenberg H (1983b) Thiazide diuretic effect on medullary collecting duct function in the rat. Kidney Int 23:711–716
Windhager E, Frindt F, Yang JM, Lee CO (1986) Intracellular calcium ions as regulators of renal tubular sodium transport. Klin Wochenschr 64:847–852
Wingo CS (1989) Active proton secretion and potassium absorption in the rabbit outer medullary collecting duct. J Clin Invest 84:361–365
Wright SH (1985) Transport of N^1-methylnicotinamide across brush-border membrane vesicles from rabbit kidney. Am J Physiol 249 (Renal Fluid Electrolyte Physiol 18):F903–F911
Wright FS, Giebisch G (1992) Regulation of potassium excretion. In: Seldin DW, Giebisch G (eds) The kidney. Physiology and pathophysiology, 2nd edn. Raven, New York, pp 2209–2248
Yang JM, Lee CO, Windhager EE (1988) Regulation of cytosolic free calcium in isolated perfused proximal tubules of Necturus. Am J Physiol 255 (Renal Fluid Electrolyte Physiol 24):F787–F799
Yoshitomi K, Burckhardt B-C, Frömter E (1985) Rheogenic sodium-bicarbonate cotransport in the peritubular cell membrane of rat renal proximal tubule. Pflugers Arch 405:360–366
Yoshitomi K, Koseki C, Taniguchi J, Imai M (1987) Functional heterogeneity in the hamster medullary thick ascending limb of Henle's loop. Pflugers Arch 408: 600–608
Yoshitomi K, Kondo Y, Imai M (1988) Evidence for conductive Cl^- pathways across the cell emebranes of the thin ascending limb of Henle's loop. J Clin Invest 82:866–871
Zeidel ML (1990) Renal action of atrial natriuretic peptide: regulation of collection duct sodium and water transport. Annu Rev Physiol 52:747–760
Zelikovic I, Chesney RW (1989) Sodium-coupled amino acid transport in renal tubule. Kidney Int 36:351–359

CHAPTER 3

Renal Energy Metabolism

W.G. GUDER and M. SCHMOLKE

A. Introduction

The kidney uses 20% of whole body oxygen, which is comparable to the figure for the heart muscle (COHEN and BARAC-NIETO 1973), indicating that this organ has a high rate of oxidative metabolism. Since the early discovery that oxygen uptake paralleled glomerular filtration (KRAMER and DEETJEN 1960) and sodium reabsorption (KIIL et al. 1961) under experimental conditions, renal oxygen uptake has been seen as a measure of renal energy turnover. Only 20% of renal oxygen uptake has been calculated to be used for the basal requirements of the kidney. These calculations were based on the assumption that sodium reabsorption was the main energy-requiring function of the kidney and that oxygen was used exclusively for mitochondrial oxidative phosphorylation to provide ATP for transport ATPases. A quantity of 28–30 mol sodium was calculated to be transported per mol O_2 consumed (DEETJEN and KRAMER 1961). Since then this basic relationship has been confirmed in studies with isolated perfused kidney (SILVA et al. 1980), tubule suspensions (GSTRAUNTHALER et al. 1985; BALABAN et al. 1980) and microdissected tubule segments (JUNG and ENDOU 1991). On the other hand, many more energy-consuming transport ATPases and metabolic functions have been attributed to defined segments of the nephron. Therefore any calculation on energy metabolism may be performed at the defined nephron level. This also holds for the sites of renal substrate turnover, which were originally thought to serve exclusively to cover renal energy demands (PITTS 1976; COHEN 1986). In the present review we seek to provide some basic information about the mechanisms of renal energy metabolism and summarize recent literature on segment-specific metabolic pathways in relation to transport function. Possible interference with the action of diuretics will also be considered.

B. Mechanisms of Renal ATP Formation

The γ-phosphate bond of adenosine triphosphate (ATP) is the main source of chemical energy used for renal transport processes. Active transport systems using this kind of energy are called ATPases. Under standard conditions the cleavage of this bond is equivalent to 7.3 kcal (30.6 kJ)/mol

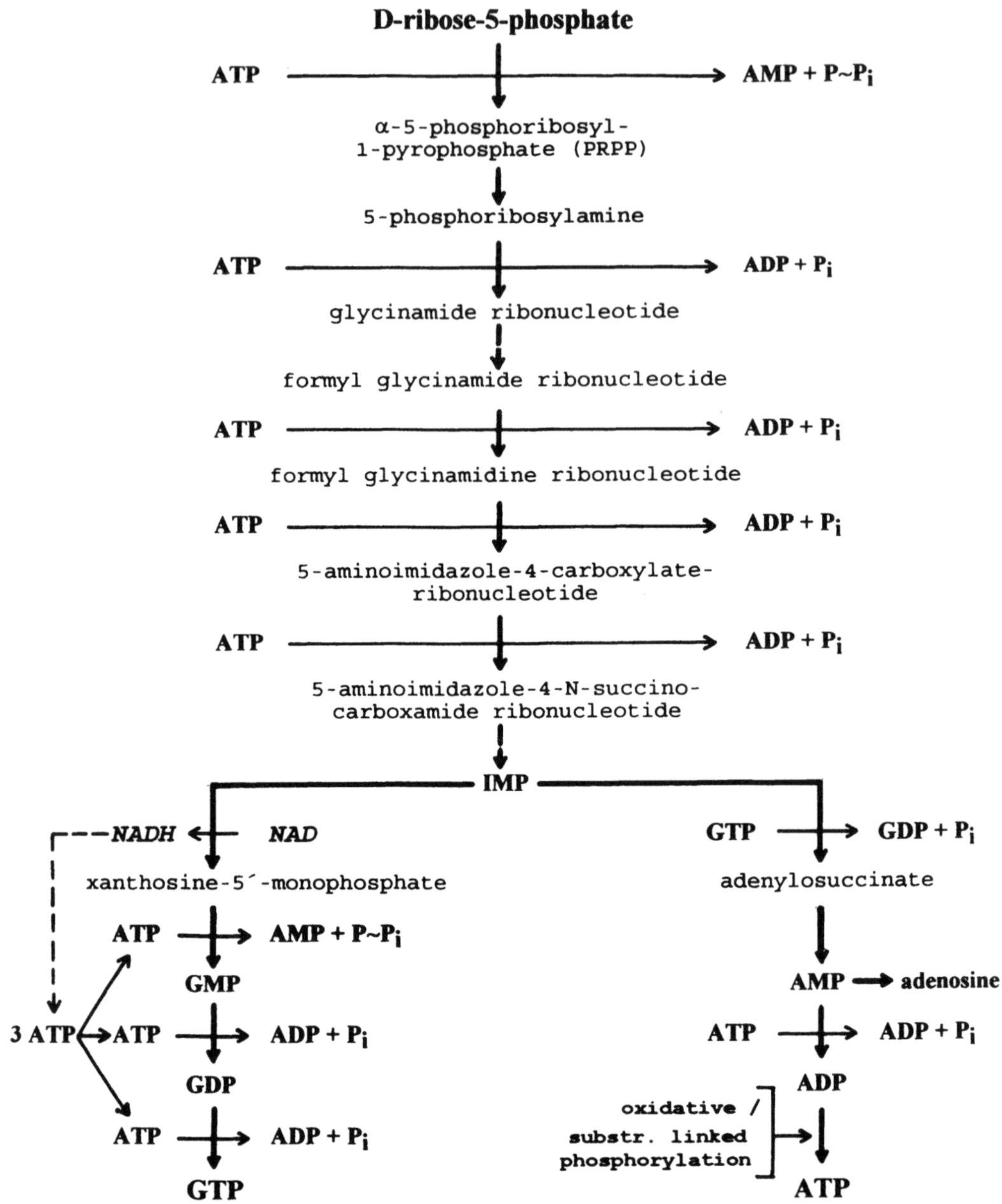

biosynthesis of	nucleotides needed	~P-bonds used
ATP	6 ATP + 1 GTP	8
GTP	8 ATP (5)	10 (6)

Fig. 1. Pathway of ATP and GTP synthesis from their precursors with respect to the consumption of energy-rich P_i-bonds and their corresponding nucleotides. *Numbers in parentheses* are the required ATP and energy-rich phosphate bonds, respectively, assuming oxidation of NADH during xanthosine-5′-monophosphate synthesis

ATP at 37°C and pH = 7. The product of ATP hydrolysis, adenosine diphosphate (ADP), has to be resynthesized in the same cell that uses the energy, because ATP does not penetrate the cells and its concentration is negligible in the extracellular space. The renal concentrations of adenine nucleotides are in the range of 2–3 mmol/l cellular water. These nucleotides are synthesized from their precursors ribose, phosphate and adenine in a multistep reaction sequence, yielding inosine monophosphate, which can be converted to adenosine monophosphate. This in turn can form ADP by the adenylate kinase reaction (Fig. 1). Six ATPs are needed to form 1 mol AMP and 7 to build ADP by de novo synthesis (LÖFFLER and PETRIDES 1992). It is therefore understandable that the cell cannot afford de novo synthesis as a continuous source of ATP. Only in hypoxia and in nephrotoxic situations is ADP degraded to AMP and adenosine, inosine, hypoxanthine and uric acid (OSSWALD et al. 1977; ENDOU and JUNG 1991). Under physiological conditions more than 90% of ATP is synthesized from ADP. This ATP resynthesis from ADP takes place in a "substrate-linked phosphorylation" or in the mitochondrial oxidative phosphorylation, respectively.

I. Substrate-Linked ATP Formation

1. Glycolysis

The glycolytic pathway is the major source of substrate-linked ATP synthesis in mammalian kidney. As shown in Fig. 2, the glycolytic pathway needs 2 mol ATP to metabolize 1 mol glucose to 1 mol fructose-1,6-bisphosphate, but releases 4 mol ATP upon further metabolism of the 2 mol triose phosphate formed by fructose 1,6-bisphosphate aldolase. ATP is formed in the phosphoglycerate kinase and pyruvate kinase steps, which use the energy-rich substrates 1,3-diphosphoglycerate and phosphoenolpyruvate (PEP), respectively. The phosphate bonds in these metabolites have energies of 49.6 and 62.7 kJ (15 kcal) under standard conditions, which are far above that of ATP (30.6 kJ = 7.3 kcal). ATP synthesis by this mechanism is independent of the presence of oxygen and can therefore replace oxidative ATP synthesis under conditions of reduced oxygen supply. In oxidative cells like those of the thick ascending limbs, this pathway is even stimulated under anaerobic conditions by the Pasteur effect, which couples oxidative and glycolytic ATP synthesis to each other (BROWN 1992).

2. Other Mechanisms

Succinate thiokinase is an enzyme of the citric acid cycle which can form 1 mol guanosine triphosphate (GTP)/mol succinyl-CoA used. This step is present in renal mitochondria, where GTP is rapidly equilibrated with ATP by their respective phosphotransferases. This pathway seems to be of minor importance under physiological conditions, since this step is tightly coupled to

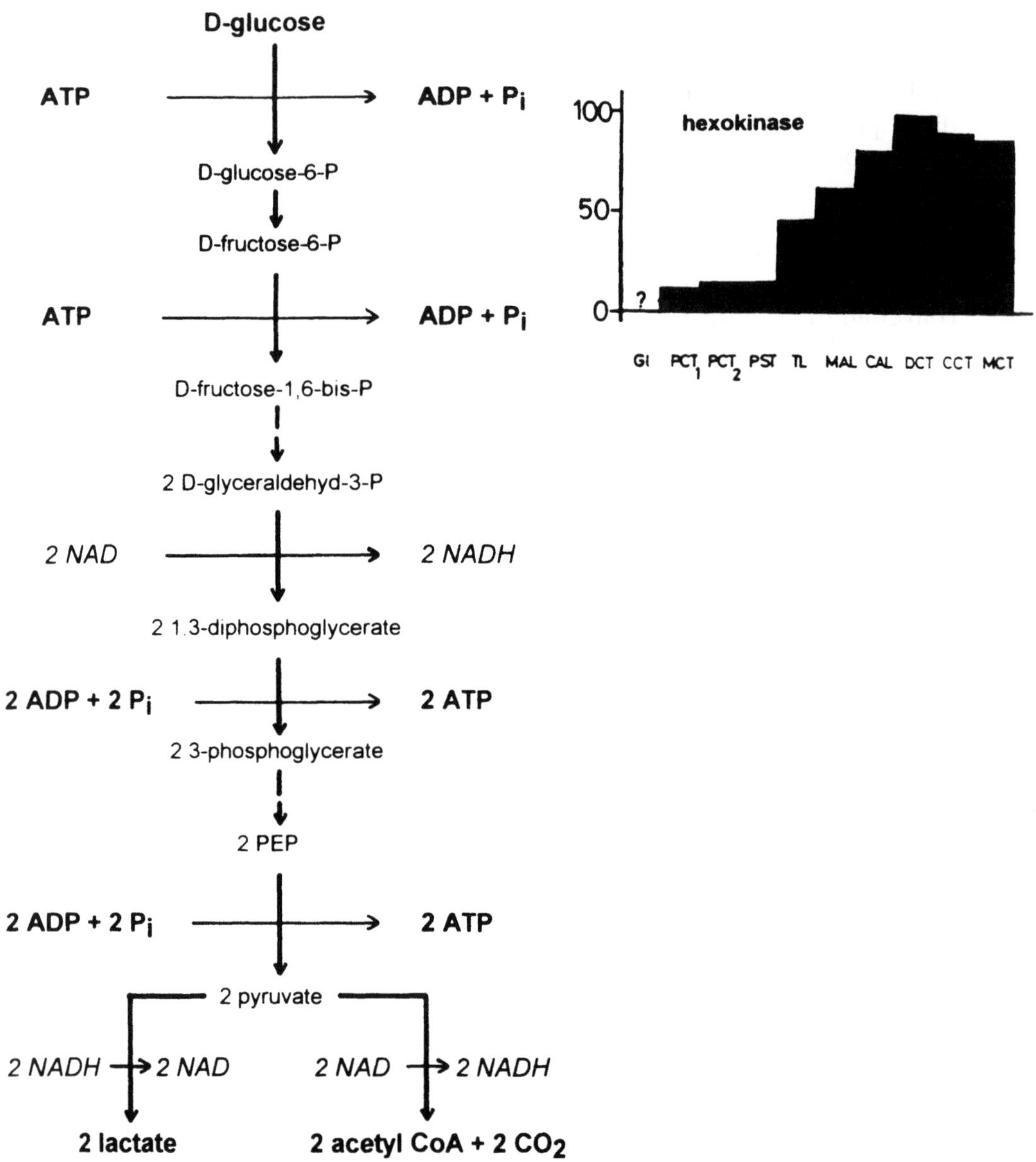

energy balance:

anaerobic: **1 glucose** $\rightarrow$ 2 lactate + **2 ATP**

aerobic: **1 glucose** $\rightarrow$ 2 acetyl CoA + **2 ATP** + **4 NADH** + $2\ CO_2$

4 NADH + $2\ O_2$ + $4\ H^+$ $\rightarrow$ $4\ H_2O$ + $4\ NAD^+$ + **12 ATP**

Fig. 2. Metabolic pathway and energy balance of glycolysis leading to lactate or acetyl-CoA. *Inset* shows the distribution of the glycolytic key enzyme hexokinase along the rabbit nephron (Vandewalle et al. 1981). *Abbreviations*: *PCT_1*, *PCT_2*, proximal convoluted tubule S1 and S2; *PST*, proximal straight tubule (S3); *TL*, thin limb; *MAL*, medullary thick ascending limb; *CAL*, cortical thick ascending limb; *DCT*, distal convoluted tubule; *CCT*, cortical collecting tubule; *MCT*, medullary collecting tubule

the respiratory chain and cannot therefore replace defects in oxidative ATP synthesis. Under experimental conditions, however, it can serve as a major mechanism of tubular ATP synthesis (COHEN and BARAC-NIETO 1973).

Creatine Kinase (CK). Creatine phosphate serves as a buffering system in tissues performing continuous work such as heart muscle, skeletal muscle and brain. Recently it has been described that the mitochondrial and cytosolic isoenzymes of CK are involved in an energy-buffering system which helps to keep ADP low during intense activity of ATP-consuming reactions (WALLIMANN et al. 1992). Although less active than in muscle, CK in kidney has been found with highest activity in the thick ascending limb of Henle's loop (FRIEDMAN and PERRYMAN 1991) and the distal convoluted tubule (BASTIN et al. 1987). These activities have been immunologically characterized as the mitochondrial CK and the brain type CK-B, respectively, indicating a tight coupling of this system to the mitochondrial ATP synthesis as shown for other organs. FRIEDMAN and PERRYMAN (1991) pointed to the close relationship between ATP turnover, oxygen consumption, Na^+ reabsorption and CK along the nephron. This system might help to maintain ATP levels constant in the thick ascending limb segment under situations of pulsatile oxygen supply as occurring in vivo and in isolated perfused rat kidney (SCHUREK et al. 1992). BASTIN et al. (1987) have shown that creatine phosphate decreases rapidly in anoxia in distal convoluted tubule but remains fairly stable in thick ascending limb cells and proximal tubules.

II. Oxidative Phosphorylation

1. Coupling to Oxygen Consumption

Renal active transport processes are tightly coupled to mitochondrial oxygen consumption (BALABAN et al. 1980; ROSS et al. 1984; BOGUSKY et al. 1986). The site of ATP synthesis is the inner mitochondrial membrane. Here "oxidative phosphorylation" is driven by a proton gradient formed by respiratory chain oxidation (GULLANS and MANDEL 1992). Figure 3 depicts the present view of the respiratory chain, which has also been reported to prevail in tubular cells. The hydrogen for oxidation is either provided in the form of reduced nicotinamide adenine dinucleotide ($NADH+H^+$) or flavine adenine dinucleotide ($FADH_2$). The coupling of oxygen consumption to ATP formation leads to 3 and 2 mol ATP/NADH and FADH oxidized, respectively. Both nucleotides consume 0.5 mol O_2/mol. This mechanism is the major source of renal ATP synthesis and is responsible for the high rate of oxygen consumption. Oxygen is provided by the arterial blood supply whereas inorganic phosphate, ADP and "hydrogen equivalents" are derived from the cellular metabolism. Calculating the ATP formed from the oxygen used, the following result can be obtained:

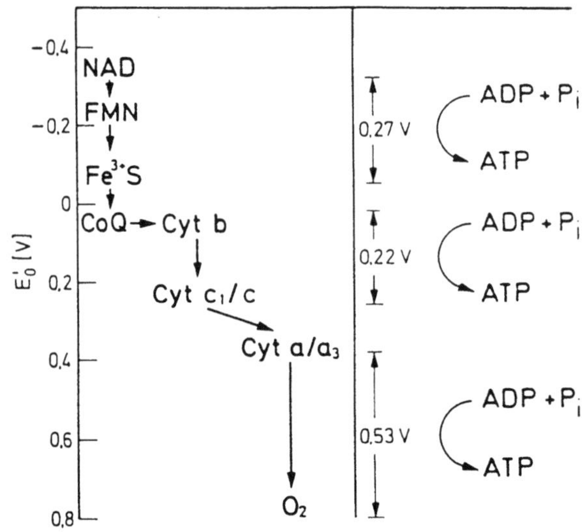

Fig. 3. Localization of coupling sites between electron transfer and ATP formation in the respiratory chain. (From LÖFFLER and PETRIDES 1992, with permission)

$$NADH + H^+ + ½O_2 + 3\ ADP + 3\ P_i = 3\ ATP + H_2O + NAD^+ \quad (1)$$

$$FADH_2 + ½O_2 + 2\ ADP + 2\ P_i = 2\ ATP + H_2O + FAD \quad (2)$$

The "hydrogen" needed for these reactions stems from various metabolic pathways, the major part deriving from the citric acid cycle.

2. Citric Acid Cycle

The citric acid cycle is the main source of mitochondrial "reducing equivalents" in the form of $NADH+H^+$ and $FADH_2$. One cycle leads to the formation of 3 molecules NADH and 1 molecule FADH, resulting in 3 × 3 and 1 × 2 molecules ATP upon oxidation in the respiratory chain (Fig. 4). In addition, 1 GTP is formed in the succinate thiokinase step. Assuming that oxygen consumption is driven exclusively by the citric acid cycle, the speed of the cycle may be calculated by the following formula:

Citric acid cycle flux = 0.5 × oxygen consumption
(μmol acetyl-CoA/min) (μmol O_2/min)

Taking the oxygen consumption figure of 2.86 μmol/min per gram obtained in the isolated perfused rat kidney (Ross et al. 1984), this results in a maximal flux rate through the citric acid cycle of 1.43 μmol acetyl-CoA/min per gram. This calculation is, however, limited by several factors, which must be considered:

1. A considerable proportion of the "reducing equivalents" are derived from sources other than the citric acid cycle (see below).
2. Part of the oxygen consumed may be due to reactions other than the respiratory chain.
3. The citric acid cycle is heterogeneously distributed across the kidney. Therefore numbers calculated for the whole kidney do not hold for specific tubule segments.
4. Under some conditions renal substrates use only parts of the citric acid cycle. Thus the proximal tubule forms glucose from glutamine by use of the ketoglutarate to oxaloacetate pathway (VINAY et al. 1978). Under such conditions the oxygen/citric acid cycle turnover rate ratio may be considerably higher (VINAY et al. 1978).

Several authors have attempted to calculate the citric acid cycle flux rate in kidney tubules using radioactive precursors (VINAY et al. 1978; JANS et al. 1989; NISSIM et al. 1990). VINAY et al. (1978) calculated a maximal rate of 2.16 in control rat proximal tubules and 2.24 μmol/min per gram in acidotic tubules at an oxygen consumption of 5–10 μmol/min per gram under gluconeogenic conditions. Studies using ^{13}C-NMR (NISSIM et al. 1990; JANS et al. 1989; JANS and LEIBFRITZ 1989) give only relative numbers for different experimental conditions and confirm the difficulties in interpreting isotopic labeling data pointed out previously (KREBS et al. 1966; KATZ 1989).

The situation in vivo is further complicated by the fact that proximal tubules reabsorb metabolites of the citric acid cycle, namely citrate and oxoglutarate (COHEN and BARAC-NIETO 1973) and amino acids, using parts of the citric acid cycle to be converted to glucose (WIRTHENSOHN and GUDER 1986).

The distribution of citric acid cycle enzymes along the nephron segments is summarized in Fig. 4. Citrate synthase, oxoglutarate dehydrogenase, succinate dehydrogenase and acetoacetyl-CoA transferase all exhibit a similar distribution pattern, with highest activities in DCT and PCT segments and low activities in segments of the inner renal medulla (HÖHMANN 1973; MARVER and SCHWARTZ 1980; LEHIR and DUBACH 1982; GUDER et al. 1983; BURCH et al. 1984). The enzyme activities (LEHIR and DUBACH 1982; ROSS and GUDER 1982) nicely correlate with the inner membrane surface areas quantitated by PFALLER (1982), suggesting that the citric acid cycle enzymes are formed in coordinated way, keeping constant the proportions of the various enzymes. Interestingly, citric acid cycle activities also correlate with that of (Na^++K^+)-ATPase in distal nephron segments during postnatal development and under steroid hormones, pointing to a close functional link between energy-producing and energy-consuming pathways (MARVER 1986; BASTIN et al. 1990; DJOUADI et al. 1993).

On the other hand, segment-specific and species-specific differences should be mentioned. Thus the step from succinyl CoA to succinate can be circumvented by acetoacetyl-CoA transferase, which is highly active in rat

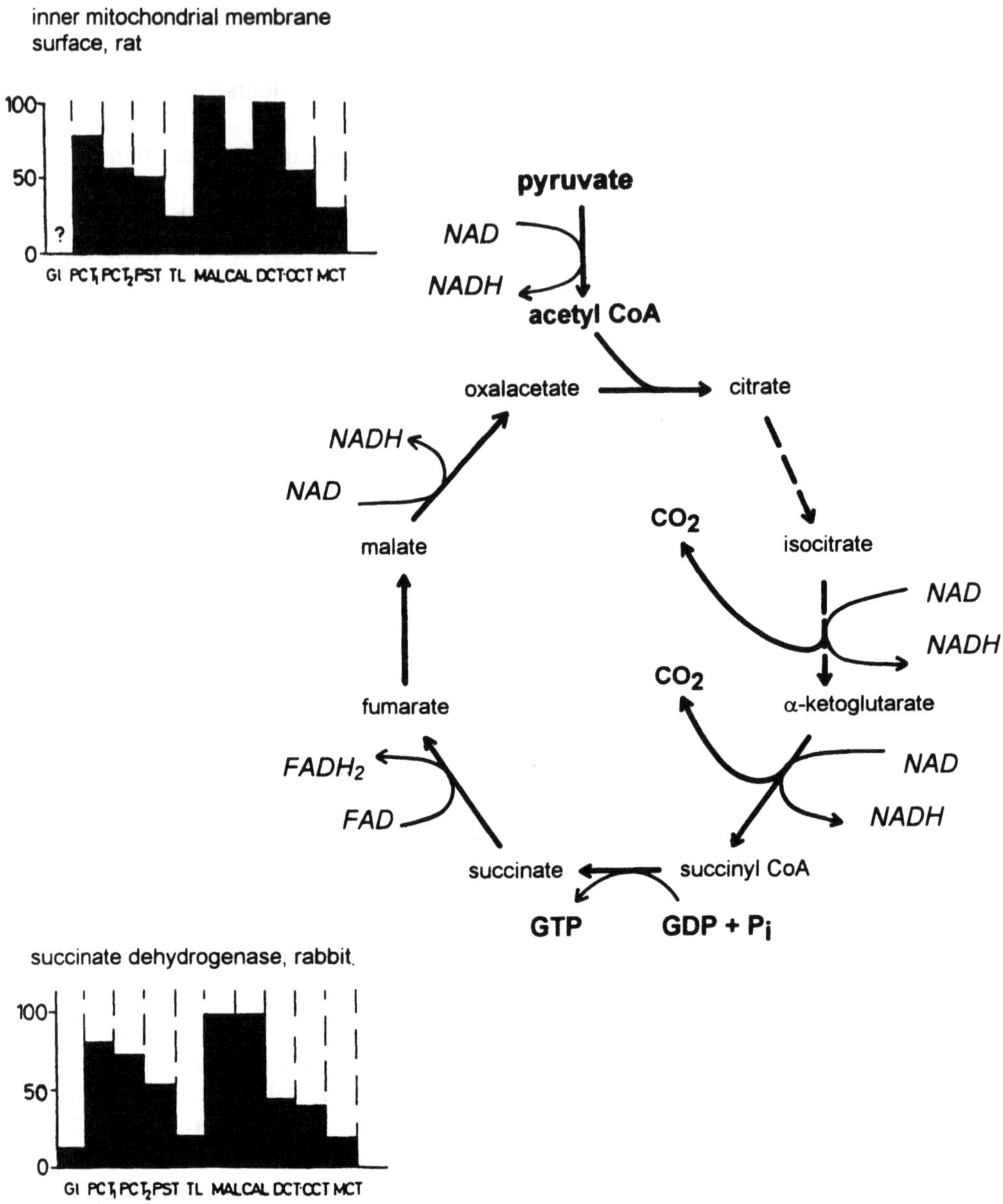

energy balance:

$$1 \text{ pyruvate} \rightarrow 3\,CO_2 + 4\,NADH + 2\,FADH_2 + 1\,GTP$$

$$1 \text{ pyruvate} \cong 17\text{ ATP} \qquad 4\,NADH \cong 12\text{ ATP} \qquad 2\,FADH_2 \cong 4\text{ ATP} \qquad 1\,GTP \cong 1\text{ ATP}$$

Fig. 4. Metabolism of pyruvate in the citric acid cycle and its energy balance. *Insets* demonstrate the distribution of the inner mitochondrial membrane surface of the rat (PFALLER 1982) and the key enzyme succinate dehydrogenase of the rabbit along the nephron (HÖHMANN 1973). For abbreviations of nephron segments see legend to Fig. 2

and mouse but not in rabbit kidney mitochondria (see Sect. C.II). This step makes the renal citric acid cycle different from that of the liver, which lacks this enzyme. Moreover, rabbit kidney exhibits this enzyme activity only in the thick ascending limb but not in proximal tubule mitochondria (BURCH et al. 1985).

In conclusion, the citric acid cycle is the major source of "hydrogen equivalents" for oxidative ATP synthesis in the proximal tubule, the thick ascending limb of Henle's loop, the distal convoluted tubule and the collecting tubule, where it resides mainly in intercalated (dark) cells. In the proximal tubule the citric acid cycle can also mediate intermediary substrate fluxes which are not completely oxidized (see Sect. B.I below). Due to its close link to the respiratory chain, the speed of the citric acid cycle is closely linked to oxidative phosphorylation and renal ATP turnover. Substrate availability can limit the capacity of this pathway when energy consumption is not met by a sufficient substrate supply. The main substrate acetyl-CoA can be provided by degradation of carbohydrates, fatty acids, ketone bodies and certain amino acids. The relative contributions of these metabolic fuels and their respective interactions will be covered in the next section.

C. Metabolic Substrates of Renal Energy Metabolism

Like glycolysis, the oxidative pathway can only form ATP if sufficient "reducing equivalents" are available. Besides the citric acid cycle, metabolic pathways form $NADH+H^+$ and $FADH_2$ before entering this cycle. The following section covers the major energy-providing pathways documented for tubule cells, their relative distribution among tubule segments and their mechanisms of interaction.

I. Glucose and Lactate

As shown in Sect. A.I.1, glycolysis not only provides ATP but 2 mol NADH/mol glucose are also produced. In the presence of oxygen, NADH can be channelled to the mitochondrial matrix by respective substrate shuttles, where it is oxidized in the respiratory chain. Pyruvate can be transported into mitochondria via a specific carrier, where it is either oxidized to acetyl-CoA or carboxylated to oxaloacetate (KREBS et al. 1966; GUDER and WIELAND 1974). The latter pathway is restricted to the proximal tubule. Pyruvate dehydrogenase mediates the formation of acetyl-CoA, leading to formation of additional NADH. This enzyme is regulated by phosphorylation dephosphorylation and by changes in the ATP/ADP, NADH/NAD and CoA/acetyl-CoA ratios, forming the crucial step in carbohydrate lipid interaction (see Sect. B.V). Assuming that glucose-derived acetyl-CoA enters the citric acid cycle quantitatively, as can be expected from experimental data in distal nephron segments (HUS-CITHAREL and MOREL 1986), 1 mol glucose provides 4 mol NADH up to acetyl-CoA and an additional 6 mol NADH and 2 mol $FADH_2$ in the citric acid cycle (Figs. 2, 4). The total energy formed would thus be 28 mol ATP from the respiratory chain plus 6 mol ATP in substrate-linked phosphorylation steps.

In contrast glucose provides only 4 mol ATP when metabolized to lactate, as occurring in anoxia in mitochondria-containing cells and in inner medullary cells under normal conditions (MEURY et al. 1994; SCHMOLKE and GUDER 1989). Thus eighttimes more glucose is needed to cover the same energy demands in anoxic as in normoxic conditions, provided that the glycolytic pathway does not become limiting. From Fig. 2 it can be seen that this is clearly the case in proximal tubules, where glycolysis cannot provide sufficient energy to sustain tubular energy demands.

In certain tubule segments lactate can replace glucose as a metabolic fuel. Indeed lactate is taken up by the kidney under all circumstances studied so far (COHEN and BARAC-NIETO 1973; WIRTHENSOHN and GUDER 1986). One mole lactate when completely oxidized results in 6 mol "reducing equivalents", which account for 17 mol ATP plus 1 mol GTP. Pyruvate can replace lactate in that respect, resulting in 1 NADH less, corresponding to 3 ATPs less. Proximal tubules, thick ascending limb segments and to a lesser extent cortical collecting tubules can keep their ATP levels and work on lactate or pyruvate as the sole substrate (HUS-CITHAREL and MOREL 1986; ENDOU and JUNG 1991).

However, the relative contribution of lactate to renal energy balance from arteriovenous differences is difficult to calculate because (a) lactate can be converted to glucose and other products mainly by the proximal tubule, (b) lactate is simultaneously formed by medullary segments and (c) the relative contribution of lactate-forming to lactate-using cells in an in vivo state is not known. Because of the large heterogeneity of metabolic pools in experiments with isotopic lactate in vivo (PITTS 1976) and in the isolated kidney (JANSSENS et al. 1980; COHEN 1986), the results are likewise difficult to interpret. On the level of defined nephron cells, however, a clear relationship between CO_2 formation from lactate and renal energy turnover can be demonstrated in the thick ascending limb of Henle's loop and cortical collecting tubule segments (HUS-CITHAREL and MOREL 1986).

II. Fatty Acids

Fatty acids have long been known to be a preferred fuel of the rat renal cortex (WEIDEMANN and KREBS 1969). As depicted in Fig. 5, fatty acids taken up have to be converted to their respective CoA-esters before they can be metabolized. This requires 2 mol ATP/mol fatty acid irrespective of its chain length, since the product of the reaction AMP has to be rephosphorylated to ATP by adenylate kinase before reentering the ADP pool of the cell (see Fig. 1). Acyl-CoA can either be oxidized in the mitochondrial

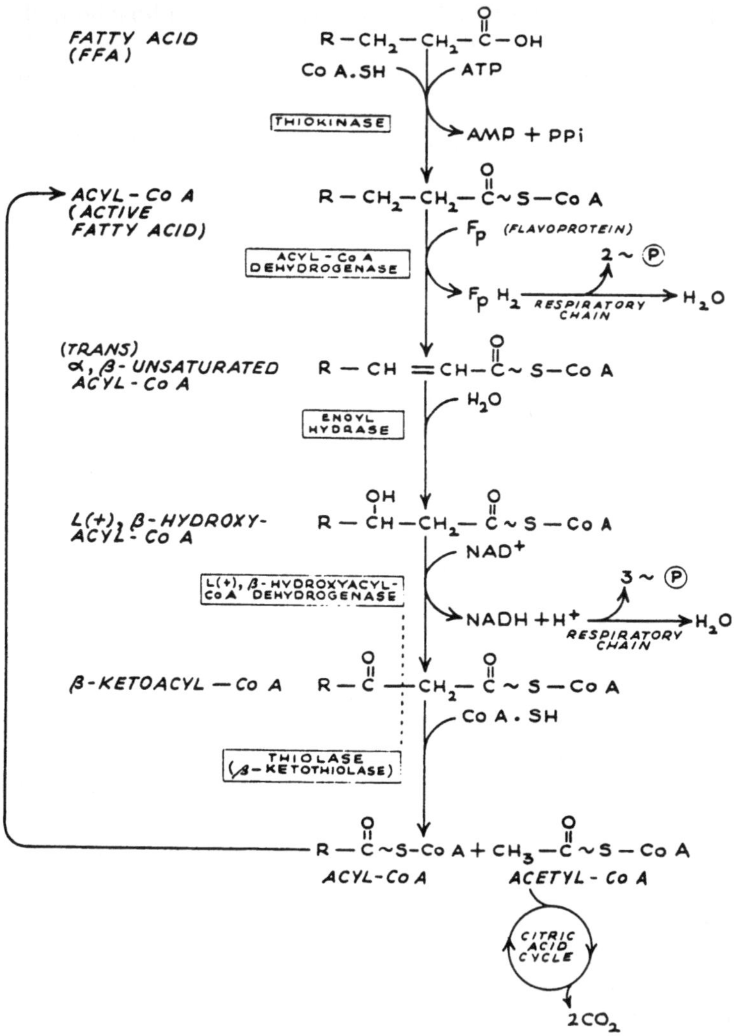

energy balance:	
palmitate + CoASH + **ATP**	→ **steaoryl-CoA + AMP + PP_i**
palmitoyl-CoA + 7 CoA SH + **7 NAD + 7 FAD**	→ **8 acetyl-CoA + 7 $FADH_2$ + 7 NADH**
7 $FADH_2$ + 7 NADH+ 7 O_2 + 35 P_i + **35 ADP**	→ **8 FAD + 8 NAD** + 42 H_2O + **35 ATP**
8 acetyl-CoA + 8 P_i + **8 GDP + 24 NAD + 8 FAD**	→ **8 CoASH + 24 NADH + 8 $FADH_2$ + 8 GTP** + 16 CO_2
24 NADH + 8 $FADH_2$ + 88 ADP + 8 P_i + 16 O_2	→ **24 NAD + 8 FAD** + 104 H_2O + **88 ATP**
1 palmitate	≅ **130 ATP (8 GTP included)**

Fig. 5. Metabolic pathway of the β-oxidation of fatty acids. The energy balance was calculated for the oxidation of palmitic acid

β-oxidative pathway or esterified to form triglycerides, phospholipids and other more complex lipids (WIRTHENSOHN and GUDER 1983). One mole of a saturated fatty acid with a chain length of 16 carbons (palmitate) results in 8 mol acetyl-CoA. This degradative pathway reveals 7 mol NADH and 7 mol $FADH_2$. Taken together with the reducing equivalents obtained from the citric acid cycle, 31 mol $NADH+H^+$ and 15 mol $FADH_2$ are formed from 1 mol palmitate. This would result in 93 plus 30 mol ATP, which together with the 8 GTPs formed results in 131 mol ATP/mol palmitate oxidized. Since 1 mol ATP is needed for activation of palmitate, 130 mol ATP can be obtained from 1 mol palmitate when completely oxidized.

As already concluded from the low respiratory quotient of 0.75 (CO_2 formed divided by O_2 consumed) by DICKENS and SIMER in 1930, fatty acids are the major fuel of the kidney under normal in vivo conditions. This has been confirmed in cortical and outer medullary segments (for literature see GUDER et al. 1986). The distribution of an enzyme of the β-oxidative pathway mirrors that of the citric acid cycle enzyme (Fig. 4), indicating a close link between fatty acid oxidation and respiratory functions. Fatty acids cannot form any ATP in the absence of oxygen. FONTELES et al. (1983) were able to show that renal Na^+ reabsorption was strikingly suppressed when fatty acid oxidation was inhibited by tetradecylglycidic acid, confirming the earlier observation that fatty acid uptake was correlated with Na^+ reabsorption in the dog kidney in vivo (BARAC-NIETO and COHEN 1968). On the other hand, the calculated Na^+ reabsorbed per mol fatty acid taken up was much smaller than expected from the above calculation. This was explained by the finding that the fatty acids taken up were not completely oxidized (BARAC-NIETO and COHEN 1971; GUDER and WIRTHENSOHN 1979).

Triacylglycerol was found to be the major product of fatty acid metabolism not accounted for by oxidation in cortical and outer medullary tubules (WIRTHENSOHN et al. 1980). The conversion of 1 mol fatty acids to triglycerides would need 1.3 mol ATP, assuming that 0.33 mol glycerophosphate have to be formed from glycerol or glucose. This pathway seems to be used under conditions of excess fatty acid supply and explains why earlier calculations on the relative contribution of fatty acids to renal energy balance resulted in too high numbers (PITTS 1976). Fatty acid uptake of rat kidney in vivo was recently measured to be 0.54 μmol/kg per minute under fed and 2.77 μmol/kg per minute under starving conditions (ELHAMRI et al. 1993) with no difference in oxygen consumption and Na^+ reabsorption under either set of conditions. As measured in vivo and in vitro the relative contribution of triglyceride synthesis to renal fatty acid metabolism increases linearly with arterial supply (BARAC-NIETO and COHEN 1968; GUDER et al. 1992). It seems that the switch is regulated by the cellular energy demand, leaving stored triglycerides as a source of endogenous energy supply to be used under conditions of substrate depletion (WIRTHENSOHN and GUDER 1980).

III. Ketone Bodies

Ketone bodies are formed in the liver and used as metabolic fuel by heart, kidney, muscle and brain. The metabolic pathway includes 3-hydroxybutyrate dehydrogenase, which equilibrates acetoacetate with 3-hydroxybutyrate and an acetoacetyl-CoA-transferase, which links ketone body degradation to the citric acid cycle (Fig. 6). The latter enzyme exhibits the highest activity in the kidney, when compared with the other organs.

In contrast to fatty acids, ketone bodies do not enter biosynthetic pathways at appreciable rates in the kidney (WIRTHENSOHN and GUDER 1986). Although acetyl-CoA formed during metabolism of acetoacetate and 3-hydroxybutyrate can be incorporated into lipids, amino acids and even glucose (KREBS et al. 1966), the relative contribution of these pathways to net ketone body turnover seems to be negligible in kidney.

Ketone bodies are taken up at rates which are linearly dependent on their arterial concentration with no saturation at physiological levels. Thus ELHAMRI et al. (1993) found uptake rates of acetoacetate in rat kidney of 1.9 μmol/kg body wt. per minute at an arterial concentration of 0.36 mmol/l in fed animals compared with 5.5 μmol/min at 1.05 mmol/l in starved animals. For 3-hydroxybutyrate the respective numbers were 0.57 μmol/min at 0.18 mmol/l and 4.8 μmol/min at 1.8 mmol/l (Table 1), confirming the earlier studies in dog (LITTLE and SPITZER 1971), rat (BARAC-NIETO 1985) and human kidney (OWEN et al. 1969).

The distribution of enzymes of ketone body metabolism indicates a close correlation with the distribution of mitochondria along the nephron (BURCH et al. 1984; GUDER et al. 1983; Fig. 6). Assuming that the activity of the transferase correlates with oxidative capacity of the respective segments, the thick ascending limb, distal convoluted tubule, and proximal tubule are the major users of ketone bodies as energy fuel. This has been confirmed in studies with tubule suspensions prepared from rat kidney cortex (GUDER et al. 1986) and outer medulla (BAVEREL et al. 1980).

IV. Amino Acids

Of the 21 amino acids present in mammalian blood, only a few are taken up by the kidney in net amounts (TIZIANELLO et al. 1980; OWEN and ROBINSON 1963; SHALOUB et al. 1963; LEMIEUX et al. 1974). Of particular relevance for the present review are those amino acids which can provide net energy during their tubular turnover.

Alanine, which is formed from pyruvate and is released by dog and human kidneys (PITTS and STONE 1967; OWEN et al. 1969), is metabolized to glutamate and glutamine. Both provide ammonia and NADPH when metabolized to oxoglutarate (for review see TANNEN 1978; ROSS and LOWRY 1981; GUDER et al. 1987). The metabolism of oxoglutarate to glucose in

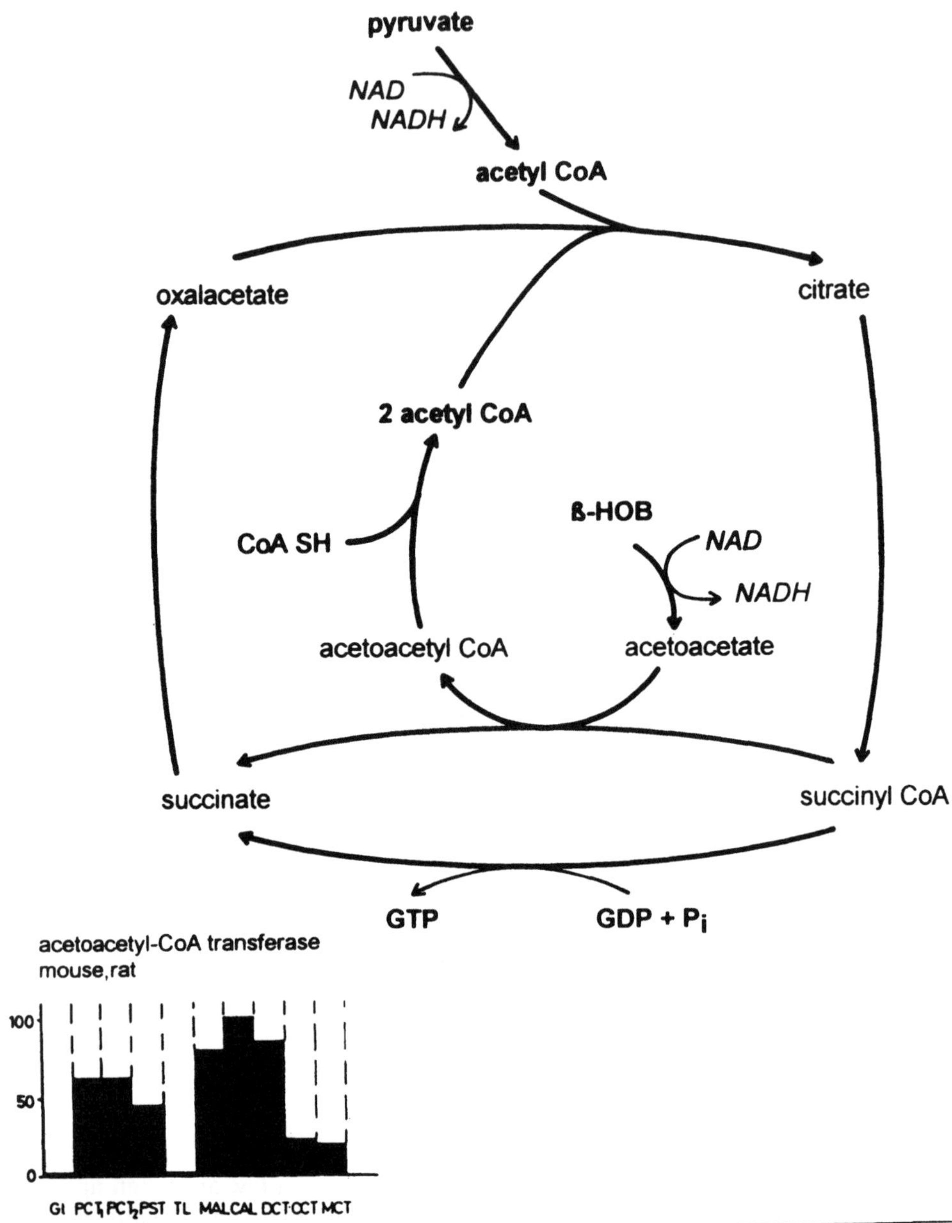

energy balance:

1 ß-HOB + succinyl CoA + CoA SH + **NAD** → **2 acetyl CoA** + succinate + **NADH**
2 acetyl CoA → 4 CO_2 + **6 NADH** + **2 $FADH_2$**

≅ ≅ ≅

1 ß-HOB ≅ **25 ATP** **18 ATP** **4 ATP** **3 ATP**

Fig. 6. Ketone body metabolism and its relation to the citric acid cycle. The energy balance was calculated from the complete metabolization of 3-hydroxybutyrate (β-HOB). *Inset* shows the distribution of acetoacetyl-CoA transferase along the mouse and rabbit nephron which catalyzes an alternative pathway in the citric acid cycle, coupling it to ketone body metabolism

proximal tubule cells reveals additional ATP, and even more is formed when this dicarboxylic acid is degraded to CO_2 (VINAY et al. 1979).

Proline, which uses the same metabolic pathways, is degraded to glutamate by mitochondrial proline oxidase and pyrroline-5-carboxylate dehydrogenase (SCHORN 1982; GUDER and SCHORN 1991). This pathway is localized mainly in the proximal tubule, thus linking proline metabolism to gluconeogenesis and glutamine formation (GUDER and SCHORN 1991).

Of special interest seems the metabolism of branched-chain amino acids. Leucine, isoleucine and valine use a common branched-chain amino acid dehydrogenase after being transaminated to the respective α-oxoacid. These enzymes have been shown to exhibit a specific distribution pattern with highest activity in the thick ascending limb of Henle's loop (BURCH et al. 1985; LEVILLAIN 1993). Whereas leucine metabolism results in the formation of three molecules of acetyl-CoA, which can be oxidized in the citric acid cycle (Fig. 4), valine is converted to succinyl-CoA, entering the cycle at a site where part of the carbon can be transformed to glucose in the proximal tubule. This implies that "ketogenic" and "gluconeogenic" branched-chain amino acids have site-specific pathways, making the calculation of a quantitative energy balance difficult (TRINH-TRANG-THAN et al. 1988).

Other amino acids taken up are converted to products without net energy gain. Thus citrulline is metabolized to equimolar amounts of arginine in the proximal tubule (LEVILLAIN et al. 1993) and released into the blood stream (TIZIANELLO et al. 1980; KETTNER and SILBERNAGL 1985). Arginine on the other hand is converted to ornithine and urea in the nephron segments of several species (LEVILLAIN et al. 1992).

Calculating the contribution of amino acids to renal energy balance reveals that the role of these substrates seems to be minimal compared to the previously discussed metabolites. This is even more so under conditions of limited nitrogen supply as in starvation (OWEN et al. 1969).

V. Substrate Interactions

Under in vivo conditions all metabolic substrates are taken up simultaneously. The relative contribution of each of these substrates to tubular energy turnover changes with arterial supply and is specific for each nephron segment. Since the early work of Krebs (WEIDEMANN and KREBS 1969), fatty acids were thought to be the preferred energy fuel in the cortex. The mechanism by which oxidation of carbohydrates and some amino acids is inhibited was shown to be at the site of pyruvate dehydrogenase (GUDER and WIELAND 1974). More recently, ketone bodies were shown to have a similar effect in tubules of renal cortex (LEMIEUX et al. 1978; GUDER et al. 1986) and outer medulla (BAVEREL et al. 1980).

In addition, ketone bodies inhibited oleate oxidation (GUDER et al. 1986). Thus a hierarchy was established indicating that ketone bodies were nearly completely oxidized under all conditions, whereas the metabolic fate of all other substrates was hard to predict for the individual nephron segment from in vivo data. Recently, ELHAMRI et al. (1993) measured substrate uptake rates of all major fuels in rat kidney under fed and starving conditions. As summarized in Table 1, we have tried to calculate whole kidney energy turnover from these data.

VI. Contribution of Individual Substrates to Whole Kidney Energy Turnover

In trying to calculate the contribution of each individual substrate to total energy turnover of the kidney, one is faced with several limitations, which have to be considered when calculating these metabolic rates. Taking an oxygen consumption of about 50 μmol/min per kilogram body wt. and the textbook number of 6 ATPs/oxygen consumed, a total renal energy turnover of 300 μmol/min per kilogram body wt. would result. This may be compared with the sum of ATP provided by complete oxidation of major substrates taken up (Table 1). As can be seen the numbers coincide quite well in fed, but are twice as high in starved, animals. In contrast, oxygen consumption does not change under these conditions. From the studies on substrate interaction in tubules it may be assumed that ketone bodies are the only substrate listed which may be completely oxidized. This would imply that the contribution of acetoacetate and 3-hydroxybutyrate rises from 19% in fed to more than 80% in starved animals. Considering the increase in fatty acid uptake under conditions of starvation, this substrate contributes less than 20% assuming that all other substrates are not completely oxidized.

The calculated contribution of lactate to total energy turnover of nearly 37% may be overestimated, because even under fed conditions part of the lactate taken up is converted to glucose, a pathway which would need 3 mol ATP/mol lactate consumed. On the other hand, lactate is formed in the papillary region, thus underestimating total lactate uptake in the whole kidney (MEURY et al. 1994; SCHMOLKE and GUDER 1989). Likewise glycerol, citrate and glutamine may be assumed to be mainly converted to glucose, thus obscuring the net uptake of glucose, which cannot be estimated from in vivo data. In conclusion, ketone bodies, lactate and fatty acids are the major energy fuels in the rat kidney, with an increasing percentage covered by ketone bodies under hyperketotic conditions. Fatty acids, although important fuels of all mitochondria-containing segments, seem to be converted to triacylglycerol under conditions of ketosis (GUDER et al. 1986, 1992).

D. Regulation of ATP in Tubule Cells

I. Compartmentation

Tubule cells contain several intracellular compartments, which are separated by phospholipid-containing membranes (see Chap. 1, this volume). With

Table 1. Substrate turnover and calculated energy balance in rat kidney

Substrate	Arterial concentration (mmol/l)	Renal uptake (μmol/min)	ATP/mol (mol/mol)	ATP turnover (μmol/min)	ATP turnover (%)
Fed rat					
Oxygen		50	6	300	100
Fatty acids	0.91	0.54	136	73.4	24.0
Acetoacetate	0.36	1.9	22	41.8	13.9
3-HO-butyrate	0.18	0.57	25	14.2	4.7
Lactate	1.23	6.13	18	110.3	37.0
Glutamine	0.45	0.85	10[a]	8.5	2.8
Glycerol	0.10	1.06	2[a]	2	0.7
Citrate	0.17	1.5	14[a]	21	7.0
Total					90.1
Starved rat					
Oxygen		50	6	300	100
Fatty acids	1.57	2.77	136	376.0	[126.0]
Acetoacetate	1.05	5.5	22	121.0	40.3
3-HO-butyrate	1.81	4.8	25	120.0	40.0
Lactate	0.57	2.2	18	39.0	[13.0]
Glutamine	0.43	2.26	10[a]	22.6	7.5
Glycerol	0.11	1.8	2[a]	3.6	1.2
Citrate	0.14	1.26	14[a]	17.6	5.9
Total (without fatty acid and lactate)					94.9

Oxygen consumption was estimated from data given by COHEN and BARAC-NIETO (1973) and by WEINSTEIN and SZYJEWICK (1974). All substrate concentrations and uptake rates were measured by ELHAMRI et al. (1993). Rates are given as μmoles per kilogram body weight per minute.

[a] Glutamine, glycerol and citrate taken up were assumed to be transformed to glucose. % ATP turnovers are relative contributions calculated from the ATP turnover rates taken from oxygen consumption. Results in brackets mean that the contribution is probably overestimated, because the relative amount of triglyceride (fatty acids) and glucose (lactate) formation cannot be estimated.

regard to ATP the mitochondrial matrix and the cytosolic compartment are the most relevant. By the use of digitonin, the cholesterol-containing plasma membrane can be permeabilized without major disruption of the mitochondrial membrane (GUDER and PÜRSCHEL 1980). This technique has been applied to the measurement of cytosolic and mitochondrial ATP in proximal tubule suspensions (PFALLER et al. 1984). The mitochondrial ATP was found to be 2.6 mmol/l and the cytosolic ATP concentration 4.3 mmol/l, the latter representing 70% of total cellular ATP (PFALLER et al. 1984).

II. ATP Turnover

From measured ATP concentrations and oxygen consumption or metabolic rates the turnover of tubular ATP may be calculated. ATP can turn over

Table 2. ATP turnover in tubules of kidney cortex and outer and inner medulla

Tubules (species)	Oxygen consumption ($\mu mol\ g^{-1}\ min^{-1}$)	ATP		Ref.
		Content ($mmol\ l^{-1}$)	Turnover ($\mu mol\ min^{-1}\ g^{-1}$)	
Cortex (rabbit)	63	2.66	15	1
Outer medulla (rabbit)	3.5	2–3	16	2, 3
Inner medulla (dog)	1.5	2.6	7	4

References: 1, NOEL et al. (1992); 2, CHAMBERLIN et al. (1984); 3, EVELOFF et al. (1980); 4, MEURY et al. (1994).

several times its total tubular amount per minute. Thus Ross et al. (1984) calculated for the isolated perfused kidney an ATP turnover rate of 16.6 μmol/min per gram wet wt. at an ATP concentration of 2.42 mmol/kg. These data have been confirmed in isolated cortical and outer and inner medullary tubules (see Table 2).

Similar results were obtained in other species (for review see GULLANS and MANDEL 1992; GUDER and MOREL 1992). In all studies cited ATP turnover was increased by the addition of oxidative substrates and inhibited by metabolic and transport inhibitors. These maneuvers showed, however, that tubular ATP concentrations did not parallel ATP turnover. Therefore determination of ATP seems of little relevance for calculating the energy state of the tubule.

The same holds true for the addition of diuretics (see below). These studies confirm that renal ATP turnover is largely regulated by ATP-using reactions, which may be defined in more detail.

III. Energy-Consuming Mechanisms

1. Transport ATPases

Transport ATPases form the major energy-using reactions in tubular ATP turnover. The distribution of various transport ATPases along the nephron has been the subject of numerous studies (Fig. 7) and enables the possible contribution of each species to energy turnover in a defined nephron segment to be calculated. Since their activities are measured under optimal conditions in vitro, the measured activities cannot directly be compared with the respective state of the tubule under in vivo conditions. Therefore, additional studies to measure oxygen consumption or metabolic rates and using specific inhibitors are of particular help in calculating the contribution of each transport to the total energy turnover.

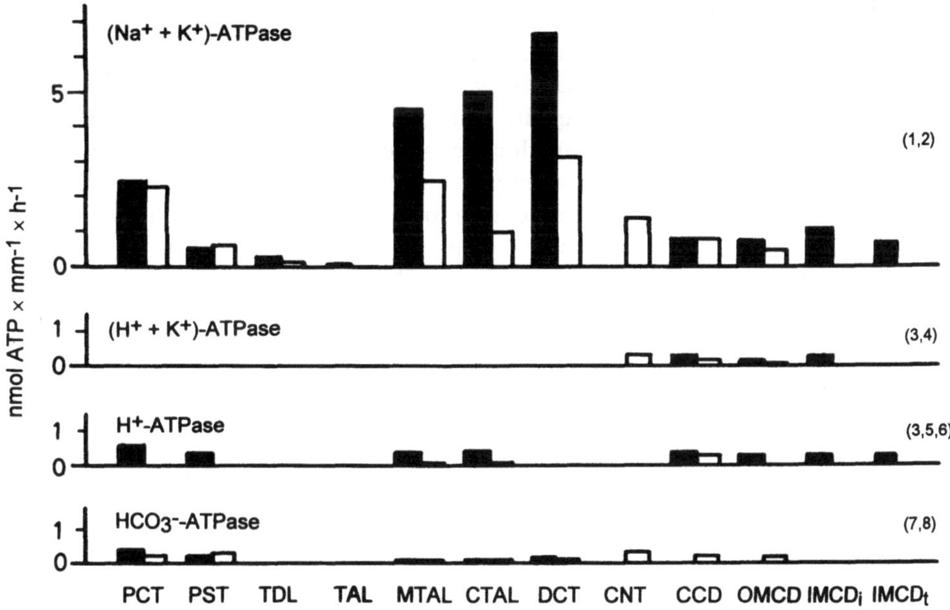

Fig. 7. Distribution of different primary active transport ATPases – $(Na^+ + K^+)$-ATPase, $(H^+ + K^+)$-ATPase, H^+-ATPase and HCO_3^--ATPase – along the rat (*closed bars*) and rabbit (*open bars*) nephron. *Abbreviations*: *PCT*, proximal convoluted tubule; *PST*, proximal straight tubule; *TDL*, thin descending limb; *TAL*, thin ascending limb; *MAL*, medullary thick ascending limb; *CAL*, cortical thick ascending limb; *DCT*, distal convoluted tubule; *CCD*, cortical collecting duct; *OMCD*, outer medullary collecting duct; *IMCD*$_i$, initial inner medullary collecting duct; *IMCD*$_t$, terminal inner medullary collecting duct. References: *1*, KATZ et al. (1979); *2*, EL MERNISSI and DOUCET (1984); *3*, DOUCET (1994); *4*, DOUCET and MARSY (1987); *5*, AIT–MOHAMED et al. (1986); *6*, KHADOURI et al. (1991); *7*, BEN ABDELKHALEK et al. (1986); *8*, DOUCET, unpublished

$(Na^+ + K^+)$-ATPase is by far the most active transport ATPase along the nephron. Its activity is highest in the distal convoluted tubule and the thick ascending limb, with less than 50% of these activities in proximal tubules and the remaining structures (measured per millimeter length, Fig. 7). In the proximal tubule brush border the H^+-ATPase can account for nearly 30% of energy turnover (NOEL et al. 1993) in dog kidney but only one-third of that in rabbit proximal tubules (NOEL et al. 1992). This enzyme has also been found in rat but not in rabbit and mouse thick ascending limbs (Fig. 7). Likewise bicarbonate-dependent ATPase is mainly found in the proximal and the collecting tubule (BEN ABDELKHALEK et al. 1986). Like $(H^+ + K^+)$-ATPase, its activity and relative contribution changes with acid-base and K^+ state of the animal (GARG and NARANG 1989, 1990, 1991; DOUCET 1994; DOUCET and MARSY 1987; PLANELLES et al. 1991). The interaction of renal transport ATPases, diuretics and energy metabolism is discussed in Sect. D.

2. Metabolic Processes

Besides transport ATPases, metabolic energy-using processes have been shown to contribute to tubular oxygen consumption and ATP turnover. Hence these reactions interfere with transport processes. Most prominent in this respect is proximal tubular gluconeogenesis, which has been shown to contribute markedly to proximal energy turnover. As shown by GUDER et al. (1971), the increase in oxygen consumption due to an added substrate largely parallels the increase in energy needed to account for the additional glucose produced. Other but minor metabolic contributors to proximal tubular energy balance are glutathione turnover, triglyceride synthesis, drug metabolism, and lysosomal protein reabsorption (for review see GUDER and SCHMIDT 1976; GUDER and MOREL 1992). In all other tubule segments metabolic contribution to energy turnover seems to be of minor relevance. Basal metabolic rate, as measured after inhibiting all known transport processes, is below 10% of the maximal rates in thick ascending limb and collecting duct cells.

E. Interaction of Diuretic Substances with Renal Energy Metabolism

It has long been known that diuretics interact with renal metabolism in both direct and indirect ways (COHEN and BARAC-NIETO 1973; SIEGENTHALER et al. 1977). Studies with diuretics thus confirm the close link between renal energy metabolism and renal transport function. This will be shown with a few examples only, since extensive studies are documented in other chapters of this volume.

I. Proximal Tubule

Ouabain has been shown to increase proximal tubular gluconeogenesis from lactate (FRIEDRICHS and SCHONER 1973) but not from other substrates (GUDER and GRUBER-STUKOWSKI 1977). This effect was explained by assuming a link between the ATP pool needed to drive $(Na^{+}+K^{+})$-ATPase and the energy-consuming metabolic pathway. In fact oxidation of lactate was reduced in the same experiment, indicating that the energy-sparing effect of ouabain was responsible for its stimulating effect on gluconeogenesis.

Azetazolamide and ethoxzolamide affected proximal tubular gluconeogenesis by a different mechanism. By inhibiting carboanhydrase, these drugs inhibited intracellular bicarbonate/CO_2 equilibration, leading to inhibition of glucose formation from pyruvate (DODGSON and CHERIAN 1989) and ammonia formation from glutamine (GOUGOUX et al. 1987; TANNEN and ROSS 1983). DODGSON and CHERIAN (1989) gave evidence that the mitochondrial isoenzyme of carboanhydrase was responsible for these effects.

Mercurials have long been known to be responsible for proximal tubular intoxication, as recently confirmed by direct measurement of ATP in S1–S3 segments (JUNG et al. 1989).

II. Thick Ascending Limb of Henle's Loop

Furosemide and other derivatives of similar structure are well-known diuretics acting at the thick ascending loop's luminal membrane site (see Chap. 7, this volume). Their inhibitory effect on $Na^{\pm}$ entry into the cells is paralleled by an inhibition of oxygen consumption (CHAMBERLIN and MANDEL 1986), glucose and lactate metabolism (HUS-CITHAREL and MOREL 1986) and oxidation of branched-chain amino acids (TRINH-TRANG-TAN et al. 1988). A similar energy-sparing effect was observed by directly measuring ATP in dissected nephron segments in thin descending short loop segments (JUNG and ENDOU 1990).

Less is known regarding the effect of *ethacrynic acid* on tubular metabolism (FÜLGRAFF et al. 1972). Its inhibitory effect on glucose metabolism seems to be related to distal tubule segments. It is not clear whether the inhibitory effects of *chlorothiazide* as measured in vivo (COHEN 1986) can be explained by its energy-sparing effect in the distal nephron. Several studies indicate that the early distal convoluted tubule may be the site of action of this drug (Chap. 8, this volume).

III. Collecting Tubule

Several diuretic substances act on the late distal convoluted tubule, the connecting tubule and the collecting duct system (see Chaps. 9–11, this volume). Therefore *spironolactone* decreases (Na^{+}+K^{+})-ATPase in the distal nephron site, thus opposing the stimulatory effect of aldosterone in the collecting tubule (DOUCET 1992; RAFESTIN-OBLIN et al. 1991). The changes in transport ATPase are paralleled by changes in citrate synthase, which is a flux-generating step of the citric acid cycle (MARVER 1986). Little is known about the metabolic effects of *amiloride* and *triamterene*, which are thought to inhibit the late distal $Na^{\pm}$ channel.

References

Ait-Mohamed AK, Marsy S, Barlet C, Khaduri C, Doucet A (1986) Characterization of N-ethylamide-sensitive proton pump in the rat kidney. Localization along the nephron. J Biol Chem 261:12526–12533

Balaban RS, Mandel LJ, Soltoff StP, Storey JM (1980) Coupling of active ion transport and aerobic respiratory rate in isolated renal tubules. Proc Natl Acad Sci USA 77:447–451

Barac-Nieto M (1985) Renal hydroxybutyrate and acetoacetate reabsorption and utilisation in the rat. Am J Physiol 249:F40–F48

Barac-Nieto M, Cohen JJ (1968) Nonesterified fatty acid uptake by dog kidney: effects of probenecid and chlorothiazide. Am J Physiol 215:98–107

Barac-Nieto M, Cohen JJ (1971) The metabolic fates of palmitate in the dog kidney in vivo. Evidence for incomplete oxidation. Nephron 8:488–499

Bastin J, Cambon N, Thompson M, Lowry OH, Burch HB (1987) Change in energy reserves in different segments of the nephron during brief ischemia. Kidney Int 31:1239–1247

Bastin J, Djouadi F, Geloso JP, Merlet-Benichou C (1990) Postnatal development of oxidative enzymes in various rat nephron segments: effect of weaning on different diets. Am J Physiol 259:E895–E901

Bavarel G, Fourissier M, Pellet M (1980) Lactate and pyruvate metabolism in dog renal outer medulla. Effects of oleate and ketone bodies. Int J Biochem 12:163–168

Ben Abdelkhalek M, Barlet C, Doucet A (1986) Presence of an extramitochondrial anion-stimulated ATPase in the rabbit kidney: localization along the nephron and effect of corticosteroids. J Membr Biol 89:225–240

Bogusky RT, Garwood M, Matson GB, Acosta G, Cowgill LD, Schleich T (1986) Spatial localisation of phosphorus metabolites and sodium in the in vivo rat kidney. Magn Reson Med 3:251–261

Brown GC (1992) Control of respiration and ATP synthesis in mammalian mitochondria and cells. Biochem J 284:1–13

Burch HB, Bross TE, Brooks CA, Cole RB, Lowry OH (1984) The distribution of six enzymes of oxidative metabolism along the rat nephron. J Histochem Cytochem 32:731–736

Burch HB, Brooks CA, Cambon N, Lowry OH (1985a) Metabolic systems of nephron segments from hydronephrotic rabbit kidney. In: Dzurik R, Lichardus B, Guder WG (eds) Kidney metabolism and function. Nijhoff, Dordrecht, pp 251–255

Burch HB, Cambon N, Lowry OH (1985b) Branched-chain amino acid aminotransferase along the rabbit and rat nephron. Kidney Int 28:114–117

Chamberlin M, LeFurgey A, Mandel LJ (1984) Suspension of medullary thick ascending limb tubules from the rabbit kidney. Am J Physiol 247:F955–F964

Chamberlin ME, Mandel LJ (1986) Substrate support of medullary thick ascending limb oxygen consumption. Am J Physiol 251:F758–F763

Cohen JJ (1986) Relationship between energy requirements for Na reabsorption and other renal functions. Kidney Int 29:32–40

Cohen JJ, Barac-Nieto M (1973) Renal metabolism of substrates in relation to renal function. In: Windhager E (ed) Handbook of physiology. Renal physiology. Oxford University Press, New York, pp 909–1001

Deetjen P, Kramer K (1961) Die Abhängigkeit des O_2-Verbrauchs der Niere von der Natriumrückresorption. Pflugers Arch 273:636–650

Dickens B, Simer F (1930) The metabolism of normal and tumor tissue. II. The respiratory quotient and the relationship of respiration to glycolysis. Biochem J 24:1301–1326

Djouadi F, Wijkhuisen A, Vilar J, Bastin J, Merlet-Benichou C (1993) Effect of glucocorticoids on mitochondrial oxidative enzyme and Na-K-ATPase activities in the rat proximal tubule and thick ascending limb of Henle. Renal Physiol Biochem 16:249–256

Dodgson SJ, Cherian K (1989) Mitochondrial carbonic anhydrase is involved in rat renal glucose synthesis. Am J Physiol 257:E791–E796

Doucet A (1992) Na-K-ATPase in the kidney tubule in relation to natriuresis. Kidney Int 41 [Suppl 37]:118–124

Doucet A (1994) Proc Conf Sodium Pumps (in press)

Doucet A, Marsy S (1987) Characterization of K-ATPase activity in distal nephron: stimulation by potassium depletion. Am J Physiol 253:F418–F423

El Mernissi G, Doucet A (1984) Specific activity of Na-K-ATPase after adrenalectomy and hormone replacement along the rabbit nephron. Pflugers Arch 402:258–263

Elhamri M, Martin M, Ferrier B, Bavarel G (1993) Substrate uptake and utilization by the kidney of fed and starved rats in vivo. Renal Physiol Biochem 16:311–324
Endou H, Jung KY (1991) Heterogeneity of nephron energy metabolism: implications for response to hypoxic insult. In: Hatano M (ed) Nephrology. Proceedings of the 11th international congress of nephrology, vol 1. Springer, Berlin Heidelberg New York, pp 665–672
Eveloff J, Bayerdorffer E, Hasse W, Kinne R (1980) Biochemical and physiological studies on cells isolated from medullary thick ascending limb of Henle's loop. Int J Biochem 12:55–59
Fonteles MC, Cohen JJ, Black AJ, Wertheim SJ (1983) Support of kidney function by long-chain fatty acids derived from renal tissue. Am J Physiol 244:F235–F246
Friedman DL, Perryman MB (1991) Compartmentation of multiple forms of creatine kinase in the distal nephron of the rat kidney. J Biol Chem 266:22404–22410
Friedrichs D, Schoner W (1973) Stimulation of the renal glucogenesis by inhibition of the sodium pump. Biochim Biophys Acta 304:142–160
Füllgraf G, Münemann H, Sudhoff D (1972) Effects of the diuretics furosemide, ethacrynic acid and chlorothiazide on glucogenesis from various substrates in rat kidney cortex slices. Naunyn Schmiedbergs Arch Pharmacol 273:86–98
Garg LC, Narang N (1989) Suppression of ouabain-insensitive K-ATPase activity in rabbit nephron segments during hyperkalemia. Renal Physiol Biochem 12:295–301
Garg LC, Narang N (1990) Effects of low-potassium diet on N-ethylmaleimide-sensitive ATPase in distal nephron segments. Renal Physiol Biochem 13:129–136
Garg LC, Narang N (1991) Changes in H-ATPase in distal nephron segments during metabolic acidosis and alkalosis. In: Endou H, Schoolwerth AC, Bavarel G, Tizianello A (eds) Molecular aspects in ammoniagenesis. Karger, Basel, pp 39–45 (Contributions to nephrology, vol 92)
Gougoux A, Vinay P, Zizian Tejedor A, Noel J (1987) Effect of acetazolamide on renal metabolism and ammoniagenesis in the dog. Kidney Int 31:1279–1290
Gstraunthaler G, Pfaller W, Kotanko P (1985) Interrelation between oxygen consumption and Na-K-ATPase activity in rat renal proximal tubule suspension. Renal Physiol 8:38–44
Guder W, Wiesner W, Stukowski B, Wieland O (1971) Metabolism of isolated kidney tubules. Oxygen consumption, gluconeogenesis and the effect of cyclic nucleotides in tubules from starved rats. Hoppe Seylers Z Physiol Chem 352:1319–1328
Guder WG, Morel F (1992) Biochemical characterization of individual nephron segments. In: Windhager EE (ed) Handbook of physiology, section 8: renal physiology, vol II. Oxford University Press, New York, pp 2119–2164
Guder WG, Pürschel S (1980) The intracellular compartmentation of metabolites in isolated kidney cortex tubules. Int J Biochem 63–67
Guder WG, Schmidt U (1976) Substrate and oxygen dependence of renal metabolism. Kidney Int 10:532–538
Guder WG, Schorn T (1991) Metabolic interactions between renal proline and glutamine metabolism. In: Endou H, Schoolwerth AC, Bavarel G, Tizianello A (eds) Molecular aspects in ammoniagenesis. Karger, Basel, pp 46–51 (Contributions to nephrology, vol 92)
Guder WG, Wieland OH (1974) Metabolism of isolated kidney tubules: regulation of pyruvate dehydrogenase by metabolic substrates. Eur J Biochem 42:529–538
Guder WG, Wirthensohn G (1979) Metabolism of isolated kidney tubules. Interactions between lactate, glutamine and oleate metabolism. Eur J Biochem 99:577–584
Guder WG, Gruber-Stukowski B (1977) Studies on the effect of ouabain on renal carbohydrate metabolism. In: Siegenthaler W, Beckerhoff R, Vetter W (eds) Diuretics in research and clinics. Thieme, Stuttgart, pp 17–20

Guder WG, Pürschel S, Wirthensohn G (1983) Renal ketone body metabolism. Distribution of 3-oxoacid CoA-transferase and 3-hydroxybutyrate dehydrogenase along the mouse nephron. Hoppe Seylers Z Physiol Chem 364:1727–1737

Guder WG, Wagner S, Wirthensohn G (1986) Metabolic fuels along the nephron: pathways and intracellular mechanisms of interaction. Kidney Int 29:41–45

Guder WG, Häussinger D, Gerok W (1987) Renal and hepatic nitrogen metabolism in systemic acid base regulation. J Clin Chem Clin Biochem 25:457–466

Guder WG, Schmolke M, Wirthensohn G (1992) Carbohydrate and lipid metabolism of the renal tubule in diabetes mellitus. Eur J Clin Chem Clin Biochem 30:669–674

Gullans StR, Mandel LJ (1992) Coupling of energy to transport in proximal and distal nephron. In: Seldin E, Giebisch G (eds) The kidney: physiology and pathophysiology. Raven, New York, pp 1291–1370

Höhmann B (1973) Einführung biochemischer Ultramikromethoden bei zellphysiologischen Fragestellungen im Nephron. Habilitationsschrift, Heidelberg

Hus-Citharel A, Morel F (1986) Coupling of metabolic CO_2 production to ion transport in isolated rat thick ascending limbs and collecting tubules. Pflugers Arch 407:421–427

Jans AWH, Leibfritz D (1989) A ^{13}C-NMR study on fluxes into the Krebs cycle of rabbit renal proximal tubular cells. Biomedicine 1:171–176

Jans AWH, Winkel C, Buitenhuis L, Lugtenburg J (1989) ^{13}C-NMR study of citrate metabolism in rabbit renal proximal tubule cells. Biochem J 257:425–429

Janssens P, Hems R, Ross BD (1980) The metabolic fate of lactate in renal cortical tubules. Biochem J 190:27–37

Jung KY, Endou H (1990) Furosemide acts on short loop of descending thin limb, but not on long loop. J Pharmacol Exp Ther 253:80–100

Jung KY, Endou H (1991) Species differences in cellular ATP turnover within the nephron. In: Koide H, Endou H, Kurokawa K (eds) Cellular molecular biology of the kidney. Karger, Basel, pp 149–154

Jung KY, Uchida S, Endou H (1989) Nephrotoxicity assessment by measuring cellular ATP content. I. Substrate specificities in the maintenance of ATP content in isolated rat nephron segments. Toxicol Appl Pharmacol 100:369–382

Katz AJ, Doucet A, Morel F (1979) Na-KATPase activity along the rabbit, rat, and mouse nephron. Am J Physiol 237:F114–F120

Katz J (1989) An NMR study of the tricarboxylic acid cycle. The fate of [3-^{13}C] citrate in renal cells monitored by ^{13}C-NMR. Biochem J 263:997–999

Kettner A, Silbernagl S (1985) Renal handling of citrulline. In: Dzurik R, Lichardus R, Guder WG (eds) Kidney metabolism and function. Nijhoff, Dordrecht, pp 51–60

Khadouri C, Cheval L, Matsy S, Barlet-Bas C, Doucet A (1991) Characterization and control of proton-ATPase along the nephron. Kidney Int 40:S71–S78

Kiil F, Aukland K, Refsum HE (1961) Renal sodium transport and oxygen consumption. Am J Physiol 201:511–516

Kramer K, Deetjen P (1960) Beziehungen des O_2-Verbrauchs der Niere zur Durchblutung und Glomerulumfiltrat. Pflugers Arch 271:782–796

Krebs HA, Hems R, Weidemann MJ, Speake RN (1966) The fate of isotopic carbon in kidney cortex synthesizing glucose from lactate. Biochem J 101:242–249

LeHir M, Dubach UC (1982) Activities of enzymes of the tricarboxylic acid cycle in segments of the rat nephron. Pflugers Arch 395:239–243

Lemieux G, Vinay P, Cartier P (1974) Renal hemodynamics and ammoniagenesis. Characteristics of the antiluminal site for glutamine extraction. J Clin Invest 53:884–894

Lemineux G, Gougoux A, Vinay P, Bavarel G, Cartier P (1978) Relationship between the renal metabolism of glutamine, fatty acids and ketone bodies. In: Guder WG, Schmidt U (eds) Biochemical nephrology. Huber, Basel, pp 379–386

Levillain O (1993) Valine oxidation in the rat medullary thick ascending limb. Pflugers Arch 424:398–402

Levillain O, Hus-Citharel A, Morel F, Bankir L (1992) Localization of urea and ornithine production along mouse and rabbit nephrons: functional significance. Am J Physiol 263:F878–F885

Levillain O, Hus-Citharel A, Morel F, Bankir L (1993) Arginine synthesis in mouse and rabbit nephron: localization and functional significance. Am J Physiol 264:F1038–F1045

Little JR, Spitzer JJ (1971) Uptake of ketone bodies by dog kidney in vivo. Am J Physiol 221:679–683

Löffler G, Petrides E (1992) Physiologische Chemie, 3rd edn. Springer, Berlin Heidelberg New York

Marver D (1986) Models of aldosterone action on sodium transport-emerging concepts. Adv Exp Med Biol 196:153–172

Marver D, Schwartz MJ (1980) Identification of mineralocorticoid target sites in the isolated rabbit cortical nephron. Proc Natl Acad Sci USA 77:3672–3676

Meury L, Noel J, Tejedor A, Sénécal J, Gougoux A, Vinay P (1994) Glucose metabolism in dog inner medullary collecting ducts. Renal Physiol Biochem 17:246–266

Nissim I, Nissim I, Yudkoff M (1990) Carbon flux through tricarboxylic acid cycle in rat renal tubules. Biochim Biophys Acta 1033:194–200

Noel J, Tejedor A, Vinay P, Laprade R (1992) Substrate-induced modulation of ATP turnover in dog and rabbit proximal tubules. J Membr Biol 128:205–218

Noel J, Vinay P, Tejedor A, Fleser A, Laprade R (1993) Metabolic cost of bafilomycin-sensitive H^+ pump in intact dog, rabbit, and hamster proximal tubules. Am J Physiol 264:F655–F661

Osswald H, Schmitz HJ, Kemper R (1977) Tissue content of adenosine, inosine and hypoxanthine in the rat kidney after ischemia and postischemic recirculation. Pflugers Arch 371:45–49

Owen EE, Robinson RR (1963) Amino acid extraction and ammonia metabolism by the human kidney during the prolonged administration of ammonium chloride. J Clin Invest 42:263–276

Owen OE, Felig P, Morgan A, Wahren J, Cahill GF Jr (1969) Liver and kidney metabolism during prolonged starvation. J Clin Invest 48:574–583

Pfaller W (1982) Structure function correlation on rat kidney. Adv Anat Embryol Cell Biol 70:1–106

Pfaller W, Guder WG, Gstraunthaler G, Kotanko P, Jehart I, Pürschel S (1984) Compartmentation of ATP within renal proximal tubular cells. Biochim Biophys Acta 800:152–157

Pitts RF (1976) Metabolism of the kidney of the dog in vivo. In: Giovanetti S, Bonomini V, D'Amico V (eds) Proceedings of the 6th international congress of nephrology. Florence, Karger, Basel, pp 159–170

Pitts RF, Stone WJ (1967) Renal metabolism of alanine. J Clin Invest 46:530–538

Planelles G, Anagnostopoulos T, Cheval L, Doucet A (1991) Biochemical and functional characterization of H^+-K^+-ATPase in distal amphibian nephron. Am J Physiol 260:F806–F812

Rafestin-Oblin ME, Couette B, Barlet-Bas C, Cheval L, Viger A, Doucet A (1991) Renal action of progesterone and 18-substituted derivates. Am J Physiol 260: F828–F832

Ross B, Lowry M (1981) Recent developments in renal handling of glutamine and ammonia. In: Greger R, Lang F, Silbernagl S (eds) Renal transport of organic substances. Springer, Berlin Heidelberg New York, pp 78–92

Ross BD, Guder WG (1982) Heterogeneity and compartmentation in the kidney. In: Sies H (ed) Metabolic compartmentation. Academic, London, pp 363–409

Ross BD, Freeman DM, Chan L (1984) Phosphorus metabolites by NMR. Adv Exp Med Biol 178:455–464

Schmolke M, Guder WG (1989) Metabolic regulation of organic osmolytes in tubules from rat renal inner and outer medulla. Renal Physiol Biochem 12:347–358

Schorn T (1982) Studien zum Prolinstoffwechsel der Niere. Thesis, Ludwig Maximilians University, Munich
Schurek HJ (1992) Oxygen shunt diffusion in renal cortex and its physiological link to erythropoietin production. In: Pagel H, Weiss C, Jelkmann W (eds) Pathophysiology and pharmacology of erythropoietin. Springer, Berlin Heidelberg New York, p 53
Shalhoub R, Webber W, Glabman S, Canessa-Fischer M, Klein J, DeHaas J, Pitts RF (1963) Extraction of amino acids from and their addition to renal blood plasma. Am J Physiol 204:181–186
Siegenthaler W, Beckerhoff R, Vetter W (1977) Diuretics in research and clinics, Thieme, Stuttgart
Silva P, Hallac R, Swartz R, Epstein FH (1980) Competition between different metabolic demands for oxygen consumption in the kidney. Int J Biochem 12:251–255
Tannen RL (1978) Ammonia metabolism. Am J Physiol 235:F265–F277
Tannen RL, Ross BD (1983) The impact of acetazolamide on renal ammoniagenesis and gluconeogenesis. J Lab Clin Med 102:536–542
Tizianello A, DeFerrari G, Garibotto G, Guerri G, Robano C (1980) Renal metabolism of amino acids and ammonia in subjects with normal renal function and patients with chronic renal insufficiency. J Clin Invest 65:1162–1173
Trinh-Trang-Than M-M, Levillain O, Bankir L (1988) Contribution to leucine to oxidative metabolism of the rat medullary thick ascending limb. Pflugers Arch 411:676–680
Vandewalle A, Wirthensohn G, Heidrich H-G, Guder WG (1981) Distribution of hexokinase and phosphoenolpyruvate carboxykinase along the rabbit nephron. Am J Physiol 240:F492–F500
Vinay P, Mapes JP, Krebs HA (1978) Fate of glutamine carbon in renal metabolism. Am J Physiol 234:F123–F129
Vinay P, Lemieux G, Gougoux A (1979) Characteristics of glutamine metabolism by rat kidney tubules: a carbon and nitrogen balance. Can J Biochem 57:346–356
Wallimann T, Wyss M, Brdiczka D, Nicolay K, Eppenberger HM (1992) Intracellular compartmentation, structure and function of creatine kinase isoenzymes in tissues with high and fluctuating energy demands: the "phosphocreatine circuit" for cellular energy homeostasis. Biochem J 281:21–40
Weidemann MJ, Krebs HA (1969) The fuel of respiration of rat kidney cortex. Biochem J 112:149–166
Weinstein SW, Szyjewicz J (1974) Individual nephron function and renal oxygen consumption in the rat. Am J Physiol 227:171–177
Wirthensohn G, Guder WG (1986) Renal substrate metabolism. Physiol Rev 66: 469–489
Wirthensohn G, Guder WG (1980) Triacylglycerol metabolism is isolated rat kidney cortex tubules. Biochem J 180:317–324
Wirthensohn G, Guder WG (1983) Renal lipid metabolism. Miner Electrolyte Metab 9:203–211
Wirthensohn, G, Gerl M, Guder WG (1980) Triacylglycerol metabolism in kidney cortex and outer medulla. Int J Biochem 12:157–161

CHAPTER 4

Discovery and Development of Diuretic Agents

H.-J. LANG and M. HROPOT

A. Introduction

The history of diuretics goes back a long way. It is believed that paleolithic man discovered the caffeine-containing plants. Beverages containing caffeine were prepared from the seeds and bark of different plants. Xanthine derivatives and osmotic diuretics were also used due to their clinically important diuretic effects before the emergence of modern diuretics.

The development and introduction of mercurial diuretics into therapy, at the beginning of this century, was a decisive step towards new discoveries in the field of modern diuretics. In the mid-1950s, the first modern diuretic, acetazolamide, a powerful carbonic anhydrase inhibitor, was developed (first generation). In the late 1950s and in the 1960s, a significant breakthrough was achieved with the discovery of chlorothiazide, furosemide and ethacrynic acid (second generation). This era also includes the development and introduction of potassium-retaining diuretics. The late 1970s and 1980s were dominated by the search for so-called polyvalent diuretics which possess adjunctive renal and/or extrarenal effects such as influence on uric acid, serum lipids, blood glucose and blood pressure (third generation). In recent years, the very new non-diuretic ion transport modulators, which have been derived from structures of the diuretic family, have opened exciting new avenues in the treatment of various diseases. In the present review, diuretics have been classified into different generations according to their chronological appearance on the market and according to their chemical structure, although there is no single system which effectively classifies all the available diuretics (Table 1; HROPOT and MUSCHAWECK 1992; KAU 1992).

B. Xanthine Derivatives

Theophylline, *caffeine* and *theobromine* are methylated xanthines which exhibit clinically important diuretic effects; however, their duration of action is short-lived. They are closely related alkaloids that occur in plants widely distributed geographically (CUSHNY and LAMBIE 1921). Xanthine itself is a dioxypurine and is structurally related to uric acid (formula I, Fig. 1).

Xanthine derivatives stimulate the central nervous system, act on the kidney to produce diuresis, stimulate the myocardium by increasing the

Table 1. Classification of diuretics

A. First generation (discovery to mid-1950s)
  I. Xanthine derivatives
  II. Osmotic diuretics
  III. Mercurial diuretics
  IV. Carbonic anhydrase inhibitors
B. Second generation (mid-1950s to late 1970s)
  I. Sulfonamide diuretics
    1. Benzothiadiazines and related heterocyclics
    2. Sulfamoylbenzoic acid derivatives
  II. Nonsulfonamide diuretics
    1. Phenoxyacetic acid derivatives
      a) Ethacrynic acid
  III. Potassium-retaining diuretics
    1. Aldosterone antagonists
      a) Spironolactone
      b) Potassium canrenoate
      c) CGP-33033
      d) Mespirenone
    2. Pteridines and pyrazine derivatives
      a) Triamterene
      b) Amiloride
C. Third generation (1980s to present)
  I. Polyvalent diuretics
    1. Diuretics with prolonged duration of action (torasemide, muzolimine)
    2. Saluretics with eukalemic properties CRPH-2823, ICI-206970)
    3. Diuretics improving renal function (ibopamine, fenoldopam, docarpamine)
    4. Diuretics with uricosuric activity (indacrinone, tienilic acid, S-8666, A-56234, DR-3438)
    5. Diuretics without adverse effects on serum lipids and blood glucose (indapamide, etzolin, HOE-708)
    6. Diuretics with predominant cardiovascular activity (cicletanine, 6-iodo-amiloride, AY-31906, M-17055, UP-788-42, PD-116948, KW-3902)
D. Aquaretica
  I. Vasopressin antagonists
  II. *K*-Agonists
E. New aspects
  I. Ion transport modulators
    1. MK-417, MK-507
    2. Piprozolin
    3. MSD compounds for treatment of brain edema
    4. NPPB
    5. EIPA, HOE-694

heart rate and contractility, and relax bronchial smooth muscle. Theophylline possesses distinct diuretic activity, and its urine and electrolyte excretion profile is very similar to that of the thiazides. The diuretic activity of xanthine derivatives results from three single effects: (a) a positive inotropic and chronotropic effect on cardiac muscle which results in an increase in glomerular filtration rate (GFR) and renal blood flow, (b) inhibition of

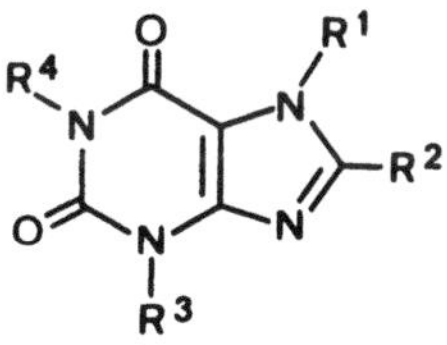

I

R¹	R²	R³	R⁴	
H	H	H	H	Xanthine
H	H	CH_3	CH_3	Theophylline
CH_3	H	CH_3	H	Theobromine
CH_3	H	CH_3	CH_3	Caffeine
H	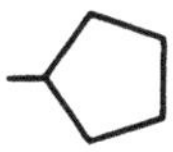	C_3H_7	C_3H_7	PD-116948
H	1-adamantyl	C_3H_7	C_3H_7	KW-3902

Fig. 1. Xanthine derivatives

tubular NaCl reabsorption and (c) an increase in medullary blood flow which reduces the functional level of the urine-concentrating mechanism. Theophylline preparations are primarily employed to relax bronchial smooth muscle in the treatment of asthma and chronic obstructive pulmonary diseases. Xanthine derivatives are no longer in use as primary diuretics because of the greater efficacy of modern agents.

C. Osmotic Diuretics

Osmotic diuretics do not belong to any one specific chemical group. In general, they are solutes that have the following properties in common: (a) they are filtered freely through the glomeruli, (b) their reabsorption by the renal tubules is low or easily saturated and (c) they are not metabolized. In the proximal renal tubules Na^+ ions are actively reabsorbed, whereas water diffuses passively (see Chap. 2, this volume). In the presence of nonreabsorbable solute, the diffusion of water is reduced relative to that of sodium. Thus, the urine flow increases with a relatively smaller increment in the excretion of electrolytes. The two osmotic diuretics most commonly used intravenously are mannitol and urea. Mannitol, which is the most widely used osmotic diuretic, is a metabolically inert carbohydrate that is freely filtered and poorly absorbed by the renal tubules. It is poorly permeant into cells and is distributed mainly in the extracellular fluid compartment; thus it has been used in conditions requiring dehydration of cells, such as cerebral edema, but also as a prophylactic agent of acute renal failure. The main site of action of the osmotic diuretics is the proximal convoluted tubule (Gennari and Kassirer 1974).

Fig. 2. Structural formulas of urea and mannitol

D. Mercurial Diuretics

Because of the clinical importance of mercurial diuretics for over 40 years and of the impetus these drugs have given to the development of modern diuretics, they will be considered briefly. Shortly before the First World War, Bayer introduced the new mercurial drug Novasurol (merbaphen) against syphilis (Fig. 3).

In Wenckenbach's Clinic in Vienna, a nurse and the medical student A. Vogel accidentally observed in 1919 that Novasurol increased urine excretion in a young female patient with congenital syphilis without edema. This was the beginning of an exciting period during which antibacterially effective organomercurial drugs were shown to be useful for the treatment of edema (Saxl and Heilig 1920). However, Novasurol, a mercury derivative, also induced serious adverse effects, such as stomatitis and diarrhea. Thus, the search for mercurial diuretics with fewer side effects continued. In 1924, Salyrgan (mersalyl) was launched onto the market by Hoechst (Fig. 3). Mersalyl was the most effective mercurial diuretic, and it had tolerable side effects. Mersalyl became a standard diuretic for more than 30 years. However, mercurial diuretics had to be administered by the parenteral route because of their poor enteral absorption. Moreover, during treatment with mercurial diuretics, patient resistance to these agents often occurred after only a few injections. Hypochloremia has been cited as the cause of patient resistance to mercurial diuretics. Elevation of plasma chloride by administration of sodium chloride or ammonium chloride has been shown to restore the response of such patients to mercurial diuretics rather than increased acidity of the urine (Axelrod and Pitts 1952). Moreover, it was also shown that mercurial diuretics significantly increased urinary excretion of uric acid in man (Engel and Epstein 1931) and in dogs (Miller et al. 1951). The phenoxyacetic moiety of the mercurial diuretics was responsible for their uricosuric effect and was later the basis for development of ethacrynic acid and other phenoxyacetic derivatives with uricosuric activity.

E. Carbonic Anhydrase Inhibitors

The era of carbonic anhydrase inhibitors began to emerge in the mid-1930s, when it was discovered that the antimicrobial agent sulfanilamide increased bicarbonate excretion and caused metabolic acidosis (STRAUSS and SOUTHWORTH 1938). These effects were attributed to inhibition of carbonic anhydrase, an enzyme that was found to be abundant in the kidney (MANN and KEILIN 1940; DAVENPORT and WILHELMI 1941). These facts and conceptualizations started a long series of experimental investigations that led to the era of modern oral diuretics. In 1949, for the first time, sulfanilamide was given to a woman with congestive heart failure and edema who was refractory to mercurial diuretics (SCHWARTZ 1949). An unexpectedly high rate of urine excretion was observed. This finding was followed by the synthesis of several sulfonamide derivatives which inhibit carbonic anhydrase, the prototype of which was acetazolamide (MAREN et al. 1954).

Acetazolamide (2-acetylamino-1,3,4-thiadiazole-5-sulfonamide) is a powerful carbonic anhydrase inhibitor, 100 times more active than sulfanilamide on a molar basis. Acetazolamide was the first sulfonamide diuretic widely used in the treatment of congestive heart failure from 1954 to 1958 (Fig. 3). However, a limiting factor in the use of acetazolamide is its excessive bicarbonate excretion. This is accompanied by a comparably less severe chloruresis. This disadvantage of acetazolamide was overcome by the development of disulfonamides. Moreover, sulfanilamide was the starting point for structural relationship studies leading to the development (a) of

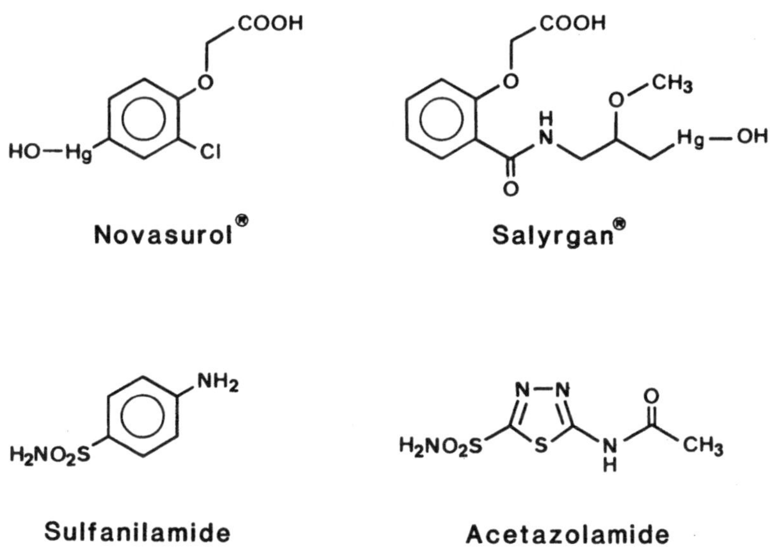

Fig. 3. Mercurial diuretics and the first sulfonamide diuretic acetazolamide

antidiabetic sulfonyl ureas (tolbutamide, glibenclamide) and (b) of thiazide diuretics by substitution variations of the sulfanilamide benzene ring.

F. Sulfonamide Diuretics

I. Benzothiadiazines and Related Compounds

The decisive breakthrough to the new-type diuretics was achieved by structural variations of sulfanilamide and by the introduction of a chlorine atom in the ortho position of the sulfonamide structure, whereby the inhibition of carbonic anhydrase was still present, although less prominent (NOVELLO and SPRAGUE 1957). The carbonic anhydrase inhibitory activity of the new diuretic chlorothiazide (6-chloro-7-sulfamoyl-2*H*-1,2,4-benzothiadiazine 1,1-dioxide) was only about 100th that of acetazolamide; however, chlorothiazide produced a remarkable sodium and chloride excretion with relatively little bicarbonate output (Fig. 4).

During the subsequent years chlorothiazide represented the lead structure for numerous series of diuretics with remarkably enhanced potency. The substitution of the double bond in the chlorothiazide by a single bond led to the largest subgroup of modern diuretics, the benzothiadiazine diuretics (formula III, Fig. 5), its most prominent representative being hydrochlorothiazide.

The replacement of R^2 = H in the hydrochlorothiazide against more lipophilic substituents resulted in increased potency of benzothiadiazines (Fig. 5). Moreover, the replacement of carbonyl-like SO_2 function in formula III by a carbonyl group led to quinazolinone diuretics (formula IV, Fig. 5). Later it was possible to synthesize highly potent hydrochlorothiazide-like acting diuretics by opening the heterocyclic ring system of benzothiadiazines (III) and quinazolinones (IV) or by replacement with a carbocyclic ring system, respectively. This yielded the corresponding 3-sulfamoylbenzenesulfonamides (formula V, Fig. 6) and 3-sulfamoylbenzamides (formula VI, Fig. 6).

The term "thiazide diuretics" was introduced as a general name for these compounds.

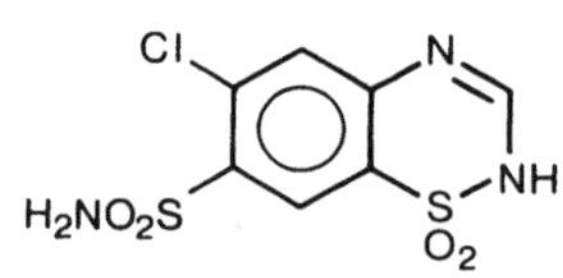

II

Fig. 4. Structural formula of the first modern sulfonamide diuretic chlorothiazide

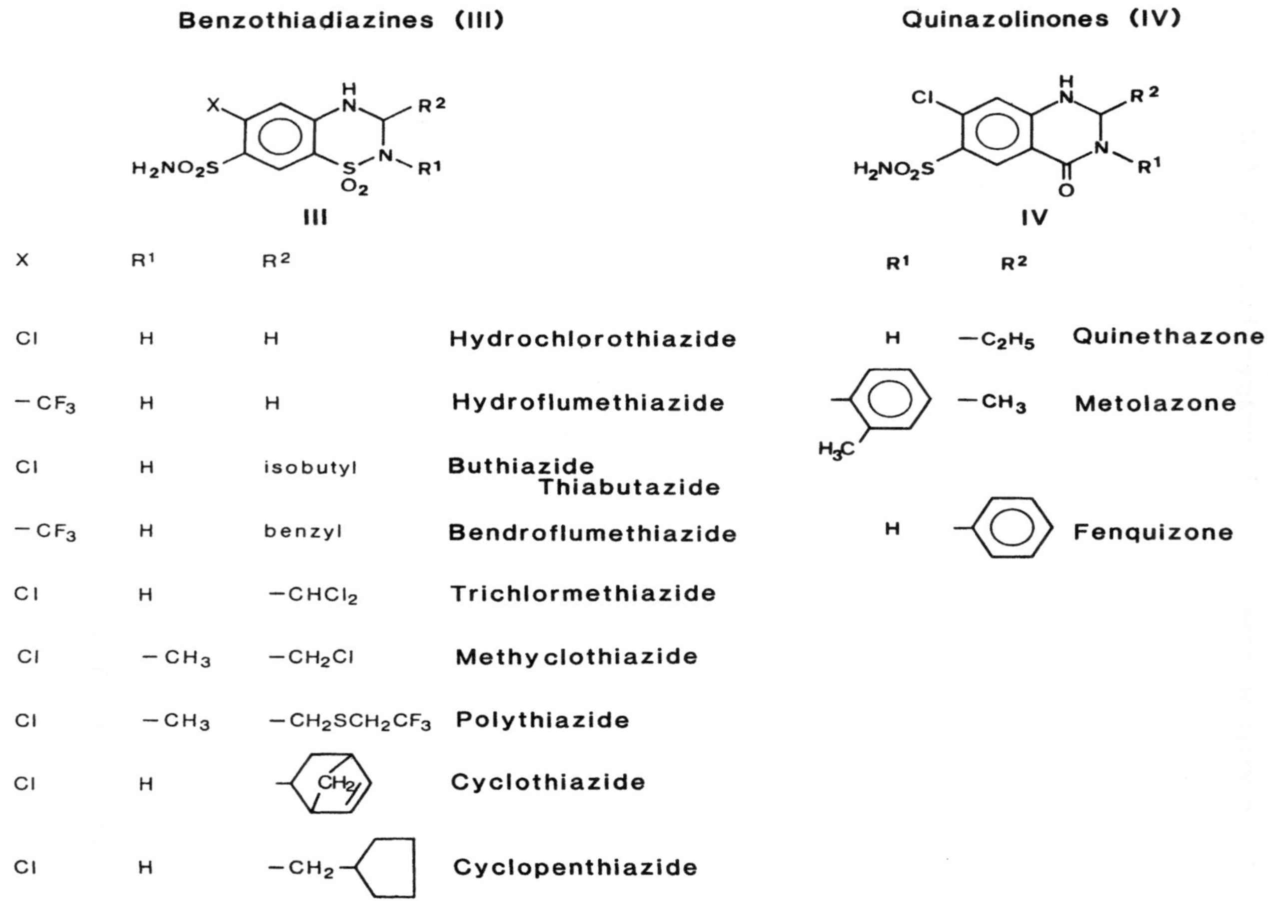

Fig. 5. Benzocondensed heterocyclic sulfonamide diuretics

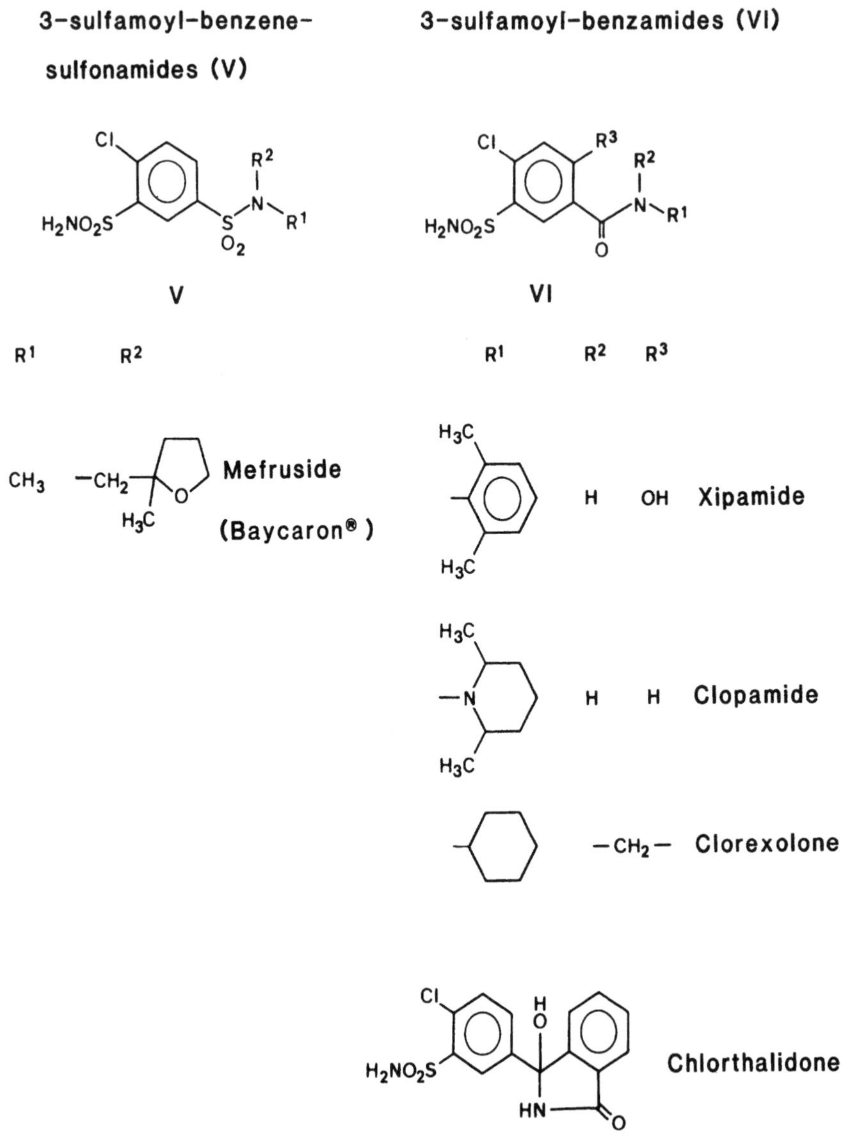

Fig. 6. Ring-opened sulfonamide diuretics

Although a large amount of different analogs were synthesized as part of the benzothiadiazine series, it should be emphasized that all thiazides have parallel dose-response curves and comparable natriuretic effects. This indicates that they have a similar mechanism of action (BEYER 1958; BEYER and BAER 1961; LANT 1985a,b). The era of sulfonamide diuretics or thiazides brought new concepts into the therapy of congestive heart failure, edema and hypertension. In most indications where diuretics had to be used, thiazide diuretics replaced mercurial diuretics. However, in cases where retained water was to be removed quickly from the body, mercurial diuretics still had to be used. The reason for this was the fact that thiazide diuretics induced the excretion of only 5% of the filtered fluid load, whereas the maximum effect of mersalyl amounted to about 20% of the filtered fluid load. Therefore, there was still a need for a highly effective diuretic.

II. Sulfamoylbenzoic Acid Derivatives

Although the free carbonic acid derivatives of the sulfamoylbenzamide type (formula VI) showed no major diuretic effects, it was possible to synthesize highly effective sulfamoylbenzoic acid derivatives by specific substitution of the benzene ring. In 1959, STURM and colleagues succeeded in synthesizing the compound Salu 58, later known as furosemide (Lasix) (STURM et al. 1966). The substitution of the NH_2 group of sulfamoylanthranilic acid by furfuryl was the crucial step (Fig. 7). Furosemide (4-chloro-*N*-furfuryl-5-sulfamoylanthranilic acid) was recognized to be the most effective diuretic ever known (MUSCHAWECK and HAJDÚ 1964). Through furosemide, new trends and new therapeutic principles were created. Due to its excellent tolerance and rapid onset of effect, furosemide became the tool of two

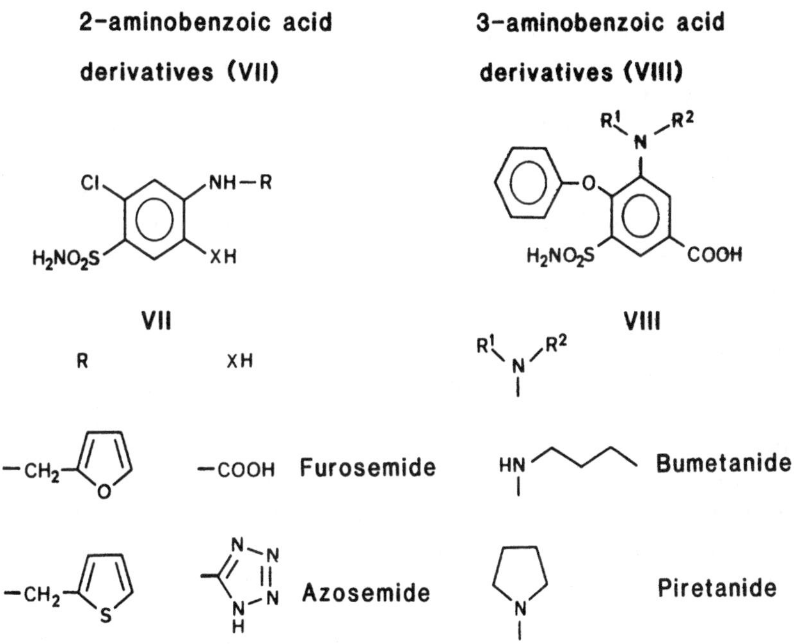

Fig. 7. 2- and 3-Sulfamoylbenzolic acid derivatives

generations of researchers and clinicians involved in basic research, especially in renal physiology. Subsequently, it was well established that furosemide specifically inhibited ion reabsorption in the loop of Henle (BURG et al. 1973) through its interaction with the specific furosemide-sensitive $Na^+2Cl^-K^+$ cotransporter (GECK et al. 1980; GREGER et al. 1983). Sulfamoylbenzoic acid diuretics can be subdivided into two distinct structural classes, 2- and 3-aminobenzoic acid derivatives for formulas VII and VIII, respectively (Fig. 7).

The common structural feature of furosemide-like derivatives was shown to be an acid group (carboxyl or tetrazole moiety) in the meta position of the sulfonamide group. Consequently, corresponding sulfonic acid derivatives (formula VII, X = SO_3^-) were synthesized and were shown to be extremely potent loop diuretics (STURM et al. 1983). Surprisingly, corresponding phosphonic acid derivatives revealed no diuretic activity (SEURING et al. 1984, unpublished results). All diuretics of this group also inhibit $Na^+2Cl^-K^+$ cotransport in the thick ascending limb of the loop of Henle (GREGER et al. 1985).

Bumetanide (3-butylamino-4-phenoxy-5-sulfamoylbenzoic acid) resembles furosemide in chemical structure and diuretic profile; however, on a weight basis it is 40–60 times more potent than furosemide in dog and man (Fig. 7). A phenoxy group was introduced at position 4 and a butylamino substitution made at position 3 (FEIT 1971).

Piretanide [4-phenoxy-3-(1-pyrrolidinyl)-5-sulfamoylbenzoic acid] differs from bumetanide in that the butylamino substituent at position 3 has been replaced by a pyrrolidine ring (Fig. 7). Piretanide causes a high-ceiling salidiuretic effect similar to that of furosemide and bumetanide in animals and man. However, the dose-response curves for diuretic and natriuretic effects of piretanide in the intact animal are flatter than those of furosemide and bumetanide. Piretanide has been claimed to evoke less potassium loss than furosemide and bumetanide (MERKEL et al. 1976; LAWRENCE et al. 1978). It was shown in patients with congestive heart failure that the immediate decrease in intrapulmonary arterial pressure which followed administration of 12 mg piretanide was induced by the vasodilating prostaglandins PGE_2 and PGI_2. The increase in plasma prostaglandins that occurs within the first 15 min after intravenous piretanide administration may be part of the prediuretic mechanism, leading to a dilating effect on the pulmonary vasculature and to increased venous pooling (KIRSTEN et al. 1986).

Azosemide [5-(4-chloro-5-sulfamoyl-2-thenylaminophenyl)-tetrazole] is a diuretic chemically related to furosemide in which the carboxy group present in classical sulfamoylbenzoate derivatives is replaced by a tetrazolyl moiety displaying a comparable acid character (Fig. 7). In comparison to furosemide, the 2-furylmethyl group is replaced by a thienyl group. It is a loop diuretic which has a diuretic profile similar to that of furosemide, but it has a slower onset of effect (BRATER 1979). Azosemide is poorly absorbed from the gastrointestinal tract, with a bioavailability of only 10%.

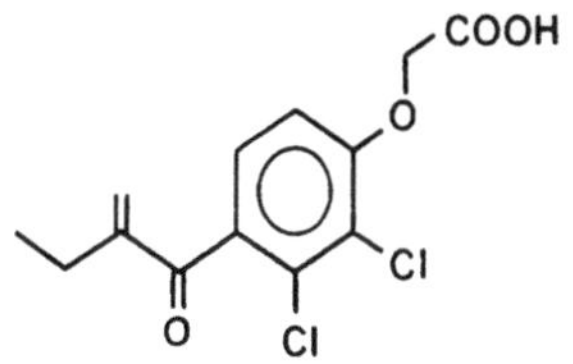

IX

Fig. 8. Structural formula of the phenoxyacetic acid derivative ethacrynic acid

G. Nonsulfonamide Diuretics

I. Phenoxyacetic Acid Derivatives

As mentioned above, the phenoxyacetic moiety in the structure of mercurial diuretics contributed to their diuretic and uricosuric activity. By searching for mercury-free derivatives, ethacrynic acid was found to be a lead compound of the phenoxyacetic acid derivatives. Moreover, ethacrynic acid contains an unsaturated ketone moiety with a high reactivity to sulfhydryl groups similar to mercurial diuretics (Fig. 8).

Thus, ethacrynic acid was synthesized in an attempt to mimic the sulfhydryl reactivity of the mercurial diuretics (BEYER et al. 1965). However, it remains unclarified whether such reactivity is crucial to the action of ethacrynic acid, which is a loop diuretic, although it also inhibits proximal fractional sodium reabsorption. Ethacrynic acid [2,3-dichloro-4-(2-methylenebutyryl)-phenoxyacetic acid] was developed and introduced onto the market in the mid-1960s. The search for diuretics with uricosuric activity resulted in an extensive research for novel phenoxyacetic acid derivatives (see Sect. H).

II. Potassium-Retaining Diuretics

1. Aldosterone Antagonists

Spironolactone and *potassium canrenoate* are used as aldosterone antagonists (DODSON and TWEIT 1959). Aldosterone stimulates sodium and water reabsorption in the late part of the distal tubule and the collecting ducts. Aldosterone antagonists, on the other hand, enhance sodium excretion and diminish potassium secretion. Aldosterone antagonists act only in the presence of enhanced aldosterone levels. The response to these drugs is slow in onset and reveals a complete natriuretic activity only after several days of treatment. Both spironalactone and canrenoate are rapidly converted to the active metabolite canrenone, which has a half-life of about 20 h.

CGP-33033 is an aldosterone antagonist from a series of 7-acylthio-9,11-epoxy-20-spiroxan derivatives. Its affinity for the mineralocorticoid receptor

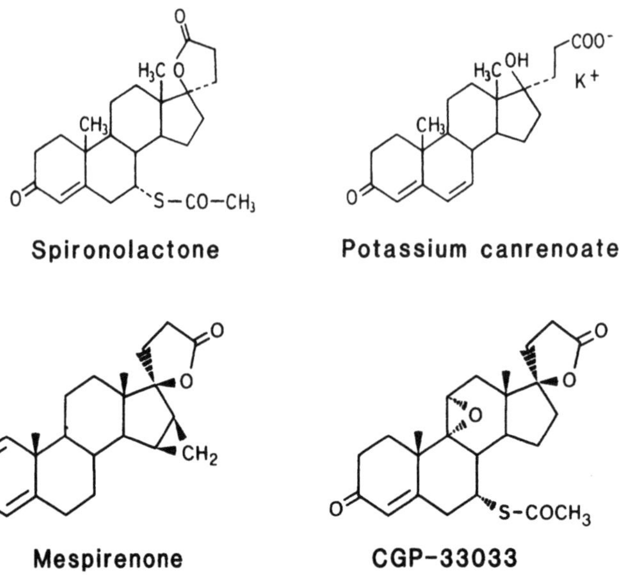

Fig. 9. Structural formulas of aldosterone antagonists

was comparable to that of spironolactone; however, the affinity of CGP-33033 for the androgen and progesterone receptor revealed 10000- and 1000-fold lower values than those for spironolactone (DEGASPARO et al. 1989).

Mespirenone is another investigational aldosterone antagonist with increased potency and reduced endocrinological side-effects compared to spironolactone (HABEREY et al. 1986). Studies in volunteers indicated that mespirenone is about six times more potent than spironolactone as a natriuretic agent without progestogenic activity.

2. Pteridines and Pyrazine Derivatives

Triamterene and *amiloride* (formulas X, XI, Fig. 10) are the most important members of these groups (CRAGOE et al. 1967; WIEBELHAUS et al. 1967).

Triamterene and amiloride decrease renal potassium excretion. However, their mechanism of action differs from that of spironolactone. They do not antagonize aldosterone, but block sodium channels in epithelia and act in the late part of the distal tubules and cortical collecting duct of the kidney, thereby inducing natriuresis and antikaliuresis. Triamterene and amiloride are used mainly in combination with hydrochlorothiazide and other saluretics in order to prevent diuretic-induced hypokalemia.

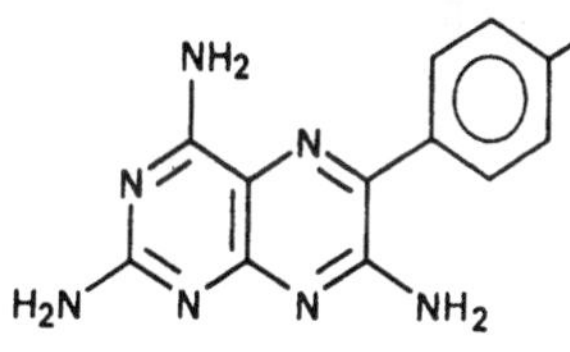

X

R = H Triamterene

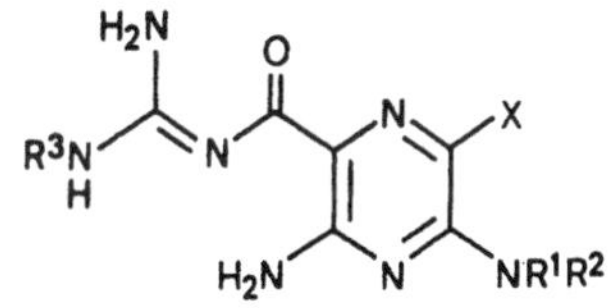

XI

$R^1, R^2, R^3 = H$, X = Cl Amiloride

$R^1, R^2, R^3 = H$, X = I 6-iodo-amiloride

$R^1 = C_2H_5$, $R^2 = CH(CH_3)_2$, $R^3 = H$, X = Cl Ethyl-isopropyl-amiloride (EIPA)

Fig. 10. Potassium-retaining diuretics

H. So-called Polyvalent Diuretics

The modern "second-generation" diuretics described above have been well established in therapy of hypertension, edema, and congestive heart failure. Consequently, there has been no medical need for further diuretics with higher potency. Therefore, during the past decade, research has focused on the development of diuretics with new pharmacological qualities, such as prolonged diuration of action, improvement of renal function or reduction of adverse effects, and it has also concentrated on diuretics displaying further favorable therapeutic properties.

I. Loop Diuretics with Prolonged Duration of Action

Attempts have been made in the past to overcome the disadvantages of brisk diuresis and short duration of action of classical loop diuretics. Especially elderly patients may suffer from side effects such as weakness and tiredness. A retard formulation might be advantageous over thiazide diuretics, particularly in hypertensives and in patients with impaired renal function. Thus, efforts have been made to develop a retard formulation for furosemide to prevent the peak diuretic effect. Furthermore, transdermal therapeutic systems (TTSs) have also been evaluated for their ability to change the time-response curve of furosemide (Hropot et al., unpublished results). However, no real breakthrough has been achieved so far. On the other hand, new loop diuretics have been synthesized and introduced onto the market with prolonged duration of action, mainly as a result of extensive metabolism of parent compounds. In addition to galenic manipulations of classical loop diuretics some novel classes of agent have been developed during the past decade which combine the high efficacy of loop diuretics with prolonged duration of action.

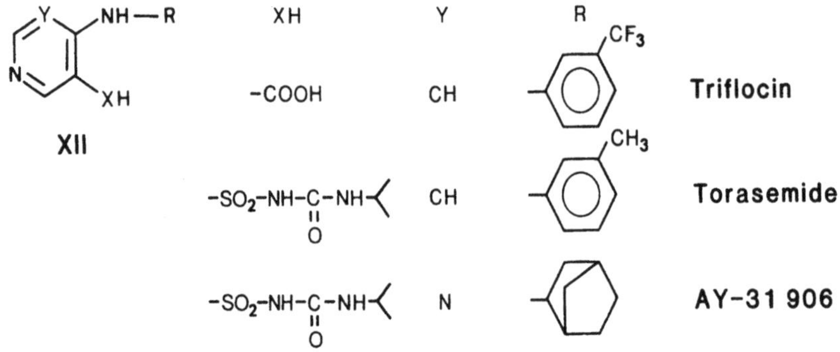

Fig. 11. Pyridine and pyrrazole derivatives and related compounds

Torasemide, 1-isopropyl-3-[(4-m-toluidino-3-pyridyl)sulfonyl] urea, is a loop diuretic which produces salidiuretic effects similar to those of furosemide, bumetanide and piretanide, but it seems that the duration of action of torasemide is longer. A striking feature of the loop diuretic torasemide is the sulfonylurea side chain (DELARGE 1988). The sulfonylurea group of torasemide has less acidity (pK 6.8) than a carboxyl group. Torasemide corresponds to triflocin, in which the methyl group is exchanged for a trifluoromethyl group and the sulfonylurea group is exchanged for a carboxyl group (formula XII, Fig. 11). Triflocin is a loop-acting diuretic which had been withdrawn from clinical use for toxicological reasons. As with furosemide, the main site of action of torasemide is the loop of Henle, where it inhibits the luminal $Na^+2Cl^-K^+$ cotransport.

In the further search for loop diuretics with prolonged duration of action, muzolimine, [3-amino-1-(3,4-dichloro-α-methylbenzyl)-2-pyrazolin-5-one], was synthesized. It is a loop diuretic similar in action to furosemide, but with a less abrupt onset of action and more prolonged duration of action (LOEW and MENG 1977). However, only 10% of muzolimine is excreted in unchanged form. The bile is the major route of elimination of metabolites. It was later shown that muzolimine acts by means of an active metabolite (WANGEMANN et al. 1987). In 1987, shortly after the introduction of muzolimine onto the market in Germany, it was withdrawn for toxicological reasons (polyneuropathy).

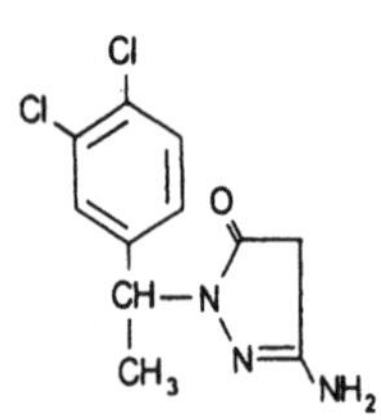

Fig. 12. Structural formula of muzolimine

II. Saluretics with Eukalemic Properties

Structural variations of triamterene (Fig. 10) led to the biologically active derivative RPH-2823 [formula X, R = $-O-CH_2-CH(OH)-CH_2-N(CH_3)_2$] with a hydrophilic side chain (MUTSCHLER et al. 1987). RPH-2823 is a potent potassium- and magnesium-retaining diuretic with a rapid onset of action which was sustained 3 h after dosing. Moreover, its physicochemical properties are superior to those of triamterene, it is water soluble and a stable solution for intravenous administration can be produced. Furthermore, it has been shown by MUTSCHLER et al. (1987) that, in volunteers who received the combination of furosemide plus RPH-2823, there was an additional change in the excretion of chloride, calcium and magnesium, whereas the furosemide-induced potassium excretion was reduced to control values. Due to its favorable pharmacodynamic and pharmacokinetic properties, RPH-2823 seems to be a suitable partner for diuretics which may provoke kaliuresis in excess. Some further structural variations of amiloride (Fig. 10) which resulted in new biological activities will be outlined below.

ICI-206970, 3,5-diamino-*N*-[2-(*N*-[2-(3-bromo-5-tert. butyl-2-hydroxybenzylamino)ethyl]-*N*-methylamino)ethyl]6-chloropyrazine-2-carboxamide, is an effective diuretic and natriuretic agent with eukalemic properties. Furthermore, it has been shown that this diuretic effectively reduces blood pressure in spontaneously hypertensive rats and that it also possesses some

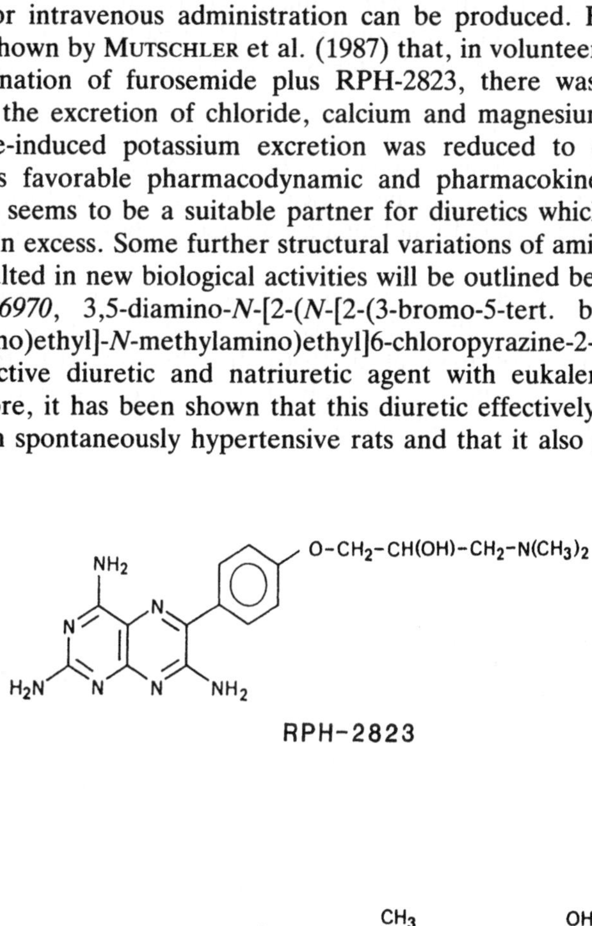

RPH-2823

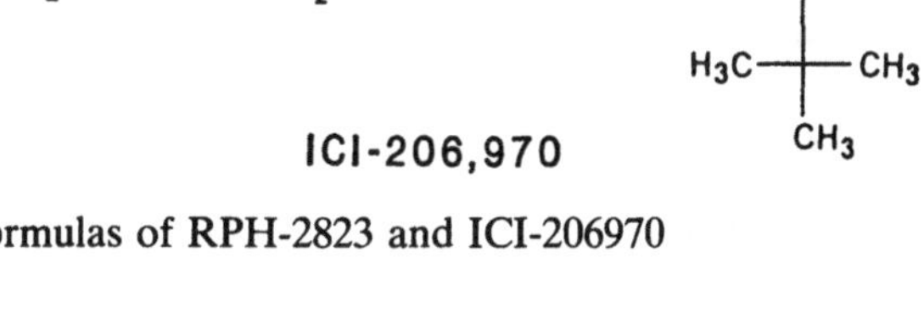

ICI-206,970

Fig. 13. Structural formulas of RPH-2823 and ICI-206970

of the pharmacologic features of calcium channel blockers (JOHNSTON and KAU 1993).

III. Diuretics Improving Renal Function: Dopamine Agonists

The search for peripheral dopaminergic drugs that would produce vasodilation, an increase in renal blood flow and natriuresis resulted in two types of investigational diuretics, those which functioned as dopamine prodrugs and others which are themselves dopaminergics. The first type is represented by *ibopamine*, the 3,4-diisobutyryl ester of *N*-methyldopamine (formula XIII, Fig. 14).

The second type is represented by *N*-allyl substituted *fenoldopam*, showing D_1 and D_2 agonistic activity. This exemplifies the influence of *N*-allyl substitution on the pharmacological action of the benzazepine, since the parent N-unsubstituted fenoldopam possesses only D_1 agonistic activity (see below).

Ibopamine is an orally active derivative of dopamine with diuretic activity. Ibopamine is rapidly and extensively metabolized after oral administration to the rat (POCCHIARI et al. 1986). It is hydrolyzed to free epinine and to conjugated epinine. It is likely that the effects of ibopamine are mediated by arteriolar vasodilatation.

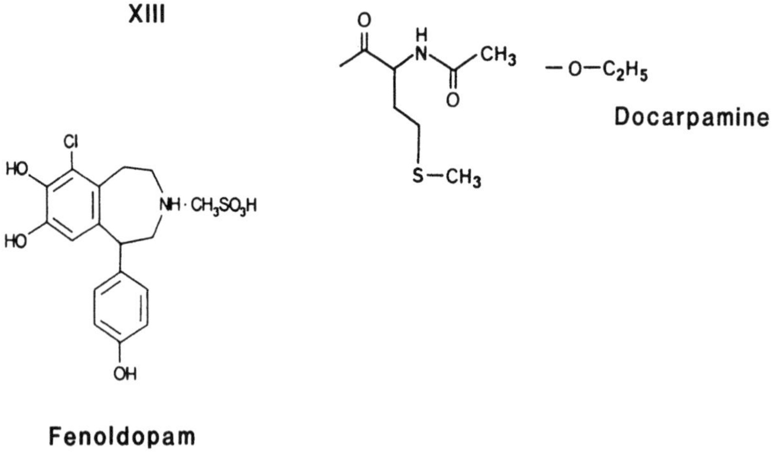

Fig. 14. Dopamine agonists

Docarpamine (TA-870) is an orally active dopamine pro-drug, *N*-(*N*-acetyl-L-methionyl)-*O*,*O*-bis-(ethoxycarbonyl) dopamine (Fig. 14). After oral administration, docarpamine is rapidly absorbed and then converted to dopamine to cause renal vasodilation, diuresis and positive inotropic effects (YAMAGUCHI et al. 1989). In dogs, it increased glomerular filtration rate, urine and sodium excretion. The duration of action was more than 4 h. In patients with renal dysfunction, docarpamine (600 mg p.o.) increased glomerular filtration rate, renal blood flow and urinary sodium excretion (OZAWA and MORI 1986). Furthermore, in patients with severe congestive heart failure, docarpamine improved left ventricular fractional shortening and mean circumferential velocity (KUBOTA et al. 1989). Docarpamine seems to be a useful alternative to dopamine due to its effects on heart and kidney.

Fenoldopam, [6-chloro-2,3,4,5-tetrahydro-1-(4-hydroxyphenyl)-1*H*-3-benzazepine-7,8-diol methanesulfonate], is a selective dopamine D_1-receptor agonist which increases the renal plasma flow, whereas the glomerular filtration rate is unchanged. These effects are consistent with decreased renovascular resistance (ALLISON et al. 1987). The mechanism of fenoldopam-induced natriuresis is not completely understood. However, it has been shown that fenoldopam decreases sodium reabsorption in proximal and distal nephron segments due to changes in renal hemodynamics, but not due to the compound's direct tubular effects (ALLISON et al. 1987).

IV. Diuretics with Uricosuric Activity

The uricosuric effect of mercurial diuretics containing a phenoxyacetic acid residue mentioned above induced a search for diuretics without hyperuricemic properties. Since the discovery of ethacrynic acid, many aryloxyacetic acid derivatives have been synthesized and evaluated for their diuretic properties. The uric acid retention caused by thiazides and furosemide diuretics contrasted sharply with the uricosuric nature of indacrinone.

Indacrinone (MK-196) is chemically related to ethacrynic acid and ticrynafen, containing a dichlorophenoxyacetic acid structure (Fig. 15).

Indacrinone has a chiral center at position 2 of the indanone ring and exists as a racemic mixture. Both enantiomers have uricosuric activity, but natriuretic activity resides predominantly in the (−)-enantiomer. In contrast to furosemide, the diuretic and saluretic response to indacrinone was reported to be slower in onset and of longer duration (FANELLI et al. 1977). Indacrinone was assigned for treatment of mild to moderate hypertension and congestive heart failure. However, the clinical development of indacrinone was discontinued in the mid-1980s due to an imbalance between diuretic and uricosuric activity resulting in indacrinone-induced hyperuricemia as with other diuretics.

Ticrynafen (tienilic acid = [2,3-dichloro-4-(2-thienylcarbonyl)-phenoxy]acetic acid), is a phenoxyacetic acid derivative (Fig. 15), a salidiuretic with a thiazide-like profile and with additional uricosuric activity

(MAASS et al. 1978). Replacement of the α, β-unsaturated ketone side chain of ethacrynic acid by a thienoyl substituent led to tienilic acid. Ticrynafen was a very promising drug; however, it was dropped from the market in most countries because of its liver toxicity in man. Structure-toxicity experiments with different phenoxyacetic acid diuretics have demonstrated that hepatotoxic effects seem to be attributed to the 2-thienylcarbonyl moiety of tienilic acid (HIGAKI et al. 1989).

S-8666 is a newly developed antihypertensive uricosuric diuretic which possesses an asymmetric carbon and exists as a racemic mixture (Fig. 15). It can be recognized from the formulas in Fig. 15 that, similarly to A-56234 and A-53385, the essential phenoxyacetic acid moiety in S-8666 is a con-

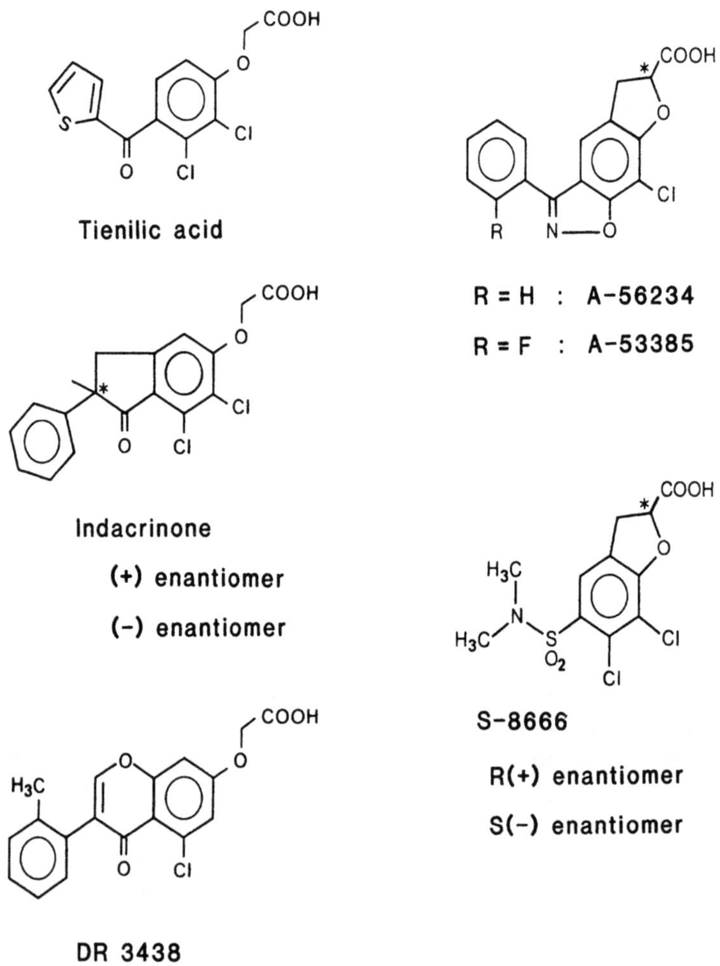

Fig. 15. Phenoxyacetic acid derivatives with uricosuric activity

stituent of the benzofuran-2-carboxylic acid structure. The (−)-enantiomer and racemic S-8666 showed a high-ceiling natriuretic effect like that of furosemide, whereas the (+)-enantiomer caused no significant change in sodium excretion (NAKAMURA et al. 1990). However, both enantiomers inhibit urate transport in the rabbit proximal tubule in vitro.

A 56234 (Fig. 15) is a new investigational diuretic with high-ceiling saluretic and diuretic properties and uricosuric activity (PLATTNER et al. 1984). Evaluation of the enantiomers of A-56234 revealed that only the (+)-enantiomer caused diuretic activity, whereas both enantiomers showed uricosuric activity.

DR-3438, a benzopyran derivative which does not contain an asymmetric carbon atom, represents another phenoxyacetic acid diuretic with uricosuric activity (Fig. 15, TANAKA et al. 1990). The main site of action of DR-3438 concerning salidiuretic effect is the loop of Henle and the distal tubule. Furthermore, it inhibits urate reabsorption in the proximal tubule. The effect of DR-3438 on hepatocytes has been compared with that of other uricosuric diuretics in primary cultured hepatocytes from rats (TAKAGI et al. 1990).

V. Avoidance of Adverse Effects on Serum Lipids and Blood Glucose

Diuretics were primarily introduced for the treatment of edema of diverse etiologies, whereas today the main indications for diuretics are hypertension and congestive heart failure. In the past, it was found in long-term studies that thiazides cause a moderate elevation of triglycerides and total cholesterol in the majority of patients. In a few studies, this could account, for the failure of long-term thiazide therapy to reduce the incidence of myocardial infarction, although the incidence of stroke was reduced (HELGELAND A 1980). Recently, it has been shown that lowering the diuretic dose is the most effective maneuver for reducing the unwanted side effects of diuretics. However, in the past, attempts were made to find diuretics with less or no adverse effects on serum lipids and carbohydrate metabolism. In long-term studies in man, indapamide was found not to have significant changes in serum lipids and serum glucose levels (CAMPBELL and MOORE 1981; GERBER et al. 1985), but this was primarily due to the low dosage of the drug (see Chap. 13, this volume). *Indapamide* (Fig. 16) belongs to a special type of ring-opened sulfonamide diuretic of general formula VI (Fig. 6) characterized by a benzoylhydrazine structure (formula XIV, Fig. 16).

Subsequently, further thiazide-like diuretics with a benzoylhydrazide structure such as the well-known *clopamide* (Fig. 6) and the novel indapamide-related compounds *tripamide* and *isodapamide* have been investigated (Fig. 16). All three exhibit salidiuretic effects comparable to those of indapamide.

Etozolin is a long-acting loop diuretic of the thiazolidone type, which lacks the benzene ring and the sulfonamide group (SATZINGER et al. 1978).

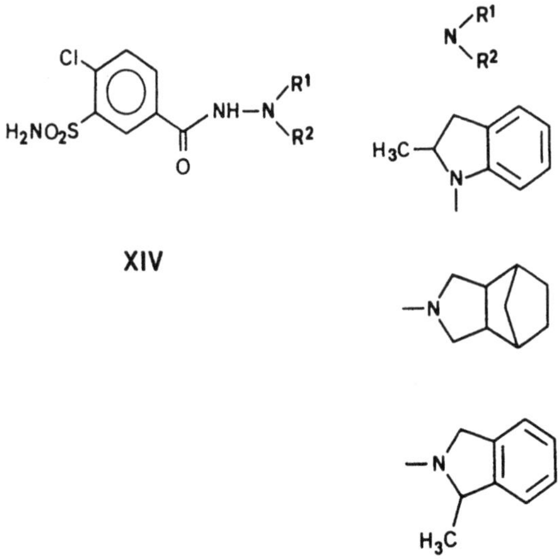

XIV

Indapamide

Tripamide

Isodapamide

Fig. 16. Sulfamoylbenzoylhydrazide derivatives

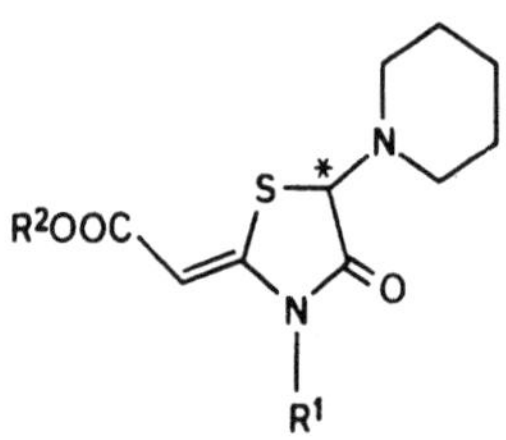

XVI

R^1	R^2	
$-CH_3$	$-C_2H_5$	Etozolin (racemate)
$-CH_3$	$-H$	Ozolinone (−)-enantiomer (diuretic) (+)-enantiomer (antihypertensive)
$-C_2H_5$	$-C_2H_5$	Piprozolin (racemate)

Fig. 17. Thiazolidone-type agents

Etozolin was synthesized in 1978 by the substitution of a methyl for an ethyl group at position 1 of the thiazolidone ring of the choleretic substance *piprozolin* (Fig. 17).

Etozolin is well absorbed from the gastrointestinal tract in animals and man and undergoes extensive first-pass metabolism. Hydrolysis of the pro-drug etozolin yields the active metabolite ozolinone (GREVEN and HEIDENREICH 1978), which has an asymmetric carbon at position 5 of the thiazolidone ring. After separation of the optical isomers, it was shown that

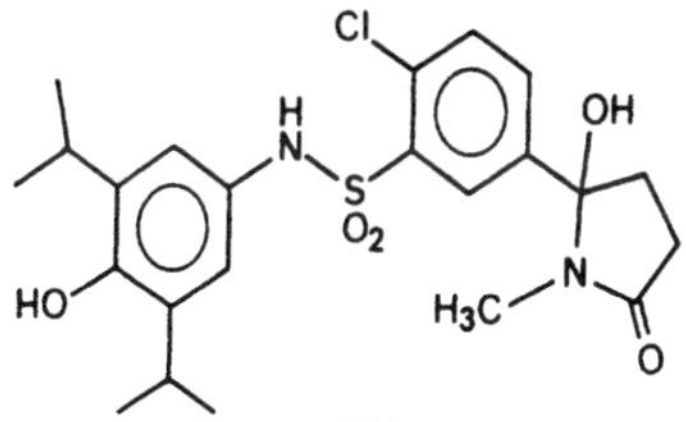

XV

Fig. 18. Structural formula of the polyvalent diuretic sitalidone (HOE-708)

(−)-ozolinone acts in the thick ascending limb of the loop of Henle. Furthermore, it has been shown in long-term experiments in rats and dogs that, in contrast to most other diuretics, etozolin did not cause glucose intolerance. Diuretic-induced glucose intolerance has long been attributed to potassium loss, which is an inherent quality of most diuretics. It has, however, been shown that a relationship between kaliuresis and hyperglycemia does not really exist, since triamterene and the nondiuretic analogue of chlorothiazide, diazoxide, also increase the blood glucose concentration (Wales et al. 1968). Moreover, it has recently been shown that low concentrations of furosemide reduce the glucose-induced insulin release, whereas high doses increase basal and glucose-induced insulin release (Sandström and Sehlin 1988). These authors suggested that the diabetogenic action of furosemide may be due to direct inhibition of insulin release from the pancreatic β-cells by the inhibition of an inwardly directed chloride flux. This leads to a reduced outward chloride transport, a factor associated with stimulated calcium uptake and insulin release.

In contrast to the sulfamoylbenzoylhydrazides, *sitalidone* (HOE-708, Fig. 18) represented a novel thiazide-like diuretic structurally related to chlorthalidone with antiatherosclerotic properties (Hropot et al. 1990).

The molecule of sitalidone comprises three moieties: the 2-chlorobenzenesulfonamide moiety shows salidiuretic activity, the benzenesulfanilide moiety shows hypolipidemic properties and the diisopropylphenol moiety may be responsible for antioxidant activity. In addition, the pronounced salidiuretic and antihypertensive effects of sitalidone caused a distinct decrease in serum LDL fraction without major changes in HDL fraction in a 7-day experiment in rats (Hropot et al. 1990). Furthermore, sitalidone inhibited malonyldialdehyde production of rat liver microsomes, indicating its antioxidant properties.

VI. Diuretics with Predominant Cardiovascular Activity

Because of the importance of diuretics in cardiovascular diseases efforts have been made to search for diuretics with additional activity on blood pressure and heart. *Cicletanine*, a dihydrofuropyridine derivative, is a potent

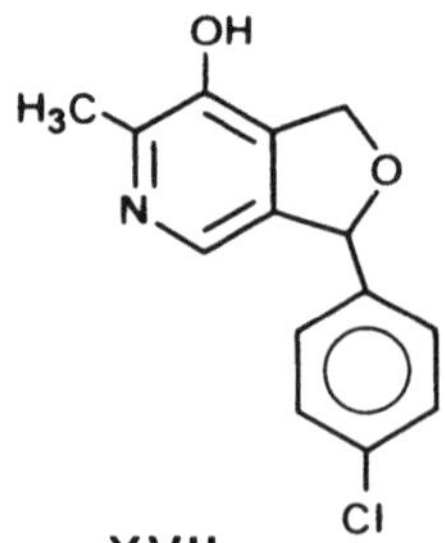

XVII

Fig. 19. Structural formula of the dihydrofuropyridine derivative cicletanine

diuretic and antihypertensive drug (Fig. 19). Comparison of the acute renal effects of cicletanine (300 mg), furosemide (40 mg) and hydrochlorothiazide (50 mg) after oral administration to healthy volunteers indicated that cicletanine is nearly equieffective to standard formulations of the two reference drugs and that it exhibits thiazide-like diuretic activity. It was reported that cicletanine increases blood and urine 6-keto-$PGF_{1\alpha}$ and urinary PGE_2 levels in hypertensive patients, and that it enhances the production of prostacyclin from exogenously added arachidonic acid in cultured smooth muscle cells of the rat aorta (GUINOT and FRÖLICH 1985).

During a 2-year study in hypertensives, cicletanine decreased serum sodium and potassium, but did not change serum urate, glucose, cholesterol, triglycerides and creatinine. Furthermore, a dissociation between plasma renin activity and aldosterone concentration was shown, suggesting an effect of cicletanine on intracellular calcium mobilization (FAGARG et al. 1987). The antihypertensive effect of this drug is clearly dissociated from the natriuretic effect, which appears at a higher dose. A direct effect on vascular smooth muscle mediated by an increase in prostacyclin synthesis has also been suggested to explain the vasodilator and antihypertensive action of cicletanine.

6-Iodo-amiloride (formula XI: R^1, R^2, R^3 = H; X = I in Fig. 10) was shown to be particularly active relative to amiloride concerning its antihypertensive effect. An intravenous infusion of 6-iodo-amiloride to spontaneously hypertensive rats produced a prompt and sustained dose-dependent decrease in arterial blood pressure. This antihypertensive effect of 6-iodo-amiloride was independent of excretory function, but correlated with peripheral vasodilatation subsequent to hyperpolarization of vascular smooth muscle cells (PAMNANI et al. 1986).

AY-31906 (Fig. 11) is a pyrimidine sulfonamide diuretic, the profile of which resembles that of furosemide, but with less action on potassium excretion. Moreover, its sodium/potassium ratio and antihypertensive activity are also favorable (ACKERMAN et al. 1990).

In the past, attempts have been made to combine diuretic effects with beta-blocking activity in order to reduce the adverse effects of diuretics and

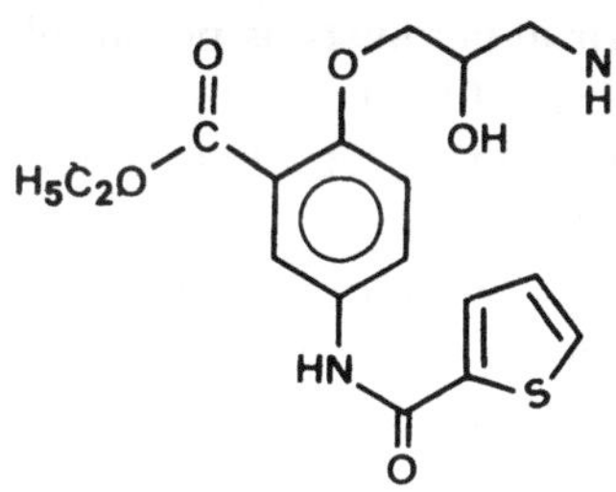

Tienoxolol

(UP788-42)

XVIII

M-17055

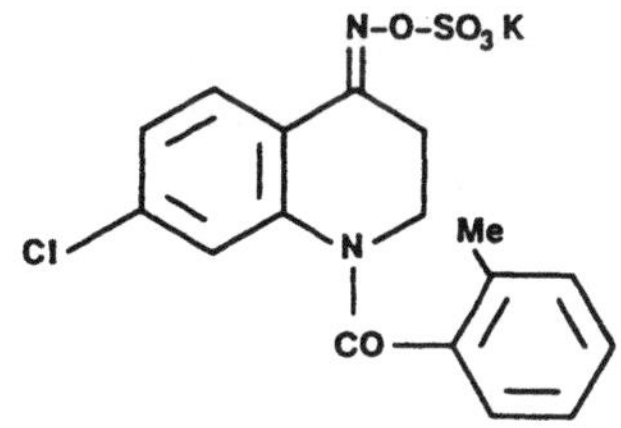

Fig. 20. Diuretics UP-788-42 and M-17055

to increase their blood pressure-lowering activity. An example par excellence of such a compound is tienoxolol (UP-788-42, formula XVIII, Fig. 20), a thienylcarbonylaminophenoxy propanolamine derivative, which possesses both diuretic and beta-blocking activity (BOULEY et al. 1986). Tienoxolol is a cardioselective beta-blocking agent and salidiuretic in animals and man. It also lowers plasma renin activity and increases the creatinine clearance, exibiting the profile of a moderate diuretic.

M-17055, 7-chloro-2,3-dihydro-1-(2-methyl-benzoyl)-4(1*H*)-quinolinone-4-oxime-*O*-sulfonic acid potassium salt (Fig. 20), is a potent loop diuretic with relative potassium-saving properties compared with furosemide. The main sites of action are the loop of Henle and also the distal nephron segments. The antihypertensive activity of M-17055 in spontaneously hypertensive rats lasted significantly longer than that of furosemide (YAMASAKI et al. 1990).

PD-116948, a xanthine derivative, has been shown to be an A_1-selective adenosine antagonist in rat whole-brain membranes (formula I in Fig. 1). In the isolated rat heart, it has been demonstrated that PD-116948 selectively antagonized the negative chronotropic activity of an adenosine agonist and that it was free of intrinsic cardiac or vascular activity. Furthermore, PD-116948 also caused significant thiazide-like salidiuretic activity after its oral administration to rats (KEDDIE and COLLINS 1989).

KW-3902 represents another xanthine derivative (formula I: R^1 = H, R^2 = 1-adamantyl, R^3, R^4 = *n*-propyl in Fig. 1) with A_1-selective adenosine antagonistic properties resulting in favorable diuretic and antihypertensive activity (KOBAYASHI et al. 1992).

I. Aquaretics

Aquaretics are agents which selectively increase the renal excretion of osmotically free water without major changes in electrolyte output. Thus, they dilute urine osmolality distinctly below the values for plasma osmolality and therefore produce hypotonic urine. Since vasopressin is the primary hormonal regulator of total body water, the discovery of diuretic vasopressin analogs has initiated interest in vasopressin receptor antagonists as potential specific water diuretic agents (MANNING et al. 1981). On the basis of the mechanism of action and chemical structures, one can differentiate three main types of aquaretics:

(a) vasopressin analog V_2 receptor antagonists such as SK&F-105494,
(b) nonpeptide V_2 receptor antagonists such as OPC-31260 and
(c) nonpeptide κ-opioid receptor agonists such as U-50488 and RU-51599 (Fig. 21).

SK&F-105494, a cyclic arginine vasopressin antagonist (Fig. 21), was shown to be the most potent in vitro and in vivo antagonist of renal V_2 receptors. It selectively increased the renal excretion of water and diluted urine osmolality to levels one-third or less that of plasma osmolality in conscious monkeys (BROOKS et al. 1989). Demonstration of SK&F-105494's inhibition of the antidiuretic response to the selective V_2 agonist desamino-8-D-arginine vasopressin indicated that the primary target is renal V_2 receptor and not the suppression of endogenous vasopressin release. Thus, the aquaretic effect of SK&F-105494 is strikingly different from the effects of the standard diuretics hydrochlorothiazide or furosemide, and possesses minimal antidiuretic agonist activity.

OPC-31260, a benzazepine derivative (Fig. 21), is a selective nonpeptide V_2 receptor antagonist (YAMAMURA et al. 1992). It has been shown that OPC-31260 inhibited AVP binding to V_1 and V_2 receptors in a competitive manner, whereby it was 100 times more selective for V_2 than for V_1 receptors. Furthermore, OPC-31260 does not possess partial agonistic activity and, after oral administration to rats, caused a dose-dependent diuretic effect, significantly reducing the urine osmolality. Moreover, it seems to be the first orally active aquaretic. OPC-31260 has already entered phase I studies.

Of the opioid receptors three main types can be differentiated: μ, κ and δ. It has been shown that the activation of κ-opioid receptors produced a marked diuresis in normally hydrated rats. Furthermore, it has been demonstrated that κ-opioid agonists produce marked diuretic effects in rats, mice and rhesus monkeys (SLIZGI and LUDENS 1982). Diuresis was due to a decrease in plasma vasopressin levels.

U-50488, *MR-2034* and *RU-51599* are potent κ-agonists producing marked diuretic effects in different animal species and man (Fig. 21). MR-2034 [(−)-(1R, 5R, 9R, 2″S)-5,9-dimethyl-2′-hydroxy-2-tetrahydrofurfuryl-6,7-benzomorphan-D-tartrate] produced an inverted U-shaped dose-effect curve for diuresis in rats, also indicating μ-opioid agonist activity in higher doses. Furthermore, the diuretic effect of MR-2034 was shown to be stereospecific, since its opposite *d*-isomer MR-2035 was markedly less potent. On the other hand, U-50488 and RU-51599 are full κ-agonists without μ-receptor activities (LEANDER et al. 1987; ROUSSEL UCLAF INVESTIGATOR'S BROCHURE 1992). They have the efficacy to suppress vasopressin release

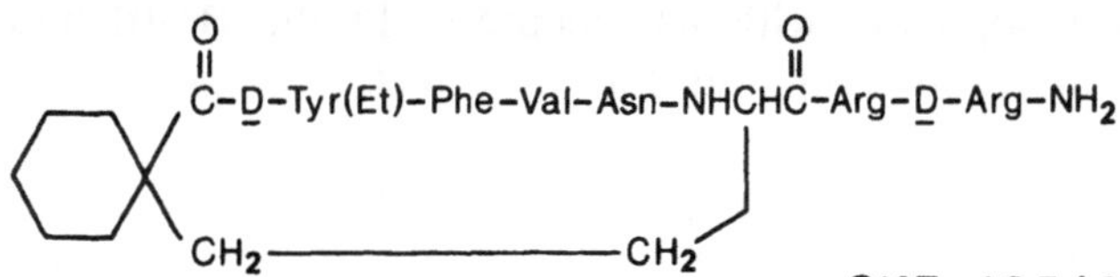

SKF-105494

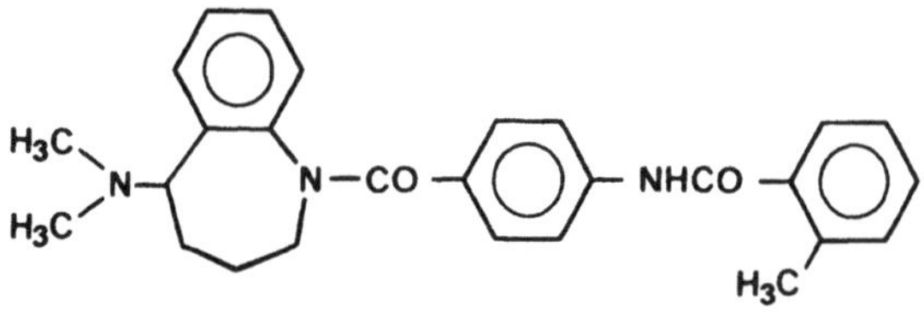

OPC-31260

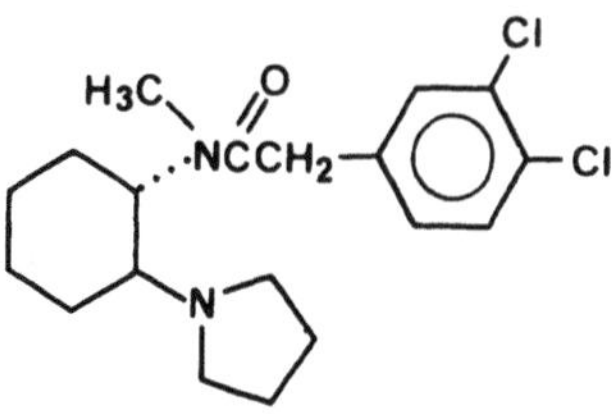

U-50 488

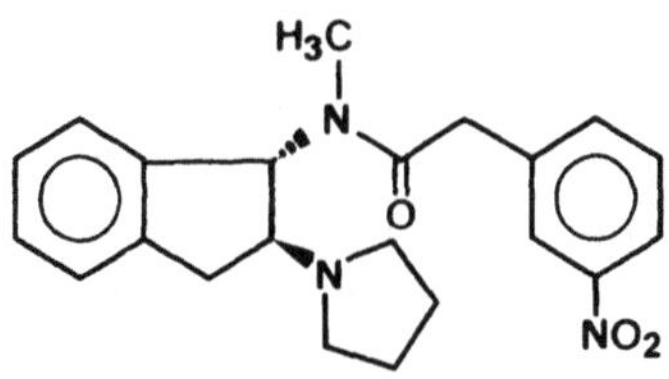

HCl

RU-51 599

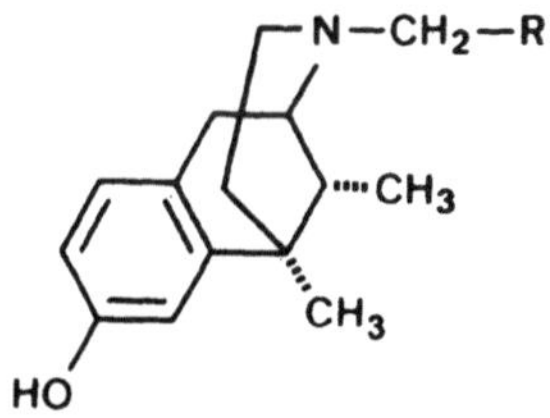

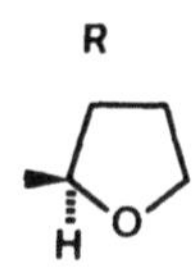

MR-2034

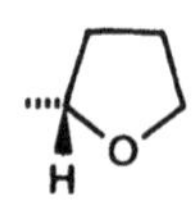

MR-2035

Fig. 21. Aquaretics

from the pituitary to the point of producing a marked diuresis even in a water-deprived condition, which is a natural stimulus for increased plasma vasopressin levels. Moreover, the diuretic effect of κ-agonists can be prevented by exogenously administered vasopressin, indicating that the diuresis is due to reduced plasma levels of vasopressin rather than to any effect on the interaction of vasopressin with its receptors. In the Brattleboro rat, which is deficient in vasopressin, the diuretic effect of κ-agonists was not observed.

Aquaretics may be therapeutically useful for the treatment of diseased states associated with excessive vasopressin and/or excessive body water, e.g., the inappropriate secretion of ADH (SIADH) syndrome, hyponatremia and congestive heart failure.

J. New Aspects: Ion Transport Modulators

In the past decade, the introduction of new sophisticated methods into electrophysiology has revealed that diuretics display their activity primarily by a modulation of renal ion transport systems. Moreover, it has also been demonstrated that diuretics modulate ion transport systems even in extrarenal tissues such as other epithelia, endothelium and smooth muscle cells of diverse origin. Consequently, diuretics per se and their structural congeners have been shown to open new approaches to the therapy of diseases caused by pathological dysregulation of existing ion transport systems in different target organs. Thus, it has recently been demonstrated that systemic effects of some diuretics can be avoided by their topical administration and that inhaled furosemide prevented exercise- and allergen-induced bronchoconstriction in asthmatics (BIANCO et al. 1988). Furthermore, molecular variation of acetazolamide resulted in the novel potent carbonic anhydrase

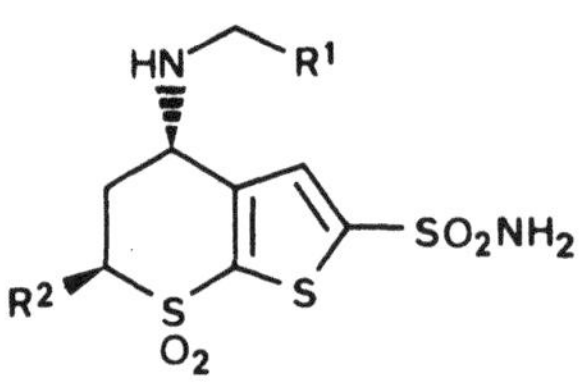

R^1	R^2	
H	$CH(CH_3)_2$	MK-417
CH_3	CH_3	MK-507

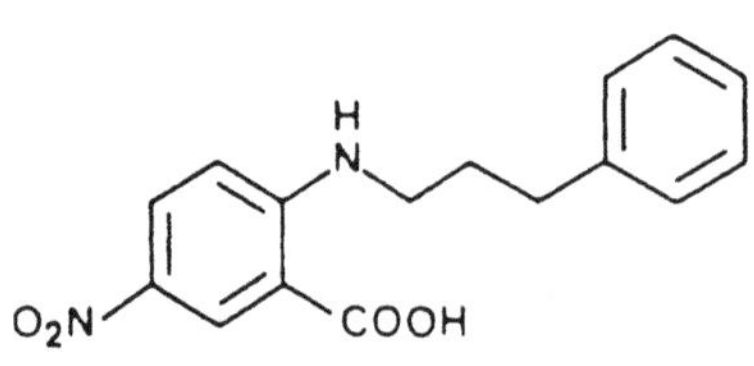

5-nitro-2-(3-phenylpropylamino)-benzoic acid (NPPB)

Fig. 22. Ion transport modulators derived from diuretics

inhibitors *MK-417* and *MK-507*, both exerting ocular pressure lowering activity after topical treatment of glaucoma (Baldwin et al. 1989).

Moreover, ion transport modulators have been derived from diuretic structures which are devoid of renal effects after systemic administration. Replacement of $R^1 = -CH_3$ in etozolin (formula XVI in Fig. 17) by an ethyl group resulted in enhanced excretion of the new compound in the bile. The increased hepatic excretion of the corresponding compound *piprozolin* in the bile causes an elevated production of bile due to inhibition of reabsorption in the bile duct without altering biliary electrolyte concentrations. Piprozolin was introduced onto the market as a choleretic agent with no diuretic activity (Satzinger et al. 1978).

Structural variations of indacrinone led to phenoxyacetic acid derivatives which possess extrarenal ion transport modulating properties (Cragoe et al. 1986). It has been postulated that head injuries result in cellular swelling of brain tissue due to dysregulation of potassium and chloride transport in astrocytes. Prevention of the cellular edema was viewed as a major objective of therapeutic intervention. Moreover, certain indanyloxyalkanoic acid derivatives exhibited marked chloride transport inhibitory activity in brain slices, but exerted little or no effect in the kidney. Furthermore, these compounds exhibited a dose-dependent effect in the in vivo acceleration/deceleration brain edema assay and abolished the perivascular astroglial swelling.

The blockers of epithelial chloride channels have been obtained from structural variations of furosemide derivatives. The chloride channel blockers, for example, *5-nitro-2-(3-phenylpropylamino) benzoic acid* (NPPB) and its congeners, have been proven to be potential antidiarrheal and antiasthmatic drugs by blocking the chloride channels in colon and trachea (Greger et al. 1991; Hropot et al. 1991). There are now indications that recently discovered nondiuretic chloride channel blockers are a novel class of ion-modulating drugs which may provide new therapeutic approaches to bronchial asthma, brain edema and/or diarrhea.

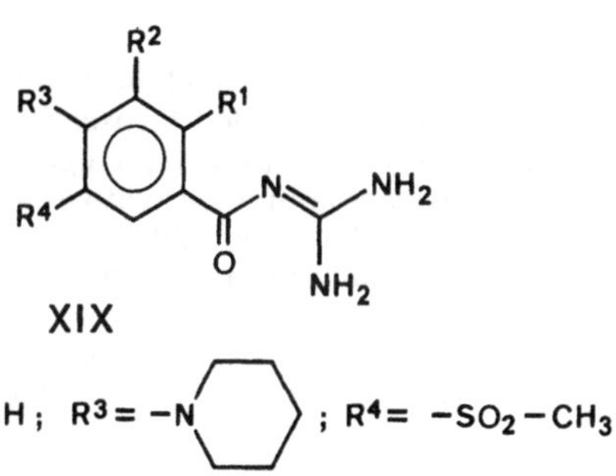

Fig. 23. Na^+/H^+ exchange-inhibiting benzoylguanidine HOE-694

Amiloride (formula XI in Fig. 10) acts as a potassium-retaining agent by blocking the epithelial sodium channels in the late distal tubules of the kidney. The alkylated product of amiloride, e.g., *ethyl-isopropyl-amiloride* (EIPA), formula XI: $R^1 = C_2H_5$, $R^2 = CH(CH_3)_2$, $R^3 = H$, $X = Cl$, was shown to be a highly potent and selective inhibitor of the Na^+/H^+ exchanger as compared to amiloride (VIGNE et al. 1983).

Further structural variations led very recently to the benzoylguanidines with Na^+/H^+ inhibitory activity, of which HOE-694 (Fig. 23) exerted very favorable protective effects against ischemically induced heart damage (SCHOLZ et al. 1993).

Although the importance of diuretics in the physician's armamentarium is obvious, the zenith of diuretic research seems to have been crossed. However, the development of very new chemical agents derived from traditional diuretics has opened the door to new horizons of tissue-selective ion transport modulation.

Acknowledgments. We wish to thank Peter Hainz for excellent technical assistance and Grace McConaghy for correcting the manuscript.

References

Ackerman DM, MacLean AG, Brady MM, Kraml M (1990) AY-31906, a novel high ceiling diuretic with potassium sparing properties. In: Puschett JB, Greenberg A (eds) Diuretics III: chemistry, pharmacology, and clinical applications. Elsevier Science, New York, pp 94–96

Allison NL, Dubb JW, Ziemniak JA, Alexander F, Stote RM (1987) The effect of fenoldopam, a dopaminergic agonist, on renal hemodynamics. Clin Pharmacol Ther 41:282–288

Axelrod DR, Pitts RF (1952) The relationship of plasma pH and anion pattern to mercurial diuresis. J Clin Invest 31:171–179

Baldwin JJ, Ponticello GS, Sugrue MF (1989) MK-417. Drugs Future 14:636–638

Beyer KH (1958) The mechanism of action of chlorothiazide. Ann NY Acad Sci 71:363–379

Beyer KH, Baer JE (1961) Physiological basis for the action of newer diuretic agents. Pharmacol Rev 13:517–562

Beyer KH, Baer JE, Michaelson JK, Russo HF (1965) Renotropic characteristics of ethacrynic acid: a phenoxy-acetic saluretic-diuretic agent. J Pharmacol Exp Ther 147:1–22

Bianco S, Vaghi A, Robuschi M, Pasargiklian M (1988) Prevention of exercise-induced bronchoconstriction by inhaled furosemide. Lancet ii:252–255

Bouley E, Teulon JM, Cazes M, Cloarec A, Deghenghi R (1986) p-(Thienylcarbonyl)amino-phenoxy propanolamine derivatives as diuretic and beta-adrenergic receptor blocking agents. J Med Chem 29:100–103

Brater DC (1979) Renal sites of action of azosemide. Clin Pharmacol Ther 25:428–434

Brooks DP, Koster PF, Albrightson CR, Huffman WF, Moore ML, Stassen FL, Schmidt DB, Kinter LB (1989) Vasopressin receptor antagonism in rhesus monkey and man: stereochemical requirements. Eur J Pharmacol 160:159–162

Burg M, Stoner L, Cardinal J, Green N (1973) Furosemide effect on isolated perfused tubules. Am J Physiol 225:119–124

Campbell DB, Moore RA (1981) The pharmacology and clinical pharmacology of indapamide. Postgrad Med J 57 [Suppl 2]:7–17
Cragoe EJ Jr, Woltersdorf OW, Bicking JB, Kwong SF, Jones JH (1967) Pyrazine diuretics. II. N-amidino-3-amino-5-substituted 6-halopyrazines. J Med Pharm Chem 10:66–75
Cragoe EJ Jr, Woltersdorf OW Jr, Gould NP, Pietruszkiewicz AM, Ziegler C, Sakurai Y, Stokker GE, Anderson PS, Bourke RS, Kimelberg HK, Nelson LR, Barron KD, Rose JR, Szarowski D, Popp AJ, Waldman JB (1986) Agents for the treatment of brain edema 2. [(2,3,9,9a-tetrahydro-3-oxo-9a-substituted-1H-fluoren-7-yl)oxy]alkanoic acids and some of their analogues. J Med Chem 29:825–841
Cushny AR, Lambie CG (1921) The action of diuretics. J Physiol (Lond) 55:276–286
Davenport H, Wilhelmi AE (1941) Renal carbonic anhydrase. Proc Soc Exp Biol Med 48:53–56
DeGasparo M, Whitebread SE, Preiswerk G, Jeunemaitre X, Corvol P, Menard J (1989) Antialdosterones: incidence and prevention of sexual side effects. J Steroid Biochem 32:223–227
Delarge J (1988) Chemistry and pharmacological properties of the pyridine-3-sulfonylurea derivative torasemide. Arzneimittelforschung/Drug Res 38(I):144–150
Dodson RM, Tweit RC (1959) Addition of alkanethiolic acids to delta^{1,4}-3-oxo- and delta^{4,6}-3-oxosteroids. J Am Chem Soc 81:1224
Engel K, Epstein T (1931) Die Quecksilberdiurese. Ergebn Inn Med Kinderheilkd 40:187–261
Fagard R, Lijnen P, Moerman E, Staessen J, Amery A (1987) Acute haemodynamic and humoral responses to felodipine and metoprolol in mild hypertension. Eur J Clin Pharmacol 32:71–75
Fanelli GM Jr, Bohn DL, Scriabine A, Beyer Jr DH (1977) Saluretic and uricosuric effects of (6,7-dichloro-2-methyl-1-oxo-2-phenyl-5-indanyloxy) acetic acid (MK-196) in the chimpanzee. J Pharmacol Exp Ther 200:402–412
Feit PW (1971) Aminobenzoic acid diuretics 2,4-substituted-3-amino-5-sulfamyl-benzoic acid derivatives. J Med Chem 14:423–439
Geck P, Pietrzyk C, Burckhardt BC, Pfeiffer B, Heinz E (1980) Electrically silent cotransport of Na^+, K^+ and Cl^- in Ehrlich cells. Biochim Biophys Acta 600:432–447
Gennari FJ, Kassirer JP (1974) Osmotic diuresis. N Engl J Med 291:714–720
Greger R, Schlatter E, Lang F (1983) Evidence of electroneutral sodium chloride cotransport in the cortical thick ascending limb of Henle's loop of rabbit kidney. Pflugers Arch 396:308–314
Gerber A, Weidmann P, Bianchetti MG, Ferrier C, Laederach K, Mordasini R, Riesen W, Bachmann C (1985) Serum lipoproteins during treatment with the antihypertensive agent indapamide. Hypertension 7 [Suppl 2]:II164–II169
Greger R, Nitschke RB, Lohrmann E, Burhoff I, Hropot M, Englert HC, Lang HJ (1991) Effects of arylaminobenzoate-type chloride channel blockers on equivalent short-circuit current in rabbit colon. Pflugers Arch 419:190–196
Greven J, Heidenreich O (1978) Effects of ozolinone, a diuretic active metabolite of etozolin, on renal function. I. Clearance studies in dogs. Naunyn Schmiedebergs Arch Pharmacol 304:283–287
Guinot Ph, Frölich JC (1985) Study of the effects of cycletanine on prostanoids. Arzneimittelforschung/Drug Res 35(11):1714–1716
Haberey M, Buse M, Losert W, Nishino Y (1986) Mespirenone a novel aldosterone antagonist. Naunyn Schmiedebergs Arch Pharmacol 334 [Suppl]: Abstr 109
Helgeland A (1980) The Oslo Study. Am J Med 69:725–732
Higaki J, Harada H, Tonda K, Hirata M (1989) Chemical structure and toxicity of diuretics in isolated hepatocytes. Pharmacol Toxicol 65(1):21–24

Hropot M, Muschaweck R (1992) Chemistry and chemical classification of diuretics. In: Reyes AJ (ed) Progress in pharmacology and clinical pharmacology, vol 9. Fischer, Stuttgart, p 3
Hropot M, Englert HC, Granzer E, Kerékjártó von B, Klaus E, Lang HJ, Scholz W (1990) HOE 708, a novel diuretic with hypolipidemic properties. In: Puschett JB, Greenberg A (eds) Diuretics III: chemistry, pharmacology, and clinical applications. Elsevier Science, New York, pp 117–123
Hropot M, Lang HJ, Alpermann HG, Hainz P (1991) Neurokinin A-induced chloride secretion is inhibited by a chloride channel blocker in canine tracheal epithelium. Naunyn Schmiedebergs Arch Pharmacol 343 [Suppl]:R46
Johnston PA, Kau ST (1993) A micropuncture study on the renal site of action of ICI 206,970, a unique eukalemic diuretic. J Pharmacol Exp Ther 264:604–608
Kau ST (1992) Basic pharmacology and pharmacological classification of diuretics. In: Reyes AJ (ed) Progress in pharmacology and clinical pharmacology, vol 9. Fischer, Stuttgart, pp 33–113
Keddie JR, Collins MG (1989) The diuretic and saliuretic properties of PD 116948, a potent adenosine receptor antagonist. Br J Pharmacol 97 [Suppl]:502P
Kirsten R, Alexandridis T, Heintz B, Hopf R, Nelson K, Pooth R, Sievert H (1986) Acute influence of piretanide on hemodynamics and vasoactive hormones in patients with congestive heart failure. In: Puschett JB, Greenberg A (eds) Diuretics II: chemistry, pharmacology, and clinical applications. Elsevier Science, New York, p 365
Kobayashi T, Mizumoto H, Karasawa A, Kubo K (1992) Diuretic and antihypertensive properties of KW-3902, a novel adenosine A1 receptor antagonist. Jpn J Pharmacol 58 [Suppl 1]:195p
Kubota J, Kubo S, Nishimura H, Ureyama M, Kino M, Nakayama A, Hara M, Kawamura K (1989) Cardiorenal effects of an orally active dopamine prodrug (T-870) in patients with congestive heart failure. J Cardiovasc Pharmacol 14:53–57
Lant A (1985a) Diuretics, clinical pharmacology and therapeutic use, part I. Drugs 29:57–87
Lant A (1985b) Diuretics, clinical pharmacology and therapeutic use, part II. Drugs 29:162–188
Lawrence JR, Ansari AF, Elliot HL, Sumner DJ, Brunton GF, Whiting B, Whitesmith R (1978) Kinetic and dynamic comparison of piretanide and furosemide. Clin Pharmacol Ther 23:558–565
Leander JD, Hart JC, Zerbe RL (1987) κ agonist-induced diuresis: evidence for stereoselectivity, strain differences, independence of hydration variables and a result of decreased plasma vasopressin levels. J Pharmacol Exp Ther 242:33–39
Loew D, Meng K (1977) The renal mechanism of Bay-g-2821. Pharmatherapeutica 1:333–340
Maass AR, Snow IB, Erickson R (1978) Ticrynafen: an antihypertensive, diuretic, uricosuric agent. In: Cragoe EJ Jr (ed) Diuretic agents. ACS, Washington DC, p 84 (ACS symposium series 83)
Mann T, Keilin D (1940) Sulfanilamide as a specific inhibitor of carbonic anhydrase. Nature 146:164–165
Manning M, Lammek B, Kolodziejczyk AM, Seto J, Sawyer WH (1981) Synthetic antagonists of in vivo antidiuretic and vasopressor responses to arginine vasopressin. J Med Chem 24:701–706
Maren TH, Mayer E, Wadsworth BC (1954) Carbonic anhydrase inhibition. I. The pharmacology of Diamox; 2-acetylamino-1,3,4-thiadiazole-5-sulfonamide. Bull Johns Hopkins Hosp 95:199–243
Merkel W, Bormann D, Mania D, Muschaweck R, Hropot M (1976) Piretanide (Hoe 118) a new high-ceiling salidiuretic. Eur J Med Chem 11(5):399–406
Miller GE, Danzig LS, Talbott JH (1951) Urinary excretion of uric acid in the dalmatian and non-dalmatian dog following administration of diodrast, sodium salicylate and a mercurial diuretic. Am J Physiol 164:155–158

Muschaweck R, Hajdú P (1964) Die salidiuretische Wirksamkeit der Chlor-N-(2-furylmethyl)-5-sulfamyl-anthranilsäure. Arzneimittelforschung/Drug Res 14(I): 44–47
Mutschler E, Knauf H, Finke M, Kraft H, Möhrke W, Priewer H, Ullrich F, Völger KD, Vollmer G (1987) Pharmacodynamics and pharmacokinetics of investigational triamterene derivatives. In: Puschett JB, Greenberg A (eds) Diuretics II: chemistry, pharmacology, and clinical applications. Elsevier Science, New York, p 25
Nakamura M, Kawabata T, Itoh T, Miyata K, Harada H (1990) Stereoselective saluretic effect and localization of renal tubular secretion of enantiomers of S-8666, a novel uricosuric antihypertensive diuretic. Drug Dev Res 19:23–36
Novello FC, Sprague JM (1957) Benzothiadiazine dioxides as novel diuretics. J Am Chem Soc 79:2028–2029
Ozawa N, Mori N (1986) The effects of a dopamine prodrug TA 8704 on improvement of kidney function. Jpn J Pharmacol 40 [Suppl]:127
Pamnani MB, Haddy FJ, Bryant HJ, Swindall BT, Hom GJ, Johnston J, Cragoe EJ Jr (1986) Effects of 6-iodo-amiloride, a sodium channel blocker, on cardiovascular parameters in spontaneously hypertensive Wister-Kyoto rats. J Hypertens 4 [Suppl 3]:S491–S493
Plattner JJ, Fung AKL, Parks JA, Pariza RJ, Crowley SR, Pernet AG, Bunnell PR, Dodge PW (1984) Substituted 5,6-dihydrofurol(3,2-f)-1,2-benzisoxazole-6-carboxylic acids: high-ceiling diuretics with uricosuric activity. J Med Chem 27:1016–1026
Pocchiari F, Pataccini R, Castelnovo P,Longo A, Casagrande C (1986) Ibopamine, an orally active dopamine-like drug: metabolism and pharmacokinetcs in rats. Arzneimittelforschung/Drug Res 26(IIa):334–340
Roussel Uclaf Investigators' Brochure (1992) Internal report. Roussel Uclaf, Romainville
Sandström PE, Sehlin J (1988) Furosemide reduces insulin release by inhibition of Cl^- and Ca^{2+} fluxes in β-cells. Am J Physiol 255:E591–E596
Satzinger G, Herrmann M, Vollmer KO, Merzweiler A, Gomahr H, Heidenreich O, Greven J (1978) Etozolin: a novel diuretic. In: Cragoe EJ Jr (ed) Diuretic agents. ASC, Washington DC, pp 155–189 (ASC symposium series 83)
Saxl P, Heilig R (1920) Über die diuretische Wirkung von Novasurol und anderen Quecksilberinjektionen. Wien Klin Wochenschr 42:943–944
Scholz W, Albus U, Lang HJ, Linz W, Martorana PA, Englert HC, Schölkens BA (1993) Hoe 694, a new Na^+/H^+ exchange inhibitor and its effects in cardiac ischaemia. Br J Pharmacol 109:562–568
Schwartz WB (1949) The effect of sulfanilamide on salt and water excretion in congestive heart failure. N Engl J Med 240:173–177
Slizgi GR, Ludens JH (1982) Studies on the nature and mechanism of diuretic activity of the opioid analgesic ethylketazocine. J Pharmacol Exp Ther 220:585–591
Strauss MB, Southworth H (1938) Urinary changes due to sulfanilamide administration. Bull Johns Hopkins Hosp 63:41–45
Sturm K, Siedel W, Weyer R, Ruschig H (1966) Zur Chemie des Furosemids. I. Synthesen von 5-sulfamoyl-anthranilsäure-Derivaten. Chem Ber 99:328–344
Sturm K, Muschaweck R, Hropot M (1983) 5-Sulfamoylorthanilic acids, a sulfonamide series with salidiuretic activity. J Med Chem 26:1174–1187
Takagi S, Takayama S, Onodera T (1990) Effects of DR-3438, tienilic acid, indacrynone and furosemide on primary cultured hepatocytes from rats. In: Puschett JB, Greenberg A (eds) Diuretics III: chemistry, pharmacology, and clinical applications. Elsevier Science, New York, p 800
Tanaka M, Hashimoto H, Tanaka S, Masura M, Shirasaki Y, Akashi A (1990) Pharmacological profile of DR-3438, a new uricosuric diuretic. In: Puschett JB, Greenberg A (eds) Diuretics III: chemistry, pharmacology, and clinical applications. Elsevier Science, New York, p 150

Vigne P, Frelin C, Cragoe EJ Jr, Lazdunski M (1983) Ethylisopropyl-amiloride: a new and highly potent derivative of amiloride for the inhibition of the Na^+/H^+ exchange system in various cell types. Biochem Biophys Res Commun 116:86–90

Wales JK, Grant A, Wolff FW (1968) Studies on the hyperglycemic effects of nonthiazide diuretics. J Pharmacol Exp Ther 159:229–235

Wangemann Ph, Braitsch R, Greger R (1987) The diuretic effect of muzolimine. Pflugers Arch 410:674–676

Wiebelhaus VD, Brennan FT, Sosknowski G, Maas AR, Weinstock J, Bender AD (1967) The natriuretic and diuretic characteristics of triamterene in the dog. Arch Int Pharmacodyn Ther 169:429–451

Yamaguchi I, Nishiyama S, Akimoto Y, Yoshikawa M, Nakajima H (1989) A novel orally active dopamine prodrug TA-870. I. Renal and cardiovascular effects and plasma levels of free dopamine in dogs and rats. J Cardiovasc Pharmacol 13:879–886

Yamamura Y, Ogawa H, Yamashita H, Chihara T, Miyamoto H, Nakamura S, Onogawa T, Yamashita T, Hosokawa T, Mori T, Tominaga M, Yabuuchi Y (1992) Characterization of a novel aquaretic agent, OPC-31260, as an orally effective, nonpeptide vasopressin V_2 receptor antagonist. Br J Pharmacol 105: 787–791

Yamasaki F, Notsu T, Kitamoto A, Shinkawa T, Inhabe H, Mochida E, Yoshitomi K, Imai M (1990) A novel potent diuretic, M 17055, having both loop and distal actions. In: Puschett JB, Greenberg A (eds) Diuretics III: chemistry, pharmacology, and clinical applications. Elsevier Science, New York, p 111

CHAPTER 5

Metabolism of Diuretics

W. MÖHRKE and F. ULLRICH

A. Introduction

The pharmacodynamics and pharmacokinetics of a substance are often closely interrelated. This applies especially to diuretic agents, which have a common site of action and elimination. In addition to renal elimination, hepatobiliary excretion and biotransformation play an important role in determining the pharmacodynamic effects of diuretics. Generally biotransformation of diuretics is not desirable unless the drug concerned is administered as a pro-drug and thus needs converting to the active compound. This chapter reviews the metabolism of the major diurectic agents presently used in clinical medicine. Diuretics can be divided into four groups according to their site of action (see Chaps. 7–9, this volume):

1. Carbonic anhydrase inhibitors, which inhibit $Na^+HCO_3^-$ reabsorption in the proximal tubule
2. Loop diuretics, which block the $Na^+2Cl^-K^+$ cotransporter in the ascending limb of Henle's loop
3. Thiazides or thiazide-like diuretics, which block Na^+Cl^- reabsorption in the early distal convoluted tubule
4. Potassium-retaining diuretics, which block Na^+ channels and thereby reduce K^+ secretion in the late distal convoluted tubule and the collecting duct

Figures 1–19 show the chemical structures of the diurectics included in this review and in Table 1 the most relevant physicochemical and pharmacokinetic data are included. Before discussing the metabolism of those compounds, however, we give a brief overview of the most relevant metabolic reactions in general.

B. Biotransformation

The lipophilic properties of drugs allow their rapid permeation through the lipid bilayer of cell membranes. This guarantees their absorption by the intestinal wall and uptake by other cells. At the same time, a high degree of lipophilia causes slow excretion from the body because, after filtration at the renal glomerulus, the absorbed drugs are reabsorbed by diffusion through

Table 1. Physicochemical and pharmacokinetic data

Drug	Mol. weight	pK_a	Log P	Bioavailability (%)	Protein binding (%)	Elimination half-life (h)	Distribution volume (l/kg)	Cl_R (ml/min)
Acetazolamide	222.2	7.4			92	4		
Furosemide	330.7	3.8	−1.60	50–65	98	1	0.17–0.27	94
Azosemide	370.7	4.0		18	96	2.7		20–40
Bumetanide	364.6	3.6	−0.37	95	95	1–1.5	0.13	97–126
Piretanide	362.4	4.1		90	96	1.5		90–100
Etozolin	284.2		2.3	>90	35	2.5(Ozo 9)		
Ethacrynic acid	303.1	3.5	−0.11		95			
Torasemide	348.4	≈7.0	0.57	80–90	98	3–4		12
Bendrofluazide	421.7	8.5		>90	94	3–9	1.2–1.5	
Chlorothiazide	295.7	6.7/9.5	−1.10	33–65	70	15–27	0.3	
Hydrochlorothiazide	297.7	7.9/9.2	−0.43	60	65	7		340
Chlorthalidone	338.8	9.4		64	75	24–50	3–5	69
Indapamide	365.8	8.3	1.51	94	79	18–24	1.60	5
Mefruside	382.9	9.5	1.54	>90	64	7	4–7	
Xipamide	354.8	4.8/10.0	1.32	73	99	5	0.2–0.28	12
Amiloride	302.1	8.7		50	>40	6–9	5	9.7
Triamterene	253.3	6.2		52	55	3–5	13.4	220
Spironolactone	416.1			60–90	98(Can)20	14		

the renal tubular cells, which may lead to an accumulation of drugs in the body.

Accumulation in the body is prevented by metabolic reactions which transform exogenous and endogenous substances to hydrohilic, most often nontoxic, compounds that can be rapidly eliminated into urine or bile. Thus, biotransformation not only accounts for drug elimination but also often results in inactivation of the compound. However, many metabolites possess pharmacological activity. These metabolites may exert effects similar to or different from the parent molecule. Sometimes, a metabolic pathway is used for drug excitation and a pro-drug is administered.

C. Patterns of Biotransformation

For most substances biotransformation occurs in two steps: Phase I reactions usually convert the parent drug to a more polar substance by introduction of, for example, amino, hydroxy or sulfhydryl groups based on oxidizing, reducing or hydrolyzing reactions. In the second step (phase II) the resulting phase I metabolites are coupled via the introduced reactive group with an endogenous substrate, such as acetic acid, amino acid, glucuronic acid or sulfuric acid.

Phase I. The most important phase I reaction is the oxidation of lipophilic drugs, which are catalyzed by the microsomal monooxygenase system. This system consists of a family of hydroxylating hemoproteins, named cytochrome P450 because of an absorption maximum at 450 nm when exposed to carbon monoxide, and of a flavoprotein (NADPH-cytochrome C reductase).

The electron transport scheme involved in microsomal drug oxidation is shown in Fig. 1. At first, a drug (substrate, SH) binds with oxidized cytochrome P450. In the next step the resulting substrate-cytochrome complex is reduced by the reductase. The reduced complex then combines with molecular oxygen. A donor system, which has not yet been fully investigated, provides a second electron and two hydrogen ions to produce an oxidized substrate-cytochrome complex and water. At the end of this cycle, the complex dissociates to cytochrome P450 and an oxidized metabolite.

About 70 biotransformation hemoproteins have been characterized (Nebert et al. 1989). In contrast to other enzymes, they have no marked substrate specificity but this guarantees that different substrates can be oxidized, e.g., aliphatic and aromatic hydrocarbons are converted to the corresponding alcohols or phenols, primary and secondary amines to hydroxylamines, sulfhydryl groups to sulfoxides, sulfones or disulfides, and methyl groups to hydroxymethyl groups or split off as formaldehyde.

Phase I reducing reactions are not as important as oxidation. For example, ketones and aldehydes are converted to the corresponding alcohols and nitro groups to amines, respectively. The reductases concerned are not

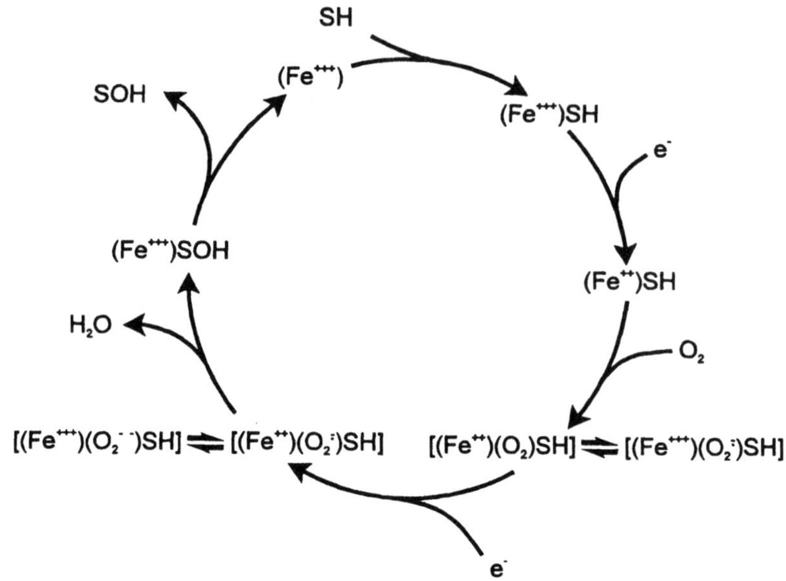

Fig. 1. Oxidation of a substrate by the microsomal monooxygenase system

only located in the microsomes and the cytosol, but part of the reductive metabolism is carried out in the intestine by bacteria.

Phase II. The second step of biotransformation is characterized by conjugation reactions. A phase I metabolite or any other endogenous or exogenous substrate is coupled via a reactive group (e.g., hydroxy, amino, sulfhydryl) to an endogenous compound. The conjugation can occur in two ways: An endogenous substance (e.g., glucuronic acid, acetic acid) has to be activated (e.g., with uridine diphosphate, UDP, coenzyme A, CoA) to be conjugated with an exogenous substrate, or a phase I metabolite has to be activated by forming an activated CoA derivate to be coupled with an endogenous compound (e.g., glycine or glutamine).

The most important phase II reaction in the human body is the conjugation of glucuronic acid, which is transferred from UDP-glucuronic acid, the activated form, to its substrate. The catalyst is a glucuronyltransferase, which is a highly active enzyme in the liver, but which is also found in numerous other organs and tissues. Comparable to cytochrome P450, different subtypes exist in the liver of different individuals. Its substrate specificity, however, is much higher than that of the hemoproteins. Glucuronic acid can be coupled with carbonic acids, aliphatic or phenolic alcohols, amines and sulfonamides. The resulting conjugates are relatively strong acids with pK_a values of 3–4, which are completely dissociated at physiological blood and tissue pH.

The enzymes of both biotransformation phases are located closely together in the endoplasmic reticulum, so that both steps can occur in sequence. The final metabolite is transported to the cavity of the endoplasmic reticulum and then transferred to the outside of the cell.

D. Biotransformation of Diuretics

I. Carboanhydrase Inhibitors

1. Acetazolamide

Although carboanhydrase inhibitors have only limited usefulness as diuretic agents, the most important substance of this class, acetazolamide (Fig. 2), has made a great contribution to the development of modern diuretics and to the understanding of renal physiology. Moreover, recent investigations have indicated the usefulness of acetazolamide combined with other potent diuretics in overcoming Na^+ hyperreabsorption in the state of reduced effective arterial blood volume (KNAUF and MUTSCHLER 1991). In rats acetazolamide is metabolized of nearly 70% of the administered drug (MAREN et al. 1954), and the structure of the metabolites is still unknown. In humans the drug is rapidly absorbed from the gastrointestinal tract. Nearly 90% of the dose is found in the urine and is excreted virtually unchanged.

II. Loop Diuretics

1. Furosemide

The biotransformation of this drug has been widely investigated in amimals and man (MIKKELSEN and ANDREASEN 1977; SMITH et al. 1980). However, due to its relative instability to light and acidic conditions, the study of its metabolism is very complicated and the data reported so far have been controversial.

HAJDU and HAEUSSLER (1964) reported that 2-amino-4-chloro-5-sulfamoyl-anthranilic acid (CAS, see Fig. 3) is the only metabolite of furosemide in man, dogs, and rats. These authors used paper chromatography and spectrofluorometry as analytical methods to determine unlabeled furosemide and its metabolites in the urine. YAKATAN et al. (1976) confirmed these results after administering single and repeated doses of ^{35}S-labeled

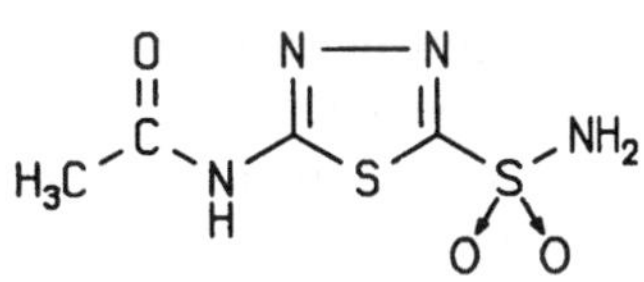

Fig. 2. Acetazolamide

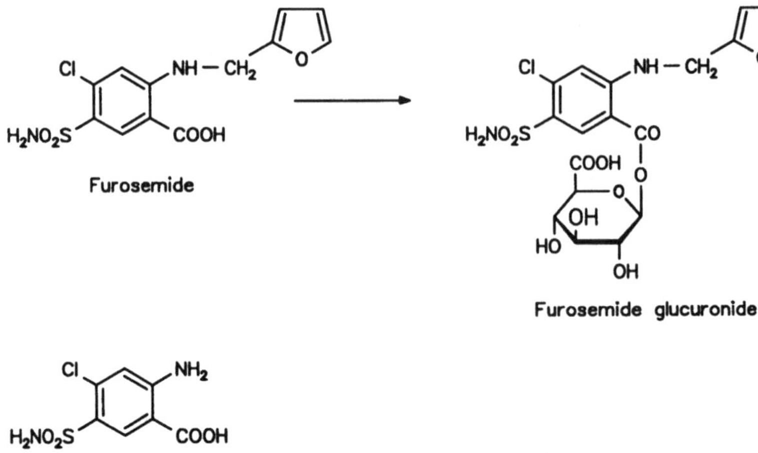

4-Chloro-5-sulfamoyl-anthranilic acid (CSA)

Fig. 3. Metabolism of furosemide

furosemide to dogs and monkeys. They demonstrated the similarity in metabolism of these two species, with 80% of the drug excreted unchanged in the urine, 7% excreted as 4-chloro-5-sulfamoyl-anthranilic acid and another 7% as two unidentified metabolites. Other investigators have also reported CSA to be a metabolite of furosemide (Andreasen et al. 1978; Perez et al. 1979). However, other authors using paper chromatography or thin layer chromatography (TLC) have been unable to support these findings (Calesnick et al. 1966; Kindt and Schmid 1970; Beermann et al. 1975b). Moreover, Smith et al. (1980) showed that CSA might be formed due to the analytical method used. By using high-performance liquid chromatography (HPLC) without prior extraction or derivatization they found no evidence of CSA. In man, the major metabolic pathway has been suggested to be the formation of an inactive glucuronic ester (see Fig. 3). There is no evidence of liver-dependent oxidation but only a phase II reaction. More than 50% of the parent compound is excreted in the urine; the glucuronic conjugate is excreted in urine and bile, the percentage of urinary excretion being 7%–15%. However, it should be mentioned that direct proof of the structure of this glucuronide conjugate is still lacking, but was only deduced from hydrolysis with β-glucuronidase.

In patients with liver cirrhosis this glucuronidation of furosemide is not affected. This can be explained by the fact that in severe hepatic disease phase II reactions are not reduced to the same extent as in phase I reactions. Animal experiments have also supported this thesis. It was shown by Verbeeck et al. (1981) that functional hepatectomy by devascularization did not influence the amount of glucuronidation. Minor differences were found with respect to elimination half-life, volume of distribution and serum protein binding. Allgulander et al. (1980) showed that the elimination half-life was increased (from 0.6 to 4.3h). In addition, the serum half-life was correlated with the volume of distribution. The serum protein binding was decreased to 96.9% in patients compared to 98.1% in healthy subjects. However, renal and nonrenal clearance and consequently total clearance did not change due to the intact glucuronidation.

In contrast to patients with liver cirrhosis, in patients with hepatitis the pharmacokinetics of this drug is significantly altered. However, the prolongation of its half-life is not based on an alteration in metabolism but is due to the impaired biliary excretion of these compounds (KELLER et al. 1981). It is suggested that the carrier-mediated transfer through the canicular cell membrane into the biliary tract of either the native drug or its metabolites is limited (KNAUF et al. 1992).

2. Bumetanide

As with furosemide, the biotransformation of bumetanide has been intensively investigated in animals and man. Pronounced species differences have been found in its metabolism.

In rats, bumetanide possesses only weak diuretic activity due to its extensive metabolism (HALLADAY et al. 1978a). The drug is nearly completely metabolized by oxidization of the side chain (Fig. 4). All the metabolites identified so far have been devoid of any diuretic activity. After intravenous administration, more than 50% of the dose is excreted as the 4′ alcohol and the corresponding acid. Interestingly, inhibition of the microsomal drug metabolism with piperonyl butoxide leads to a significant prolongation of serum half-life and increases diuretic activity twofold. Further evidence of the metabolism of bumetanide was obtained from a study investigating the uptake of bumetanide in rat liver cells (PETZINGER et al. 1989). Besides investigating the mechanism of uptake the authors reported two hydroxylated metabolites and at least one conjugate which were separated by TLC. Some of the metabolites found in the rat such as the *N*-desbutyl derivative could also be detected in mouse, rabbit and dog. However, especially the dog excretes large amounts of bumetanide unchanged (FREY 1975; COHEN 1981).

In man and dogs, early studies with ^{14}C-labeled bumetanide suggested an insignificant metabolism (DAVIES et al. 1974). Other investigators, however, also using radiolabeled drug, showed that about 80% of the administered dose can be recovered in urine, with 40% being metabolized and the rest excreted as unaltered drug (HALLADAY et al. 1978b; PENTIKAINEN et al. 1977). Another 14% was excreted in the bile, most of this appearing as metabolites and very little as unchanged drug. Hydroxylation at each of the carbon atoms in the *N*-butyl side chain is the primary metabolic pathway (Fig. 4). Furthermore, phase II conjugation with glucuronic acid was demonstrated. The primary urinary metabolite was the 3′ alcohol of the *N*-butyl chain; the primary biliary metabolite was the 2′ alcohol.

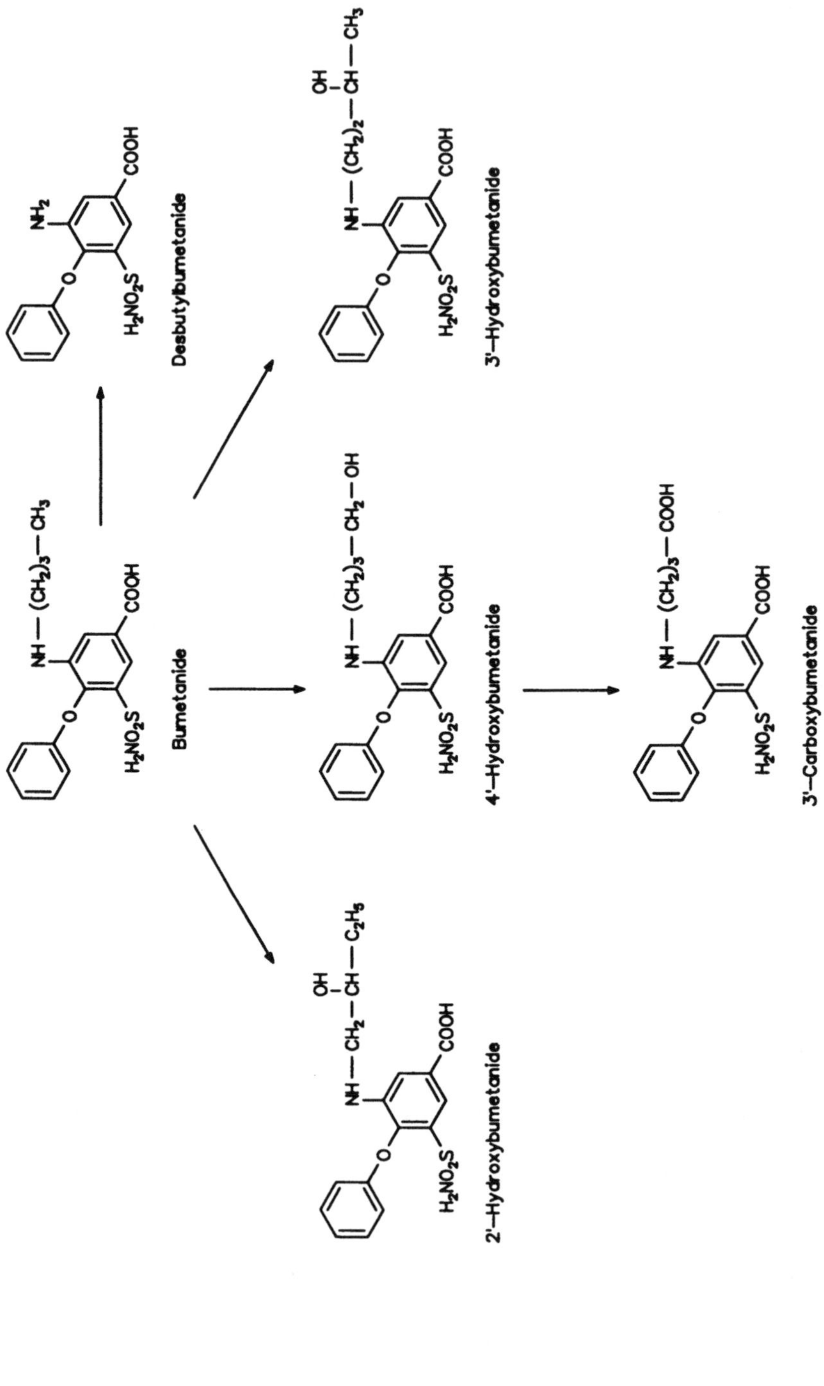

Fig. 4. Metabolism of bumetanide

In patients with impaired kidney function, the serum levels of native bumetanide are increased up to twice those in healthy subjects. In addition, the volume of distribution is elevated and the plasma half-life is prolonged (LAU et al. 1986; PENTIKAINEN et al. 1984, 1985). Biliary excretion becomes the major route of elimination, so up to 90% of the drug is eliminated by this route (MARCANTONIO et al. 1983). In hepatic disease both renal and nonrenal clearance is decreased.

3. Piretanide

Little information on the metabolism of this drug has been published (BRATER et al. 1983). In healthy subjects the bioavailability of piretanide is at least 80%. The biliary excretion accounts for about 40% of the dose, while up to 50% is recovered in the urine as unchanged drug (MEYER et al. 1983). CLISSOLD and BROGDEN (1985) isolated five metabolites using TLC. One of these metabolites was identified and used in clinical studies by HEPTNER et al. (1984) (Fig. 5). Two more metabolites were quantified by SPAHN-LANGGUTH et al. (1993). In contrast to furosemide, no glucuronic acid derivate was detected.

4. Azosemide

Pharmacokinetic studies with azosemide have been conducted in rat, rabbit and man. However, most of the studies performed did not focus on the metabolism of this drug.

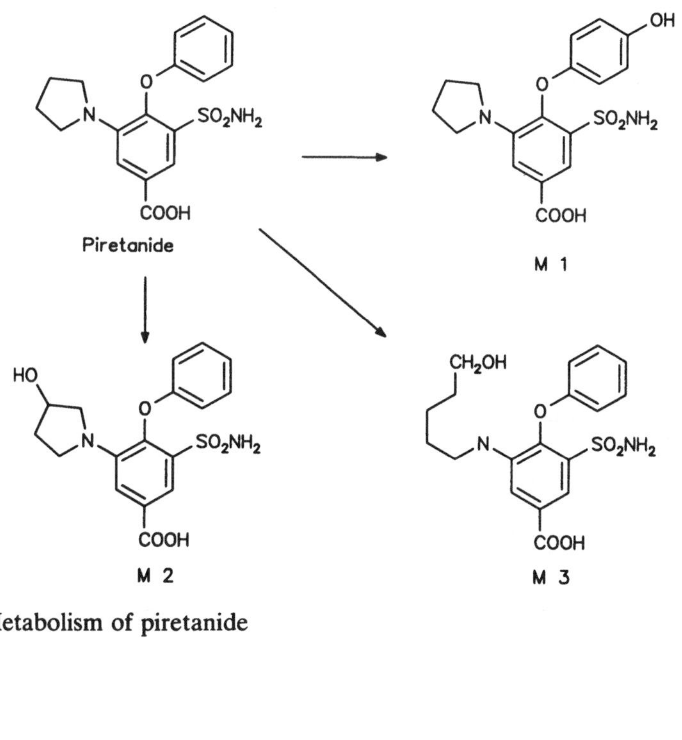

Fig. 5. Metabolism of piretanide

Studies in man have shown that azosemide has a low bioavailabilty of 10%–20% (BRATER et al. 1983). A marked first-pass effect may partly account for this effect. In the liver the drug is metabolized in two pathways. One part of the absorbed amount is coupled to glucuronic acid, the other part being metabolized to desthenyl-azosemide after previous cleavage of the thenyl moiety (Fig. 6). In order to further investigate the metabolism of azosemide, SCHUCHMANN et al. (1992) orally administered the drug to patients with T-drain after cholecystectomy. Azosemide was shown to be excreted mainly unchanged. Of the given dose, 20% was eliminated via the bile and 40% via the urine, respectively. Only a small amount was excreted as azosemide glucuronide. No other metabolites could be detected. Also, no evidence was found for substances such as thiophenemethanol, thiophene-carboxylic acid and its corresponding glycinate, which were described as metabolites in rats.

5. Etozolin

The metabolism of etozolin, which in its ester form is a pro-drug, has been investigated in rats, dogs and man using radiolabeled drug and TLC or HPLC as analytical methods (VOLLMER et al. 1977; GLADIGAU and VOLLMER 1977). Etozolin is well absorbed and undergoes an extensive first-pass effect (HENGY et al. 1980). The metabolism of this compound has been found to be nearly the same in all species including man. The main biotransformation step is the enzymatic cleavage of the ester group to give as the major

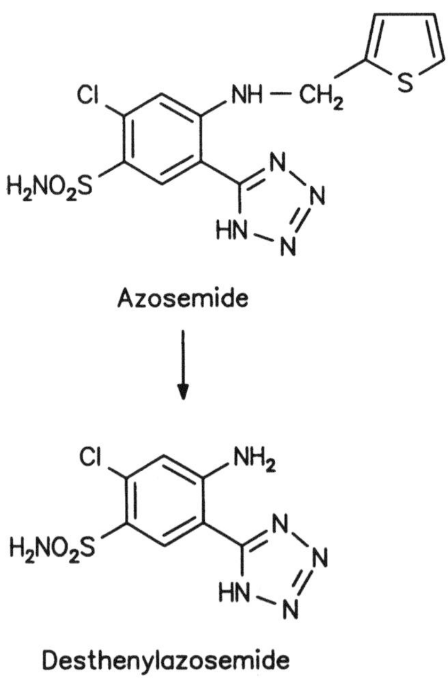

Fig. 6. Metabolism of azosemide

metabolites the two optical isomers of ozolinone (Fig. 7). It is of great significance for the understanding of the mode of action that only the (−)-enantiomer possesses diuretic activity. Both enantiomers are subsequently coupled with glucuronic acid to give the corresponding ester conjugates.

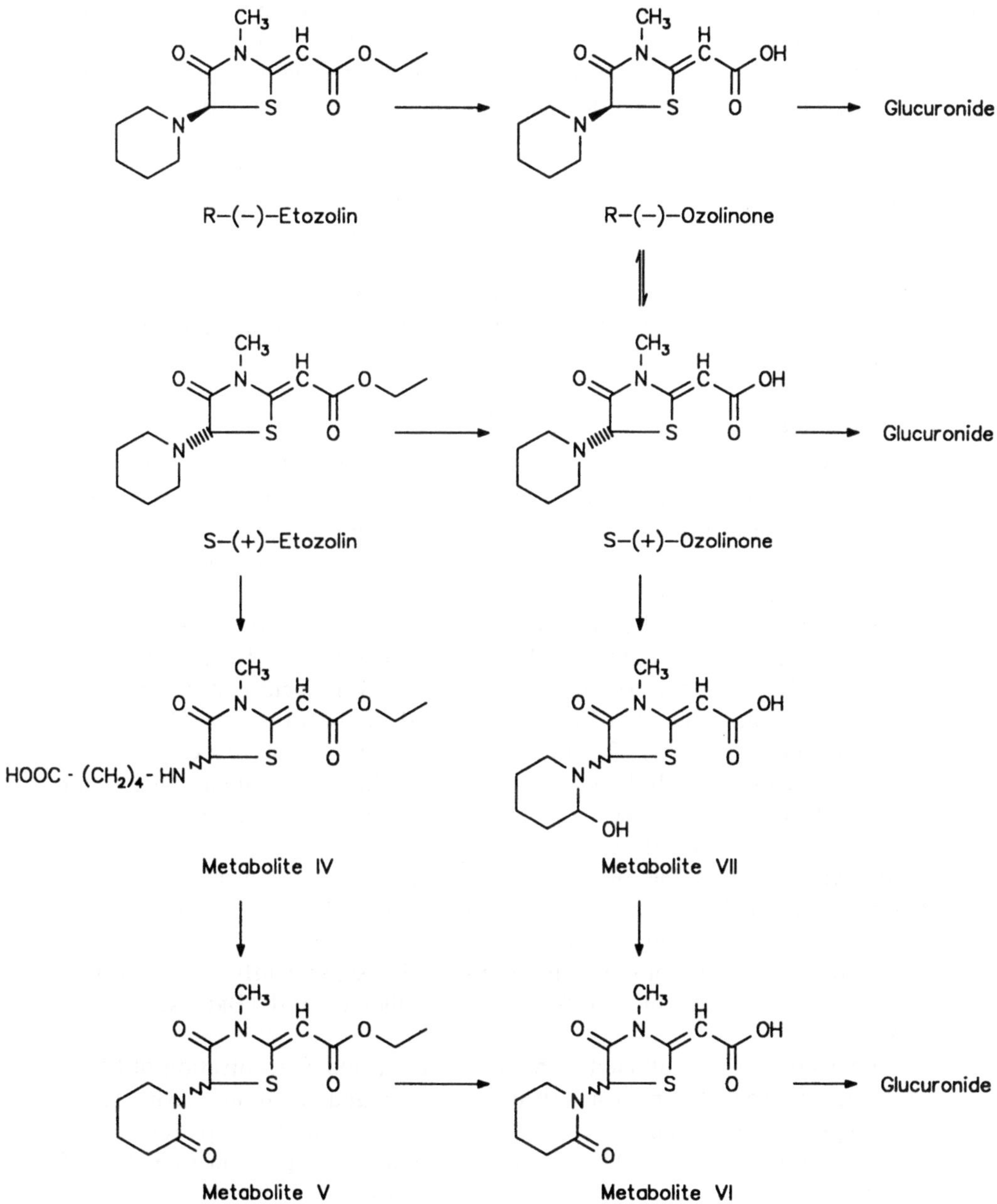

Fig. 7. Metabolism of etozolin

Less than 5% of etozolin or ozolinone can be recovered in the urine, whereas 75% is excreted as the two diastereoisomeric ester conjugates. Of the administered dose, 12% was eliminated via biliary excretion (VON HODENBERG et al. 1977). In addition, some other metabolites were identified as oxidative products of the piperidine moiety (Fig. 7). In liver cirrhosis the hydrolysis of etozolin is markedly reduced, resulting in an elevation of the plasma half-life by a factor of 5. The area under the plasma concentration time curve (AUC) increases by a factor of 3 compared with normal subjects. The C_{max} value of ozolinone decreases to one-sixth that of controls. Despite the prolongation of plasma half-life of ozolinone, the calculated AUC is significantly reduced (KNAUF et al. 1987). Based on these findings the use of etozolin in patients suffering from liver cirrhosis appears difficult.

In contrast to patients with hepatic diseases, in patients with impaired kidney function the biotransformation of etozolin is not altered. KNAUF et al. (1980, 1984) found no increase in plasma levels of ozolinone and no correlation between the plasma half-life of either etozolin or ozolinone and creatinine clearance.

6. Torasemide

Torasemide is a potent new diuretic agent. Its biotransformation has been elucidated in rats, dogs and man. In healthy subjects the drug is mainly metabolized at the methyl group of the phenyl ring to metabolite M1 and then further oxidized to the corresponding carboxylic acid M5 (Fig. 8). A second pathway which leads to the production of small amounts of metabolite M3 involves hydroxylation in the para position of the phenyl ring. Interestingly, M3 is diuretically as active as the parent compound. M1 possesses only weak diuretic activity. The glucuronic acid conjugate M5 is inactive. The total amount of drug recovered in the urine was 83%, with 25% torasemide, 11% M1, 3% M3 and 44% M5 (NEUGEBAUER et al. 1988). In rats two further metabolites M2 and M4, in which the aliphatic side chain had been transformed, have been found. However, M2 and M4 have not been detected in man (GHYS et al. 1985).

In liver cirrhosis the time of maximal plasma concentration, t_{max}, is higher than in healthy subjects, and the elimination half-life is prolonged to 4.8 h (3 h for healthy volunteers). The AUC values are 2.5 times higher in these patients. In parallel the formation of M5 is significantly slowed, which demonstrates that the biotransformation is also altered (BRUNNER et al. 1988).

In chronic renal insufficiency the maximal plasma concentration of M5 is elevated, and the elimination half-life is prolonged to about 7.5 h. Thus, with long-term treatment and high doses of torasemide in patients with impaired kidney function, M5 might accumulate. The pharmacokinetics of torasemide and the saluretically active metabolite M3, however, has not been found to differ from those in healthy subjects (SPAHN et al. 1990; KNAUF et al. 1991).

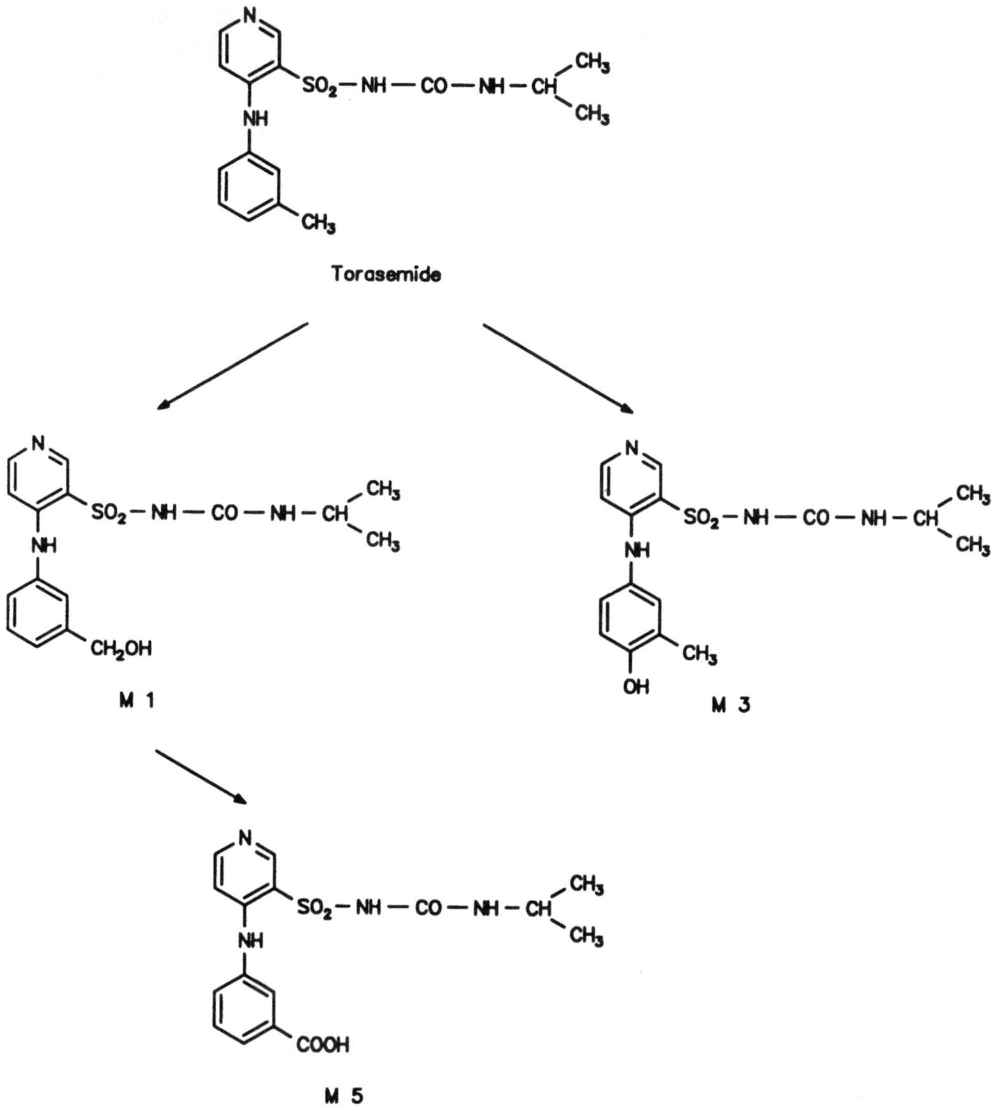

Fig. 8. Metabolism of torasemide

7. Ethacrynic Acid

Although ethacrynic acid is the first loop diuretic introduced onto the market, only little information is available with respect to its pharmacokinetics, especially in man. In rats and dogs ethacrynic acid has been shown to be extensively metabolized (Fig. 9). About 35% of the unchanged drug was recovered in the urine, and a further 20%–30% of the dose was excreted as a cysteine adduct (Beyer et al. 1965). However, a large fraction

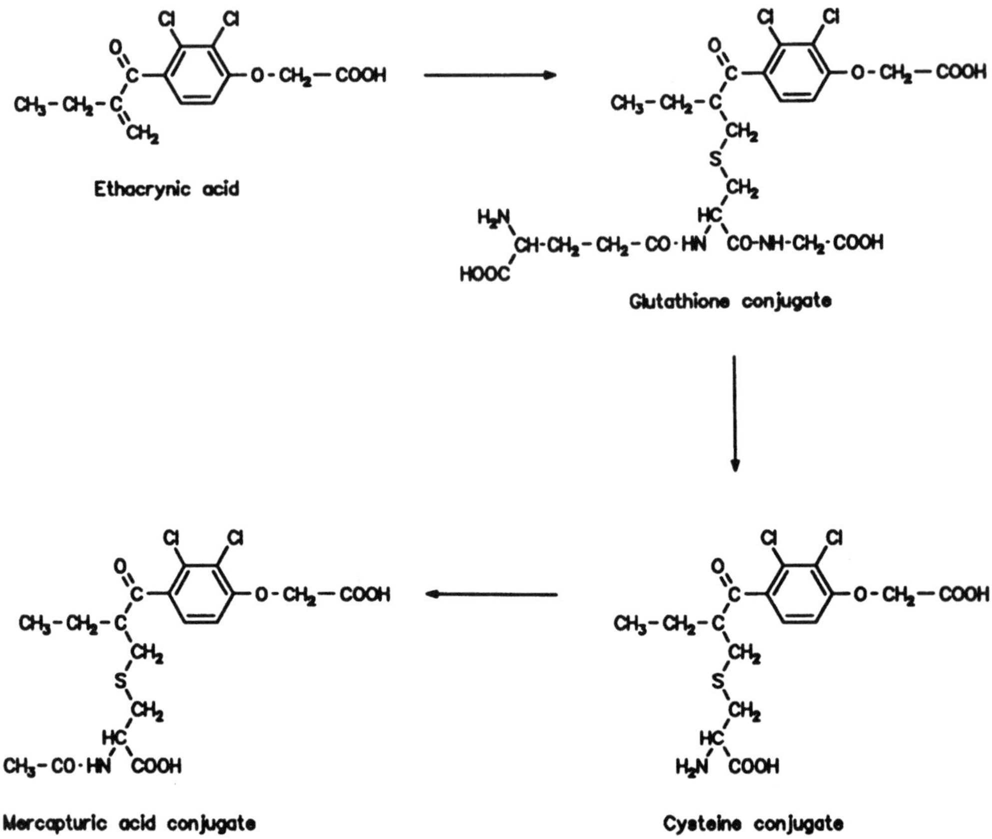

Fig. 9. Metabolism of ethacrynic acid

(about 50%) was also found in feces. In rat bile the major metabolites were identified as a glutathione conjugate and the corresponding mercapturic acid.

About 40% of the dose was found as the glutathione conjugate and about 20% as the mercapturic acid metabolite. An additional metabolite found in dogs was identified as the above-mentioned cysteine conjugate. Whereas the mercapturic acid conjugate was the major metabolite in dog bile (25%), the cysteine conjugate was the major metabolite in monkey bile (26%) (KLAASSEN and FITZGERALD 1975).

In man less than 1% of the dose is renally excreted in unchanged form after i.v. or oral administration. Bioavailability after oral administration is variable, with an average value of 32% ± 19% (VOITH et al. 1993). From this and the clearance characteristics it may be hypothesized that a significant fraction of the dose is eliminated by first-pass extraction. A close correlation of urine volume or Na^+ excretion to ethacrynic acid excretion rate was not found. Compared with the results obtained from isolated loop studies (SCHLATTER et al. 1983) and from investigations on $Na^+2Cl^-K^+$ cotransport in avian erythrocytes (PALFREY and LEUNG 1993), a significantly higher potency would be expected for ethacrynic acid. It was concluded (VOITH et al. 1994) that the diuretic effect was not based solely on the parent drug. A likely candidate for the active metabolite would be the L-cysteine adduct, although other conjugates cannot be ruled out.

Ethacrynic acid has been shown to cause a dose-dependent ototoxic effect in anmials and man (KOECHEL 1981). Its ototoxicity may be due to the formation of the cysteine conjugate, which is suggested to act on the cochlea. The cochlear site of action is the stria vascularis, where an active K^+ transport system is inhibited by the cysteine adduct (BROWN 1973).

III. Thiazide and Thiazide-Type Diuretics

Some of the thiazide diuretics are eliminated virtually unchanged, e.g., chlorothiazide, hydrochlorothiazide and methychlothiazide, or despite its widespread use in clinical practice form as yet unknown metabolites, e.g., bendrofluazide, polythiazide, cyclopenthiazide and chorthalidone.

1. Bendrofluazide

The metabolism of bendrofluazide (Fig. 10) has not been elucidated. BRETTELL et al. (1964) found more than 98% of a ^{35}S-radiolabeled drug in urine, approximately 30% as unchanged drug.

2. Chlorothiazide

As mentioned above, chlorothiazide (Fig. 11) is not metabolized. There is no evidence for in vivo hydrolysis to 6-amino-4-chlorobenzene-1,3-disulfonamide (BRETTELL et al. 1960). The gastrointestinal absorption of the drug is low and dose dependent (ADEBAYO and MABADEJE 1985; OSMAN et al. 1982), which should be considered in bioavailability studies (WELLING et al. 1982). Chlorothiazide was found to be taken up by erythrocytes, resulting in a blood/plasma concentration ratio of about 1.5 (SHAH et al. 1984).

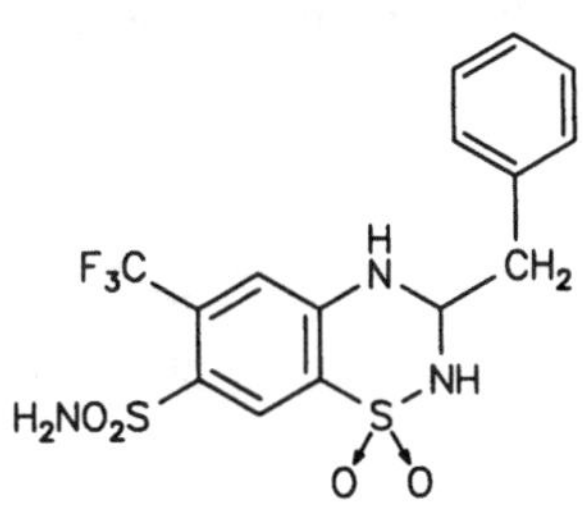

Fig. 10. Bendrofluazide

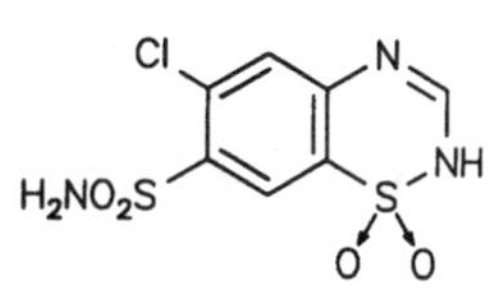

Fig. 11. Chlorothiazide

3. Chlorthalidone

Studies with radiolabeled chlorthalidone in man have shown by TLC that approximately 10% of the material recovered in urine was different from the parent compound (BEERMANN et al. 1975a). In rats 2-(4-chloro-3-sulfamoylbenzoyl)benzoic acid was tentatively identified in the bile (BEISENHERZ et al. 1966). Chlorthalidone is subject to saturable binding to carbonic anhydrase (DIETERLE et al. 1976; FLEUREN and VAN ROSSUM 1980), which is responsible for the accumulation of the drug in erythrocytes and its long and variable half-life. Its concentration in red cells is 50–80 times higher than that in plasma (BEERMANN et al. 1975) and the erythrocyte compartment accounts for roughly 30% of the total amount in the body after multiple dosing (COLLSTE et al. 1976). Although chlorthalidone is a racemic mixture of two enantiomers, its stereopharmacokinetics has been ignored to date.

4. Hydrochlorothiazide

This drug (Fig. 12) is by far the most used diuretic, and numerous pharmacokinetic studies have been published. The concentration of hydrochlorothiazide in red blood cells is at least three times its concentration in plasma (BEERMANN et al. 1976a). Carbonic anhydrase has not been found to be a major binding protein and acetazolamide has been found unable to displace hydrochlorothiazide but not chlorthalidone from binding to erythrocytes (DIETERLE et al. 1976). Therefore, the mechanism of the accumulation of hydrochlorothiazide into erythrocytes remains to be elucidated. Hydrochlorothiazide is excreted virtually unchanged in urine and feces. After intravenous administration more than 90% of the drug is excreted in urine and as little as 1.0% and 4.3% is found in feces (BEERMANN et al. 1976). As biliary excretion appears to be negligible (CALESNICK et al. 1961) it must be considered that the observed amount of drug in feces was secreted into the gastrointestinal tract.

5. Hydroflumethiazide

This drug (Fig. 13) is incompletely absorbed after oral administration. Binding to plasma proteins has been estimated to be 95% (AGREN and BÄCK 1973) and uptake by red blood cells may occur but the binding properties

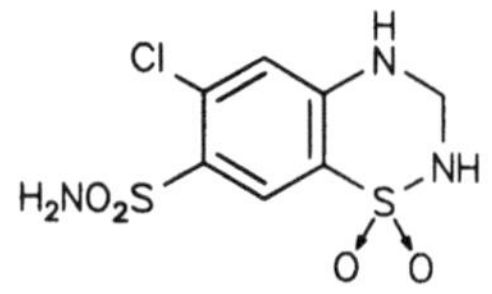

Fig. 12. Hydrochlorothiazide

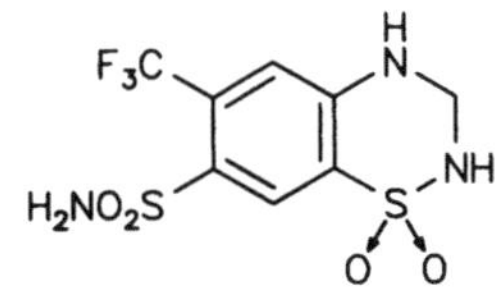

Fig. 13. Hydroflumethiazide

have not yet been reported. In healthy subjects less than 2% of the drug is recovered as 2,4-disulfamoyl-5-trifluoromethiazide (BROERS et al. 1979). Hydrolysis upon storage can account only in part for the formation of this compound. Biliary excretion of hydroflumethiazide is less than 0.1% in patients with T-drain after cholecystectomy.

6. Indapamide

The oral absorption of indapamide is high and the drug is widely distributed in the body. Binding to plasma proteins is 76%–79% and it is also preferentially bound to erythrocytes. Less than 5% of the drug (Fig. 14) is excreted unchanged in the urine and up to 20% is found in the feces. The metabolism of indapamide is complex and 19 different metabolites have been identified (CAMPBELL et al. 1977). Two major pathways characterize the extensive degradation of indapamide. Dehydrogenation yields an indole derivative which partially possesses diuretic activity but this pathway accounts for only 5% of urinary metabolites. Hydrolysis of the amide bond degrades the compound to 2-methylindoline and to chlorosulfamoylbenzoic acid, which in turn is methylated to the corresponding ester. Indapamide is a chiral drug but the metabolism of the enantiomers has so far not been analyzed.

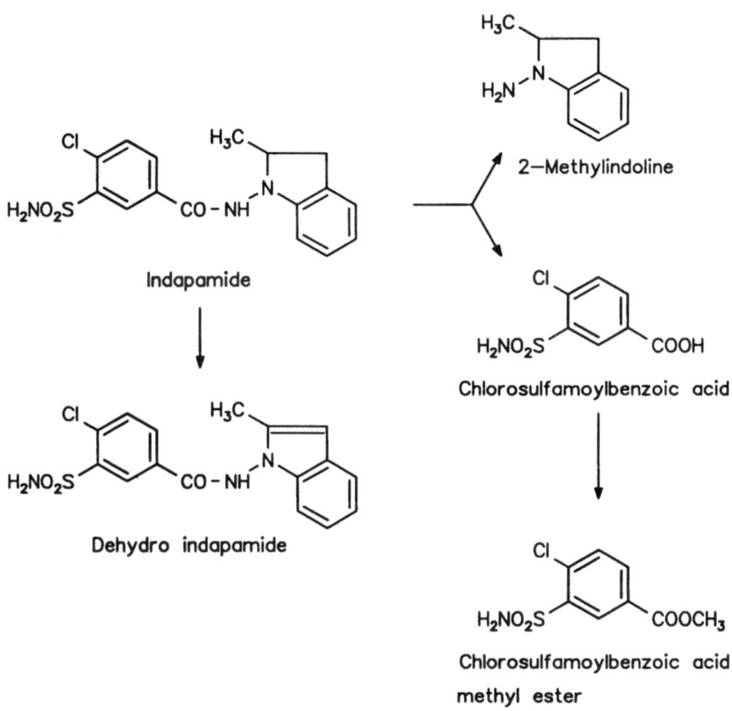

Fig. 14. Metabolism of indapamide

7. Mefruside

Mefruside (Fig. 15) is characterized by extensive oxidation of the furan ring. The resulting lactone metabolite undergoes hydrolytic ring opening to a hydroxy acid metabolite and a substantial part of this carboxylic acid is conjugated with glucuronic acid (DUHM et al. 1967). The interconversion of the lactone and hydroxycarboxylic acid metabolite is pH dependent and catalyzed by components of rat plasma and liver homogenates (PÜTTER and SCHLOSSMANN 1972; SCHLOSSMANN and PÜTTER 1973). The two unconjugated metabolites possess the same diuretic activity in rats as the parent compound (KRONEBERG and MENG 1967). Mefruside and the lactone metabolite accumulate in red blood cells as indicated by an erythrocyte to plasma ratio of about 30 and 20, respectively, and in vitro the metabolite can displace mefruside from erythrocyte-binding sites (FLEUREN et al. 1980). In contrast the hydroxycarboxylic acid metabolite exhibits a blood to plasma ratio of only 0.1. A minor unidentified metabolite was suggested to result from N-demethylation as some radioactivity was given off into the air after administration of mefruside labeled with ^{14}C in the methyl group. In urine

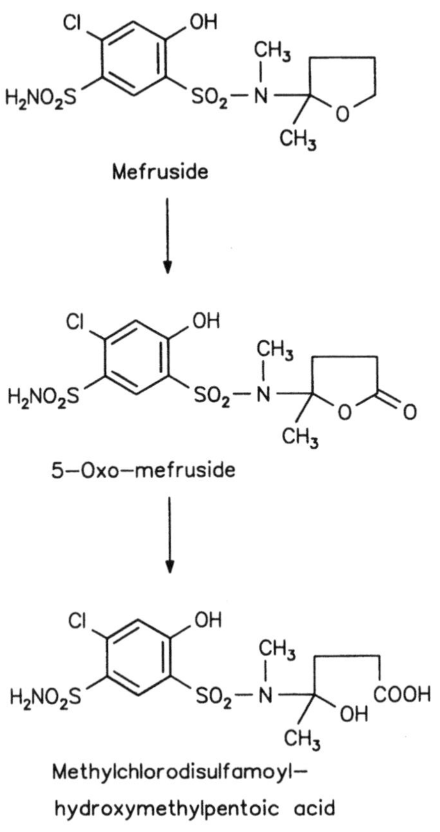

Fig. 15. Metabolism of mefruside

less than 1% of a dose is excreted unchanged and approximately 13%, 46% and 15% are recovered as the lactone metabolite, the hydroxy carboxylic acid and the corresponding glucuronic acid, respectively. A substantial amount of unchanged drug and metabolites are eliminated into bile and reabsorbed. The enterohepatic circulation may contribute to the long diuretic action. Mefruside is another example of a diuretic which possesses a chiral carbon atom, and again the metabolic pathways of the enantiomers have not been elucidated and the racemate has not been analyzed by stereospecific methods.

8. Xipamide

In man approximately 30% of the drug is converted to xipamide-*O*-glucuronide (HEMPELMANN and DIEKER 1977). Apparently no other metabolites are formed and hydrolysis of the amide bond is unlikely because of steric hindrance by the two *O*-methyl groups (Fig. 16).

IV. Potassium-Sparing Diuretics

1. Amiloride

Due to its low lipophilicity the uptake of amiloride (Fig. 17) after oral administration is assumed to be low and variable. Plasma protein binding has been determined to be 40%. Most of the drug is excreted unchanged in the urine. In plasma but not in urine of healthy volunteers 24 h after dosing an additional peak has been observed and tentatively identified as 3,5-diamino-6-chloropyrazinecarboxylic acid (REUTER 1984). This metabolite can be formed by hydrolytic cleavage of the amide bond. The pharmacokinetic properties and the clinical significance of this presumably minor metabolite are unknown.

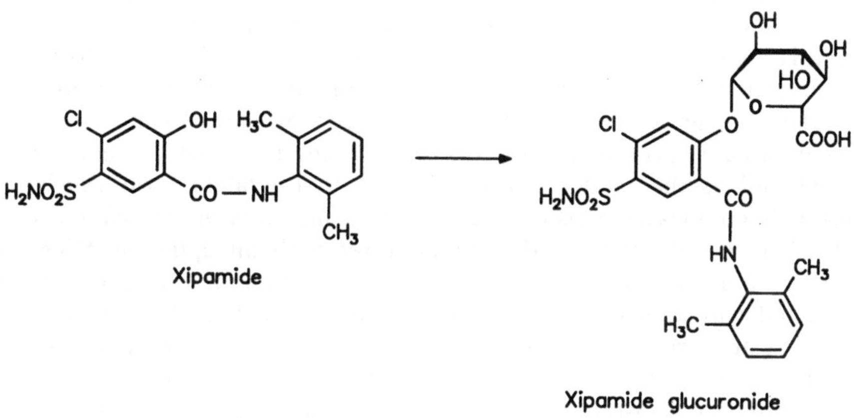

Fig. 16. Metabolism of xipamide

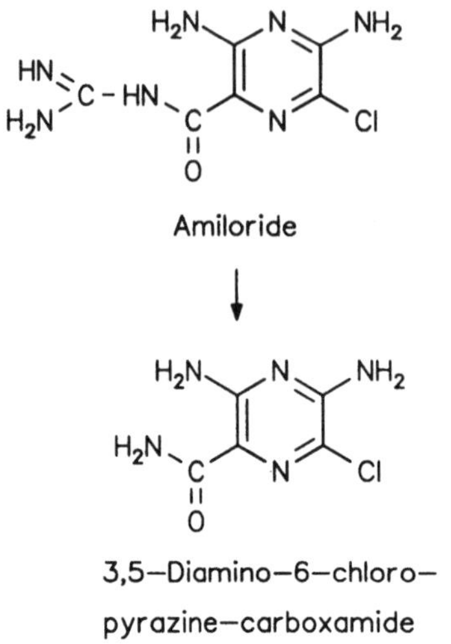

Fig. 17. Metabolism of amiloride

2. Triamterene

Considerable efforts have been made to elucidate the characteristics of triamterene (TA) (Fig. 18) in man (Mutschler et al. 1983). The uptake of the drug in the gastrointestinal tract has been shown to be 83%. As absolute bioavailability was only 52%, a substantial first-pass effect must be assumed (Gilfrich et al. 1983). The main metabolic pathway is the hydroxylation to *p*-hydroxytriamterene (OH-TA), which is then very rapidly conjugated with active sulfate to form the phase II metabolite *p*-hydroxytriamterene sulfuric acid ester (OH-TA-ester). The minor metabolite triamterene-N-glucuronide was identified by Lehmann (1965). The urinary excretion ratio of TA to OH-TA-ester exhibits a unimodal distribution which indicates that the metabolism of TA is not subject to polymorphism (Sörgel et al. 1982). The total plasma clearance has been calculated to be approximately 4.5 l/min, which is higher than normal liver blood perfusion. As the blood to plasma concentration ratio is close to unity, extrahepatic metabolism must be considered. The phase II reaction is catalyzed by phenol sulfotransferase. This enzyme is present in two isoforms, a thermostable protein with *p*-nitrophenol as the model substrate and a thermolabile form which preferentially conjugates dopamine. It has been shown in human platelets that OH-TA is transformed to the sulfate ester with an apparent Michaelis-Menten value of 26 μM. The conjugation reaction of OH-TA is highly correlated with that of *p*-nitrophenol but not with that of dopamine. The OH-TA-ester excretion is not associated with the phenolsulfotransferase activity measured with *p*-nitrophenol and dopamine. These results show that OH-TA is the substrate of the thermostabile isoenzyme and that this metabolic step is not rate limiting for the excretion of the phase II metabolite

Fig. 18. Metabolism of triamterene

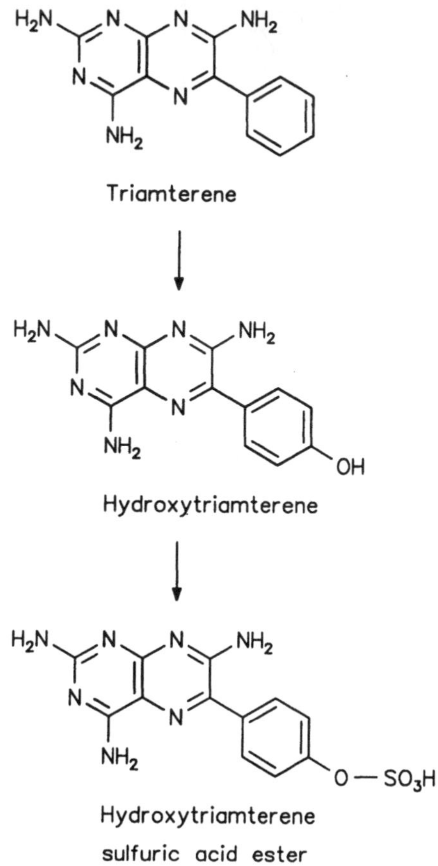

(Reiter et al. 1983). In patients with cirrhotic or inflammatory liver disease the concentration ratio of OH-TA-ester to TA in plasma and urine was decreased but the total urinary recovery (parent drug and metabolite) was increased to approximately 100%, whereas OH-TA was not detected in significant amounts (Antonin et al. 1982). These results suggest that the elimination by phase I but not phase II reactions is rate limiting even in hepatic failure and that, most interestingly, in liver disease the biliary excretion of TA and OH-TA-ester is diminished. After repeated oral administration of TA the AUC ratios of OH-TA-ester to TA are higher than after the first dose in healthy subjects and patients with hepatitis and compensated liver cirrhosis but not in patients with decompensated liver cirrhosis, indicating a metabolic self-induction ability except in advanced disease states.

3. Spironolactone and Potassium Canrenoate

Although both drugs (Fig. 19) are chemically similar and available under the same brand name, one for oral use and the other for intravenous administra-

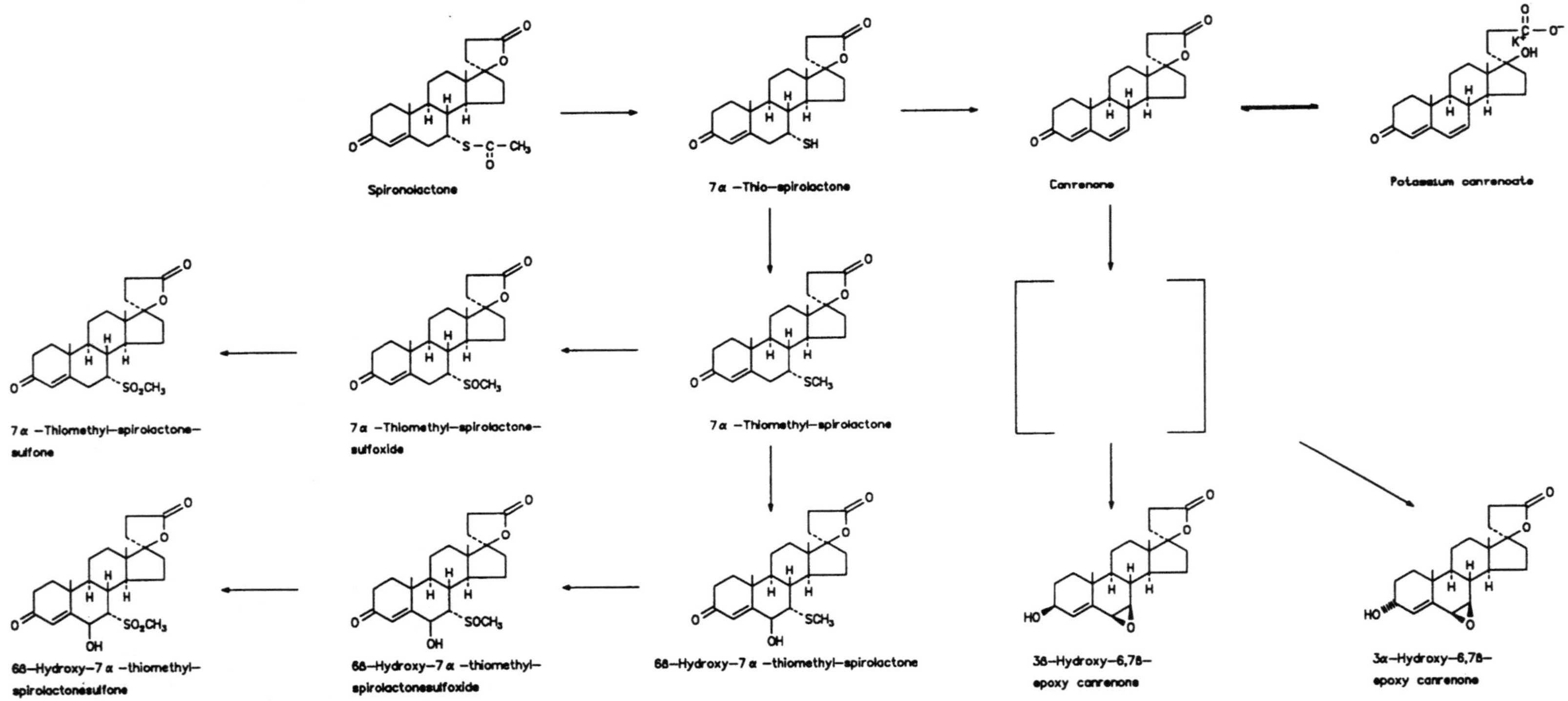

Fig. 19. Metabolism of spironolactone and potassium canrenoate

tion, they differ in their metabolic pathways, with significantly different toxicological findings in animals. Evaluation of the metabolism of both compounds is complicated by its complexity and the fact that unspecific methods of determination have sometimes been employed, the majority in the early studies.

The metabolism of spironolactone can be divided into two routes, one independent of canrenone and the other via canrenone. As shown below, this does not necessarily mean that canrenone formed from spironolactone is biotransformed similarly to potassium canrenoate administered intravenously. With specific HPLC assays canrenone has been found to account for only 20%–25% of the fluorigenic substances of spironolactone in man (DAHLOF et al. 1979; MERKUS et al. 1983; ABSHAGEN et al. 1979; KOJIMA et al. 1985) in contrast to an earlier report (KARIM et al. 1975) where canrenone was found to be the main metabolite circulating in the serum besides appreciable amounts of the 6β-hydroxythiomethyl derivative and traces of the 6β-hydroxysulfoxide. The predominant metabolites of spironolactone are those retaining the sulfur moiety (OVERDIEK et al. 1985). The initial step is deacetylation, leading to formation of a 7α-thiol derivative. This metabolite can be methylated to form the thiomethyl compound. Ring hydroxylation leads to 6β-hydroxy-7α-thiomethylspirolactone in animals and man. The thiomethyl moiety is subject to oxidation to form the corresponding sulfoxide and sulfone derivatives.

Potassium canrenoate readily equilibrates with canrenone. A major metabolic pathway leads to the canrenoate ester glucuronide. Other pathways involve hydroxylation to 15β-hydroxy-, 21-hydroxy- (KARIM et al. 1977) or 20-hydroxycanrenone (FINN et al. 1976). Metabolites formed by reduction of the double bonds yield 6,7-dihydrocanrenone, 4,5β,6,7-tetrahydrocanrenone, and 4,5α,6,7-tetrahydrocanrenone. Further reduction of the 3-oxo group yields 3α- and 3β-hydroxy-4,5β,6,7-hexahydrocanrenone in rat, dog and monkey (KARIM et al. 1977).

In vitro studies in rat liver preparations have shown that ^{3}H-canrenone is extensively metabolized to form both hydrophilic and lipophilic metabolites and that the former was not hydrolyzed by β-glucuronidase. The metabolism is dependent on NADPH and oxygen and inhibited by carbon monoxide and SKF-525-A. Glutathione content is simultaneously depleted and dihydrodiol metabolites detected. Hence, it was concluded that an intermediate epoxide was formed (BOREHAM et al. 1981). OPPERMANN et al. (1988), using rat and human liver S-9 preparations, have shown by GC-MS that potassium canrenoate but not spironolactone was converted in a first step to 6,7β-epoxy-canrenone and then to 3α- and 3β-hydroxy-6,7β-epoxy-canrenone. The latter two compounds are direct mutagens in the Mouse Lymphoma Assay and the former only after metabolic activation. The simultaneous incubation of potassium canrenoate and spironolactone greatly reduced the number and amount of canrenone-derived metabolites presumably by inactivation of cytochrome P450 as described by MENARD et al.

(1979) and Decker et al. (1986). Therefore, the finding of myelocytic leukemia in toxicological studies in rats after administration of potassium canrenoate (Lumb et al. 1978) may not be applicable to the clinical use of spironolactone in man.

References

Abshagen U, Besenfelder E, Endele R, Koch K, Neubert B (1979) Kinetics of canrenone after single and multiple doses of spironolactone. Eur J Clin Pharmacol 16:255–262

Adebayo GI, Mabadeje AF (1985) Chlorothiazide absorption in humans – possible example of Michalis-Menten kinetics. Pharmacology 31:181–188

Agren A, Bäck R (1973) Complex formation between macromolecules and drugs. VII. Binding of saccharine, N-ethyl saccharine, and the diuretic drugs hydroflumethiazide and bendroflumethiazide to human serum albumin. Acta Pharm Suec 10:223–228

Allgulander C, Beermann B, Sjoegren A (1980) Furosemide pharmacokinetics in patients with liver disease. Clin Pharmacokinet 5:570–575

Andreasen F, Hansen HE, Mikkelsen E (1978) Pharmacokinetics of furosemide in anephric patients and in normal subjects. Eur J Clin Pharmacol 13:41–48

Antonin KH, Antonin C, Möhrke W, Mutschler E, Völger KD (1982) Veränderungen der Pharmakokinetik von Triamteren und Hydrochlorothiazid nach ein- und mehrmaliger Gabe an Patienten mit akuten und chronischen Lebererkrankungen. Therapiewoche 32:3905–3917

Beermann B, Hellstrom K, Lindstrom B, Rosen A (1975a) Binding site interaction of chlorthalidone and azetazolamide, two drugs transported by red blood cells. Clin Pharmacol Ther 17:424–432

Beermann B, Dalen E, Lindstrom B (1975b) On the fate of furosemide in man. Eur J Clin Pharmacol 9:57–61

Beermann B, Groschinsky-Grind M, Rosen A (1976) Absorption, metabolism and excretion of hydrochlorothiazide. Clin Pharmacol Ther 19:531–537

Beisenherz G, Koss FW, Klatt L, Binder B (1966) Distribution of radioactivity in the tissues and excretory products of rats and rabbits following administration of ^{14}C-Hygroton. Arch Int Pharmcodyn Ther 161:76–93

Beyer KH, Baer JE, Michaelson JK, Russo HF (1965) Renotropic characteristics of ethacrynic acid: a phenoxyacetic saluretic-diuretic agent. J Pharmacol Exp Ther 147:1–22

Boreham DR, Cheeseman K, Palmer RF, Slater TF, Vose CW (1981) Metabolism of canrenone in vitro: evidence for the formation of a glutathione conjugate via epoxidation. Br J Pharmacol 73:181P

Brater DC, Anderson S, Baird B, Kaojarern S (1983) Effects of piretanide in normal subjects. Clin Pharmacol Ther 34:324–330

Brettell HR, Aikawa JK, Gordon GS Harms DR (1960) Studies with chlorothiazide tagged with radioactive carbon (C^{14}) in human beings. Arch Intern Med 106: 109–115

Brettell HR, Smith JG, Aikawa JK (1964) S^{35} labelled bendroflumethiazide in human beings. Arch Intern Med 113:373–377

Broers O, Jacobsen S, Arnesen E (1978) Pharmacokinetics of a single oral dose of hydroflumethiazide in health and cardiac failure. Eur J Clin Pharmacol 14:29–37

Broers O, Haffner JFW, Jacobsen S (1979) Excretion of hydroflumethiazide in bile and urine of man. Eur J Clin Pharmacol 15:287–289

Brown RD (1973) Ototoxicity caused by ethacrynic acid and furosemide. Minerva Otorinolaringol 23:213–221

Brunner G, von. Bergmann K, Haecker W, von Moellendorf E (1988) Comparison of diuretic effects and pharmacokinetics of torasemide and furosemide after a

single oral dose in patients with hydropically decompensated cirrhosis of the liver. Drug Res 38:176–183
Calesnick B, Sheppard H, Bowen N (1961) Direct comparison of C^{14}-chlorothiazide and T^{3}-hydrochlorothiazide in man. Fed Proc 20:409
Calesnick B, Christensen JA, Richter M (1966) Absorption and excretion of furosemide-S^{35} in human subjects. Proc Soc Exp Biol Med 123:17–22
Campbell DB, Taylor AR, Hopkins YW, Williams JRB (1977) Pharmacokinetics and metabolism of indapamide: a review. Curr Med Res Opin 5 [Suppl 1]:13–24
Clissold SP, Brogden RN (1985) Piretanide: a preliminary review of its pharmacodynamic and pharmacokinetic properties and therapeutic efficacy. Drugs 29:489–530
Collste P, Garle M, Rawlins MD, Sjoquist F (1976) Interindividual differences in chlorthalidone concentration in plasma and red cells of man after single and multiple doses. Eur J Clin Pharmacol 9:319–325
Cohen M (1981) Pharmacology of bumetanide. J Clin Pharmacol 21:537–542
Dahlof CG, Lundborg P, Persson BA, Regardh CG (1979) Re-evaluation of the antimineralocorticoid effect of the spironolactone metabolite, canrenone, from plasma concentrations determined by a new high-pressure liquid chromatographic method. Drug Metab Dispos 7:103–107
Davies DL, Lant AF, Millard NR (1974) Renal action, therapeutic use, and pharmacokinetics of the diuretic bumetanide. Clin Pharmacol Ther 15:141–155
Decker C, Sugiyama K, Underwood M, Correia MA (1986) Inactivation of rat hepatic cytochrome P-450 by spironolactone. Biochem Biophys Res Commun 136:1162–1169
Dieterle W, Wagner J, Faigle JW (1976) Binding of chlorthalidone (Hygroton) to blood components in man. Eur J Clin Pharmacol 10:37–42
Duhm B, Maul W, Medenwald H, Patzschke K, Wegner LA (1967) Untersuchungen mit ^{14}C-markiertem N-(4′-chloro-3′-sulfamoyl-benzol-sulfonyl)-N-methyl-2-aminomethyl-2-methyl-tetrahydrofuran. Stoffwechesel und Kinetik. Drug Res 17:672–687
Finn AM, Finn C, Sadée W (1976) Chemical structure and aldosterone receptor affinity of canrenoate-potassium metabolites in rabbits. Res Commun Chem Pathol Pharmacol 15:613–625
Fleuren HLJ, van Rossum JM (1977) Nonlinear relationship between plasma and red blood cell pharmacokinetics of chlorthalidone in man. J Pharmacokinet Biopharm 5:359–375
Fleuren HLJ, Verwey-van Wissen CPW, van Rossum (1980) Pharmacokinetics of mefruside and two active metabolites in man. Eur J Clin Pharmacol 17:59–69
Frey HH (1975) Pharmacology of bumetanide. Postgrad Med J 51 [Suppl 6]:14–18
Ghys A, Denef J, de Suray JM, Gerin M, Georges A, Delarge J, Willems J (1985) Pharmacological properties of the new potent diuretic torasemide in rats and dogs. Drug Res 35:1520–1526
Gilfrich HJ, Kremer G, Möhrke W, Mutschler E, Völger KD (1983) Pharmacokinetics of triamterene after i.v. administration to man: determination of bioavailability. Eur J Clin Pharmacol 25:237–241
Gladigau V, Vollmer KO (1977) Beschreibung des pharmakokinetischen Verhaltens von Etozolin und dessen Hauptmetaboliten. Drug Res 27:1786–1799
Hajdu P, Haeussler A (1964) Untersuchungen mit dem Salidiuretikum 4-Chlor-N-(2-furylmethyl)-5-sulfamyl-anthranilsaeure. Arzneimittel Forsch 14:709–710
Halladay SC, Carter DE, Sipes EG (1978a) A relationship between the metabolism of bumetanide and its diuretic activity in the rat. Drug Metab Dispos 6:45–49
Halladay SC, Sipes EG, Carter DE (1978b) Diuretic effect and metabolism of bumetanide in man. Clin Pharmacol Ther 22:179–187
Hempelmann FW, Dieker P (1977) Untersuchungen mit Xipamid (4-Chlor-5-sulfamoyl-2′,6′-salicyloxylidid). Teil II: Pharmakokinetik beim Menschen. Drug Res 27:2143–2151

Hengy H, Vollmer KO, Gladigau V, Koelle EU (1980) Assay of etozolin and its main metabolite ozolinone in plasma by high performance liquid chromatography. Drug Res 30:1788–1790
Heptner W, Baudner S, Dagrosa EE, Hellstern C, Irmisch R, Strecker E, Wissmann H (1984) A radioimmunoassay to measure piretanide in human serum and urine. J Immunoassay 5:13–27
Karim A, Hribar J, Aksamit W, Doherty M, Chinn LJ (1975) Spironolactone metabolism in man studied by gas chromatography-mass spectrometry. Drug Metab Dispos 3:467–478
Karim A, Hribar J, Doherty M, Aksamit W, Chappelow D, Brown E, Markos C, Chinn LJ, Liang D, Zagarella J (1977) Spironolactone: diversity in metabolic pathways. Xenobiotica 7:585–600
Keller E, Hoppe-Seyler G, Mumm R, Schollmeyer P (1981) Influence of hepatic cirrhosis and end-stage renal disease on pharmacokinetics and pharmacodynamics of furosemide. Eur J Clin Pharmacol 20:27–33
Kindt H, Schmid E (1970) Ueber die Harnausscheidung von Furosemid bei Gesunden und Kranken mit Leberzirrhose. Pharmacol Clin 2:221–226
Klaassen CD, Fitzgerald TJ (1975) Metabolism and biliary excretion of ethacrynic acid. J Pharmacol Exp Ther 191:548–556
Knauf H, Mutschler E (1991) Pharmacodynamic and kinetic considerations on diuretics as a basis for differential therapy. Klin Wochenschr 69:239–250
Knauf H, Hasenfuß G, Schollmeyer P, Wais U, Mutschler E (1980) Independence of etozolin elimination of kidney function. Single dose experiments in patients with renal insufficiency. Drug Res 30:1791–1793
Knauf H, Liebig R, Schollmeyer P, Rosenthal J, Koelle EU, Mutschler E (1984) Pharmacodynamic and kinetics of etozolin/ozolinone in hypertensive patients with normal and impaired kidney function. Eur J Clin Pharmacol 26:687–693
Knauf H, Missmahl M, Schoelmerich J, Gerok W, Mutschler E (1987) Altered kinetics of etozolin and its active metabolite ozolinone in hepatitis and hepatic cirrhosis with ascites. Drug Res 37:1385–1388
Knauf H, Spahn H, Mutschler E (1991) The loop diuretic torasemide in chronic renal failure. Pharmacokinetics and pharmacodynamics. Drugs 41 [Suppl 3]: 23–34
Knauf H, Gerok W, Mutschler E (1992) Pharmakokinetik von Diuretika. In: Knauf, H, Mutschler (eds) Diuretika – Prinzipien der klinischen Anwendung. Urban and Schwarzenberg, Munich
Koechel DA (1981) Ethacrynic acid and related diuretics: relationship of structure to beneficial and detrimental actions. Annu Rev Pharmacol Toxicol 21:265–293
Kojima K, Yamamoto K, Fujioka H, Kaneko H (1985) Pharmacokinetics of spironolactone and potassium canrenoate in humans. J Pharmacobiodyn 8: 161–166
Kroneberg G, Meng K (1967) Pharmakologie und Metabolismus einer neuen diuretischen Substanz. In: Heilmeyer L, Holtmeier J, Mazzei ES, Marongiu F (eds) Diureseforschung. Thieme, Stuttgart, p 24
Lau HSH, Hyneck ML, Berardi RR, Swartz RD, Smith DE (1986) Kinetics, dynamics and bioavailability of bumetanide in healthy subjects and patients with chronic renal failure. Clin Pharmacol Ther 39:635–645
Lehmann K (1965) Trennung, Isolierung und Identifizierung von Stoffwechselprodukten des Triamterens. Drug Res 15:812–816
Lumb G, Newberne P, Rust JH, Wagner B (1978) Effects in animals of chronic administration of spironolactone – a review. J Environ Pathol Toxicol 1:641–660
Marcantonio LA, Auld WHR, Murdoch WR, Purohit R, Skellern GG, Howes CA (1983) The pharmacokinetics and pharmacodynamics of the diuretic bumetanide in hepatic and renal disease. Br J Clin Pharmacol 15:245–252
Maren TH, Meyer E, Wadsworth BC (1954) Carbonic anhydrase inhibition. I. The pharmacology of diamox, 2-acetylamino-1,3,4-thiadiazole-5-sulfonamide. Bull Johns Hopkins Hosp 95:199

Menard RH, Guenthner TM, Kon H, Gillette JR (1979) Studies on the destruction of adrenal testicular cytochrome P-450 by spironolactone. J Biol Chem 254: 1726–1733

Merkus FWHM, Overdiek JWPM, Cilissen J, Zuidema J (1983) Pharmacokinetics of spironolactone after a single dose: evaluation of the true canrenone serum concentrations during 24 hours. Clin Exp Hypertens 5 [A]:239–248

Meyer BH, Müller FO, Grigoleit HG, Dagrosa EE (1983) Pharmacodynamics and urine pharmacokinetics of three doses of piretanide. Eur J Clin Pharmacol 25:783–785

Mikkelsen E, Andreasen F (1977) Simultaneous determination of furosemide and two of its possible metabolites in biological fluids. Acta Pharmacol Toxicol 41:254–262

Mutschler E, Gilfrich HJ, Knauf H, Möhrke W, Völger KD (1983) Pharmacokinetics of triamterene. Clin Exp Hypertens 5 [A]:249–269

Nebert DW, Nelson DR, Adesnik M, Coon MJ, Estabrook RW, Gonzalez FJ, Guengerich FP, Gunsalus IC, Johnson EF, Kemper B, Levin W, Phillips IR, Sato R, Waterman MR (1989) The P450 superfamily: updated listing of all genes and recommended nomenclature for the chromosomal loci. DNA 8: 1–13

Neugebauer G, Besenfelder E, von Möllendorff E (1988) Pharmacokinetics and metabolism of torasemide in man. Drug Res 38:164–166

Oppermann JA, Piper C, Gardiner P (1988) Spironolactone and potassium canrenoate. Despite chemical similarities, differing metabolism accounts for different toxicological findings in animals. In: Mutschler E (ed) Therapie mit Aldosteronantagonisten. Urban and Schwarzenberg, Munich, p 3

Osman MA, Patel RB, Irwing DS, Craig WA, Welling PG (1982) Bioavailability of chlorothiazide from 50, 100 and 250 mg solution doses. Biopharm Drug Dispos 3:89–94

Overdiek HWPM, Hermens WAJJ, Merkus FWHM (1985) New insights into the pharmacokinetics of spironolactone. Clin Pharmacol Ther 38:469–474

Pentikainen PJ, Penttilae A, Neuvonen PJ, Gothoni G (1977) Fate of ^{14}C-bumetanide in man. Br J Clin Pharmacol 14:39–44

Pentikainen PJ, Pasternack A, Lampainen E, Neuvonen PJ, Penttilae A (1984) Effect of renal failure on the pharmacokinetics of bumetanide. In: Puschett JB (ed) Diuretics. Elsevier Science, New York, p 302

Pentikainen PJ, Pasternack A, Lampainen E, Neuvonen PJ, Penttilae A (1985) Bumetanide kinetics in renal failure. Clin Pharmacol Ther 37:582–588

Perez J, Sitar DS, Ogilvie RI (1979) Biotransformation of furosemide in patients with acute pulmonary oedema. Drug Metab Dispos 7:383–387

Petzinger E, Muller N, Follmann W, Deutscher J, Kinne RKH (1989) Uptake of bumetanide into isolated rat hepatocytes and primary liver cell cultures. Am J Physiol 256:G78–G86

Pütter J, Schloßmann K (1972) The degradation of mefruside: the participation of a "lactonase" in drug metabolism. Biochim Biophys Acta 286:186–188

Reiter C, Werness PG, van Loon J, Smith LH, Weinshilboum RM (1983) Sulphate conjugation of p-hydroxytriamterene by platelet phenol sulphotransferase: assay conditions and correlation with metabolism in man. Br J Clin Pharmacol 15: 211–220

Reuter C (1984) Untersuchungen zur quantitativen Bestimmung und Human-Pharmakokinetik von Amilorid unter besonderer Berücksichtigung der Nierenfunktion. Thesis, Frankfurt am Main

Schloßmann K, Pütter J (1973) Untersuchungen zur Entstehung der Lakton- und Säureform des Hauptmetaboliten des Mefrusid. Drug Res 23:255–262

Shah VP, Walker MA, Hunt JP, Schuirman D, Prasad VK, Cabana BE (1984) Thiazides XI: partitioning of chlorothiazide in red blood cells after oral administration. Biopharm Drug Dispos 5:55–62

Smith DE, Lin ET, Benet LZ (1980) Absorption and disposition of furosemide in healthy volunteers, measured with a metabolic-specific assay. Drug Metab Dispos 8:337–342
Sörgel F, Kiefl H, Hasegawa J, Geldmacher-von Mallinckrodt M, Mutschler E, Benet LZ (1982) Variability of triamterene's fate in the human body. Clin Pharmacol Ther 31:74
Spahn H, Knauf H, Mutschler E (1990) Pharmacokinetics of torasemide in healthy controls and in chronic renal failure. Eur J Clin Pharmacol 39:345–348
Spahn-Langguth H, Langguth P, Brockmeier D (1993) The pharmacokinetics of piretanide in humans: an update. In: Piretanide, Knauf H, Mutschler E (eds) Diuretic agents. Marius Press, Carnforth, pp 11–29
Verbeeck R, Gerkens JF, Wikinson GR, Branch RA (1981) Disposition of furosemide in functionally hepatectomized dogs. J Pharmacol Exp Ther 216: 479–483
Voith B, Spahn-Laugguth H, Paliege R, Knauf H, Mutschler E (1993) Ethacrynic acid: determination of pharmacokinetic and pharmacodynamic parameters. Fundam Clin Pharmacol 7:386
Voith B, Spahn-Laugguth H, Paliege R, Knauf H, Mutschler E (1994) Ethacrynic acid: evidence for active metabolites. Naunyn-Schmiedeberg's Arch Pharamacol Suppl 349:594
Vollmer KO, von Hodenberg A, Poisson A, Gladigau V, Hengy H (1977) Resorption, Verteilung, Metabolismus und Ausscheidung von ^{14}C-Etozolin bei Ratte, Hund und Mensch. Arzneimittelforschung 27:1767–1776
von Hodenberg A, Vollmer KO, Klemisch W, Liedtke B (1977) Metabolismus von Etozolin bei Ratte, Hund und Mensch. Drug Res 27:1776–1785
Welling PG, Barbhaiya RH, Patel RB, Foster TS, Shah VP, Hunt JP, Prasad VK (1982) Thiazides X: lack of dose proportionality in plasma chlorothiazide levels following oral solution doses. Curr Ther Res 31:379–386
Yakatan GJ, Maness DD, Scholler J, Novick WJ, Dolusio T (1976) Absorption, distribution, metabolism, and excretion of furosemide in dogs and monkeys. I. Analytical methodology, metabolism, and urinary excretion. J Pharm Sci 65: 1456–1460

CHAPTER 6

Interaction of Diuretics with Transport Systems in the Proximal Renal Tubule

K.J. ULLRICH

A. Introduction

The pharmacological effect of a drug depends primarily on its concentration at the place of action. This is the plasma concentration for many drugs, but not for diuretics, which have their main site of action on the luminal side of the renal tubules. They reach this site partly by filtration. However, since most diuretics are to a large extent bound to plasma albumin (KNAUF et al. 1992) and are therefore not available for filtration, transtubular secretion of the substances plays a pivotal role. In general, transtubular transport of xenobiotics takes place in the proximal tubule, where several transport systems are involved (Fig. 1): a contraluminal transport system for hydrophobic organic anions (*para*-aminohippurate, PAH) (ULLRICH et al. 1987a,b, 1988, 1989, 1991a), luminal and contraluminal transport systems for hydrophobic organic cations (N^1-methylnicotinamide, $NMeN^+$; tetraethylammonium, TEA^+; *N*-methyl-4-phenylpyridinium, MPP^+; $choline^+$) (DAVID et al. 1995; ULLRICH et al. 1991b, 1992a; ULLRICH and RUMRICH), contraluminal and luminal transport systems for dicarboxylates (methylsuccinate, succinate) (SHERIDAN et al. 1983; ULLRICH et al. 1984), and contraluminal and luminal transport systems for sulfate (DAVID and ULLRICH 1992; ULLRICH et al. 1985a,b,c). The specificities of these transport systems have also been evaluated.

Since all luminal K_i values are obtained from the rat proximal tubule in situ with the same method and all contraluminal K_i values again from the proximal tubule in situ by a similar method, both methods are briefly described here: Transport from the tubular lumen into the tubular cell is measured with the tubular lumen microperfusion method (SHERIDAN et al. 1983). The proximal tubule is punctured with a micropuncture capillary and filled with colored castor or mineral oil. A column of equilibrium solution – preventing net transport of fluid – is then injected into the oil column. After 1–20 s contact time the injected fluid containing radiolabeled test substances and radiolabeled inulin as the reference substance is withdrawn. The time-dependent loss of the test substance relative to inulin is measured with a scintillation counter. Since the disappearance of the test substances within 4 s for sulfate, 3.5 s for methylsuccinate or succinate and 2 s for *N*-methylphenylpyridinium (MPP^+) and $choline^+$ is almost linear, a simple

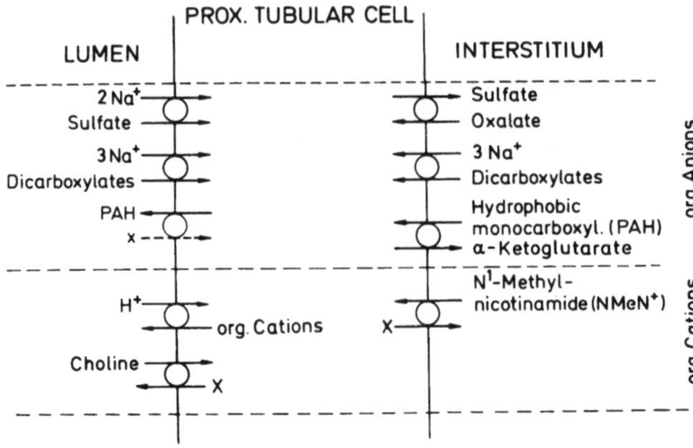

Fig. 1. Sidedness of the transporters for organic anions and organic cations in the proximal renal tubule

two-parameter kinetic is employed to calculate K_m, V_{max}, and a possible permeability term for the test substances. By adding different amounts of interfering substrates, apparent K_i values are evaluated, whereby competitive inhibition is assumed, but not tested, in all instances. For evaluation of the contraluminal transport parameters, the vessels of the exposed kidney are clamped so that the blood flow through the kidney and glomerular filtration is stopped (FRITZSCH et al. 1984). Since the tubular fluid is reabsorbed, the tubular lumen collapses, so that there are only two compartments left, the peritubular and the cellular. Then with a micropipette a peritubular star capillary is punctured and the test solution containing radiolabeled test substrate and radiolabeled inulin as reference substance is injected. Again the disappearance of the test substances within 1–4 s contact time is measured and the transport parameters of the test substances and the apparent K_i values of interfering substrates are evaluated using a computer program (FRITZSCH et al. 1984). With the many substrates tested no K_m or apparent K_i values less than 0.01 mmol/l were found. Furthermore, with the short contact times used for the determinations of the apparent K_i values, up to 20% ethanol could be added to the test solutions and the controls to dissolve hydrophobic substrates.

B. Transport System for Hydrophobic Organic Anions (*para*-Aminohippurate)

The specificity of the *para*-aminohippurate (PAH) transport system is poor and partially overlaps those of the dicarboxylate and sulfate transporters (ULLRICH and RUMRICH 1988). The PAH transport system accepts hydrophobic molecules with a negative ionic or a partial negative charge of electron-attracting side groups, i.e., the substrate need not necessarily be an anion. The hydrophobic domain requires a minimal length of 4 Å (FRITZSCH et al. 1989). Thus, in the fatty acid series interaction starts with molecules longer than valerate (ULLRICH et al. 1987b). The PAH transport system also interacts with substrates which have two ionic or partial negative charges, preferentially with a charge distance of 6–7 Å (ULLRICH et al. 1987a, 1988). The hydrophobic domain can be up to 10 Å and might be located at least partially outside a line connecting the two charges (Fig. 2) (FRITZSCH et al. 1989). The role of hydrophobicity in the interaction with the transporter has not only been shown for substituted benzonates (ULLRICH et al. 1988), but also for a large group of dipeptides (ULLRICH et al. 1989) and imidazoles (ULLRICH et al. 1993a). Furthermore, the strength of the ionic charge is a determinant for the interaction with the PAH transporter. Thus, the interaction of substituted benzoates and phenols with the PAH transporter

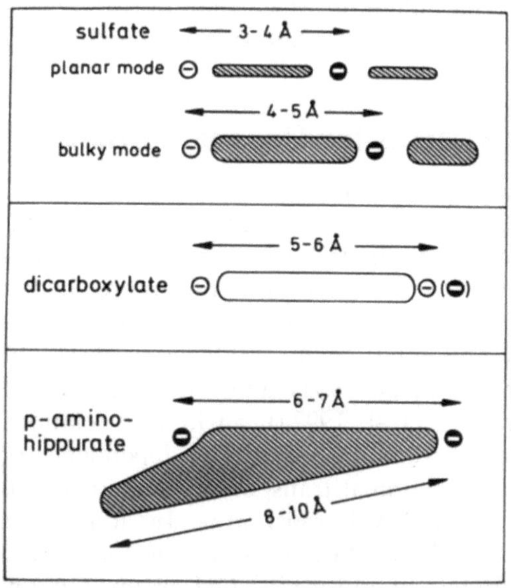

Fig. 2. Structural models of substrates for the three types of contraluminal anion-transporting systems in the rat kidney. The sulfate system accepts short mono- or bivalent anions: planar molecules with COO^- residues, flat hydrophobic domains, and a second ionic or partial negative charge at 3–4 Å distance and bulky molecules with SO_3^- residues, bulky or flat hydrophobic domain and a charge separation of 4–5 Å. The dicarboxylate system accepts bivalent anions with a charge distance of 5–6 Å, whereby one charge might be a partial electronegative charge. The PAH system accepts monovalent anions if they bear a hydrophobic moiety of a minimal length of 4 Å and substrates that have only electron-attracting side groups. It also accepts bivalent anions with a charge distance of 6–7 Å. For some inhibitors the hydrophobic area is longer than the spacing between the two charges. ▨, hydrophobic domain; ⊖, negative ionic charge obligatory; ●, negative ionic charge favorable, but electron-attracting or hydrogen bond forming groups sufficient. (Adapted from FRITZSCH et al. 1989)

increases in inverse relation to their pK_a values (ULLRICH et al. 1988). As side groups with electron-attracting properties serve Cl, Br, NO_2, and CHO. However, OH, NH-CO-CH_3, and O-C_2H_5 groups also interact with the carrier presumably by hydrogen bond formation. This is also indicated by the interaction of many corticosteroid hormones with the contraluminal PAH transporter (ULLRICH et al. 1991a), whereby the spatial orientation of the OH groups is important. Thus, hydrophobicity, negative charge strength (electron-attracting power), and hydrogen bond formation in favorable spatial orientation determine interaction with the PAH transporter. Metabolic transformations of xenobiotics and drugs, such as hydroxylation, N-acetylation, N-benzoylation, conjugation with sulfate, glucuronate, taurine, or glutathione (with further processing), render a molecule acceptable by the PAH transporter (ULLRICH et al. 1990). Specific inhibitors for the contraluminal PAH transport system are apalcillin, benzoylbutyrate, and 2-nitro-4-azido-phenylalanine, while probenecid and tienilic acid are only relatively specific inhibitors since they also interact – albeit weakly – with the contraluminal sulfate/oxalate transporter (ULLRICH et al. 1992b).

C. Transport Systems for Organic Cations

The contraluminal transport system for organic cations accepts positively charged molecules according to their hydrophobicity. Thus, primary, secondary, and tertiary aliphatic, and aromatic amines as well as quaternary N-compounds interact more strongly the more hydrophobic they are (BRÄNDLE et al. 1991; ULLRICH et al. 1991b). This "rule" does not, however, hold for anilines and their analogs (ULLRICH et al. 1992a). With local anesthetics (BRÄNDLE et al. 1991) and with heterocyclic compounds and their analogs (ULLRICH et al. 1992a), an inverse relationship between pK_a and apparent K_i can be demonstrated. The dependence on pK_a and hydrophobicity of the contraluminal transport system for organic cations closely resembles those of the PAH transporter. In addition, steroid hormones interact with the contraluminal transport system for organic cations, whereby the pattern of hydrogen bond forming OH groups favorable for interaction is different from that for the PAH transporter (ULLRICH et al. 1993b). Recently it was shown that even strong carboxylic acids such as ethacrynic and tienilic acid, which have many hydrogen bond forming groups, also interact weakly with the organic cation transporters (DAVID et al. 1994). Thus, similarly to the contraluminal PAH transporter, interaction with the contraluminal organic cation transporter depends on hydrophobicity, charge strength (electron-donating power), and hydrogen bond formation (ULLRICH et al. 1993a).

The luminal transport system for H^+/organic cations is an exchange system and is usually driven by H^+-ion countertransport (for review see ULLRICH 1994). The determinants for this transport system are the same as for the contraluminal transport system for organic cations: hydrophobicity, basicity and size of the substrate molecules (DAVID et al. 1995). The slope of the K_i values against both basicity and hydrophobicity is steeper for the luminal H^+/organic cation exchanger than for the contraluminal $NMeN^+$ transporter. This means that the differences between luminal and contraluminal K_i values become larger at high K_i values, whereas in the range of low K_i values the differences disappear. The luminal membrane of the proximal renal tubule also contains a $choline^+$ transporter, which has similar transport parameters for $choline^+$ (K_m, 0.18 mmol/l; V_{max}, 0.42 pmol/cm per second) as the H^+/organic cation transporter for MPP^+ (K_m, 0.21 mmol/l; V_{max}, 0.42 pmol/cm per second). The specificity of this transporter is under investigation (ULLRICH and RUMRICH, to be published). At the moment it can only be said that the test substrates for the contraluminal organic cation transporter TEA^+, $NMeN^+$ and cimetidine have 160, 27 and 12 times higher K_i values against the luminal $choline^+$ transporter than against the contraluminal cation transporter. MPP^+, the test substrate for the luminal H^+/organic cation exchanger, has a K_i value of 1.4 mmol/l against the luminal $choline^+$ transporter, which is seven times higher than its K_m value against the luminal H^+/organic cation transporter (0.21 mmol/l). Amiloride and its analogs have the same affinity against both luminal organic cation transporters, which will be discussed below. Whether the different affinities play a role in the direction of net transport of organic cations has to be established. A high-affinity, although not specific, inhibitor for the luminal H^+/organic cation transport system is decynium 22 ($K_{i,MPP^+,l}$, 0.013 mmol/l), for the luminal $choline^+$ transporter 4-(4-dimethylamino-styryl)-*N*-methyl-pyridinium, ASP^+ ($K_{i\ choline^+,l}$, 0.014 mmol/l), while a high-affinity inhibitor of the contraluminal organic cation transport system is 4-(perfluorobutoxy)-*N*-methyl-2-methyl-pyridinium (NMR5) ($K_{i,NMeN^+,cl}$, 0.04 mmol/l).

D. Transport Systems for Sulfate

A considerable number of diuretics which interact with the contraluminal PAH transporter also interact with the contraluminal and – very weakly – with the luminal sulfate transporter. The sulfate transport system in the luminal membrane of proximal tubule cells is an Na^+-cotransport system; the respective contraluminal transporter is, however, a sodium-independent exchanger (LÜCKE et al. 1979; ULLRICH et al. 1980; PRITCHARD and RENFRO 1983; LÖW et al. 1984; TURNER 1984; BÄSTLEIN and BURCKHARDT 1986). Some substrates are accepted luminally and contraluminally with almost identical affinities (i.e., sulfate, thiosulfate, selenate, molybdate, phosphate, or 1,3-benzene sulfonates, carboxylates, or sulfocarboxylates). Substituted benzene carboxylates (salicylates), however, interact preferentially with the contraluminal transporter (DAVID and ULLRICH 1992). A closer analysis of substrates for the contraluminal sulfate transport system reveals that the system accepts molecules with two positive charges at a distance of less than 5 Å (Fig. 2) (FRITZSCH et al. 1989). If the substrate is of the "bulky mode" (SO_3H^-), compared to the "planar mode" (COO^-), it can also contain an apolar domain. A specific inhibitor of the contraluminal sulfate transport system has not been found. The potent inhibitors 8-anilinonaphthalene-sulfonate, bromophenol blue, and orange G inhibit contraluminal sulfate and PAH transport equally well (ULLRICH et al. 1992b). SH reagents (complexed Hg) apparently interact with SH groups of the contraluminal sulfate transporter, but not with those of the luminal transport system (ULLRICH et al. 1991c; DAVID and ULLRICH 1992).

E. Transport Systems for Dicarboxylates

Although of the diuretics listed in this review only furosemide interacts weakly with the contraluminal dicarboxylate transport systems, the specificity requirements of this system should also be mentioned. Dicarboxylates are transported both on the luminal and on the contraluminal cell side by an Na^+-dependent transport system (WRIGHT et al. 1980; SHEIKH et al. 1982; JØRGENSEN et al. 1983; SHERIDAN et al. 1983; BURCKHARDT 1984; ULLRICH et al. 1984; WRIGHT and WUNZ 1987). The specificities of luminal and contraluminal transport are very similar, but not identical (ULLRICH et al. 1985d). The charge distance within the substrates has to be between 5 and 6 Å, whereby one charge might be a partial charge of at least −0.5 elementary charge (Fig. 2) (FRITZSCH et al. 1989). Since solely contraluminal sulfate transport is inhibited by acidic pH and only luminal sulfate transport is stimulated by starvation, it is possible to reverse net active secretion of methylsuccinate into net active reabsorption. The intracellular concentration of transported dicarboxylates is high, especially in starvation. High intracellular accumulation may pose a problem with substrates that are potentially nephrotoxic. The dicarboxylate transporters are relatively specifically inhibited by tetrafluorosuccinate (ULLRICH et al. 1992b).

F. Interaction of Diuretics with the Different Proximal Transport Systems

Most diuretics show net secretion through the contraluminal PAH-transport system because of their affinity for this system. Other diuretics such as amiloride and analogs have equal affinities to the contraluminal and luminal organic cation transport systems and can show net secretion or net reabsorption depending on the driving forces. Many diuretics interact with organic anion and organic cation transport systems with only one type, however, and usually with low affinity, so that transport through this system is hardly detectable. Both the PAH and the NMeN/TEA transport systems seem to run best with substrates with a K_m, or K_i, respectively, of 0.1 mmol/l.

Substrates with much lower K_m or K_i values face the risk that their transport rate is low despite high affinity, because their off rate from the carrier is low. This holds for probenecid ($K_{i,PAH}$, 0.03 mmol/l) and apalcillin ($K_{i,PAH}$, 0.02 mmol/l), which are recommended inhibitors of the PAH transport system and possibly for 4-(4-dimethylaminostyryl)-N-methylpyridinium (ASP^+) for the luminal $choline^+$ transport system ($K_{i,choline+,l}$ 0.014 mmol/l). Below the apparent K_i values of the different diuretics against the contraluminal and luminal transport systems for organic anions and organic cations are given. This will enable the reader to judge which transport system might be used by the respective diuretic. Since the apparent K_i values of a large number of drugs are known, it is also possible to predict drug interactions. However, one must keep in mind that some drug interactions seen in the whole animal are not due to the parent drugs but rather to quickly appearing metabolites (see Chap. 5, this volume). Such metabolic transformation can shift a substrate to another transport system (ULLRICH et al. 1990). It will also be discussed what part of the diuretic molecule is responsible for the interaction with the different transporters.

I. Sulfonamide/Thiazide Derivatives

In this group of diuretics only furosemide and piretanide have a high affinity to the contraluminal PAH transporter (Table 1). with $K_{i,PAH,cl}$ values of 0.04 and 0.07 mmol/l. This is clearly due to their COOH group with a pK_a of ≈3.9. Other substances within this group, acetazolamide, hydrochlorothiazide, cyclopenthiazide, bendroflumethiazide, and indapamide, undergo an acidic dissociation at the sulfonamide group, with a pK_a value of between 7.4 and 8.8. These diuretics have a $K_{i,PAH,cl}$ value of between 0.6 and 1.7 mmol/l, i.e., they interact moderately with the PAH transporter. However, two diuretics within this group, chlorthalidone and mefruside, with higher pK_a values than the others, 9.4 and 9.5, do not interact with the PAH transporter. Our suggestion that only sulfonamides with a pK_a below 9.5 interact with the PAH transporter was confirmed in experiments with benzenesulfonamides and their para-NH_2 and NO_2 derivatives. Thus the sulfonamides behave like phenols, where direct correlation between their pK_a values and their apparent $K_{i,PAH}$ was observed. It must be emphasized, however, that the PAH carrier interacts with the ionized and the nonionized form of these substrates equally well (ULLRICH et al. 1992c). Inhibition measurements on tissue/bath (PAH) ratio in isolated rabbit S2 segments after preincubation with thiazide and loop diuretics gave IC_{50} values which are in accordance with the data in Table 1 (BARTEL et al. 1993).

Two-thirds of the tested sulfonamide/thiazide diuretics interacted with the contraluminal sulfate transporter, but only furosemide with good affinity ($K_{i,SO_4^{2-},cl}$, 0.9 mmol/l; for comparison the K_m of sulfate is 1.4 mmol/l). Acetazolamide, hydrochlorothiazide, cyclopenthiazide, mefruside, and piretanide had only a moderate affinity toward the contraluminal sulfate transporter, with $K_{i,SO_4^{2-},cl}$ values between 2.9 and 6.5 mmol/l. Furosemide

Table 1. Interaction of diuretics with transport systems in the proximal renal tubule: chemical structure of dirutetics; apparent K_i values in millimoles per liter as evaluated with the method described for luminal (l) transporters by SHERIDAN et al. (1983) and for contralateral (cl) transporters by ULLRICH et al. (1984)

		PAH (cl)	Succinate (cl)	Sulfate (cl)	Sulfate (l)	NMeN$^{\oplus}$ (cl)	MPP$^{\oplus}$ (l)	Choline$^{\oplus}$ (l)
I. Sulfonamide-/thiazide derivatives								
Acet-azolamide	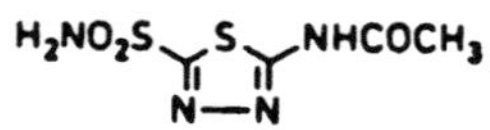	1.3 ± 0.45	NS > 10	6.4 ± 7.7		NS > 20	8.9 ± 4.0	34.0 ± 19.0
Hydrochlorothiazide	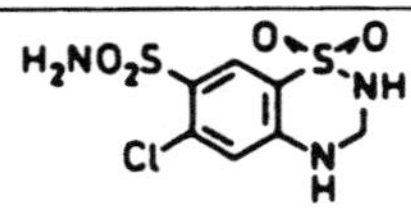	0.72 ± 0.19	NS > 10	4.4 ± 4.3	NS > 37	5.0 ± 1.3	18.5 ± 4.3	14.3 ± 7.1
Cyclopenthiazide	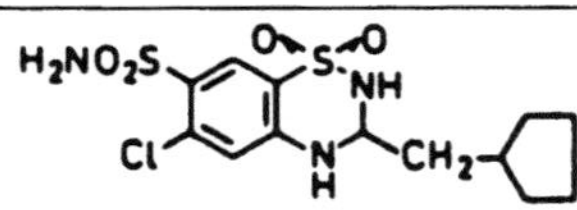	1.7 ± 0.8		4.7 ± 4.3		9.3 ± 4.2	4.7 ± 0.8	12.6 ± 3.8
Bendroflumethiazide	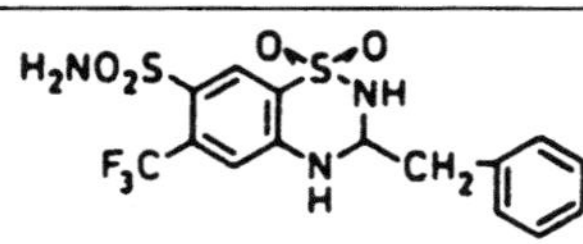	0.57 ± 0.16		NS > 10		7.2 ± 2.6	2.0 ± 0.5	22.1 ± 12.2
Indapamide	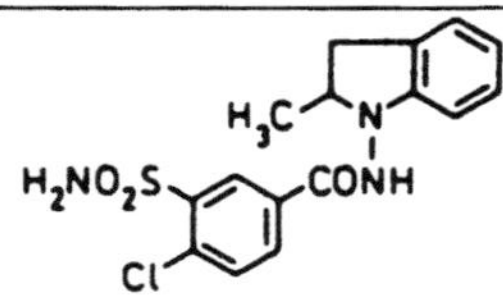	1.0 ± 0.3		NS > 10		15.5 ± 8.2	4.7 ± 1.6	18.0 ± 10.5
Chlorthalidone	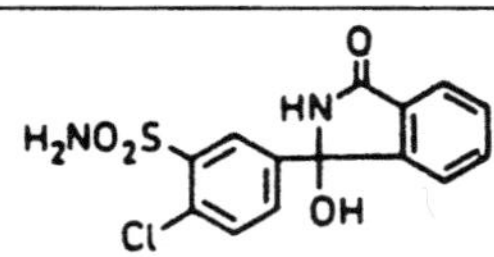	NS > 5		NS > 10		8.0 ± 2.0	0.87 ± 0.25	1.8 ± 0.3

Mefruside	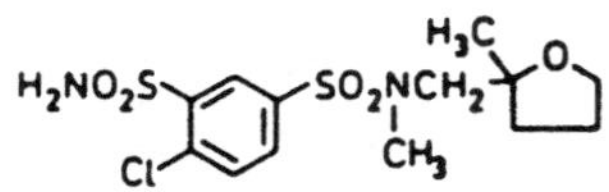	NS > 5		6.5 ± 9.5		NS > 20	16.2 ± 7.4	NS > 20
Furosemide	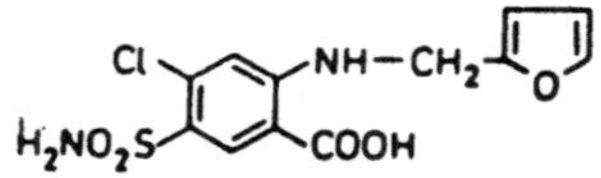	0.04 ± 0.01	5.1 ± 8.9	0.87 ± 0.3	36.0	11.5 ± 3.5	1.7 ± 0.7	17.5 ± 7.7
Piretanide	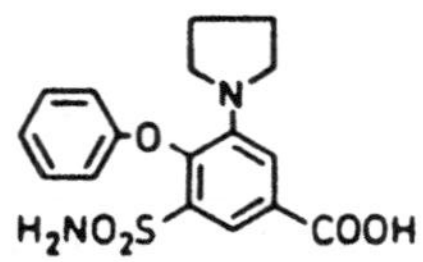	0.07 ± 0.02	NS > 10	2.9 ± 1.9	46.5	9.1 ± 2.9	10.9 ± 3.0	16.7 ± 7.9
II.a Thiazolidine derivatives								
Ozolinone	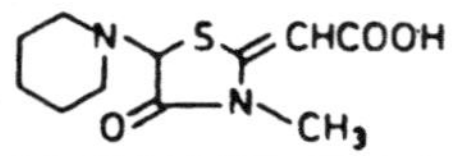	0.07 ± 0.02		NS > 10				
Etozoline	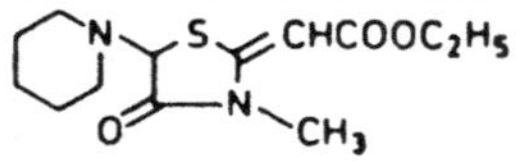	NS > 7.5				17.4 ± 5.3	NS > 40	
II.b Aminopyrazoline derivatives								
Muzolimine	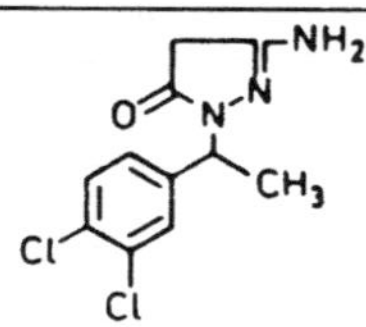	1.7 ± 0.92		NS > 10		11.3 ± 2.6	NS > 25	
II.c Pyrazolidine derivatives								
Sulfinpyrazone	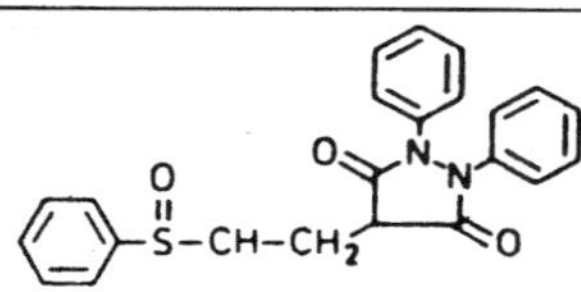	0.17 ± 0.03				3.8 ± 1.0	3.5 ± 0.7	11.9 ± 6.1

Table 1. *Continued*

		PAH (cl)	Succinate (cl)	Sulfate (cl)	Sulfate (l)	$NMeN^{\oplus}$ (cl)	$MPP^{\oplus}$ (l)	$Choline^{\oplus}$ (l)
III. Arylamine-pyridine-carboxylate/sulfonylurea derivatives								
Torasemide	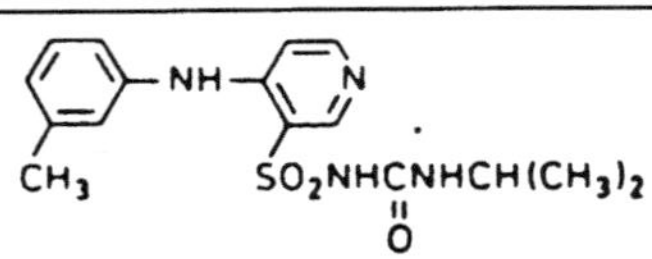	0.16 ± 0.04		NS > 10		11.2 ± 2.8	9.0 ± 2.0	18.5 ± 6.1
Triflocin	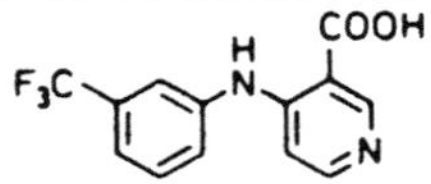	0.15 ± 0.04	NS > 20	4.1 ± 3.6		2.7 ± 0.7	2.2 ± 0.8	16.4 ± 5.2
IV. Phenoxyacetic acid derivatives								
Ethacrynic acid	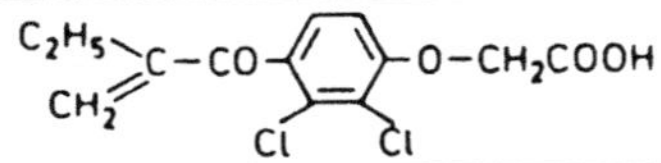	0.12 ± 0.04	NS > 10	4.7 ± 4.8	11.3	14.8 ± 6.6	15.8 ± 5.1	NS > 20
Tienilic acid	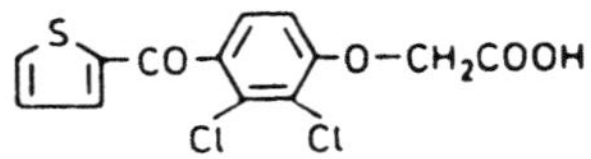	0.04 ± 0.01	NS > 10	2.7 ± 1.6	NS > 75	20.5 ± 5.3	7.7 ± 1.9	NS > 20
V. Pyrazinoyl-guanidine derivatives/Pyrazinoyl-aminomethylphenol derivatives								
Amiloride	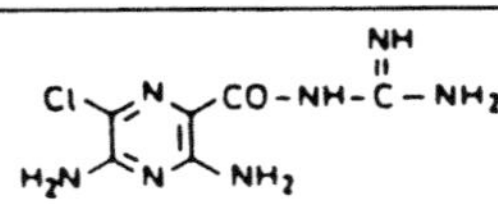	5.1 ± 3.4				0.11 ± 0.06	0.13 ± 0.04	0.14 ± 0.04
Benzylamiloride	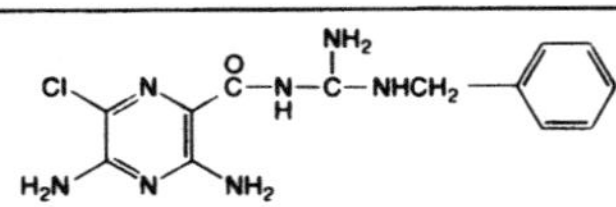	1.3 ± 0.3				0.05 ± 0.02	0.024 ± 0.001	0.025 ± 0.003
Amiloride 5-(N,N-hexamethylene)	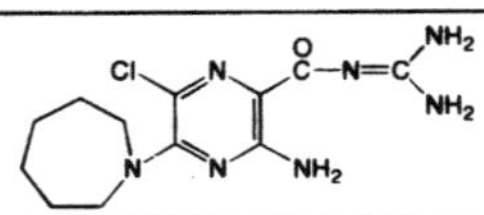	1.8 ± 0.9				0.06 ± 0.02	0.07 ± 0.03	0.12 ± 0.05

ICI 207.828	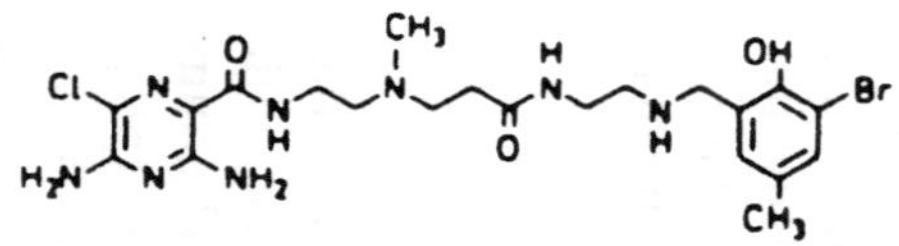	0.75 ± 0.2			NS > 12.0		
VI. Pteridine derivatives							
Triamterene	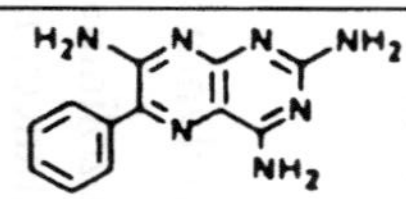	NS > 2.5			NS > 5.0	5.3 ± 1.6	NS > 20
Hydroxytriamterene-sulfuric acid ester	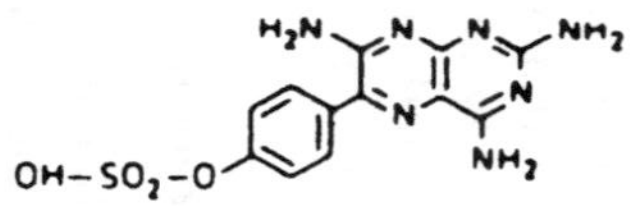	0.56 ± 0.2		NS > 10	NS > 4.0	8.3 ± 2.5	NS > 20
VII. Aldosterone antagonists							
Spironolactone	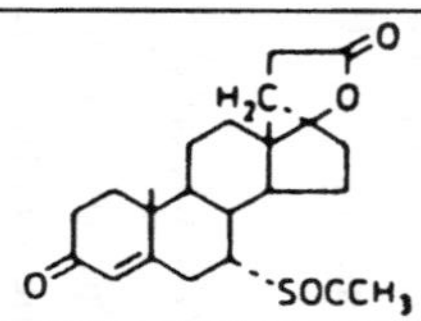	0.07 ± 0.02	NS > 6	NS > 0.7			
Canrenoate	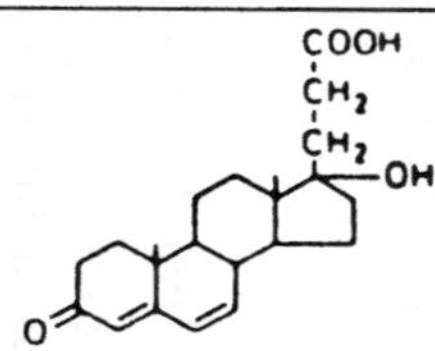	0.22 ± 0.07	NS > 15	NS > 7.2			
d-Aldosterone	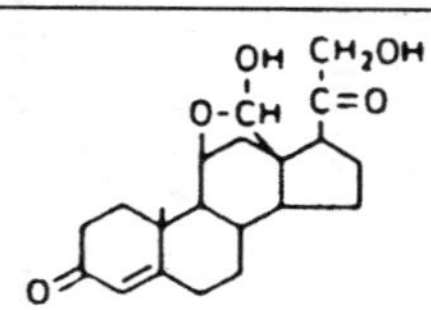	0.18 ± 0.05	NS > 25	NS > 7.2	0.9 ± 0.2	0.87 ± 0.4	NS > 17

and piretanide showed a very weak interaction with the luminal sulfate transporter. The prerequisite for interaction with the contraluminal sulfate transporter is a negative charge accumulation in the neighbourhood of an ionizable group and this seems to be fulfilled with the respective substances. The reason that bendroflumethiazide, indapamide, and chlorthalidone did not interact with the contraluminal sulfate transporter might be because these compounds have bulky and hydrophobic phenyl groups (ULLRICH et al. 1994).

Of the four substances of this group tested against the contraluminal dicarboxylate transporter (succinate), only furosemide showed a slight interaction (apparent $K_{i,succ,cl}$, 5.1 mmol/l). This might be due to the constellation of the electronegative COOH and the electronegative Cl and/or furan oxygen.

Quite unexpected all the tested substances in this group showed moderate to weak interaction with the luminal H^+/organic cation exchanger ($K_{i,MPP^+,l}$, 0.9–18.6 mmol/l) and a weak to very weak interaction with the luminal choline$^+$ transporter ($K_{i,cholin^+,l}$ >12.6 mmol/l), except chlorthalidone, which interacts moderately with both luminal transporters for organic cations (K_i, 0.9 and 1.8 mmol). Seven out of nine substances showed a weak interaction with the contraluminal organic cation transporter ($K_{i,NMeN^+,cl}$, 5.0–15.5 mmol/l). The interaction with the organic cation transporters can apparently be achieved by any N-bearing group: the sulfonamide group itself, and very likely the thiazide group, the furylamino group (in furosemide), the pyrrolidyl group (in piretanide), the indolinyl group (in indapamide), and the isoindolinyl group (in chlorthalidone). However, as already mentioned, the K_i, organic cation values are rather high, i.e., the affinities to both organic cation transporters are too low to play a significant role in the renal transport of these diuretics. This is true especially if their affinity to the PAH transport system is high. In the case of chlorthalidone, which only interacts with the organic cation transporters especially the luminal ones, a net reabsorptive component may exist, which may be responsible for the fact that chlorthalidone is found in the proximal tubular cells, even 7 h after i.v. administration (PULVER et al. 1959).

II. Thiazolidine, Aminopyrazol, and Pyrazolidine Derivatives

To this category belong the diuretics ozolinone, muzolimine, and the uricosuric drug sulfinpyrazone (Table 1). All these substances interact with the contraluminal PAH transporter, ozolinone, which bears a COOH group (pK_a, 2.5), with high affinity ($K_{i,PAH,cl}$, 0.07 mmol/l), sulfinpyrazone (pK_a, 2.8) also with high affinity ($K_{i,PAH,cl}$, 0.17 mmol/l), and muzolimine, which has no acidic dissociable group, but two electronegative chloro and one oxo side group, with moderate affinity ($K_{i,PAH,cl}$, 1.7 mmol/l). Etozolin is the ethyl ester of ozolinone and does not interact with the PAH transporter. As shown previously, esterification of a carboxylic group abolishes its interaction with the PAH transporter (ULLRICH et al. 1987b). The three substrates etozoline, muzolimine and sulfinpyrazone also interact with the contraluminal organic cation transporter and sulfinpyrazone also with both luminal transporters, although with low to very low affinity. Responsible for this interaction seems to be the pyrazol(idine) group of muzolimine and sulfinpyrazone, the thiazolidine and/or the piperidine residue in etozoline.

III. Arylamine-Pyridinecarboxylate and Arylamine-Pyridine Sulfonylurea Derivatives

Two diuretics listed in this group of compounds are the sulfonylurea derivative torasemide (acidic pK_a, 6.44) and the pyridine carboxylate compound triflocin (pK_a ~4.7). It is the carboxy and the sulfonylurea groups which interact with the PAH transporter. A similar good interaction with this transporter was also observed by other sulfonylurea compounds: tolbutamide (pK_a, 5.4), chlorpropamide (pK_a, 4.8), and glyburide (pK_a, 6.5) (ULLRICH et al. 1994). Both compounds, torasemide and triflocin, interact also moderately to weakly with the contraluminal and both luminal organic cation transporters but only triflocin reacts with the contraluminal sulfate transporter. The pyridine moiety of the two compounds might be responsible for their interaction with the organic cation transporters. But the sulfonylurea, phenylamine, and even the CF_3 group might also participate (ULLRICH et al. 1994). The COOH group together with the nearby NH group is responsible for the interaction of triflocin with the contraluminal sulfate transporter as was revealed by the similar behaviour of related compounds (ULLRICH et al. 1985d).

IV. Phenoxyacetic Acid Derivatives

Ethacrynic acid has a similar affinity to the contraluminal PAH transporter as PAH itself (K_m, 0.12 mmol/l). The affinity of tienilic acid, however, is considerably higher ($K_{i,PAH,cl}$, 0.04 mmol/l) and actually is one of the highest affinities toward the contraluminal PAH transporter observed. Clearance data reported by LEMIEUX et al. (1979) confirm this. These authors also point out that the process of renal transcellular transport of tienilic acid is slow. This is in agreement with the suggestion that high-affinity substrates block the transporter because of easy binding but difficult debinding. The affinity of ethacrynic and tienilic acid to the contraluminal sulfate transporter is likely to be caused by the negative charge accumulation of the COOH group, the neighbouring oxo group, and the chloro group in position 2 of the benzene ring (ULLRICH et al. 1989). Surprisingly ethacrynic and tienilic acid also interfere weakly with the organic cation transporters, except the choline$^+$ transporter. This might be caused by the many H^+-bond-forming groups which these two substances possess.

V. Pyrazinoyl-Guanidine Derivatives, Pyrazinoyl-Aminomethylphenol Derivatives

Unlike the above diuretics, these compounds have a high affinity to the contraluminal and both luminal organic cation transporters (K_i, 0.02–0.13 mmol/l). This is likely to be due to the guanidine group, which has a base pK_a of around 8.7 (Ullrich et al. 1994). On the other hand, the site for the weak interaction of these substances with the contraluminal PAH transporter might reside in the ring-attached electronegative chloro group. With radiolabeled amiloride it was seen in short-term flux measurements that ten times as much amiloride crossed the contraluminal cells side through the organic cation as through the organic anion transporter (Ullrich et al. 1993a). The "eukalemic" diuretic ICI 207 828 (Table 1) is a hybrid between the Na^+-channel-blocking amiloride and the Cl^--channel-blocking 2-(aminomethyl)phenol (Simchowitz et al. 1993), whereby the guanidine moiety of amiloride is missing. This compound showed a proximal tubular effect, apparently caused by the 2-(aminomethyl)phenol moiety and a distal tubular amiloride-like effect (Johnston and Kau 1993). The substance does not interact with the contraluminal organic cation transporter, which indicates that it is the guanidine part of amiloride which interacts with the $NMeN^+$ transporter. ICl 208 828 with a pK_a of 6.5, however, interacts with the PAH transporter with moderate affinity (apparent $K_{i,PAH,cl}$, 0.75 mmol/l). It also fits into the $K_{i,PAH,cl}/pK_a$ relationship established for substituted phenols (Ullrich et al. 1988).

VI. Pteridine Derivatives

The pteridine derivatives triamterene and *p*-hydroxytriamterene sulfuric acid ester are quite hydrophobic (Werness et al. 1982). Thus, we had to make the perfusates alkaline (pH 8.3) and to add 10% ethanol to obtain for each substrate a final concentration of 5 mmol/l. At this concentration triamterene neither interacted with the contraluminal $NMeN^+$ nor with the contraluminal PAH transporter, but with the luminal MPP^+ transporter (Table 1). *p*-Hydroxytriamterene sulfuric acid ester, in turn, had a good affinity to the PAH transporter (apparent $K_{i,PAH}$, 0.56 mmol/l), a weak affinity to the luminal MPP^+ transporter but no affinity to the luminal $choline^+$ and the contraluminal sulfate transporter. It has been reported that ranitidine diminished the renal clearance of triamterene and of hydroxytriamterene sulfate (Muirhead et al. 1988). Cimetidine, however, only reduced the renal clearance of triamterene (Muirhead et al. 1986). Both interfering substrates, ranitidine and cimetidine, are bisubstrates (Ullrich et al. 1993a), i.e., they interact with organic anion and organic cation transporters. Thus the data of Table 1 indicate that *p*-hydroxytriamterene sulfuric acid ester is secreted by the PAH transporter, while the transport systems by which triamterene is handled remain to be elucidated.

VII. Aldosterone Antagonists

The two compounds listed in this category, spironolactone and canrenoate, have a carboxylic group (pK_a ~7.0) which can form a γ-lactone ring. For interaction with the PAH carrier it seems to be irrelevant whether the compounds exist in the straight or the ring form. In any case their affinity to the PAH transporter is high ($K_{i,PAH,cl}$, 0.07 and 0.22 mmol/l, respectively) (ULLRICH et al. 1991c). *d*-Aldosterone, which has an aldehyde group in position 18 which can form a hemiacetal, also interacts with high affinity with the PAH transporter ($K_{i,PAH,cl}$, 0.18 mmol/l). However, transport of radiolabeled *d*-aldosterone through the PAH transporter has not been documented because of the high lipid solubility of this substance: it crosses the cell membrane through the lipid bilayer by diffusion rather than by a carrier. The interaction of *d*-aldosterone and other steroids, especially 11-dehydrocorticosterone, with a $K_{i,PAH,cl}$ of 0.04 mmol/l, show that substances which do not bear a dissociable group, but negative, electrophilic and hydrogen bond forming side groups, can also interact with the contraluminal PAH transporter.

d-Aldosterone and many other steroid hormones also interact with the contraluminal TEA^+/$NMeN^+$ transporter (ULLRICH et al. 1993b). This indicates that the contraluminal cation transporter also interacts with substrates by H-bond formation only. However, the optimal spatial arrangement of the hydrogen bond forming OH and O groups for the contraluminal $NMeN^+$/TEA^+ transporter is different to that of the contraluminal PAH transporter.

G. How Does Metabolic Transformation Change the Interaction with the Transport Systems for Organic Anions and Cations?

Of the diuretics listed in this chapter, only a few, such as acetazolamide, hydrochlorothiazide, and amiloride, are excreted by the kidney without being metabolized (GEROK 1986). The othes become less hydrophobic through metabolic transformation – hydrolysis, oxidation, conjugation with glucuronic acid, sulfuric acid, and glutathione with subsequent processing to *N*-acetylcysteine (see Chap. 5, this volume). This diminishes or abolishes in the first instance their ability to bind to plasma albumin and enhances their chances of being excreted by filtration. However, by analogy with the behavior of other xenobiotics, some rules of altered affinities toward proximal tubular transport systems can also be established. Hydrolysis and hydroxylation create OH groups, which can serve as binding sites to the contraluminal PAH transporter (ULLRICH et al. 1990). They also create, however, the prerequistes for conjugation with sulfuric and glucuronic acid. These two processes add an anionic charge to the molecule and render it a substrate for the PAH transporter, and in the case of sulfation sometimes also a substrate for the contraluminal sulfate transporter (ULLRICH et al. 1990). Conjugation with glutathione can result in interaction of the conjugated substrate with the PAH transporter (ULLRICH et al. 1989). Further processing to a cysteine and to an *N*-acetylcysteine (mercapturic acid) conjugate increases the affinity toward the PAH transporter. Thus, as a rule, biotransformation increases the chances of a diuretic being excreted by filtration and by transport through the PAH transporter.

Conclusion. Of the 25 diuretics tested, 7 have a high affinity for the contraluminal PAH transport system, namely furosemide, ozolinone, sulfinpyrazone, ethacrynic acid, tienilic acid, canrenoate, and spironolactone. Thus it is likely that these diuretics are net secreted by this system. Amiloride and its analogs on the other hand show high affinity for the contraluminal and both luminal organic cation transport systems. For these substances either net secretion or net reabsorption by these systems seems possible, depending on the prevailing H^+-ion and other countertransporteé gradient over the brush border membrane of the proximal tubule. Half of the diuretics tested showed moderate affinity, usually to several transport systems. For these substances the interaction with the different transporters and other xenobiotics may be complicated and difficult to evaluate.

References

Bästlein C, Burckhardt G (1986) Sensitivity of rat renal luminal and contraluminal sulfate transport systems to DIDS. Am J Physiol 250:F226–F234

Bartel C, Wirtz C, Brändle E, Greven J (1993) Interaction of thiazide and loop diuretics with the basolateral para-aminohippurate transport system in isolated S_2 segments of rabbit kidney proximal tubules. J Pharmacol Exp Ther 266:972–977

Brändle E, Fritzsch G, Greven J (1991) The affinity of different anaesthetic drugs and catecholamines to the contraluminal transport system for organic cations in proximal tubules of rat kidney. J Pharmacol Exp Ther 260:734–741

Burckhardt G (1984) Sodium-dependent dicarboxylate transport in rat renal basolateral membrane vesicles. Pflugers Arch 401:254–261

David C, Ullrich KJ (1992) Substrate specificity of the luminal Na^+-dependent sulphate transport system in the proximal renal tubule as compared to the contraluminal sulphate exchange system. Pflugers Arch 421:455–465

David C, Rumrich G, Ullrich KJ (1995) Luminal transport system for H/organic cations in the rat proximal tubule. Kinetics, dependence on pH, specificity as compared with the contraluminal organic cation tranport system. Pflugers Arch (in press)

Fritzsch G, Haase W, Rumrich G, Fasold H, Ullrich KJ (1984) A stopped flow capillary perfusion method to evaluate contraluminal transport parameters of methylsuccinate from interstitium into renal proximal tubular cells. Pflugers Arch 400:250–256

Fritzsch G, Rumrich G, Ullrich KJ (1989) Anion transport through the contraluminal cell membrane of renal proximal tubule. The influence of hydrophobicity and molecular charge distribution on the inhibitory activity of organic anions. Biochim Biophys Acta 978:249–256

Gerok W (1986) Biotransformation von Diuretika bei Leberkrankheiten. In: Knauf H, Mutschler E (eds) Diuretika, Prinzipien der klinischen Anwendung. Urban und Schwarzenberg, Munich, pp 73–92
Johnston PA, Kau ST (1993) A micropuncture study on the renal site of action of ICI 206970, a unique eukalemic diuretic. J Pharmacol Exp Ther 264:604–608
Jørgensen KE, Kragh-Hansen U, Roigaard-Petersen H, Sheikh MI (1983) Citrate uptake by basolateral and luminal membrane vesicles from rabbit kidney cortex. Am J Physiol 244:F686–F695
Knauf H, Gerok W, Mutschler E (1992) Pharmakokinetik von Diuretika. In: Knauf K, Mutschler E (eds) Diuretika, Prinzipien der klinischen Anwendung, 2nd edn. Urban and Schwarzenberg, Munich, pp 149–187
Lemieux G, Vinay P, Paquin J, Gougoux A (1979) The renal handling of tienilic acid (Ticrynafen), a new diuretic with uricosuric properties. Nephron 23 [Suppl 1]:7–14
Löw I, Friedrich T, Burckhardt G (1984) Properties of an anion exchanger in rat renal basolateral membrane vesicles. Am J Physiol 246:F334–F342
Lücke H, Stange G, Murer H (1979) Sulphate-ion/sodium-ion co-transport by brush-border membrane vesicles isolated from rat kidney cortex. Biochem J 182: 223–229
Muirhead M, Hons BSc, Somogyi AA, Rolan PE, Bochner F (1986) Effect of cimetidine on renal and hepatic drug elimination: studies with triamterene. Clin Pharmacol Ther 40:400–407
Muirhead M, Bochner F, Somogyi A (1988) Pharmacokinetic drug interactions between triamterene and ranitidine in humans: alterations in renal and hepatic clearances and gastrointestinal absorption. J Pharmacol Exper Ther 244:734–739
Pritchard JB, Renfro JL (1983) Renal sulfate transport at the basolateral membrane is mediated by anion exchange. Proc Natl Acad Sci USA 80:2603–2607
Sheikh MI, Kragh-Hansen U, Jorgensen KE, Roigaard-Petersen H (1982) An efficient method for isolation and separation of basolateral-membrane and luminal-membrane vesicles from rabbit kidney cortex. Biochem J 208:377–382
Sheridan E, Rumrich G, Ullrich KJ (1983) Reabsorption of dicarboxylic acids from the proximal convolution of rat kidney. Pflugers Arch 399:18–28
Simchowitz L, Textor JA, Cragoe EJ Jr (1993) Cell volume regulation in human neutrophils: 2-(aminomethyl)phenols as Cl^- channel inhibitors. Am J Physiol 265:C143–C155
Turner RJ (1984) Sodium-dependent sulfate transport in renal outer cortical brush-border membrane vesicles. Am J Physiol 247:F793–F798
Ullrich KJ (1994) Specificity of transporters for "organic anions" and "organic cations" in the kidney. Biochim Biophys Acta 1197:45–62
Ullrich KJ, Rumrich G (1988) Contraluminal transport systems in the proximal renal tubule involved in secretion of organic anions. Am J Physiol 254:F453–F462
Ullrich KJ, Rumrich G (1992) Renal contraluminal transport systems for organic anions (para-aminohippurate, PAH) and organic cations (N^1-methyl-nicotinamide, NMeN) do not see the degree of substrate ionization. Pflugers Arch 421:286–288
Ullrich KJ, Rumrich G (to be published) Luminal transport system for choline in the rat proximal tubule.
Ullrich KJ, Rumrich G, Klöss S (1980) Active sulfate reabsorption in the proximal convolution of the rat kidney: specificity, Na^+ and HCO_3^- dependence. Pflugers Arch 383:159–163
Ullrich KJ, Fasold H, Rumrich G, Klöss S (1984) Secretion and contraluminal uptake of dicarboxylic acids in the proximal convolution of rat kidney. Pfluger Arch 400:241–249
Ullrich KJ, Rumrich G, Klöss S (1985a) Contraluminal sulfate transport in the proximal tubule of the rat kidney. II. Specificity: sulfate-ester, sulfonates and amino sulfonates. Pflugers Arch 404:293–299

Ullrich KJ, Rumrich G, Klöss S (1985b) Contraluminal sulfate transport in the proximal tubule of the rat kidney. III. Specificity: disulfonates, di- and tricarboxylates and sulfocarboxylates. Pflugers Arch 404:300–306
Ullrich KJ, Rumrich G, Klöss S (1985c) Contraluminal sulfate transport in the proximal tubule of the rat kidney. IV. Specificity: salicylate analogs. Pflugers Arch 404:307–310
Ullrich KJ, Rumrich G, Klöss S (1985d) Contraluminal sulfate transport in the proximal tubule of the rat kidney. V. Specificity: phenolphthaleins, sulfonphthaleins, and other sulfo dyes, sulfamoyl-compounds and diphenylamine-2-carboxylates. Pflugers Arch 404:311–318
Ullrich KJ, Rumrich G, Fritzsch G, Klöss S (1987a) Contraluminal para-aminohippurate (PAH) transport in the proximal tubule of the rat kidney. II. Specificity: aliphatic dicarboxylic acids. Pflugers Arch 408:38–45
Ullrich KJ, Rumrich G, Klöss S (1987b) Contraluminal para-aminohippurate transport in the proximal tubule of the rat kidney. III. Specificity: monocarboxylic acids. Pflugers Arch 409:547–554
Ullrich KJ, Rumrich G, Klöss S (1988) Contraluminal para-aminohippurate (PAH) transport in the proximal tubule of the rat kidney. IV. Specificity: mono- and polysubstituted benzene analogs. Pflugers Arch 413:134–146
Ullrich KJ, Rumrich G, Wieland T, Dekant W (1989) Contraluminal para-aminohippurate (PAH) transport in the proximal tubule of the rat kidney. VI. Specificity: amino acids, their N-methyl-, N-acetyl- and N-benzoylderivates; glutathione- and cysteine conjugates, di- and oligopeptides. Pflugers Arch 415:342–350
Ullrich KJ, Rumrich G, Gemborys M, Dekant W (1990) Transformation and transport: how does metabolic transformation change the affinity of substrates for the renal contraluminal anion and cation transporters? Toxicol Lett 53: 19–27
Ullrich KJ, Rumrich G, Hierholzer K (1991a) Contraluminal p-aminohippurate transport in the proximal tubule of the rat kidney. VIII. Transport of corticosteroids. Pflugers Arch 418:371–382
Ullrich KJ, Papavassiliou F, David C, Rumrich G, Fritzsch G (1991b) Contraluminal transport of organic cations in the proximal tubule of the rat kidney. I. Kinetics of N^1-methylnicotinamide and tetraethylammonium; influence of K^+, HCO_3^-, pH; inhibition by aliphatic primary, secondary and tertiary amines and mono- and bisquarternary compounds. Pflugers Arch 419:84–92
Ullrich KJ, Rumrich G, Gemborys MW, Dekant W (1991c) Renal transport and nephrotoxicity. In: Bach PH, Gregg NJ, Wilks MF, Delacruz L (eds) Nephrotoxicity, chapter 1. Dekker, New York, pp 1–8
Ullrich KJ, Rumrich G, Neiteler K, Fritzsch G (1992a) Contraluminal transport of organic cations in the proximal tubule of the rat kidney. II. Specificity: anilines, phenylalkylamines (catecholamines), heterocyclic compounds (pyridines, quinolines, acridines). Pflugers Arch 420:29–38
Ullrich KJ, Rumrich G, Fritzsch G (1992b) Substrate specificity of the organic anion and organic cation transport systems in the proximal renal tubule. Prog Cell Res 2:315–321
Ullrich KJ, Rumrich G, David C, Fritzsch G (1993a) Bisubstrates: substances that interact with renal contraluminal organic anion and organic cation transport systems: I. Amines, piperidines, piperazines, azepines, pyridines, quinolines, imidazoles, guanidines, and hydrazines. Pflugers Arch 425:280–299
Ullrich KJ, Rumrich G, David C, Fritzsch G (1993b) Bisubstrates: substances that interact with both renal contraluminal organic anion and organic cation transport systems: II. Zwitterionic substrates: dipeptides, cephalosporins, quinolone-carboxylate gyrase inhibitors, and phosphamide thiazine carboxylates; nonionizable steroid hormones and cyclophosphamides. Pflugers Arch 425:300–312

Ullrich KJ, Fritzsch G, Rumrich G, David C (1994) Polysubstrates: substances that interact with renal contraluminal PAH, sulfate, and NMeN/TEA transport: sulfamoyl-, sulfonylurea-, thiazide-, and benzeneamino-carboxylate (nicotinate) compounds. J Pharmacol Exp Ther 269:684–692

Werness PG, Bergert JH, Smith LH (1982) Triamterene urolithiasis: solubility, pK, effect on crystal formation, and matrix binding of triamterene and its metabolites. J Lab Clin Med 99:254–262

Wright SH, Wunz TM (1987) Succinate and citrate transport in renal basolateral and brush-border membranes. Am J Physiol 253:F432–F439

Wright SH, Kippen I, Klinenberg JR, Wright EM (1980) Specificity of the transport system for tricarboxylic acid cycle intermediates in renal brush borders. J Membr Biol 57:73–82

CHAPTER 7

Loop Diuretics

R. GREGER

A. Introduction

Loop diuretics are the most potent diuretics and saluretics known. The term "loop" diuretics was introduced only after their invention and introduction into therapy when it was recognized in 1973 that they act in the thick ascending limb of the loop of Henle (BURG et al. 1973). Another term for these diuretics is 'high-ceiling" diuretics, a term based on the finding that diuresis and saluresis increase dose dependently over a wide range of dosages, i.e., diuresis and saluresis continue to increase almost linearly when dosage is increased. This is valid until a diuresis of 20%–30% of glomerular filtration rate (GFR) is achieved. Then, even for the "high-ceiling" diuretics, the dose-response curve levels off. The term was introduced to distinguish this group of diuretics from the thiazide group, which have a much lower "ceiling effect" of less than 10% of GFR (see Chap. 8, this volume). These, consequently, have been named "low-ceiling" diuretics.

Loop diuretics comprise a chemically heterogeneous group. Phenoxyacetic acid type diuretics, such as ethacrynic acid and indacrinone (Fig. 1), mercurial diuretics, furosemide-related drugs, and several others belong to this group. Of these the furosemide-derived compounds have become almost exclusively of therapeutic importance, whereas the others have been abandoned for therapy (mercurials) or are used only rarely (ethacrynic acid). The furosemide-derived diuretics comprise a specific entity of loop diuretics. Their cellular mode of action has been clarified and will be discussed in Sect. D.

Loop diuretics of the furosemide type were invented in 1964 by MUSCHAWECK and HAJDÚ in the search for new sulfonamide diuretics (see Chap. 4, this volume). Their mode of action has only been known for the 19 years since their invention and introduction into therapy. Their pharmacodynamic and pharmacokinetic properties, their usage, and their side effects will be the focus of this chapter. It will become evident that these substances all have a clearly defined mode of action, that they are very potent saluretic and diuretic substances, and that they all share the same side effects. They are indispensable therapeutic tools in the treatment of diseases with a variety of clinical indications.

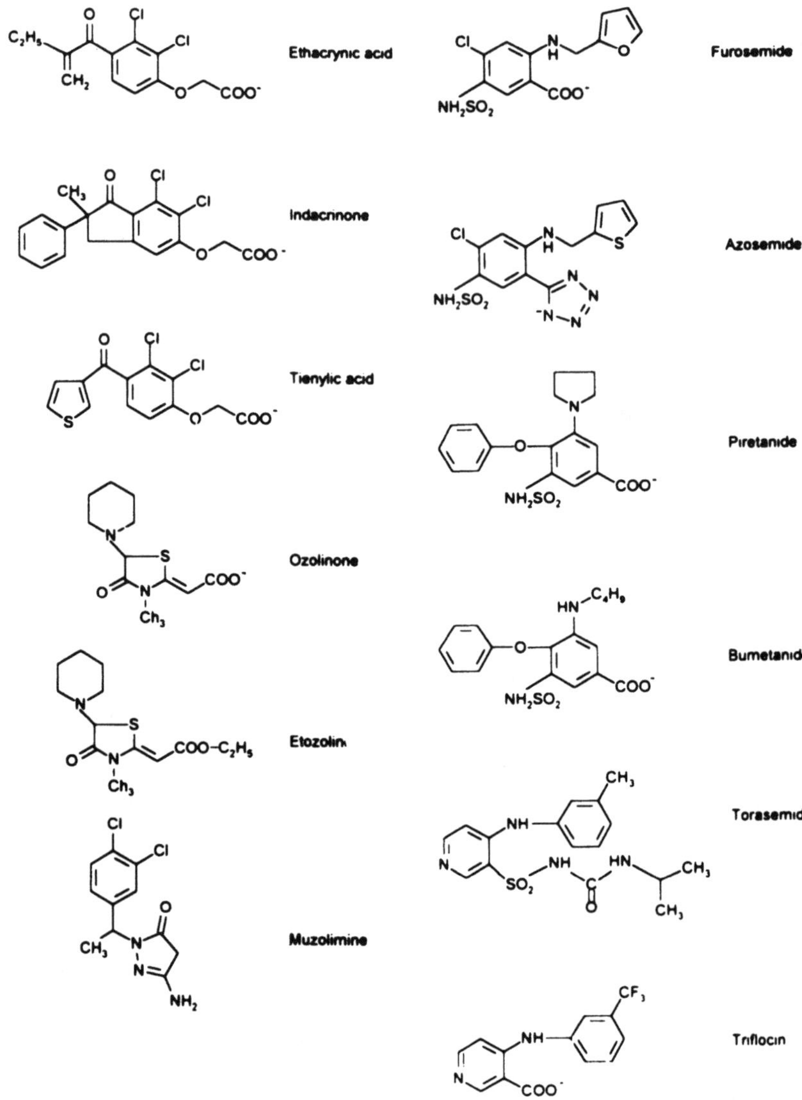

Fig. 1. Structural formulas of the loop diuretics

B. The Heterogeneous Group of Loop Diuretics

Table 1 and Fig. 1 list some loop diuretics and their general properties. They are classified into mercurials, phenoxyacetic acids, furosemide-related compounds, and others (see Chap. 4, this volume). Mercurials are no longer of pharmacological importance and will therefore not be discussed in any detail in this chapter. The most prominent examples of the phenoxyacetic acids are

Table 1. Biophysical, pharmacodynamic, and pharmacokinetic data of various loop diuretics

Loop diuretic	Distribution (l/kg)	Fractional absorption (%)	Protein binding (%)	Half-life (h)	Metabolism (%)	IC_{50} (mol/l)
Ethacrynic acid					SH-coupling	5×10^{-6}
Furosemid	0.11	60	98	1	20–30	3×1.0^{-6}
Bumetanide	0.13	90	90	1.5	40	0.2×10^{-6}
Piretanide		90	96	1.5	40	10^{-6}
Azosemide		<50	High			3×10^{-6}
Torasemide	<0.15	85	High	3	75	0.3×10^{-6}

ethacrynic acid, indacrinone, and others. Loop diuretics of the furosemide type are the sulfonamide diuretics. All compounds used, furosemide, piretanide, azosemide, bumetanide, and even torasemide and triflocin, share one general structure in common, which will be discussed in the section on structure-activity relations (Sect. D.VI). The remaining loop diuretics are largely heterogeneous. Some are pro-drugs and need to be metabolized before they become pharmacologically active. Etozolin and muzolimine belong to this group. The active metabolite of etozolin is known to be ozolinone (GREVEN and HEIDENREICH 1978), while that of muzolimine is still unknown (WANGEMANN et al. 1987).

Most loop diuretics are weak organic acids with pK_a values ranging between 7 and 4. Most are largely bound to plasma proteins. The free fraction is as low as 2% for furosemide and 4% for piretanide. Many loop diuretics are hydrophilic. Examples are furosemide and related drugs. However, some more recently developed drugs such as torasemide are also soluble in lipids. These properties are of relevance for the understanding of their pharmacokinetics (see Sect. G).

C. Organotropy of Loop Diuretics

All loop diuretics are accumulated in tubule fluid. This occurs by secretion in the proximal tubule (see Chap. 6, this volume) and by the reabsorption of water. As a result the concentration of the respective compound is much higher in the luminal fluid of the thick ascending limb of the loop of Henle (TAL) than the free concentration in plasma. This aspect is highly relevant for the organ specificity of these compounds. In other words: It will be shown that most of the loop diuretics interfere with transport systems which are present in many if not all cells of the body. The organotropy of these substances is a result of their pharmacokinetics and is not related to their pharmacodynamic effect. This has an important ramification: whenever the secretion of these compounds by the proximal tubule is inhibited because of, e.g., acute ischemia or pharmacological interference, they lose their organotropy, they are much less diuretically active, and they may produce severe side effects because they accumulate.

Compounds which interfere with the proximal secretion of loop diuretics are probenecid and many other substrates of the so-called *para*-aminohippuric acid (PAH) transporter in the basolateral membrane (ULLRICH et al. 1989; Chap. 6, this volume). Many of these substances are summarized in the respective chapter of this volume (Chap. 6). Important examples are probenecid, antibiotics such as penicillin, nonsteroidal anti-inflammatory drugs (NSAIDs), and many others. The effect of probenecid has been examined in clearance studies in the rat (BRAITSCH et al. 1990) and it was shown that probenecid shifted the dose response curve of loop diuretics such as furosemide, piretanide, bumetanide, azosemide, and torasemide to the right. An example of this is shown in Fig. 2. It is worth noting that the effect of probenecid was comparable for all diuretics when these were used at a dosage exerting a half-maximal diuretic and saluretic effect. The interference is likely to occur at the level of the secretory system localized in the basolateral membrane of the proximal tubule. The data presented in Fig. 2 suggest that this secretory system has different affinities for the various diuretics, with the highest affinities for bumetanide and piretanide and a lower affinity for azosemide and furosemide (see Chap. 6, this volume).

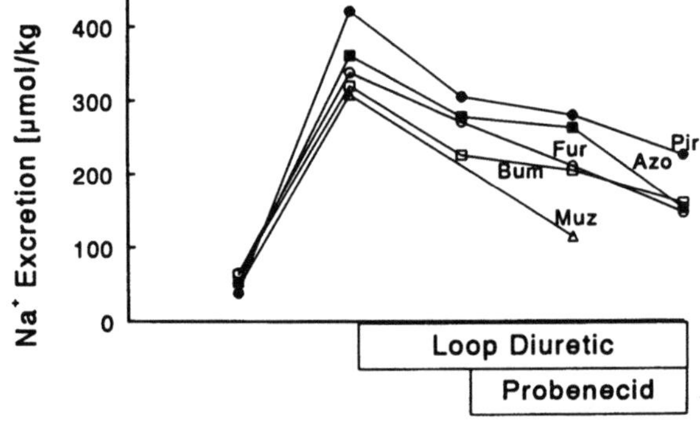

Fig. 2. Effect of probenecid on loop diuretic induced diuresis. Data taken from BRAITSCH et al. (1990) and LOHRMANN et al. (1991). Clearance experiments in rats. The animals were administered equipotent doses of diuretics, exerting approximately half-maximal natriuresis. *Muz*, muzolimine; *Bum*, bumetanide; *Fur*, furosemide; *Azo*, azosemide; *Pir*, piretanide. After stable natriuresis was established (for clarity only the mean values are given here) probenecid was administered i.v. in increasing doses of 0.07, 0.14, and 0.28 mmol/kg. Note that probenecid attenuates the saluresis dose dependently for all used loop diuretics (the effects were significant for the lowest dose). Probenecid up to 0.14 mmol/kg had no effect on GFR in these studies

It is well known that the ability to secrete, e.g., PAH, is compromised rapidly in ischemic kidney. The secretion of loop diuretics is reduced to the same extent. In view of this it is easily seen that renal ischemia compromises the secretion and hence the action of loop diuretics. In addition, ischemia may quickly diminish NaCl reabsorption in the proximal tubule and even more so in the TAL. As a consequence, the tubule concentration of the used loop diuretic is even lower, and higher dosages have to be used to induce diuresis and saluresis. In such a state it may well prove necessary to increase the dosage of a loop diuretic. Under these conditions the tubule concentration is only slightly higher than that in circulating plasma. Therefore, the loop diuretics may start to exert systemic effects which never occur under normal conditions. One such acute complication is the ototoxic effect (see Sects. F.I, J.X).

One might predict that loop diuretics due to their interaction with the PAH-secretory system, which is probably identical to the urate secretory system, might produce uricosuria. However, it is well known that the opposite is the case. Loop diuretics usually induce antiuricosuria and enhance plasma urate concentration. The predominant antiuricosuric effect is due to an increase in urate reabsorption in the late proximal tubule (LANG et al. 1977). This effect outdates by far the limited uricosuric action in the early proximal tubule.

D. Saluretic and Diuretic Effects of Loop Diuretics and Cellular Mechanisms

Shortly after their introduction it was shown that loop diuretics such as furosemide act in the proximal tubule and inhibit NaCl and water reabsorption at this site (DEETJEN 1966). This effect has been clearly demonstrated in micropuncture studies in the rat and was also entirely predicted because loop diuretics of this type were predicted to have some carbonic anhydrase inhibitory potency (for review see SUKI and EKNOYAN 1992). On the other hand, clearance studies suggested that loop diuretics such as ethacrynic acid and furosemide interfere with the concentrating ability of the kidney, an effect which would be localized in the thick ascending limb of the loop of Henle (TAL). It was not until the TAL segments were perfused in vitro (BURG and STONER 1976) that it was possible to examine the putative effect of loop diuretics in these in vitro perfused segments (BURG and BOURDEAU 1978; BURG and GREEN 1973a; BURG et al. 1973). In these experiments it immediately became evident that all loop diuretics examined mercurials (BURG and GREEN 1973b), ethacrynic acid (BURG and BOURDEAU 1978), and furosemide (BURG et al. 1973) inhibited active NaCl reabsorption in the medullary and cortical portion of the TAL segment (ROCHA and KOKKO 1973; BURG 1982). Not only were these experiments crucial to the understanding of the tubular site of action of this group of diuretics, but these same experiments were also important for the characterization of the mechanism of NaCl reabsorption in this nephron segment.

Originally it was believed that furosemide and related diuretics inhibited a Cl^- pump (Cl^--ATPase) in this nephron segment, inasmuch as the respective experiments appeared to indicate that Cl^- reabsorption was primarily active (ROCHA and KOKKO 1973; BURG and GREEN 1973b). The evidence was as follows: This nephron segment exposes a lumen-positive voltage when the composition of the solutions on the two sides of the epithelium is identical. Such a voltage would be compatible with a primary active mechanism of Cl^- reabsorption. Furthermore, experiments in which most of the Na^+ on both sides of the tubule was replaced by choline indicated that the voltage was, if anything, enhanced but not reduced (BURG and GREEN 1973b). On the other hand, replacement of Cl^- by large anions such as sulfate and methylsulfate abolished the transepithelial voltage. These experiments seemed to indicate that the active transport of NaCl in the TAL segment was relying on Cl^- but not on Na^+ (BURG and GREEN 1973b; ROCHA and KOKKO 1973).

After a few years these types of experiments were repeated with modified techniques (GREGER 1981a). It was shown then that, in contrast to previous observations, the active NaCl reabsorption was entirely dependent on Na^+. However, this dependence was only apparent at very low Na^+ concentrations. This is shown in Fig. 3 (GREGER 1981, 1985). Active NaCl reabsorption, shown here as equivalent short circuit current, was reduced sharply when the luminal Na^+ concentration was reduced to below 3 mmol/l. However, active transport was unaltered at higher Na^+ concentrations (Fig. 3). These data indicated that active NaCl reabsorption was entirely Na^+ dependent and that the affinity for Na^+ was very high. In fact, to inhibit active Cl^- reabsorption completely, Na^+-free glass had to be used in the perfusion experiments (GREGER 1981).

In similar studies the Cl^- dependence was examined and it was shown that the Cl^- dependence was S-shaped with a Hill coefficient of 2, suggesting that $2Cl^-$ ions have to be bound to the respective transporter (Fig. 3) (GREGER et al. 1983b; GREGER 1985). By that time a loop diuretic sensitive KCl transporter had already been described in Ehrlich ascites tumor cells (GECK et al. 1980) and it was postulated, on the basis of a flux analysis, that this transporter was a $Na^+2Cl^-K^+$ cotransporter (GECK et al. 1980; GECK and HEINZ 1986). Hence, it was examined whether active NaCl reabsorption in TAL segments was K^+ dependent. Respective data are shown in Fig. 3. It is evident that active NaCl reabsorption requires the presence of K^+ (GREGER and SCHLATTER 1981). The apparent affinity was around 1 mmol/l.

Active reabsorption of NaCl could be inhibited by luminal removal of one of the three ions Na^+, Cl^-, or K^+, by adding Ba^{2+} to the luminal perfusate (GREGER and SCHLATTER 1981; GREGER 1985), by adding a loop diuretic such as furosemide to the luminal perfusate (GREGER and SCHLATTER 1983a), by adding ouabain to the peritubular bath (GREGER 1985a), by adding Cl^- channel blockers such as nitrophenylpropyl-aminobenzoate

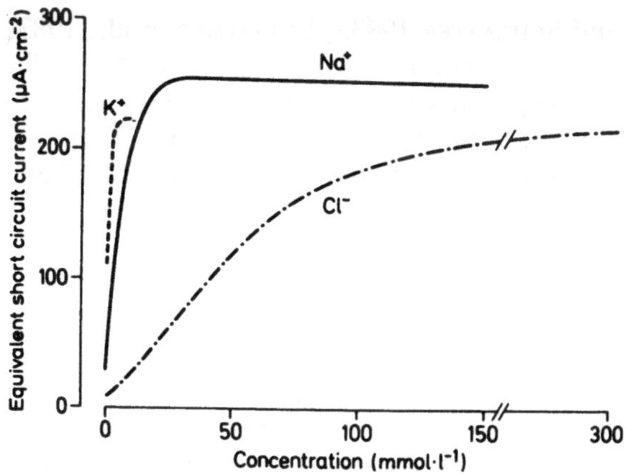

Fig. 3. Na^+, Cl^-, and K^+ dependence of active transport in TAL. Data are summarized by Greger (1985b). The equivalent short circuit current of isolated perfused rabbit cortical thick ascending limb segments (cTAL) is plotted as a function of the luminal concentration of the respective ion. Note that the $Na^+2Cl^-K^+$ cotransporter present in the luminal membrane has a very high affinity for Na^+ and K^+ but a much smaller affinity for Cl^-. The sigmoid shape of the Cl^- curve is caused by the fact the 2 Cl^- ions interact with the carrier (Greger et al. 1983b)

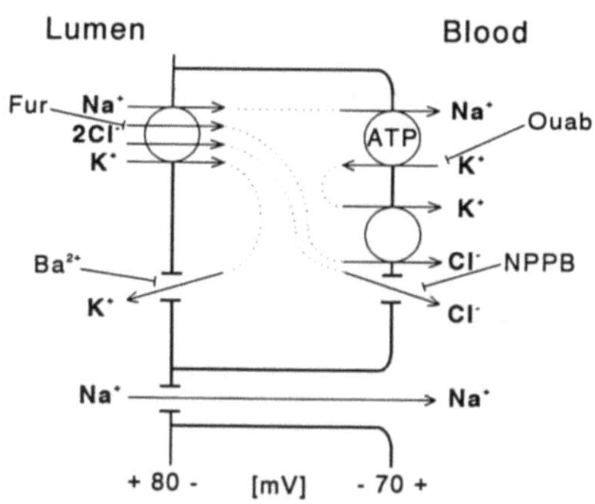

Fig. 4. Model of NaCl reabsorption in cTAL. Modified from Greger (1985b). *Circle with ATP*, ATP-driven pump; *circle*, cotransport systems; *arrows*, ion conductances; *Fur*, furosemide (or any other loop diuretic of this group); *Ouab*, ouabain; *NPPB*, 5-nitro-2-(3-phenylpropylamino)-benzoate. For further details on this model see text

(NPPB, see below) (Wangemann et al. 1986) to the bath, and by removing metabolic substrates from the bath solution (Wittner et al. 1984; Greger and Wangemann 1987). The analysis of these effects required in addition to transepithelial measurements the measurement of the basolateral membrane voltage with microelectrodes and the measurement of the cytosolic ion activities by microelectrodes. This was achieved in 1981–1984 (Greger

1981b; GREGER and SCHLATTER 1983a, b; GREGER et al., 1983, 1984a), and a model was designed to account for the various findings (Fig. 4). It is the model which was proposed originally by ourselves and which still holds today (GREGER 1985b). On the basis of this model the above inhibitory effects will be discussed below in some detail to gain further insight into the function of the TAL segment.

I. Luminal K^+ Conductance

Figure 5 shows the inhibitory effect of Ba^{2+} (GREGER and SCHLATTER 1981). The effect occurs rapidly and is fully reversible, requiring 1–3 mmol/l for half-maximal inhibition and 20 mmol/l (HEBERT and ANDREOLI 1984) for complete inhibition. Ba^{2+} is known to block K^+ channels (for review see LANG 1991). This inhibition occurs from the cytosolic side (BLEICH et al. 1990). The respective K^+ channel in the luminal membrane serves to hyperpolarize this cell membrane (Fig. 4) and allows for a recycling of K^+. This recycling is required since the $Na^+2Cl^-K^+$ cotransporter reabsorb Na^+ and K^+ at equal rates, and the luminal perfusate contains only a low concentration of K^+. Hence, Ba^{2+} inhibits K^+ recycling and reduces luminal K^+ concentration to a value which becomes rate limiting for the $Na^+2Cl^-K^+$ cotransporter.

The properties of this luminal K^+ conductance have been examined recently by patch clamp techniques (BLEICH et al. 1990; WANG et al. 1990).

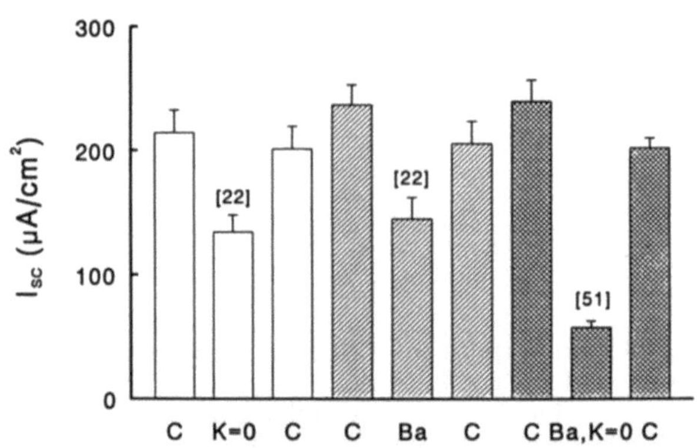

Fig. 5. Effect of Ba^{2+} and low luminal K^+ concentration on active NaCl reabsorption in rabbit cTAL segments. Data [mean values ± SEM (*n*)] are taken from GREGER and SCHLATTER (1981). Active NaCl reabsorption is shown as equivalent short-circuit current (I_{sc}). *C*, control, i.e., symmetrical Ringer's solution on both sides of the in vitro perfused tubule segment; *K = 0*, K^+ removed from luminal perfusate; *Ba*, 1 mmol/l Ba^{2+} added to luminal perfusate; *Ba, K = 0*, K^+ removed from and 1 mmol/l Ba^{2+} added to luminal perfusate. Note that the addition of Ba^{2+} to the luminal perfusate inhibits I_{sc} significantly. This effect is augmented in the apparent absence of K^+

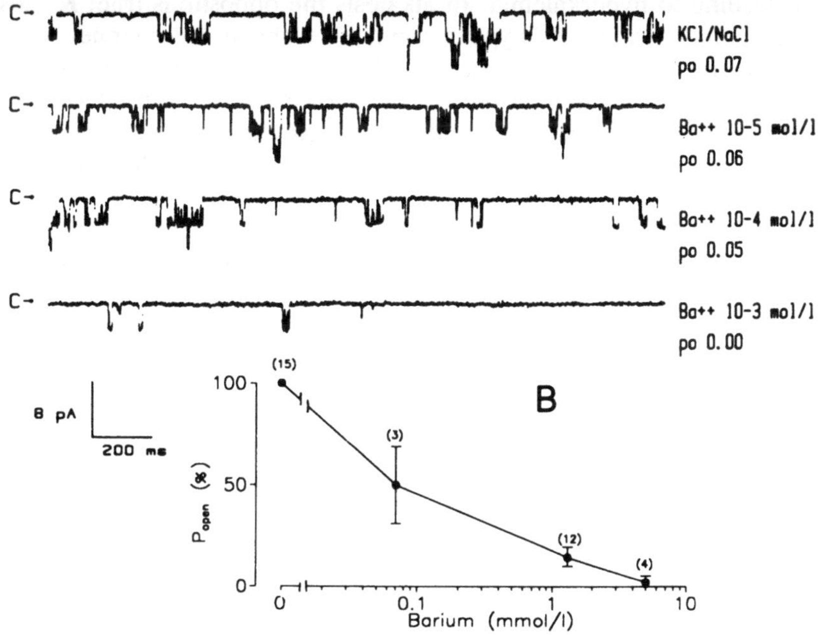

Fig. 6. K^+-channel in the luminal membrane of rat cTAL. Data taken from BLEICH et al. (1990). Patch clamp analysis of an excised inside out oriented membrane patch of the luminal membrane. Pipette, K^+-Ringer's solution; bath, Na^+-Ringer's solution. Clamp voltage, 0 mV. The K^+ gradient drives a K^+ current from the pipette to the bath. Two single K^+ channels are present in this patch (*upper three trances*). Increasing Ba^{2+} concentrations in the bath reduce the single-channel open probability (*po*). At 1 mmol/l po approaches 0. *Lower panel*, concentration response curve for Ba^{2+}. The single-channel open probability (P_{open}) is plotted versus the Ba^{2+} concentration on the cytosolic side [mean values ± SEM (*n*)]. Half-maximal inhibition of this K^+ channel by Ba^{2+} occurs at 100 μmol/l

In cultured TAL cells a large conductance K^+ channel was found. This channel was activated by Ca^{2+} and was blocked by a variety of blockers (GUGGINO et al. 1985, 1987a,b; CORNEJO et al. 1987, 1989a,b). However, an analysis for intact TAL cells indicated that the K^+ channel present in the luminal membrane of these segments was a much smaller channel with distinct properties (BLEICH et al. 1990). An example of this channel is shown in Fig. 6. It is highly selective for K^+ ions, and even distinguishes between Rb^+ and K^+ ions. This channel is active even at low cytosolic Ca^{2+} activities. In fact, a reduction in Ca^{2+} activity even enhances its open probability. It is inhibited by a variety of K^+ channel blockers, Ba^{2+}, quinine, quinidine, verapamil, lidocaine, choline, Cs^+, and is regulated by cytosolic pH (GREGER et al. 1991). Acid pH reduces and alkaline pH enhances the open probability. The pH regulation of this channel may be of particular relevance for K^+ homeostasis: In acidosis the channel is inhibited and K^+ secretion is minimized, leading to hyperkalemia. In alkalosis the opposite is true; K^+ losses are augmented, leading to hypokalemia. The above K^+ channel was the only channel found in the luminal membrane of TAL cells. This indicates that findings obtained in cultured cells cannot easily be translated into the intact tissue. Furthermore, these findings confirm the conclusion drawn from previous experiments, namely that the only relevant conductance of the luminal membrane is that for K^+ (GREGER et al. 1983b; GREGER 1985b). A K^+ channel from kidney medulla has recently been cloned (S. HEBERT 1993, personal communication). This channel might be identical to the above channel identified in patch clamp studies.

II. Furosemide-Sensitive $Na^+2Cl^-K^+$ Cotransporter

The Na^+, Cl^-, and K^+ dependence of the $Na^+2Cl^-K^+$ cotransporter has already been discussed above (Fig. 3). Furosemide-type loop diuretics bind to the same cotransporter with high or very high affinity. An example of an experiment is shown in Fig. 7. Luminal addition of furosemide abolishes the

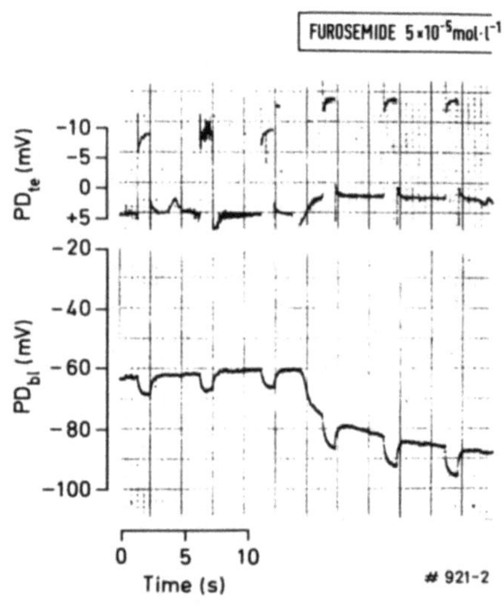

Fig. 7. Effects of furosemide on the transepithelial potential difference (PD_{te}) and on the potential difference across the basolateral membrane (PD_{bl}) of an isolated in vitro perfused rabbit cTAL segment. Data taken from GREGER and SCHLATTER (1983b). *Upper trace*, PD_{te}; *lower trace*, simultaneously measured PD_{bl}. The voltage deflections are caused by current pulses injected into the lumen. At the time indicated 50 μmol/l furosemide was added to the luminal perfusate. This abolished PD_{te} within 2 s. At the same time PD_{bl} hyperpolarized markedly until, after some 10–15 s, the equilibrium potential for K^+ (ca. −90 mV) was achieved. These data indicate that furosemide abolishes Cl^- entry from the lumen, and that hence cytosolic Cl^- activity falls until the equilibrium potential for Cl^- reaches that for K^+. Also note that the basolateral membrane resistance, as reflected by the voltage excursions in the PD_{bl} trace, increases after furosemide administration. The effects of furosemide are fully and rapidly reversible (not shown in this figure)

transepithelial voltage and hyperpolarizes the luminal membrane to values very close to the EMF for K^+. These effects occur very rapidly (within 1–2 s) and they are completely reversible (GREGER and SCHLATTER 1983c). Measurements with K^+- (GREGER et al. 1984a) and Cl^--selective (GREGER et al. 1983a) microelectrodes indicate that furosemide has no effect on cytosolic K^+ activity but reduces that for Cl^- markedly. The electromotive force (EMF) for Cl^- is around −40 mV under control conditions and approaches that for K^+ (−80 to −90 mV) in the presence of furosemide. These data have been interpreted to indicate that drugs such as furosemide bind to the $Na^+2Cl^-K^+$ cotransporter. The uptake of Na^+, Cl^-, and K^+ are inhibited. This leads to a fall in cytosolic Na^+ activity because the (Na^++K^+)-ATPase keeps pumping out Na^+. This has been documented for the rectal gland of *Squalus acanthias*, in which NaCl is secreted by the $Na^+2Cl^-K^+$ cotransporter (GREGER and SCHLATTER 1984). Subsequently it was also shown for the TAL (RAJERISON et al. 1988). Similar to that of Na^+, the cytosolic Cl^- activity falls after the administration of furosemide because Cl^- is extruded via basolateral Cl^- channels and via the K^+Cl^- symporter (GREGER 1985). These processes cease when the EMF for Cl^- approaches the EMF for K^+. The K^+ activity changes little after furosemide administration because the reduced uptake via the $Na^+2Cl^-K^+$ cotransporter in the transition phase is matched by the still ongoing uptake of K^+ via the (Na^++K^+)-ATPase (GREGER et al. 1984). The effect of furosemide has also been shown in vesicle transport and binding studies for TAL cells (KOENIG et al. 1983; KINNE et al. 1986). It was shown that furosemide or related compounds bind to one of the Cl^--binding sites of the $Na^+2Cl^-K^+$ cotransporter. For the binding to occur, all three ions need to be present, and the affinities of the three ions match closely those obtained in the above transport studies (Fig. 3).

It has long been unclear whether furosemide binds from the outside (luminal fluid) or whether it has to enter the cell to exert its effect on the cotransporter. This issue has been clarified recently by experiments in which the furosemide-related drug piretanide was conjugated to polyethylene glycol or dextran, both with a molecular mass of 5000 Da. It was shown that the concentration response curve for piretanide and the conjugated large molecular weight derivatives were very similar. Since these large molecules can hardly enter the cell, these data indicate that the binding of these diuretics to the $Na^+2Cl^-K^+$ cotransporter occurs on the luminal surface of the TAL membrane (NITSCHKE et al. 1989).

Very similar data on transport and on binding of furosemide and related compounds have been obtained with other cells and preparations containing the $Na^+2Cl^-K^+$ cotransporter (PALFREY et al. 1980; HAAS 1989; BURNHAM et al. 1990). These data suggest that the furosemide-sensitive $Na^+2Cl^-K^+$ cotransporter of various tissues is very similar. For many cells the loop diureticsensitive $Na^+2Cl^-K^+$ cotransporter has been postulated. Among these are various epithelial organs: exocrine glands, small intestine, colon crypt, trachea, cornea, epididymis, hepatocytes; and apolar cells: heart muscle, smooth muscle, endothelial cells, red blood cells, ganglion cells, glial cells, fibrobalsts (for review, e.g., GREGER 1986). The affinity towards loop diuretics is similar in all tissues. This raises the possibility that the cotransporter is similar in the various tissues. Attempts have been made for more than 10 years now to clone this protein. A final answer seems to be in prospect (S. HEBERT, personal communication 1993). Thus far the published data favor the view that the cotransporter has a molecular mass of >100 kDa (FORBUSH et al. 1987). However, much smaller molecular masses have also been proposed (DI STEFANO et al. 1986; FEIT et al. 1988). It is likely that this cotransporter has regulatory sites which are controlled by PKA- or PKC-dependent phosphorylation (HAAS 1989). With its cloning from the TAL it will also be possible to probe all the above tissues for this cotransporter.

The driving force for the cotransporter-mediated uptake of Na^+, $2Cl^-$, and K^+ amounts to approximately 60 mV for Na^+, 80 mV for Cl^-, and −90 mV for K^+, i.e., 50 mV (GREGER 1985b). Under physiological conditions the luminal concentrations of Na^+, Cl^-, and K^+ are between 30 and 150 mmol/l (NaCl) and 4–10 mmol/l (K^+). It can be seen easily from Fig. 3 that Cl^- will be the transport-limiting ion. The Cl^- transport sites will become desaturated (apparent K_M, 30–40 mmol/l), and the transport rate will be reduced accordingly. This situation will be met in the cortical TAL in volume contraction and antidiuresis. Then the early distal NaCl concentration may be around or even lower than 30 mmol/l (WIRZ 1956). In diuresis and volume expansion the situation is quite different. Then luminal Cl^- concentration stays above 50–100 mmol/l and reabsorption continues throughout the cortical TAL. It will be shown below that the luminal dilution superimposes a (diffusion) voltage on top of the active transport voltage.

III. Role of the Basolaterally Localized ($Na^+ + K^+$)-ATPase

While originally a Cl^--ATPase was assumed to account for active NaCl reabsorption in the TAL (BURG and GREEN 1973b), it soon became evident that this nephron segment was especially rich in (Na^++K^+)-ATPase (SCHURRMANS STEKHOEN and BONTING 1981). Inhibition of this pump by ouabain slowly reduced active reabsorption. This was paralleled by a depolarization of the basolateral (and luminal) membrane voltage, with an increase in cytosolic Na^+ and Cl^- concentrations and with cell swelling (GREGER 1985b; RAJERISON et al. 1988a; RAJERISON 1988b). An example of one such experiment is shown in Fig. 8. The effect of ouabain is slow and reversible, provided ouabain has not been present for too long. The effects of ouabain mentioned above can all be explained by its direct effect on the (Na^++K^+)-ATPase. Ouabain has one additional effect, namely to reduce the K^+ conductance of the luminal membrane (GREGER et al. 1984b). This effect could be caused by one of the following cytosolic second messengers:

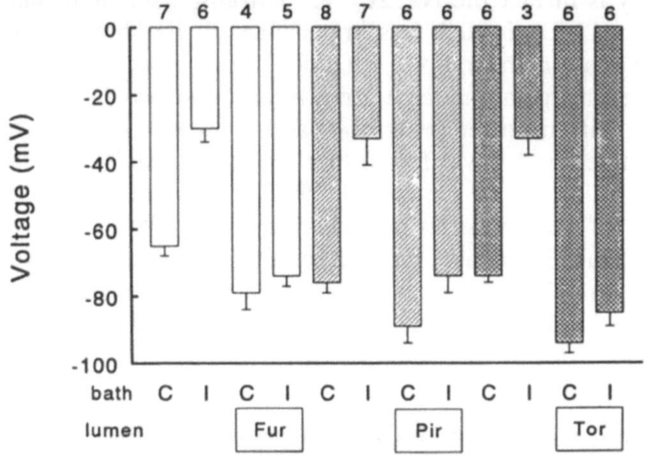

Fig. 8. The inhibitory effect (*I*) of ouabain (10 μmol/l, added to the bath) is abolished by loop diuretics. Data taken from GREGER (1985b), GREGER and WANGEMANN (1987), and LOHRMANN et al. (1993). The basolateral membrane potential [voltage, mean values ± SEM (*n*)] is given for the various conditions. *C*, control; *I*, ouabain. Note that furosemide (FUR, 50 μmol/l, lumen) hyperpolarizes the membrane voltage (see also Fig. 7). Ouabain, due to the inhibition of the (Na^+-K^+)-ATPase, depolarizes the cell. The effect of ouabain is almost completely abolished in the presence of furosemide. Due to the furosemide-induced inhibition of Na^+ uptake, the rate of the (Na^+-K^+)-ATPase is also reduced sharply. Similar observations were made for piretanide (*Pir*, 5 μmol/l, lumen) and torasemide (*Tor*, 10 μmol/l, lumen)

(a) with increased ATP (less ATP is metabolized in the presence of ouabain) the K^+ channel of the luminal membrane is inhibited by ATP (BLEICH et al. 1990; WANG et al. 1990), (b) increased cytosolic Ca^{2+}, since Ca^{2+} is probably exported by an Na^+/Ca^{2+} exchanger, increased Ca^{2+} inhibits the K^+ channel in the luminal membrane (BLEICH et al. 1990), (c) cytosolic acidification, since H^+ ions are extruded in the TAL by the Na^+/H^+ exchanger, ouabain decreases cytosolic pH, and decreased pH inhibits the K^+ channel in the luminal membrane (BLEICH et al. 1990).

The effect of ouabain is explained as the inhibition of the primary active transport mechanism, namely the ATP-driven export of Na^+. The model in Fig. 4 predicts that the most relevant entry pathway for Na^+ is the $Na^+2Cl^-K^+$ cotransporter. Inhibition of this cotransporter, e.g., by furosemide, should block the Na^+ entry and hence obliterate the ouabain effects. This type of experiment has been performed with furosemide (GREGER 1985b), piretanide (GREGER and WANGEMANN 1987), and torasemide (LOHRMANN et al. 1993). The results of all three studies were comparable. A typical experiment is shown in Fig. 8. While ouabain depolarized the basolateral voltage under control conditions, no such effect was noted in the presence of the loop diuretic. These findings indicate that the major source of Na^+ entry is in fact the $Na^+2Cl^-K^+$ cotransporter, and that other Na^+ uptake systems are of much less importance. It has been argued that this may be the case in the rabbit, but that, e.g., in the mouse Na^+/H^+ and Cl^-/HCO_3^- exchange are much more relevant, and may be as relevant as the $Na^+2Cl^-K^+$ cotransporter (FRIEDMAN and ANDREOLI 1982). This issue has been reexamined and it was found that HCO_3^- has no measurable effect on NaCl reabsorption (DI STEFANO et al. 1992). Therefore, it appears likely that the $Na^+2Cl^-K^+$ cotransporter is the major source of Na^+ entry in all species investigated thus far (mouse, rat, and rabbit). This does, however, not imply that Na^+/H^+ and Cl^-/HCO_3^- exchange are not present. In fact, the opposite appears quite likely, and these systems may serve to reclaim some HCO_3^- and to control cytosolic pH (GOOD 1985; KNEPPER et al. 1985).

IV. Metabolic Control of NaCl Reabsorption in the TAL

The TAL segment has a very high ATP turnover (see Chap. 3, this volume) and transport rates; yet the cells are small and do not possess fuel stores. As a consequence, NaCl absorption is fuel dependent (WITTNER et al. 1984). This is shown in Fig. 9. When all metabolic substrates are removed from the perfusates, active transport collapses fairly rapidly. After 3 min only 50% of the control transport rate is observed; after 10 min transport collapses altogether, and does not recover completely if the substrate free perfusion lasts for more than 20 min. It has also been shown that the key substrates are taken up across the basolateral membrane. The metabolism is obviously aerobic (WITTNER et al. 1984; GREGER 1985b).

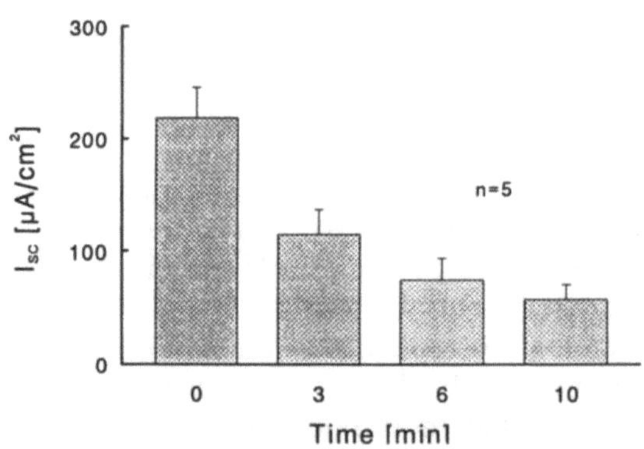

Fig. 9. Effect of substrate removal in isolated perfused rabbit cTAL segments. Data taken from WITTNER et al. (1984). The equivalent short circuit current (I_{sc}, mean values ± SEM, n) corresponding to the rate of active NaCl reabsorption, is plotted versus time. At zero time all metabolic substrates were removed from bath and lumen. I_{sc} collapsed rapidly. After 3 min I_{sc} was only half of the normal rate. This collapse in I_{sc} is paralleled by a depolarization of the basolateral membrane voltage (see also Fig. 10). These data indicate that cTAL segments are strictly dependent on continued fuel and O_2 supply

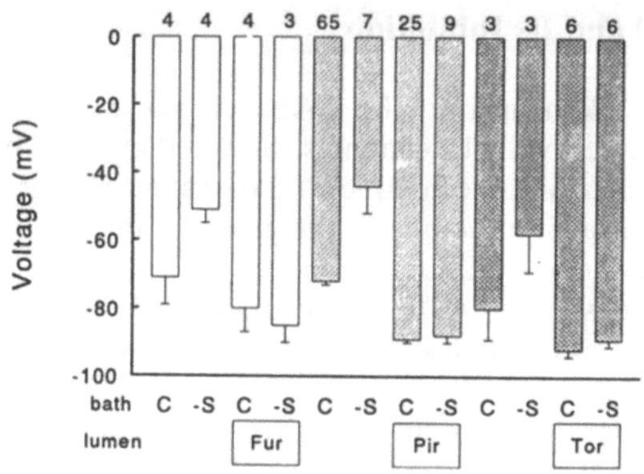

Fig. 10. Protective effect of loop diuretics. Data taken from GREGER (1985a), GREGER and WANGEMANN (1987), and LOHRMANN et al. (1993). The basolateral membrane potential [voltage, mean values ± SEM (*n* on top of respective columns)] is given for the various conditions. *C*, control; *−S*, substrate free. Note that furosemide (*Fur*, 50 μmol/l, lumen) hyperpolarizes the membrane voltage (see also Fig. 7). Substrate free perfusion, due to the inhibition of ATP generation, depolarizes the cell. The effect of substrate free perfusion is completely abolished in the presence of furosemide. Due to the furosemide-induced inhibition of Na^+ uptake, the rate of ATP consumption is also reduced sharply. Similar observations were obtained for piretanide (*Pir*, 5 μmol/l, lumen) and torasemide (*Tor*, 10 μmol/l, lumen)

These findings are highly significant for the pathophysiology of TAL transport. Any underperfusion of the peritubular capillaries will lead to a reduced transport capacity of the TAL segment. In fact, it may induce irreversible damage. The specific vulnerability of the TAL segment in renal ischemia is well known (BREZIS et al. 1984) and has been demonstrated, e.g., in the isolated in vitro perfused kidney. Together with the damage in the proximal tubule the ischemic deterioration of the TAL is of pathophysiological relevance in acute renal failure (GREGER 1987). Therefore, it was of interest to test whether the damage induced by substrate depletion or hypoxia could be prevented by reducing the workload of the TAL segment with loop diuretics. Figure 10 shows an example. It is apparent that loop diuretics can completely prevent the depolarization which would be induced by substrate removal (GREGER 1985b; GREGER and WANGEMANN 1987; LOHRMANN et al. 1993). Therefore, it is to be predicted that administration of furosemide-like diuretics can prevent the TAL damage which would ensue in acute renal failure (GREGER and HEIDLAND 1991). This is highly relevant, but cannot normally be utilized, because established acute renal failure obviously cannot be ameliorated. Furosemide or a comparable substance would have to be given prior to the ischemia to exert a beneficial effect. It is not surprising then that clinical studies in which the effect of furosemide was examined in established acute renal failure gave controversial results (GREGER and HEIDLAND 1991). Beyond this, acute renal failure certainly includes more than damage to the TAL.

V. Cl^- Channel and Its Inhibition

Most of the Cl^- taken up into the TAL cell via the $Na^+2Cl^-K^+$ cotransporter leaves the cell via Cl^- channels. So far they have been only poorly analyzed by patch clamp techniques (GREGER et al. 1991; PAULAIS and TEULON 1990) and seem to belong to the group of intermediate conductance channels. Their control by PKA-dependent phosphorylation appears to be the key activation step in ADH-induced increase in net reabsorption of NaCl (SCHLATTER and GREGER 1985; PAULAIS and TEULON 1990). Their inhibition by so-called Cl^- channel blockers has been described in comprehensive reports from our laboratory (DI STEFANO et al. 1985; WANGEMANN et al. 1986; WITTNER et al. 1987; BIJMAN et al. 1987). The most potent compounds belong to the same group as NPPB. The effect of NPPB on active transport in TAL is shown as the inhibition of the equivalent short circuit current (GREGER 1985b) in Fig. 11, which also contains a concentration response curve. NPPB induces a flicker-type block, an effect which is fully reversible. The binding of the blocker occurs at the outside mouth of the channel (facing the blood side). Therefore, the concentration response curve of NPPB is shifted to the right when it is added to the cytosolic side of excised membrane patches (TILMANN et al. 1991), and, furthermore, NPPB conjugated to polyethyleneglycol or dextran with a molecular mass of 5000 Da acts in outside out patches but not in inside out patches (TILMANN et al. 1991). In the intact TAL NPPB inhibits half-maximally at 70 nmol/l. When comparing the formula of NPPB-type Cl^- channel blockers (Fig. 11) with furosemide-type loop diuretics (Fig. 1) it becomes apparent that both kinds of compound have similar characteristics: The benzoate moiety and the secondary (or teriary) amino group are present in both types of compounds. The most striking similarity is easily apparent when comparing torasemide with NPPB or diphenylamine-2-carboxylate (DPC). In fact, torasemide has a dual effect in the TAL: at 300 nmol/l it inhibits the $Na^+2Cl^-K^+$ cotransporter, and at 30 μmol/l it inhibits the basolateral Cl^- conductance (WITTNER et al. 1987b).

In a large number of experiments it has been clarified which changes of furosemide-type diuretics alter their main target toward Cl^- channels, and which changes in Cl^- channel blockers make them inhibitors of the $Na^+2Cl^-K^+$ cotransporter (GREGER et al. 1987; GREGER and WANGEMANN 1987). The similarity of both classes of inhibitors has been suggested by these authors to indicate that the respective membrane proteins share some common motifs.

Occasionally it has been speculated that Cl^- channel blockers might be used as diuretics under certain conditions. This pharmacological avenue is highly unlikely to become successful, since these substances would have no organotropy and would hit the respective Cl^- channel in every tissue where it is present. This probably even holds for drugs such as torasemide where the dual effect has clearly been demonstrtated (WITTNER et al. 1987). Even

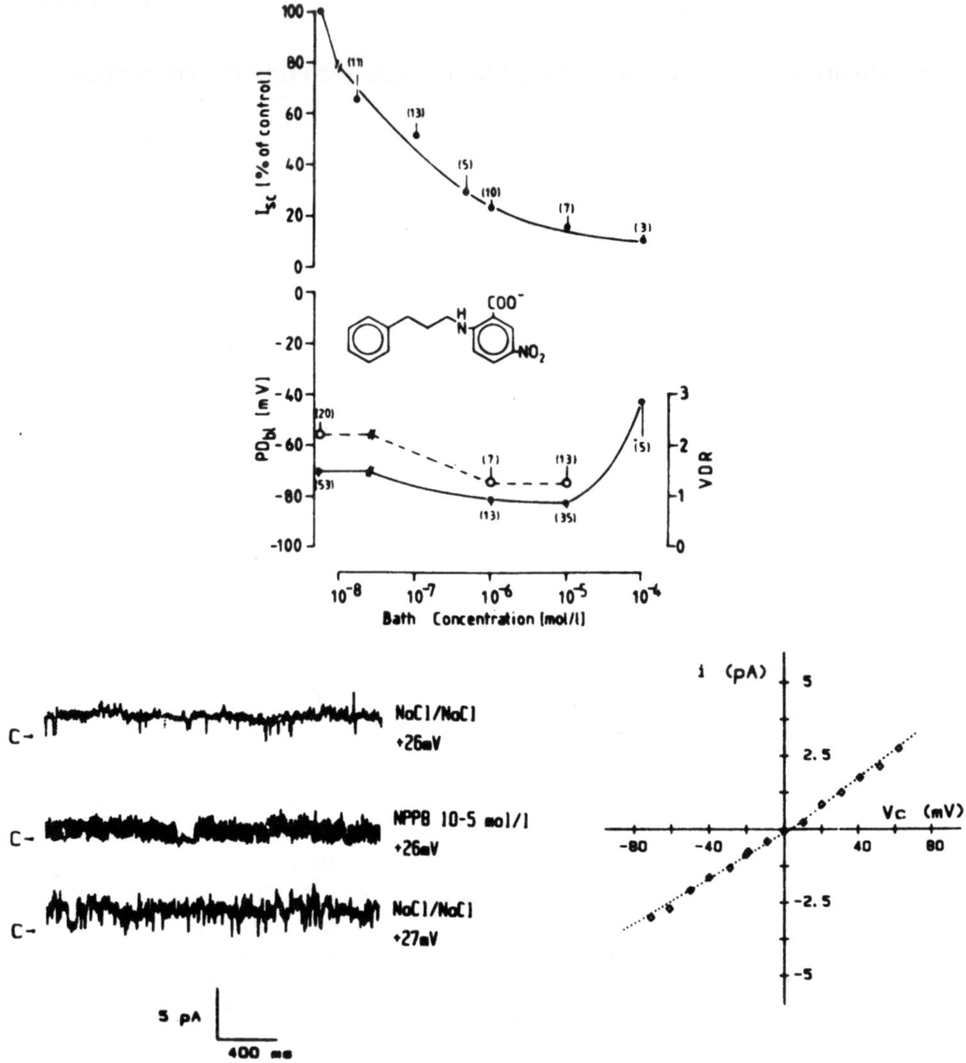

Fig. 11. Effect of the Cl^- channel blocker NPPB on Cl^- conductance and Cl^- channel in cTAL segments. Data taken from WANGEMANN et al. (1986) and GREGER et al. (1991). *Upper panel*, concentration response curves for NPPB on equivalent short circuit current (I_{sc}), on basolateral membrane potential (*PD_{bl}, solid line*), and voltage divider ratio (*VDR, broken line*, ratio of luminal membrane resistance divided by basolateral membrane resistance). All data are given as mean values ± SEM (*n*). *Inset* displays the structural formula of NPPB. Note that NPPB blocks active NaCl-reabsorption concentration dependently with a half-maximal inhibition of 70 nmol/l. As expected for the inhibition of the Cl^- conductance, this occurs with a hyperpolarization of PD_{bl} and a fall in VDR. In the *lower panel* a patch clamp experiment from an excised basolateral inside out oriented membrane patch of a rat cTAL is shown. The pipette and bath contained 150 mmol/l Cl^-. The patch was clamped at +26 (+27) mV. Individual channel openings are seen in the typical traces *on the left-hand side*. In the middle trace NPPB thas been added to the bath solution. This induces a flicker-type block, i.e., the mean open times are reduced drastically. This results in a significant inhibition of the open probability of the channel. The effect of NPPB is completely reversible. The *right-hand lower panel* shows the current (i) voltage (V_c) relationship for this channel. The single-channel conductance for this type of channel is around 30–50 pS

in patients with largely compromised torasemide secretion the concentration in the lumen of TAL will exceed that of free torasemide on the blood side by far. Consequently an inhibition of the $Na^+2Cl^-K^+$ cotransporter would prevail long before an inhibition of basolateral Cl^- channels became apparent. Even putative cellular uptake of a compound such as torasemide from the luminal fluid is not likely to produce Cl^- channel inhibition because these substances act on the outside mouth of the channel (see above).

VI. Loop Diuretics Related to Furosemide

Furosemide and related compounds act by their reversible binding to the $Na^+2Cl^-K^+$ cotransporter. It is not surprising then that all compounds of this class show concentration response curves in the isolated perfused TAL which look very similar. Figure 12 is taken from a comprehensive study by

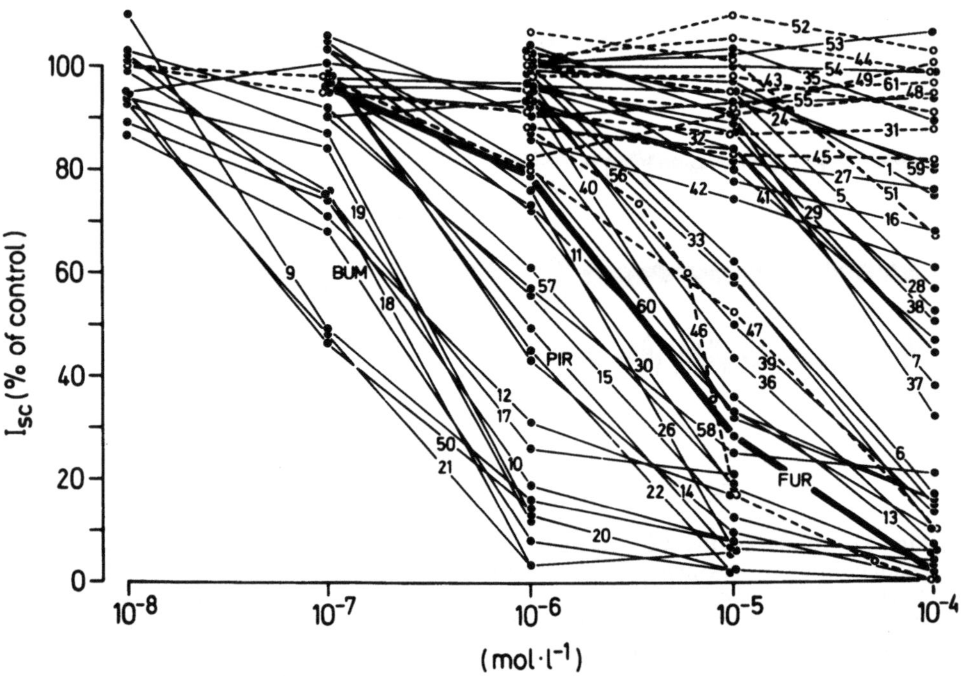

Fig. 12. Concentration-response curves for diuretics. Data taken from SCHLATTER et al. (1983). The equivalent short-circuit current [I_{sc}, mean values ± SEM, *n*], corresponding to the rate of active NaCl reabsorption, of isolated in vitro perfused rabbit cTAL segments is plotted versus the luminal concentration of the respective diuretic. *Bum*, bumetanide; *Pir*, piretanide; *Fur*, furosemide. *Solid lines*, diuretics related to furosemide; *broken lines*, diuretics structurally not related to furosemide. The data indicate that most diuretics nonrelated to furosemide have no effect. Exceptions are ethacrynate (#46) and indacrinone (#47) (see text). All compounds related to furosemide show very similar slopes (Hill coefficients around 1). From these data structure-activity relationships were deduced by SCHLATTER et al. (1983)

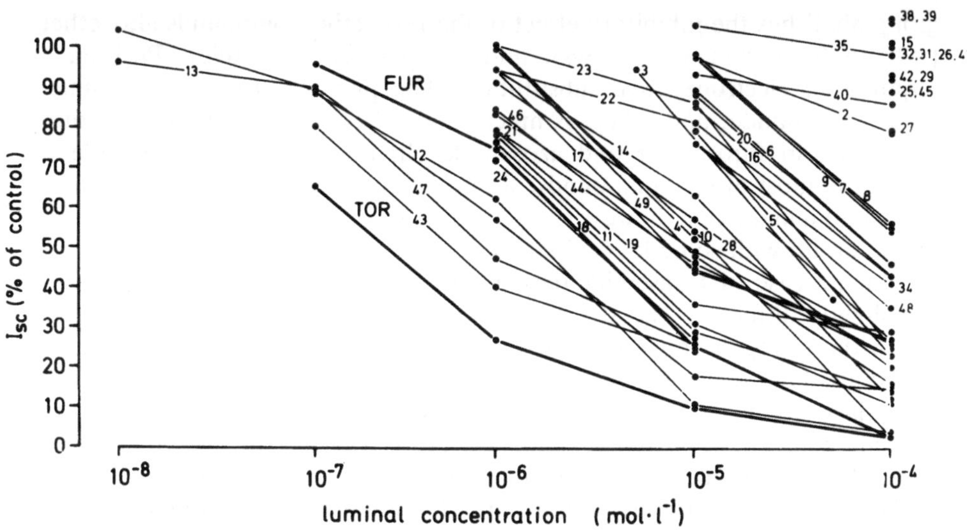

Fig. 13. Concentration-response curves for diuretics related to torasemide. Data taken from Wittner et al. (1987). The equivalent short-circuit current [I_{sc}, mean values ± SEM, n] corresponding to the rate of active NaCl reabsorption, of isolated in vitro perfused rabbit cTAL segments is plotted versus the luminal concentration of the respective diuretic. *TOR*, torasemide. Most compounds related to torasemide show very similar slopes (Hill coefficients around 1). From these data structure-activity relationships for the torasemide family have been deduced by Wittner et al. (1987)

our group in which we examined a large number of furosemide-related and nonrelated compounds (Schlatter et al. 1983). It is easily seen that all furosemide-related compounds (solid lines) in this figure show the same type of concentration response as furosemide (FUR), bumetanide (BUM), and piretanide (PIR). The Hill coefficient of all of these curves is close to 1, and the compounds only differ inasmuch as some curves are shifted to the right and others to the left. Comparable studies have also been performed for torasemide (TOR) and related substances (Wittner et al. 1987). These are summarized in Fig. 13. It is from this type of study that structure-activity relationships have been obtained.

The furosemide-related compounds (Fig. 12) all possess an aromatic ring. In position 1 they need to carry an acidic substituent such as carboxylate (furosemide, piretanide, bumetanide, and many others of. Fig. 1). This group determines the overall pKa of these compounds, which is around 4. Other acidic groups are also tolerated such as a tetrazolate group (azosemide and #21 in Schlatter et al. 1983) and sulfonate (e.g., #50 in Schlatter et al. 1983). A phosphonic acid residue, however, abolishes the effect (#45 in Schlatter et al. 1983).

Position 2 or 3 carries a secondary (furosemide, bumetanide, azosemide, and many others) or tertiary amino group (piretanide). A primary amino group abolishes the inhibitory effect of the respective compounds altogether (#42 in Schlatter et al. 1983, and a xipamide derivative). Position 4 requires an electronegative substitution. The phenoxy group as in bumetanide, piretanide, and others is much more effective than the Cl substitution, e.g., in furosemide and azosemide. In position 5 a sulfonamide is present in most of the compounds. Removal of this group destroys potency. The sulfonamide can, however, be modified by alkyl substitutions on the amino group with some loss of potency.

In the case of torasemide-derived substances (Fig. 13) the constraints are slightly different. The structures are all based on a pyrimidine ring, in which the nitrogen apparently serves the same function as the sulfonamide group in the furosemide-related compounds (Wittner et al. 1986). The secondary amino group is conserved, and it can carry larger and even aliphatic rings (#24, #L1035, #L961, etc., in Wittner et al. 1987; Lohrmann et al. 1992). The acidic function in position 1 of the furosemide group is replaced here by a sulfonylurea group. Consequently these compounds are much less acidic (e.g., torasemide has a pK_a value of 6.8, others have pK_a values of up to 9). This prompts the question of whether an anionic moiety at this site is required to bind to one of the Cl^--binding sites of the $Na^+2Cl^-K^+$ cotransporter. This has been examined recently by perfusing TAL segments with luminal solutions of alkaline normal and acidic pH and comparing the concentration response curves of furosemide, torasemide, and #BM10 (Fig. 13) (Lohrmann et al. 1991). The results are shown in Fig. 14. The concentration response curves of torasemide and #BM10 are clearly

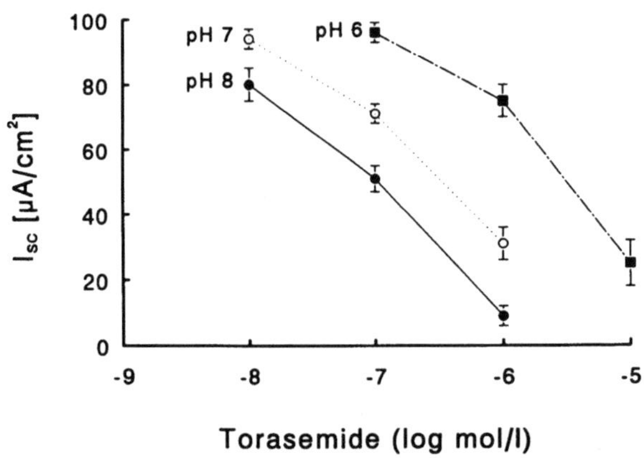

Fig. 14. pH dependence of torasemide effects in rabbit cTAL segments. The equivalent short-circuit current [I_{sc}, mean values ± SEM, *n*], corresponding to the rate of active NaCl reabsorption, of isolated in vitro perfused rabbit cTAL segments is plotted versus the luminal concentration of torasemide dissolved in luminal perfusates of different pH values. Data taken from Lohrmann et al. (1993). These data indicate that acid pH shifts the torasemide concentration-response curve to the right. Alkaline pH has the opposite effect. Hence, it can be concluded that only the anionic moiety of torasemide acts as a diuretic

pH-dependent in this pH range, while that of furosemide is not. Acid pH shifts the concentration response curve of torasemide to the right, alkaline pH to the left. A quantitative analysis of these data reveals that torasemide and related compounds such as furosemide-related compounds act only if they are deprotonated (anionic). This is possibly of clinical relevance. The pH at this nephron site is usually between 6.6 and 6.8 (ALPERN et al. 1991). Half of the concentration of torasemide present will be active, the other half will be inactive. Because the pK_a of torasemide is exactly in this range, small pH variations will produce marked changes in the two moieties. For furosemide and related diuretics such considerations are of no importance since the anionic moiety is 99.9% at pH 6.8, 99.99% at pH 7.8, and 99% at pH 5.8. For activity relationships of the torasemide group of loop diuretics this consideration will have to be taken into account.

Structure-activity relationships such as the ones given here are still of limited use, because the structures are fairly small, only very limited numbers of basic structures have been looked at, and almost nothing is known on the diuretic-binding site of the $Na^+2Cl^-K^+$ cotransporter. With the cloning of the transporter (S. HEBERT 1993, Boston, personal communication) such data will be of value and spatial modeling of the binding will possibly become particularly useful.

VII. Loop Diuretics Not Related to Furosemide

Figures 1 and 12 show several loop diuretics structurally not related to furosemide. Among these are ethacrynic acid, indacrinone, tienylic acid, etozolin, and muzolimine. They are indicated by broken lines in Fig. 12. On the basis of transepithelial measurements in isolated perfused TAL they can be subdivided into active, poorly active, and inactive compounds.

Ethacrynic acid (#46) is active, but its dose response curve is very steep, with a Hill coefficient of less than 0.5. Furthermore it was noted that it acted equally well when added from either the lumen or the blood side (SCHLATTER et al. 1983). It was also noted that the cysteine adduct of ethacrynic acid was acting in addition to ethacrynic acid in this study, but a stronger effect for the adduct was claimed by others (BURG and STONER 1976). On the other hand, it has been claimed in vesicle- and protein-binding studies that this substance binds to the $Na^+2Cl^-K^+$ cotransporter (KOENIG et al. 1983); the data in the intact in vitro perfused TAL segments render this highly unlikely. This is further supported in measurements of the luminal and basolateral membrane voltage where it was shown that ethacrynic acid, unlike furosemide and related compounds, depolarized both voltages. Inhibitors such as furosemide hyperpolarize (see above). Such data are compatible with the conclusion that ethacrynic acid interferes with mitochondrial ATP production. This concept may also explain why adducts of ethacrynic acid also show bioactivity: These compounds may enter the cell easily, are hydrolyzed, and act like ethacrynate.

Of the two other phenoxyacetic diuretics, indacrinone (#47) and tienylic acid, only indacrinone showed a reasonable IC_{50} value. However, the concentration response curve was much less steep than that for furosemide-type loop diuretics, which may be taken as an indication that indacrinone acts differently and not by binding to the $Na^+2Cl^-K^+$ cotransporter. The effect of tienylic acid was fairly small.

Etozolin and muzolimine were completely inactive in in vitro perfused TAL segments (Fig. 12). For etozolin this may not be surprising since it is well known that it has to be metabolized by hydrolysis of the esterified acetic acid group. The active compound formed is ozolinone (see Fig. 20). This compound has been shown to inhibit NaCl transport in the loop of Henle (GREVEN and HEIDENREICH 1978). It was only poorly active in isolated in vitro perfused TAL segments (SCHLATTER et al. 1983).

Muzolimine has been claimed to act directly on the TAL (FAEDDA et al. 1985) from the blood side. This has not been validated in in vitro perfusion experiments of TAL segments (SCHLATTER et al. 1983). It was found that muzolimine was entirely inactive. However, in the intact animal (e.g., rat) muzolimine acts like a loop diuretic with some delay and with a longer-lasting effect (WANGEMANN et al. 1987). To examine whether muzolimine was metabolized two types of studies were performed. Muzolimine was examined in clearance studies of intact and hepatectomized rats. It was shown that muzolimine lost its diuretic effect with acute hepatectomy (WANGEMANN et al. 1987). Furthermore, the urine obtained from intact rats after muzolimine was sampled and examined in isolated in vitro perfused TAL segments. It was shown, as indicated in Fig. 15, that the urine of

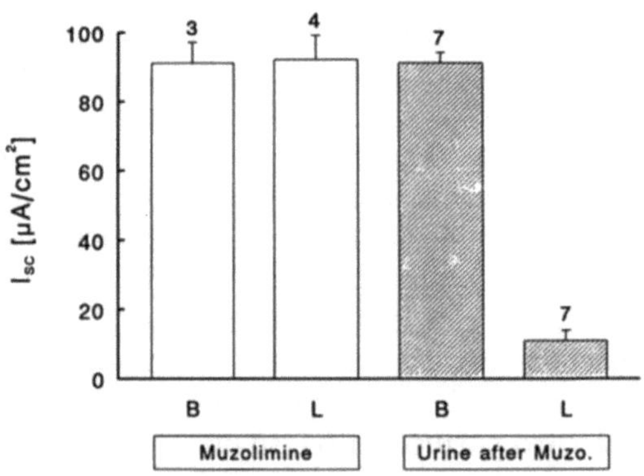

Fig. 15. Muzolimine and muzolimine derivative effects in the rabbit cTAL segment. Data taken from WANGEMANN et al. (1987). The equivalent short-circuit current [I_{sc}, mean values ± SEM, *n*], corresponding to the rate of active NaCl reabsorption, of isolated in vitro perfused rabbit cTAL segments is shown, Muzolimine (100 μmol/l) added to the bath (*B*) or lumen (*L*) had no effect on I_{sc}. However, the diluted urine of muzolimine-infused rats inhibited I_{sc} strongly and reversibly. These data indicate that muzolimine is metabolized, and that the metabolite inhibits NaCl reabsorption in TAL segments like furosemide

muzolimine-treated rats contained a compound which inhibited TAL transport just like furosemide (WANGEMANN et al. 1987). Furthermore it was shown that pretreatment with probenecid attenuated muzolimine diuresis. All this is compatible with the interpretation that muzolimine is metabolized in the liver to a diuretically active metabolite which is secreted by the proximal tubule and acts from the luminal side in the TAL. This active metabolite has thus far not been purified.

E. Effects of Loop Diuretics in the Intact Kidney

As stated above, loop diuretics produce a marked diuresis and saluresis of up to 20%–30% of the filtered load. On a qualitative basis, this is true for every species which has been looked at: man and other primates (GREGER and HEIDLAND 1991), dog, rabbit, guinea pig, rat, mouse, and many others (for review see SUKI and EKNOYAN 1992; ULLRICH and GREGER 1985). Even in nonmammals diuresis and natriuresis can be induced by, e.g., furosemide in: birds, snakes, fish, frog, and *Amphiuma* (OBERLEITHNER et al. 1983). The strong diuretic and natriuretic effect is due to the paralysis of the countercurrent concentrating system (see Chap. 2, this volume). Active NaCl absorption in the TAL is the only driving force for the concentrating ability in the outer and inner kidney medulla. The reabsorption of NaCl with no reabsorption of water leads to the diluted tubule fluid and a hyperosmolar interstitial fluid. Due to the low water permeability of the luminal membrane of this nephron segment, it has been named diluting segment by BURG (1982). The peritubular hyperosmolality is utilized to withdraw water from neighboring tubule segments, namely descending limbs of Henle loops and collecting ducts. Hence, the diuresis is not generated in the TAL itself, and flow rate does not increase in this segment when a loop diuretic is added.

In the intact kidney the effect of loop diuretics in the TAL can easily but indirectly be demonstrated by their effect on fractional NaCl clearance. In water diuresis (with ADH inhibited) urine diluted with respect to its NaCl concentration is excreted. In the presence of a loop diuretic urinary flow rate increases further and the urine NaCl concentration increases to approximately isotonic values (for review SUKI and EKNOYAN 1992). In micropuncture studies the effect of a loop diuretic can easily be shown as the increase in early distal NaCl delivery (DEETJEN 1966). An example of this is shown in Fig. 16. Normally, under moderate volume expansion, approximately 10% of the filtered load of NaCl is delivered to the early distal tubule. After systemic administration of a loop diuretic this value can increase to 30%–40% of the filtered load. However, in most of these studies an effect on the proximal tubule has also been reported for most of the loop diuretics of the furosemide type (GREGER 1985b; DILLINGHAM et al. 1993). This may not be surprising because these compounds still possess some inhibitory activity on carbonic anhydrase (GREGER 1985b). Clearly, however,

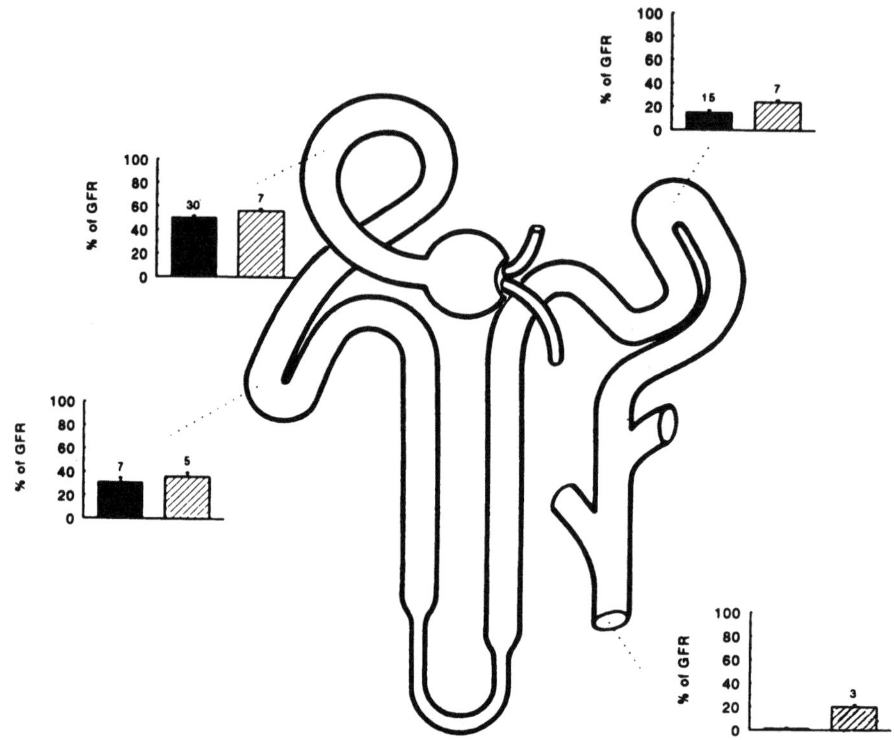

Fig. 16. Micropuncture data on the effect of furosemide in the rat. Data are taken from DEETJEN (1966). The fractional volume recovery is given for proximal and distal micropuncture sites as well as for the final urine. Mean values ± SEM (*n*). *Black bars*, control; *hatched bars*, furosemide (40 mg/kg). Furosemide has little effect in the proximal tubule if a fall in GFR is prevented. The effect is apparent as a significant difference in fluid recovered in the distal tubule and final urine

the proximal tubule effect does not account for the majority of the diuretic and saluretic response. It has also been claimed that loop diuretics, via their inhibitory effect on the $Na^+2Cl^-K^+$ cotransporter, exert an effect in other nephron segments, e.g., in the papillary collecting duct and in the kidney pelvis (SUKI and EKNOYAN 1992). These effects are probably irrelevant for the overall effects of loop diuretics. The effect on the macula densa segment will be discussed below.

I. Macula Densa Segment

Loop diuretics have been shown to interrupt the so-called tubuloglomerular feedback (TGF), i.e., single nephron filtration rate (SNGFR) becomes independent of early distal delivery of NaCl (WRIGHT and SCHNERMANN 1974). Normally, due to TGF, SNGFR falls whenever the NaCl delivery to early distal tubule or more precisely, to the macula densa, increases. This SNGFR reduction is prevented by loop diuretics such as furosemide and bumetanide. Besides, it has also been noted that loop diuretics increase plasma renin (SCHNERMANN and BRIGGS 1991), even if attention is paid to prevent any volume contraction by output-matched infusions. The rate of renin secretion from specialized renin-producing cells in the afferent arteriole is, besides the pressure-sensing and the β-sympathergic discharge, also controlled by the macula densa. Whenever the NaCl delivery to early distal tubule, or more precisely, to the macula densa, increases, renin release is decreased. It has long been postulated that the macula densa cells must possess some Na^+ or Cl^- concentration-sensing mechanism (SCHNERMANN and BRIGGS 1991). Recently this issue has been examined directly (SCHLATTER et al. 1989; SCHLATTER 1993). The voltage of macula densa cells has been measured in in vitro perfused macula densa segments by microelectrodes or with whole cell patch clamp and furosemide, or other diuretics have been added to the lumen. The results of one of these studies is shown in Fig. 17. It is evident that loop diuretics such as furosemide induce a concentration-dependent hyperpolarization of the cell membrane voltage, just like they do in TAL cells. Nonrelated diuretics such as thiazides or muzolimine have no such effect. Additional experiments indicate that a reduction in luminal NaCl concentration also hyperpolarizes these cells (SCHLATTER et al. 1989). This indicates that the macula densa cell possesses a similar or even identical $Na^+2Cl^-K^+$ cotransporter. This cotransporter serves as a chemoreceptor to sense the luminal ion concentrations at this nephron site. From the above (see Sect. D.II) it is clear that the limiting concentration will probably be that of Cl^-. The transduction steps distal to the change in TAL voltage and ion composition are largely unknown (SCHLATTER 1993; SCHLATTER et al. 1989; SCHNERMANN and BRIGGS 1991). The direct effect of furosemide-type

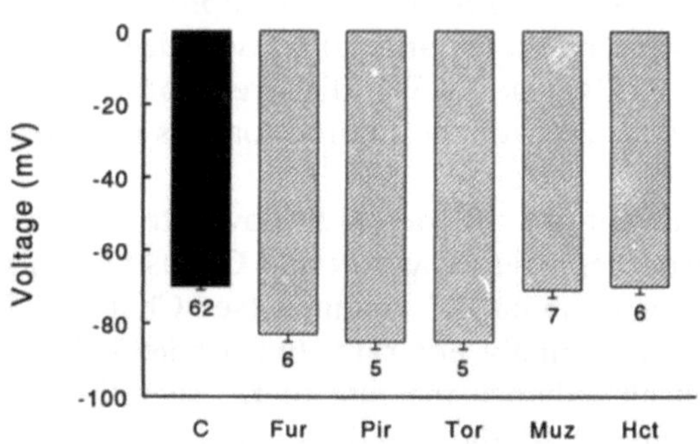

Fig. 17. Voltage of macula densa cells. Data taken from SCHLATTER (1993). Whole cell patch clamp recordings of macula densa cells. Voltage [mean values ± SEM (n)] is given in the absence (*C*, control) and presence of various diuretics (all 100 μmol/l). *Fur*, furosemide; *Pir*, piretanide; *Tor*, torasemide; *Muz*, muzolimine; *Hct*, hydrochlorothiazide. These data indicate that loop diuretics hyperpolarize macula densa cells as for TAL cells

loop diuretics in macula densa cells explains the interruption in TGF. The interruption of TGF helps to sustain the diuretic and saluretic effect, because SNGFR and GFR stay high. The macula densa effect of loop diuretics also explains the increase in renin secretion, because the macula densa cell now perceives the same message as is normally conveyed by a fall in luminal NaCl concentration.

II. Excretion of K^+

Most studies in animals and in man agree inasmuch as K^+ excretion is enhanced by loop diuretics (GREGER and HEIDLAND 1991); they differ, however, with respect to the magnitude of this loss, the interpretation of its relevance, and the loss induced by one individual compound as compared to others of this class. From a pharmacological point of view the K^+ losses are the unavoidable consequence of inhibition of the $Na^+2Cl^-K^+$ cotransporter (see below), and distinct differences between various compounds can hardly be rationalized. Rather, during peak diuresis K^+ excretion relative to natriuresis is practically the same for all loop diuretics studied thus far (KNAUF and MUTSCHLER 1990, 1992a,b).

Normally, in the TAL segment some K^+ is reabsorbed. The rate of reabsorption depends largely on the degree of luminal dilution, i.e., if the flow rate in the loop is high, and hence dilution is minimal, no additional dilution voltage will be built up. Then the driving force for K^+ reabsoption is small, and almost all of the K^+ taken up by the $Na^+2Cl^-K^+$ cotransporter is recycled via the luminal K^+ conductance. The paracellular current of K^+ is probably very small (due to its low concentration compared with NaCl). The net driving force for paracellular absorption is probably only some 10–20 mV. In antidiuresis a large dilution voltage builds up along the axis of the TAL and becomes maximal in the cortical TAL. This voltage may well be 20–30 mV lumen positive. It compromises K^+ recycling across the luminal membrane markedly and, to a lesser degree, favors paracellular reabsorption of K^+ (GREGER 1985b). This reabsorptive component, which may account for 10%–20% of the filtered load, is completely abolished by loop diuretics.

Even more relevant are the increased flow rate and the increased NaCl delivery to the distal nephron (STANTON and GIEBISCH 1992). K^+ is secreted in the distal tubule via luminal K^+ channels (see Chaps. 2, 10, this volume). Since this process is normally not rate, but gradient limited, an increased flow rate will in itself enhance the rate of K^+ secretion. Furthermore, the increased NaCl delivery to the distal tubule leads to an increased NaCl reabsorption. To the extent that this reabsorption increases, the secretion of K^+ will also be increased. This is caused by the fact that the reabsorption of Na^+ depolarizes the luminal membrane of collecting duct cells, and the depolarization increases the driving force for K^+ secretion (see also Chap. 2, this volume).

In addition, this mechanism of K^+ secretion will be augmented if, due to the loop diuretic-induced diuresis and saluresis, volume contraction enhances ADH and aldosterone concentrations. Aldosterone and ADH increase the Na^+ conductance in the luminal membrane of the collecting duct, and therefore stimulate K^+ secretion (STANTON and GIEBISCH 1992). It has also been claimed that ADH, via cAMP, might directly increase the luminal K^+ conductance of the luminal membrane and thus augment K^+ losses (WANG and GIEBISCH 1991). In the same context it is important to be aware of the acid base status when loop diuretics are administered. Firstly because they may by themselves change acid base status (see below) and secondly because the amount of K^+ lost during loop diuretic induced diuresis and saluresis will depend on the acid base status, which is large in alkalosis and attenuated in acidosis (STANTON and GIEBISCH 1992). This is due to the pH dependence of the respective luminal K^+ channels in the TAL and in the collecting duct (BLEICH et al. 1990; SCHLATTER et al. 1993). An example is shown in Fig. 18. The open probability of these channels is controlled by the pH on the cytosolic side. Respective changes of luminal pH are without effect. Alkalosis augments, acidosis reduces the open probability of these channels. The steepest portion of this curve is usually found in the physiological pH range. The respective K^+ channels can be regarded as sensors of cytosolic pH, with a very high sensitivity, and at the same time these channels are regulatory targets coupling acid base status to K^+ homeostasis.

Due to the large number of additional factors which determine K^+ excretion, it is impossible to predict the amount of K^+ lost in a certain patient in his or her specific condition, and for a given loop diuretic. It

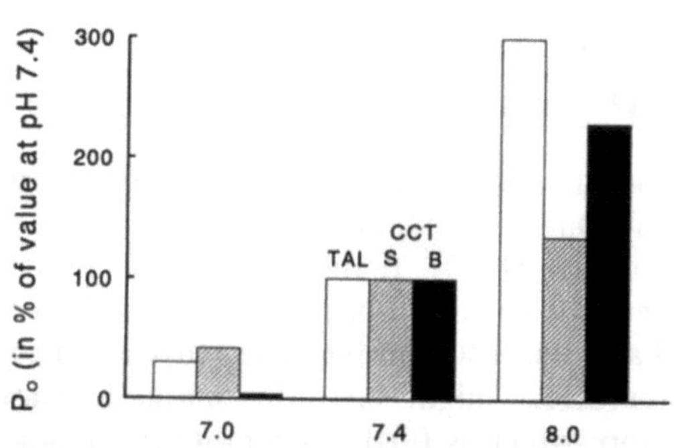

Fig. 18. pH dependence of K^+ channels in the luminal membrane of TAL segments and cortical collecing ducts (*CCT*). Data taken from SCHLATTER et al. (1993). The open probability (P_o) is normalized to pH 7.4 on the cytosolic side of the channels. For clarity only the mean values are given. *Open bars*, K^+ channel of rat TAL segments (see also Fig. 6); *hatched bars*, small (*S*) K^+ channel of rat CCT; *black bars*, big K^+ channel of CCT. Note that all three channels are pH sensitive. The open probability is reduced by acid pH in the cells, and is increased by alkaline pH. The data explain why acid base changes lead to respective changes in K^+ secretion

appears much safer to conceptualize the above mechanisms and to monitor plasma K^+ concentration. Most of the claims that one compound is superior to another with respect to its K^+ losses are bases on unpaired study designs, and specific protocols (e.g., different urine collection periods). If under such circumstances two compounds are compared and less additional K^+ loss is found for one compound than for the other, it falls short of any scientific standards to call the former substance "K^+ saving." Theoretically it might still be possible to come up with a structure which is an inhibitor both of the $Na^+2Cl^-K^+$ cotransporter and of the luminal K^+ channel in the TAL and collecting duct. No such compound has been reported thus far. It may appear wise then to prevent K^+ losses by sufficient dietary supplements or to combine the loop diuretic with a K^+-sparing diuretic (see below and Chap. 13, this volume).

III. Excretion of NH_4^+

Normally a limited amount of 20–40 mmol/day NH_4^+ is excreted in the urine. This amount is derived from intrarenal (mostly proximal tubule) production from glutamine (see Chaps. 2, 3, this volume). In metabolic acidosis this quantity may be greatly increased. NH_4^+ is reabsorbed in the TAL, enters the interstitium, and is then secreted into the collecting duct. This mechanism short circuits the passage of NH_4^+ from the proximal tubule to the final urine, keeps it in the kidney medulla, and improves its renal excretion (HALPERIN et al. 1992). Loop diuretics such as furosemide inhibit the reabsorption of NH_4^+ in the TAL (KIKERI et al. 1989). This will probably compromise the renal excretion of NH_4^+. But little attention has been paid to this as yet.

IV. Excretion of H^+ and HCO_3^-

Loop diuretics of the furosemid type have two opposing effects with respect to proton excretion. One is residual carbonic anhydrase inhibition, which slightly reduces proximal HCO_3^- reabsorption. The other effect is on the collecting duct, where the secretion of H^+ is enhanced, due to the increased distal load of NaCl and the resultant increased driving force (lumen negative voltage). The latter effect might be more marked. Therefore metabolic alkalosis may develop, which is further promoted if hypokalemia develops. The mechanism of the hypokalemia induced by alkalosis is not completely understood. Several factors may contribute: enhanced HCO_3^- reabsorption in the proximal nephron due to the hyperpolarization of the basolateral membrane voltage and due to the rheogenicity of HCO_3^- reabsorption as $Na^+(HCO_3^-)_3$ (BORON and BOULPAEP 1989); enhanced NH_4^+ production; reduced H^+ extrusion from cells; and hyperaldosteronism with the resulting further increased distal Na^+ reabsorption and H^+ and K^+ secretion.

V. Excretion of Ca^{2+} and Mg^{2+}

Loop diuretics generally increase Ca^{2+} and Mg^{2+} excretion (Greger and Heidland 1991; Greger 1985). This is due to an effect in the TAL. In this segment normally as much as 30% of the filtered load of Ca^{2+} and as much as 75% of that of Mg^{2+} are reabsorbed (Quamme 1989; Costanzo and Windhager 1992). The mechanism of reabsorption is still not completely understood. The lumen-positive voltage provides a driving force for the paracellular reabsorption of both cations (Greger 1985). However, this simple model of reabsorption is not easily reconciled with the finding that the reabsorption of both ions is under the control of PTH and calcitonin in the TAL (Greger 1985). These hormones, their second messenger cAMP, in isolated in vitro perfused tubules, and the hormones glucagon and ADH (Wittner et al. 1988) enhance the reabsorption of both cations. It is more likely that both paracellular and transcellular components contribute to Ca^{2+} and Mg^{2+} reabsorption. In fact, it has been suggested, on the basis of fura 2 measurements of cytosolic Ca^{2+} in these cells in the absence and presence of PTH, that PTH enhances cellular uptake of Ca^{2+} (Friedman and Gesek 1993).

There is little controversy that furosemide-type loop diuretics inhibit the reabsorption of Ca^{2+} and Mg^{2+} strongly. This again is easily reconciled with the paracellular concept of divalent ion reabsorption, but is difficult to envision for the transcellular mode of reabsorption. Hypercalcuria and hypermagnesuria are unavoidable consequences of loop diuretic therapy (Greger and Heidland 1991). This may be a problem in patients with kidney stone disease, in whom thiazides will be preferred over loop diuretics, because they enhance Ca^{2+} reabsorption, but have no direct effect on Mg^{2+} reabsorption (Costanzo and Windhager 1978; Chap. 8, this volume). A combination of loop diuretics and thazides may therefore also be considered. On normal diets the loop diuretic induced Ca^{2+} and Mg^{2+} losses rarely cause any major complication.

VI. Excretion of Li^+

The renal excretion of Li^+ has become a relevant issue because Li^+ is used widely in psychiatry in the therapy of depressive disorders. Furthermore, Li^+ has become a marker for proximal tubule Na^+ and water reabsorption (Thomsen et al. 1969). This is based on the findings that under many experimental conditions Li^+ is reabsorbed only in the proximal tubule with a fractional reabsorptive rate identical to that of Na^+ (Koomans et al. 1989). However, exemptions to this rule have been noted: In volume contraction and Na^+ depletion Li^+ is reabsorbed to some extent in the TAL and in the collecting duct (Greger 1990). The latter is inhibited by amiloride; the former is inhibited by loop diuretics of the furosemide type. The route, driving force, and the mechanism of Li^+ reabsorption in the TAL have been reviewed recently (GREGER 1990). Li^+ is only poorly conducted by the K^+ channels present in the luminal membrane (BLEICH et al. 1990); also it is not accepted by the furosemide-sensitive $Na^+2Cl^-K^+$ cotransporter (GREGER 1985). On the other hand, it has been shown that Li^+ can permeate the paracellular shunt (GREGER 1981c). Furthermore, especially in volume contraction the transepithelial voltage built up in the cortical TAL may be as high as +30–+40 mV (see above). This driving force enhances Li^+ absorption considerably. Loop diuretics such as furosemide abolish this dilution and the active transport voltage. As a consequence this component of Li^+ reabsorption is obliterated, and Li^+ clearance increases.

VII. Excretion of Urate

Loop diuretics have a dual effect on renal urate excretion (LANG et al. 1977). On the one hand, acutely loop diuretics can have a probenecid-like, i.e., uricosuric, effect in the proximal tubule, due to an interference with urate reabsorption. On the other hand, and much more marked, loop diuretics have an antiuricosuric effect when volume contraction occurs. This is probably mostly due to a reduced tubule flow rate especially in the pars recta of the proximal tubule, where urate is reabsorbed flow rate dependently (GREGER et al. 1974). This antiuricosuric effect of all loop diuretics leads to a marked increase in plasma urate concentration after long-term treatment with these substances. An exception to this general rule applies to some phenoxyacetic acid derivatives such as indacrinone and tienylic acid, which have been designed on the basis of their relative uricosuric response (FANELLI et al. 1974). To the extent that hyperuricemia has lost its importance as an independent risk factor in vascular disease, the efforts to design such compounds with uricosuric and diuretic effects have slowed and stopped eventually.

VIII. Phosphaturic Effect

Loop diuretics exert a moderate phosphaturic effcct, which has been shown to depend on the intactness of PTH control mechanisms, i.e., in parathyroidectomized animals furosemide no longer had a phosphaturic effect (HAAS et al. 1977). The mechanism of furosemide-induced phosphaturia has been clarified. In the presence of a carbonic anhydrase inhibitor (acetazolamide), the phosphaturic effect of furosemide is obliterated. This indicates that loop diuretics such as furosemide induce their phosphaturic effect by the proximal tubular effect to inhibit carbonic anhydrase (HAAS et al. 1977). In the absence of PTH this phosphaturic effect in the proximal tubule is compensated for by enhanced phosphate reabsorption in the straight portion of the proximal tubule. Furthermore, it should be kept in mind that the hypercalciuria induced by loop diuretics can reduce plasma-ionized Ca^{2+} and hence increase PTH, which will induce secondary hyperphosphaturia (LANG et al. 1981).

F. Effects of Loop Diuretics on Other Organs

The experimental use of loop diuretics of the furosemide type has not only led to the clarification of how they act in the TAL (GREGER and SCHLATTER 1983c), but it has also clarified that the $Na^+2Cl^-K^+$ cotransporter is present in many epithelia and in apolar cells (see above). Today it appears that this cotransporter is one of the universally used transporters of most if not all cells. It serves to import Cl^-, to regulate cell volume up, and to effect transepithelial Cl^- secretion and reabsorption (GREGER 1986). This might suggest that loop diuretics induce pleotropic effects. However, as discussed above (Sect. C), loop diuretics are organotropic due to the renal secretion of diuretics and their accumulation in TAL luminal fluid. Under certain conditions such as local administrations, reduced clearance of loop diuretics, and more importantly with high doses, loop diuretics can exert additional effects.

I. Ototoxic Effects

High doses of loop diuretics have been reported to induce acute hearing losses. These effects occurred, e.g., with 600 mg to 1 g furosemide administered i.v. (for review see GREGER and HEIDLAND 1991; KLINKE et al. 1981). Even lower doses of 200 mg/day furosemide have been reported to induce hearing losses, which were reversible in most cases. The ototoxic effect induced by high doses of ethacrynic acid may be less reversible (MAHER and SCHREINER 1965). The hearing losses are especially pronounced if loop diuretics are given together with aminoglycoside antibiotics (BRUMMET et al. 1981).

The cellular mechanism of furosemide effects on the inner ear has meanwhile been studied (MARCUS et al. 1987). It was shown that the equivalent short-circuit current of the utricle membrance of *Gerbillus* was inhibited reversibly by various furosemide-type loop diuretics, suggesting that the $Na^+2Cl^-K^+$ cotransporter is present in these cells and that it is involved in the generation of the endolymph-positive voltage which is required for inner hair cell function.

It is likely, though not easily explained, that equipotent doses of the various furosemide-type loop diuretics are not equally ototoxic. The ototoxic potential of furosemide seems to be more marked than that of piretanide or bumetanide (for review see GREGER and HEIDLAND 1991).

II. Asthma

It has been reported recently that the inhalation of loop diuretics can prevent bronchoconstriction and improve asthmatic states caused by allergen exposures (BIANCO et al. 1988, 1989). This effect of locally administered loop diuretics has not yet been clarified with respect to its cellular mechanism. An effect on respiratory epithelial cells, which possess the $Na^+2Cl^-K^+$ cotransporter in the basolateral membrane (WELSH 1987), would cause a reduction in NaCl secretion but would hardly have an effect on airway resistance. It appears more likely, but has not yet been documented, that loop diuretics have a relaxant effect on bronchial smooth muscle cells. Furthermore, the beneficial effect of furosemide inhalation has thus far been examined by one group of researchers only. Further data on this issue are badly needed.

III. Preload to the Heart

It has been shown that acute administration of loop diuretics such as furosemide reduces the preload to the heart before the diuresis has caused any detectable loss in volume (DIKSHIT et al. 1973). This has led to speculation that loop diuretics may have a relaxing effect on vascular smooth muscle cells. This appears not unlikely since these cells possess the $Na^+2Cl^-K^+$ cotransporter in their plasma membrane. Studies on isolated vascular stripes have indeed revealed that loop diuretics have a direct effect (for review see SUKI and EKNOYAN 1992). However, the concentration response curves obtained from these studies indicate that very high concentrations are required to exert these effects. The systemic administration of loop diuretics in the above studies results in systemic loop diuretic concentrations which are several orders of magnitude lower than that required for a direct relaxing effect. Furthermore, extended studies with anephric patients have revealed that loop diuretics, unlike, e.g., organic nitrates, have no effect on preload in these patients. This indicates that intact kidneys are needed to exert the vascular effects. Current speculation is that the interaction might be mediated by; e.g., prostaglandins released from the kidney in response to acute loop diuretic administration (GREGER and HEIDLAND 1991).

IV. Glucose Metabolism

The hyperglycemic complication of diuretic therapy has been well documented for thiazides and loop diuretics (for review see GREGER and HEIDLAND 1991). Hyperglycemia is probably caused by a variety of factors. Inhibition of insulin release in pancreatic β-cells may be one component, and hepatic glycogenolysis may be enhanced, as well as hepatic glucogenesis. This may be a consequence of inhibition of the $Na^+2Cl^-K^+$ cotransporter also present in hepatocytes (HÄUSSINGER and LANG 1991). Glucose utilization may also be reduced in skeletal muscle (for review see GREGER and HEIDLAND 1991). For adipocytes it has been shown that loop diuretics inhibit (directly) the carrier-mediated uptake of glucose (JACOBS et al. 1984). All in vitro findings, however, have to be viewed with some skepticism since fairly high concentrations of loop diuretics have to be used to exert these effects. It is questionable whether such concentrations are present in circulating plasma with normal chronic doses.

G. Pharmacokinetics

Even though furosemide-type loop diuretics are all similar with respect to their action on the $Na^+2Cl^-K^+$ cotransporter in the TAL, and differ only with respect to their affinity to one of the Cl^--binding sites of this carrier protein, large differences exist with respect to their pharmacokinetics. The pharmacokinetics for an individual loop diuretic comprises: (a) its absorption from the gut after oral administration. This is usually expressed as the amount taken up into the body divided by the amount ingested = fractional absorption, which is usually high for loop diuretics, being 90% for, e.g., piretanide. Only in the case of azosemide is it much smaller. (b) The volume of distribution is determined largely by the lipid solubility, which is poor for most loop diuretics. Only for torasemide is a larger volume of distribution found. For some derivatives of torasemide an even much larger volume of distribution corresponds to their high lipid solubility (Lohrmann et al. 1992). (c) In plasma most loop diuretics are bound to proteins to a large extent. Typical values are 90% and higher. (d) Some loop diuretics are excreted unaltered to a large extent (e.g., piretanide). Others are biotransformed. This occurs mostly in the liver. The type of reactions are classified into phase I biotransformation (e.g, oxidation, hydroxylation, reduction, see Chap. 5, this volume) and phase II biotransformation (conjugation with, e.g., glucuronic acid, glutathione). In some instances, which have already been mentioned above, biotransformation is essential for the activation of a pro-drug. Examples are etozolin, muzolimine, and to some extent also ethacrynic acid. (e) The excretion of most loop diuretics is by the kidney. Smaller fractions are usually excreted with the bile. Only substances such as torasemide are excreted to a large extent by the liver (bile). The importance of accumulation in the tubule fluid has already been discussed in Sect. C. In other words: what matters for the pharmacodynamics is the luminal concentration of a given loop diuretic in the tubule fluid of the TAL and not the amount administered orally or i.v. Since the tubule fluid concentration cannot be measured in a volunteer or patient, it has become customary to demonstrate the pharmacodynamic effect as a correlation of saluresis versus urinary concentration of the diuretic under study (Brater 1991).

In the following several loop diuretics will be discussed systematically. For their physical properties and structures, Fig. 1 and Table 1 should be consulted.

I. Ethacrynic Acid

After oral administration the bioavailability of ethacrynic acid is highly variable, with a mean value of 32% (Voith et al. 1993a). Ethacrynic acid is rather hydrophilic. It is metabolized to a large extent by a phase II reaction, namely conjugation to cysteine and glutathione. The cysteine adduct is active as well as ethacrynic acid itself (Burg 1982; Schlatter et al. 1983). It is very likely that both the cysteine adduct and ethacrynic acid itself contribute to the diuretic and saluretic effect. Hence it is not surprising that there was no correlation of urinary excretion of unaltered ethacrynic acid and urinary volume and Na^+ excretion (VOITH et al. 1993b).

II. Indacrinone

Indacrinone is very similar to ethacrynic acid, from which it has been derived (WOLTERSDORF et al. 1978). Its fractional absorption is high. However, indacrinone is largely metabolized in man by a phase I reaction to the 2-(*p*-hydroxyphenyl) derivative. Indacrinone is secreted into the tubule fluid and its renal clearance is strongly pH dependent. With neutral and alkaline pH, renal clearance is high and low with acid pH. One of the specific features of this compound, namely its effect as a uricosuric, has already been mentioned above.

III. Furosemide

Furosemide has a bioavailability of around 60%. Ninety-eight percent of the furosemide present in plasma is bound to proteins. The volume of distribution of furosemide is relatively small at 0.11 l/kg. This is due to its hydrophilicity and its rather acidic pKa of 3.8. Only a smaller but still sizeable portion of furosemide, 20%–40%, is biotransformed, which involves a glucuronidation. The biliary clearance of furosemide is approximately 40%. The remaining 60% is cleared by renal excretion. Due to significant tubular secretion, renal furosemide clearance is usually larger than that for creatinine. It has been mentioned above that the saluresis and diuresis induced by furosemide correlate well with the furosemide concentration present in urine (BRATER 1991). Due to the predominant role of the renal excretion, the half-life of furosemide after a single dose is markedly enhanced in chronic renal failure. On the other hand, liver insufficiency due to cirrhosis has either little impact on the half-life of furosemide or may prolong it slightly. Even under such conditions biotransformation appears to proceed almost unimpeded, perhaps also in other organs. Any fall in hepatobiliary clearance is associated with an increased urinary recovery of the drug at the cost of a prolonged half-life time, i.e., the renal clearance of the drug remains constant (KELLER et al. 1981).

IV. Piretanide

Piretanide has a very high bioavailability of 90%. The protein binding is slightly less than that of furosemide, but it is still the high figure of 96%. The biotransformation of piretanide is 40%, its bilary excretion is also 40%, and its renal clearance is 60% of the total clearance (for review see KNAUF and MUTSCHLER 1993a).

V. Bumetanide

For bumetanide as for piretanide bioavailability is very high at approximately 90% of the orally administered dose. Protein binding is 90% and biotransformation is 40%–60%, the remainder being excreted unaltered renally (for review see KNAUF et al. 1992). Biotransformation by phase I reactions (e.g., oxidation) in the butylamino residue generates inactive compounds.

VI. Torasemide

The bioavailability of torasemide is high, and its protein binding is high. Unlike other loop diuretics binding to the $Na^+2Cl^-K^+$ cotransporter, torasemide is metabolized to a large extent, its renal clearance being only 10% of the normal GFR (FRIEDEL and BUCKLEY 1991; SPAHN et al. 1990). Also, unlike furosemide and related compounds, torasemide is amphiphilic (LOHRMANN et al. 1992). Therefore, its volume of distribution is larger. This may be one of the reasons why torasemide has a half-life after a single dose approximately twice as long as that of, e.g., furosemide. The metabolism of torasemide is shown schematically in Fig. 19. The three major metabolites

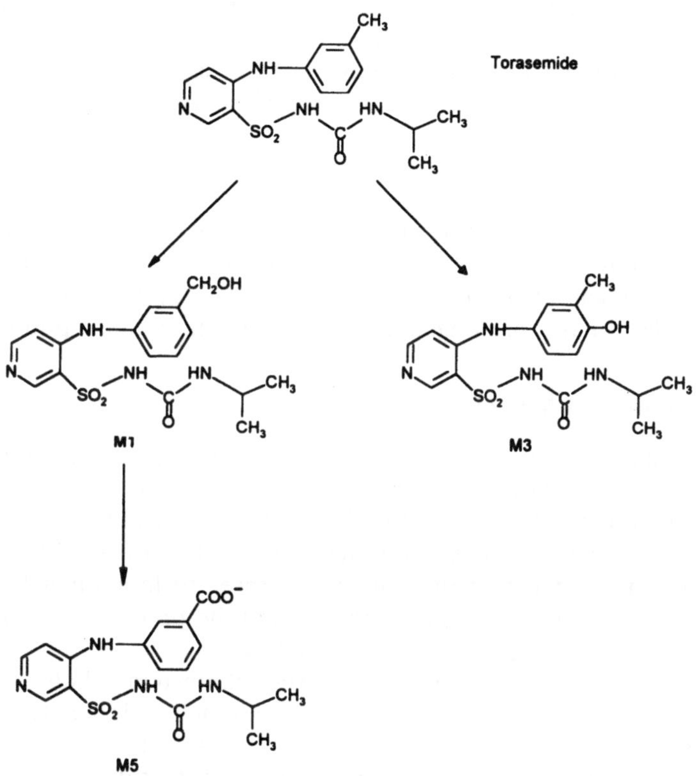

Fig. 19. Metabolism of torasemide

are produced in the liver by phase I reactions. Metabolite 1 involves a hydroxylation of the methyl substituent. This metabolite is only very poorly diuretically active, being oxidized further to the respective carboxylate, which is entirely inactive. An alternative route produces metabolite 3, which is hydroxylated in the *p*-position. This metabolite is diuretically still active (Brater 1990) and inhibits as such the $Na^+2Cl^-K^+$ cotransporter (own unpublished observations). All known metabolites of torasemide do not total 100%. Therefore, further pathways must be considered. As a consequence of its predominant role in hepatic inactivation, the half-life of torasemide is increased in liver cirrhosis. Conversely, chronic renal failure has little effect on the total clearance of torasemide (Knauf et al. 1992). However, the clearance of one specific metabolite (M5 in Fig. 19), which is normally excreted into the urine, is greatly reduced (Knauf et al. 1992). Biliary excretion of torasemide metabolites accounts for a large fraction of total excretion. The importance of tubule pH for the deprotonation of the anionic, and solely active, form of torasemide, has already been mentioned above. With torasemide as an amphiphilic compound, advantage can be taken of the fact that the volume of distribution is larger. This causes a retardation and a prolongation of the effect. However, the differences in half-life of action between torasemide and furosemide or piretanide are probably only a factor of 2 or even less. Several new substances of this class with largely increased lipophilicity have been designed. These compounds might show much more of a retarded action (Lohrmann et al. 1992).

VII. Azosemide

Azosemide is closely related to furosemide. Its bioavailability is fairly small and little is known about its metabolism (Knauf et al. 1992). It is clear, however, that it acts as such (Lohrmann et al. 1991), and that, in comparison to furosemide, it has a prolonged action due to delayed absorption.

VIII. Etozolin and Muzolimine

Both compounds are pro-drugs. Etozolin has a very high bioavailability after oral administration (90%), has a protein binding of only 30%, and is fairly lipophilic. In a first path through the liver it is hydrolyzed to the much more hydrophilic ozolinone (Greven and Heidenreich 1978) (Fig. 20). The half-life of etozolin is 2.5 h, that of ozolinone is three to four times longer. As expected etozolin accumulates in cirrhosis, and the production of ozolinone is delayed correspondingly (Knauf et al. 1987).

Muzolimine is also converted to an as yet unknown metabolite, a process which probably occurs in the liver because functional hepatectomy has been found to inhibit the diuretic and saluretic effects of this compound (Wangemann et al. 1987). Due to this metabolism the saluretic and diuretic effects of muzolimine are delayed and prolonged. After it was noticed that

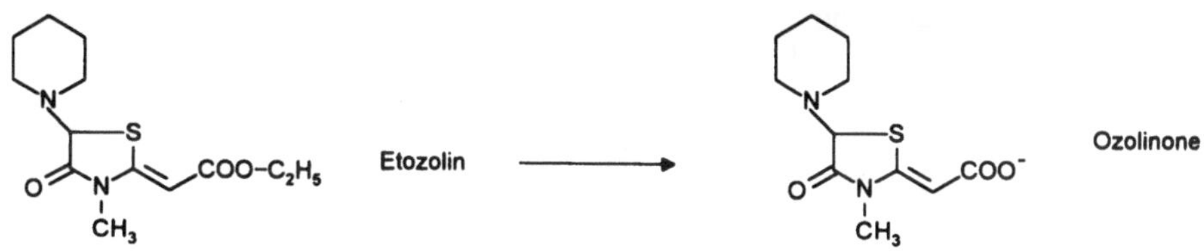

Fig. 20. Bioactivation of ozolinone from etozolin

this compound caused neuropathy, it was withdrawn from the market and no related compounds have since been developed (DAUL et al. 1987).

H. Pharmacokinetics and Pharmacodynamics

The onset of effect depends on the route of administration (p.o. versus i.v.), the volume of distribution, and on whether a diuretic requires metabolizing to become active. It is easily understandable then that the effects of furosemide, piretanide, and bumetanide are almost instantaneous (Table 1), those of torasemide are slightly delayed, and those of etozolin and muzolimine are significantly delayed.

The duration of action depends on total clearance, which is determined by renal excretion, metabolism to inactive compounds, alternative production of active metabolites, or hepatic excretion. The half-life of the effect (Table 1) is approximately 1 h for furosemide and related compounds, twice as long for torasemide, and about 7 h in the case of etozolin (KNAUF et al. 1992).

After a single dose of a loop diuretic saluresis and diuresis ensue fairly rapidly compared with other classes of diuretic. However, the effect is fairly short lived, and antidiuresis and antisaluresis develop (KNAUF and MUTSCHLER 1991, 1992a), a phenomenon called rebound (shown for torasemide in Fig. 21). It is evident that the urinary excretion of Na^+ peaks after a few hours and falls below control values after 6 h. During this rebound, Na^+ excretion is lower than that under control conditions. The total amount of Na^+ excreted with one single dose of the diuretic is the difference between the diuretic and rebound phase. The rebound phenomenon is explained by the diuretic-induced activation of the renin-angiotensin as well as the sympathergic system. It has been shown above (Sect. E.I) that loop diuretics enhance renin secretion directly via their effect on macula densa cells (SCHNERMANN and BRIGGS 1991). In addition and beyond this effect, loop diuretics produce volume contraction, leading independently to the activation of both systems mentioned above. Consideration of the rebound phenomenon has some important practical ramifications. The single dose effect of torasemide on total Na^+ excretion is larger than that of an "equivalent" dose of piretanide or furosemide. This is even more impressive for a rather slowly acting thiazide, for which the net Na^+ excretion can be even higher

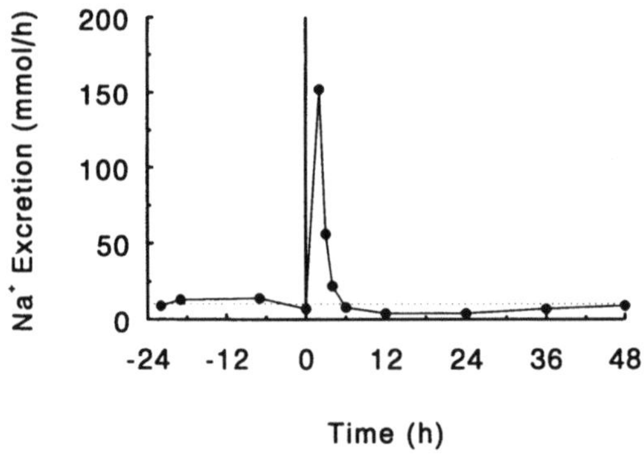

Fig. 21. Rebound phenomenon after a single dose of torasemide (20 mg i.v.). Data taken from KNAUF and MUTSCHLER (1990). Na^+ excretion is shown as a function of time. *Dotted line* shows the baseline excretion. After some 6 h Na^+ excretion falls below this line and is subnormal (rebound) for another 30–40 h

(KNAUF and MUTSCHLER 1984, 1992a). Since Na^+ excretion and that of K^+ are not strictly coupled due to independent pleotropic regulatory mechanisms (see Chap. 2, this volume), the ratios of Na^+ to K^+ excretion can very largely, especially for different types of diuretics. It is still true, though, that all loop diuretics produce comparable kaliuresis (KNAUF and MUTSCHLER 1990, 1992a,b).

Partial or total refractoriness to the "normal" oral dose of a loop diuretic can have a variety of causes (see Chap. 13, this volume). Very simply the bioavailability can be reduced. Such is the case in some patients with cardiac insufficiency (BRATER 1991). The bioactivation of the compound can be delayed, such as in patients with liver cirrhosis when they are given etozolin. The proximal tubule secretion of the respective diuretic can be inhibited, as is the case in acute renal failure, in cases of chronic tubulopathy, and after administration of other substances which are also secreted by the same secretory system (Sect. C, Chap. 6, this volume). The tubule can become refractory to the effect of loop diuretics. This is explained schematically in Fig. 22. The rebound phenomenon, which has been discussed above, is one such example. Even more pronounced are all those cases in which the circulating ADH concentrations are inappropriately high (for review see GREGER and HEIDLAND 1991). This has been noted for patients with chronic heart failure and those with liver cirrhosis. The pathophysiological key issue appears to be an enhanced proximal tubule reabsorption. Hence, the load of water and NaCl delivered to the TAL is reduced and less NaCl is subjected to the control of this nephron segment. As a consequence, the maximal effect caused by loop diuretics is blunted. This is shown in Fig. 23 as the dose response curve for furosemide in cirrhotics and normals. For equal concentrations of the diuretic in the urine, the saluresis is reduced sharply in the patient group (KELLER et al. 1981). Part of this ineffectiveness

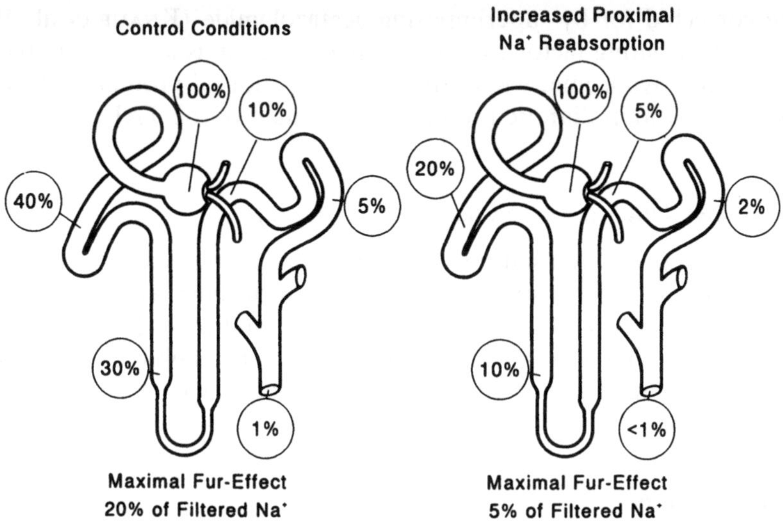

Fig. 22. Model to explain diuretic refractoriness due to increased proximal absorption. *Numbers in the circles* refer to the fractional Na^+ recovery. *Left side*, control conditions: approximately 70% of filtered Na^+ is reabsorbed in the proximal tubule. Hence, 30% is delivered to the TAL. Two-thirds of this (30% end proximal, 10% early distal) can be excreted in the presence of furosemide (maximal furosemide effect, 20% of filtered load). *Right side*, increased proximal Na^+ reabsorption. Only 10% of the filtered load is delivered to the TAL, where only 5% is reabsorbed, and only this fraction can be excreted in the presence of furosemide (maximal furosemide effect, 5% of filtered load). In this setting the coadministration of a proximally acting diuretic can strongly augment the furosemide effect

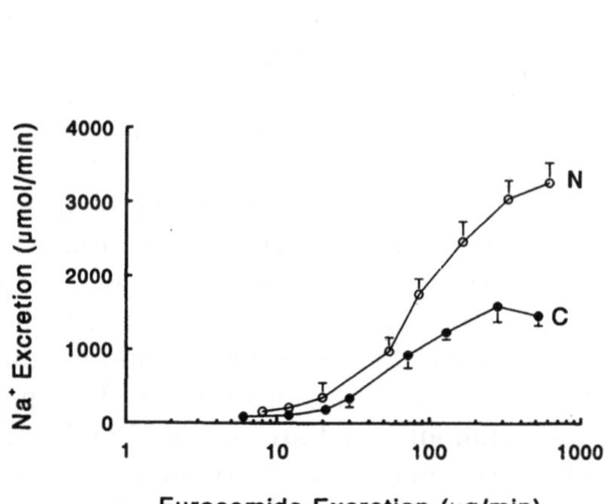

Fig. 23. Loss of diuretic effect of furosemide in cirrhotic patients with ascites (*C, closed symbols*) versus normals (*N, open symbols*). Data taken from KNAUF and MUTSCHLER (1992a). Due to reduced arterial blood volume proximal tubule reabsorption is enhanced (probably angiotensin II effect) As a result of the increased proximal reabsorption (see Fig. 22), loop diuretics lose some of their potency

can be corrected for by superimposing acetazolamide (KNAUF et al. 1990). Another contributing factor to a similar phenomenon is apparent in patients with nephrotic syndrome, where the renal excretion of proteins reduces the free fraction of loop diuretic present in the lumen of the TAL.

I. Clinical Uses (see also Chap. 13, this volume)

Diuretics in general are used for a variety of indications such as (a) essential hypertension, (b) congestive heart failure with lung edema, (c) other edematous states such as ascites, (d) nephrotic syndrome, (e) prevention of a positive Na^+ and water balance in chronic renal failure, and (f) other specific indications. Each of these clinical indications will be briefly discussed in the following.

I. Hypertension

Previously loop diuretics and even more so thiazides have been regarded as the number one medication in essential hypertension for the elderly patient (GREGER and HEIDLAND 1991; Chap. 13, this volume), and the second step in medication even for the young hypertensive, who would usually be treated first by β-blockers. This picture has changed somewhat with the appearance of converting enzyme inhibitors and calcium antagonists (FLECKENSTEIN 1983; LARAGH 1984). In most schemes, however, diuretics are still highly relevant for this indication.

The pathophysiological reasoning for the usage of diuretics is still not quite clear: Volume contraction is the initial step, which may be paralleled by a diuretic-induced increase in prostaglandins (JOHNSTON et al. 1983). Loop diuretics might act in a vasodilatating manner themselves (see Sect. F.III, above) (CANTIELLO et al. 1986). This has, however, been discarded: it appears more likely that diuretics reduce the cytosolic activity of Na^+ or Ca^{2+} (GREGER and HEIDLAND 1991; ERNE et al. 1984).

Intuitively it is clear that for chronic treatment the drug of choice should preferentially have a longer-lasting effect, since a strong short-lasting effect may be blunted by undesired compensatory mechanisms such as activation of the renin-angiotensin-aldosterone system, and catecholamine release. This may be the reason that, at least in some countries, thiazides are much more popular for this indication than loop diuretics. The appropriate dosages of the loop diuretics commonly used are provided in Table 2. All attempts to formulate retard preparations of the furosemide-type diuretics have thus far not been very successful. Torasemide might offer a small advantage in this respect, due to its longer half-life (see above). However, the differences among most respective compounds are borderline at best. It is important to note that the lowest possible effective doses should be used for this indication (for review see GREGER and HEIDLAND 1991). Otherwise the side effects of

Table 2. Usual oral doses for various loop diuretics

Name	Typical doses (mg)	See also	Comment	Renal clearance
Furosemide	20–500	Chaps. 5, 13	Fast effect	High
Bumetanide	1–15	Chaps. 5, 13	Fast effect	High
Piretanide	6–50	Chaps. 5, 13	Fast effect	High
Azosemide	40–100	Chaps. 5	Low bioavailability	
Etozolin	400	Chaps. 5	Slow effect	Small
Torasemide	2.5–25	Chaps. 5, 13	Slower effect	Small

The doses are taken from the respective literature. The safety instructions of the respective manufacturers must be strictly adhered to.

chronic diuretic treatment will partially offset their beneficial effects. The side effects of loop diuretics will be summarized in Sect. J.

II. Congestive Heart Failure and Lung Edema

Diuretics are beneficial in this condition because they reduce the preload to the heart. One might predict that this reduces, due to Frank-Starling mechanisms, the cardiac output. This, however, is not the case in most patients because diuretics reduce afterload by vasodilatation at the same time (Nishijma 1984). It has been claimed that the fall in preload in these patients occurred prior to the diuresis (Dikshit et al. 1973). This has led to the speculation that loop diuretics act directly on the venous system. However, this effect is probably not due to a direct action, because it cannot be demonstrated in anephric patients (Johnston et al. 1983; Bourland et al. 1977). This issue has been addressed in Sect. F.III.

The benefit to the patient is mostly due to improvement of respiratory performance. In such edematous states Na^+ restriction usually has to be combined with a strongly acting (loop) diuretic. The patient may even be refractory to the treatment by loop diuretics. In such a situation a low delivery of NaCl to the TAL may impede a diuretic response, and it may be useful to superimpose another diuretic such as metolazone or other thiazides (Greger and Heidland 1991). Chronic side effects are of less relevance for this group of patients. The dosages of loop diuretics (Table 2) can be much higher for this indication.

III. Ascites

Most current evidence suggests that severe ascites is concomitant with a reduction rather than an increase in extracellular volume. Hence, in its treatment NaCl and some water restriction will be the first steps (for review see Greger and Heidland 1991). Only if ascites is gross or generates additional symptoms such as discomfort, reflux, and breathing problems should diuretics be used. It is wise to volume deplete the patient only slowly. Due to the pathomechanism involved, and due to the fact that hyperaldosteronism exists in a large group of these patients, aldosterone antagonists (see Chap. 9, this volume) or distally acting diuretics (see Chaps. 10, 11, this volume) are to be preferred over loop diuretics for this indication.

IV. Edematous States in Nephrotic Syndrome

In nephrotic syndrome edematous states and even more so overhydration may justify the therapy by loop diuretics. Care should be taken in these patients to avoid volume contraction with the danger of hypercoagulation, a further fall in GFR, renal vein thrombosis, and even acute renal failure (GREGER and HEIDLAND 1991). Fairly high doses are usually required to treat these edematous states. This is probably due to the above-mentioned effect of luminally protein bound diuretics (GREEN and MIRKIN 1980). This may account for a dose three times as high as normal. In addition, there may be reduced delivery of NaCl to the TAL as a consequence of proximal tubular hyperabsorption in these states of reduced effective arterial blood volume. A normal therapeutic scheme involves NaCl restriction and administration of higher doses of, e.g., furosemide, orally. If this fails a thiazide diuretic could be added to this regimen. In refractory patients a "segmental nephron blockade" may be tried by short-term comedication with acetazolamide (KNAUF and MUTSCHLER 1993b, 1994).

V. Chronic Renal Failure

Many patients with chronic renal failure are overhydrated and hence require, besides a tolerable Na^+ restriction, diuretic therapy. Loop diuretics again are the drugs of choice here (HEIDLAND et al. 1972). For most loop diuretics it has been shown that they are effective down to a GFR of less than 10 ml/min (for review see GREGER and HEIDLAND 1991). For thiazides this has previously been refuted. However, more recent studies indicate that thiazides are effective, too, at fairly low GFR values albeit much less so than loop diuretics (KNAUF et al. 1994; KNAUF and MUTSCHLER 1993b). In fact, loop diuretics may kick up fractional excretion of Na^+ excretion to values much higher than in normals. This is due to the fact that the residual nephrons have an increased filtration rate and very high delivery rates to the TAL. In this sense, this condition is just the opposite of states of inappropriate ADH secretion and nephrotic syndrome where proximal reabsorption is enhanced.

The secretion of loop diuretics can be markedly attenuated in chronic renal failure. This is due to an inhibition of proximal secretion due to the accumulation of endogenous substrates of the proximal tubular secretory system. Only approximately 10% instead of 40%–60% of the i.v. administered furosemide may then reach the urine. As a consequence, fairly high doses may be required to exert an effect (for review see GREGER and HEIDLAND 1991). This is most relevant for the patients with very low GFRs (<10 ml/min). With a regimen of up to 1000 mg furosemide/day i.v. or twice that amount orally, side effects such as ototoxicity may occur (see Sect. F.I, above).

VI. Other Indications

The other indications comprise a heterogeneous group which will briefly be summarized here: (a) acute renal failure, (b) mountain sickness, (c) hypercalcemia, (d) renal tubular acidosis, (e) asthmatic states, and (f) intoxications.

Acute renal failure has been shown to affect mostly the proximal nephron and the TAL. This has been discussed above (see Sect. D.IV). It may suffice to repeat here that prophylactic administration of loop diuretics will enhance the ability of the TAL segment ot remain intact in spells of hypoxemia and/or ischemia (for review see GREGER and HEIDLAND 1991). On the other hand, the administration of loop diuretics after the establishment of acute renal failure has been shown to be of little value, because the damage to the kidney occurs early and has probably already been established in most respective studies prior to the administration of the loop diuretic.

Mountain sickness is not infrequent if lowlanders climb up rapidly to altitudes of 3000 m and higher. This may be concomitant with peripheral, lung, or even brain edema. The acute or prophylactic administration of diuretics of acetazolamide or furosemide and related compounds can be very valuable for such states (LARSON et al. 1982).

Hypercalcemia. It has already been mentioned that loop diuretics enhance the excretion of Ca^{2+} and Mg^{2+} (see Sect. E.V). Use can be made of this side effect for states of hypercalcemia (SUKI et al. 1970). Adequate volume and NaCl replenishment is necessary in such patients in order to prevent rebound phenomena with the danger of Ca^{2+} precipitations in the urinary tract and because volume contraction in itself, by enhancing proximal absorption, inhibits the calciuric effect.

In renal tubular acidosis type I (congenital distal defect in building up proton gradients) and in type IV (defect caused by low aldosterone or diminished aldosterone response, defect in H^+ secretion) furosemide-type loop diuretics might be helpful inasmuch as they enhance the distal NaCl load and thus increase the rate of distal H^+ secretion (for review see GREGER and HEIDLAND 1991). The proton secretion defect of some patients is refractory to furosemide.

The inhalation therapy of furosemide has already been mentioned in Sect. F.II as a possible treatment for asthma.

Loop diuretics may be of value for various intoxication states. In such cases a close monitoring of the salt and water balance is part of the routine regimen.

J. Adverse Effects

All furosemide-type loop diuretics share the same main side effects: (a) hypokalemia, (b) hyponatremia, (c) hypocalcemia, (d) hypomagnesemia, (e) metabolic alkalosis, (f) hyperlipidemia, (g) hyperglycemia, (h) hyperuricemia, (i) male impotence, and (j) ototoxicity. These side effects, to the extent that they have not previously been mentioned, will be discussed below in some detail.

I. Hypokalemia

The causes of hypokalemia have already been discussed in Sect. E.II. It suffices here to mention the therapeutic strategy. If plasma K^+ is below 3.5 but above 3.0 mmol/l, correction of the diet, if applicable, would be the first step. The clinician will be especially alerted if there is a history of cardiac arrhythmias or if the patient is on therapy with cardiac glycosides. On the other hand, rigorous treatment may lead to hyperkalemia, which is more likely in the elderly patient with reduced GFR. KCl supplements are not always tolerated well. An alternative is the combination of the loop diuretic with a K^+-sparing diuretic such as triamterene, amiloride, or the aldosterone antagonist spironolactone (GREGER and HEIDLAND 1991). Similarly, hypokalemia can be prevented by the combination of a loop diuretic with an angiotensin-converting enzyme inhibitor (GRIFFING et al. 1983).

II. Hyponatremia

Hyponatremia is a rare but possibly dangerous complication of chronic diuretic treatment. Its pathophysiology involves: (a) the inhibition of the dilution ability of the kidney by the loop diuretic; (b) excessive urinary salt loss combined with a small intake; (c) inappropriate secretion of ADH, this being especially frequent in cirrhotics, in patients with congestive heart failure, and in the elderly (GROSS 1988); and (d) unrestricted drinking and or polydipsia due to some drugs (e.g., tricyclic antidepressants).

The treatment is difficult. The diuretic has to be stopped and water intake restricted. Plasma Na^+ and K^+ concentrations have to be corrected slowly (GREGER and HEIDLAND 1991). Head-out-of-water immersion can be beneficial in these patients inasmuch as it may successfully reduce ADH and the sympathergic discharge to the kidney (EPSTEIN 1983).

III. Hypocalcemia

Loop diuretic induced hypocalcemia (the mechanism of which is discussed above in Sect. E.V) is not a serious complication for most patients. If there is a risk of kidney stone disease (Ca^{2+}/phosphate/oxalate), loop diuretics could be replaced (or combined) with thiazides. Thiazides enhance distal Ca^{2+} absorption (COSTANZO and WINDHAGER 1978; see also Chap. 8).

IV. Hypomagnesemia

The mechanisms of hypomagnesemia have already been described in Sect. E.V. Mg^{2+} losses occur both with thiazides and with loop diuretics (see Chap. 13, this volume). Therefore, unlike with Ca^{2+}, thiazides or a combination of loop diuretics and thiazides will not prevent this complication. Hypomagnesemia is frequently paralleled by hypokalemia, and both disturbances are interdependent. However, the precise mechanism of this interdependence has still not been clarified. It has been noted that the success of the K^+ replenishment of hypokalemia depends on the parallel correction of hypomagnesemia (DYKNER and WESTER 1979). K^+-sparing diuretics are helpful in this regard since they enhance distal Mg^{2+} absorption (DYKNER and WESTER 1984). The mechanism of this process is not clear (see chapters in this volume these diuretics).

V. Metabolic Alkalosis

Metabolic alkalosis is usually not a serious problem with loop diuretics (see Sect. E.IV). If metabolic alkalosis ensues, this may generate secondary problems (cf. hypokalemia); and, conversely, hypokalemia may promote metabolic alkalosis. Again, a combination of K^+-sparing and loop diuretics will largely prevent this problem, because the former diuretics reduce distal tubule H^+ secretion (HYAMS 1981).

VI. Hyperlipidemia

Hyperlipidemia is a frequently discussed complication of chronic treatment with diuretics. Thiazides and loop diuretics may cause this effect, which has been well documented in acute studies and in some but not all chronic studies (summarized in GREGER and HEIDLAND 1991). Triglycerides and low-density lipoprotein were increased in one acute study, while high-density lipoproteins were unaltered. In some long-term studies, e.g., the Veteran's Administration Cooperative study, the MRFIT study 1985, and the HAPPHY study 1987 (HAPPHY 1987; MRFIT 1985), an effect on lipids was not found; but in other studies it has been reported to occur after 1 year of diuretic treatment (WEINBERGER 1985; MIDECKE et al. 1987; SKARFORS et al. 1989). Also the MRFIT study 1985 at least supports the conclusion that the beneficial effect of a low-fat diet was partially offset by diuretics (MRFIT 1985).

The pathophysiological mechanism by which loop diuretics produce hyperlipidemia has not yet been clarified. Hyperinsulinemia together with some insulin resistance may be causally involved since insulin increases lipolysis. Even more important may be the development of hypokalemia, which itself aggravates the diabetogenic effect of diuretics. The predominant role of hypokalemia is supported by the finding that prevention of hypokalemia, e.g., by the coadministration of K^+-sparing and loop diuretics, prevents this complication. As for most complications, it also seems to hold for hyperlipidemia that a low-dose therapy may offer beneficial therapeutic effects and yet reduce the risk of hypokalemia and consequent hyperlipidemia (see also Chap. 13, this volume).

VII. Hyperglycemia and Diabetogenic Effects

The possible pathomechanisms of diuretic-induced hyperglycemia have already been discussed in Sect. F.IV. It suffices to mention here that the development of glucose intolerance and overt type II diabetes is a rather frequent complication, especially with thiazides but also with moderate to high dosages of loop diuretics. It is unclear whether loop diuretics themselves have a diabetogenic effect or whether they only unmask diabetes in patients who are predisposed to it (for review see GREGER and HEIDLAND 1991). The above-cited (Sect. F.IV) pathomechanisms, inhibition of insulin release and inhibited glucose uptake, are probably aggravated by hypokalemia. Increased sympathergic discharge will augment these effects. Consequently, close monitoring of plasma K^+ concentration will reduce the incidence of this complication (McFARLAND and CARR 1977). Diuretic-induced glucose intolerance is reversible when diuretic therapy is stopped.

VIII. Hyperuricemia

The causes of hyperuricemia have been described in Sect. E.VII. Its incidence is as high as 50%–70% in chronic diuretic therapy, while gouty arthritis is rare (BEEVERS et al. 1971). Since hyperuricemia is no longer regarded as an independent risk factor in cardiovascular disease, it is not treated.

IX. Male Impotence

This complication is probably more frequent with inappropriately high dosages of thiazides than with loop diuretics. Impotence is caused by a failure of normal erection, and it may be the reason for the patient discontining therapy. The mechanism of this rare complication is unknown.

X. Ototoxicity

The pathomechanism of loop diuretic induced ototoxicity has been discussed in Sect. F.I. This serious complication occurs only if high doses of loop diuretics are employed. It has been reported to be reversible after discontinuation of the loop diuretic (KLINKE et al. 1981; HEIDLAND and WIGAND 1970). Whether more potent loop diuretics are less harmful in this respect will have to be clarified in clinical studies.

References

Alpern RJ, Stone DK, Rector FC (1991) Renal acidification mechanisms. In: Brenner BM, Rector FC (eds) The kidney. Saunders, Philadelphia, pp 318–379
Beevers DG, Hamilton M, Harpur JE (1971) The long-term treatment of hypertension with thiazide diuretics. Postgrad Med J 47:639–643
Bianco S, Vaghi A, Robuschi M, Pasargiklian M (1988) Prevention of exercise-induced bronchoconstriction by inhaled furosemide. Lancet 2:252–255
Bianco S, Pieroni MG, Refini RM, Rottoli L, Sostini P (1989) Protective effect of inhaled furosemide on allergen-induced early and late asthmatic reactions. N Engl J Med 321:1069–1073
Bijman J, Englert HC, Lang HJ, Greger R, Frömter E (1987) characterization of human sweat duct chloride conductance by chloride channel blockers. Pflugers Arch 408:511–514
Bleich M, Schlatter E, Greger R (1990) The luminal K^+ channel of the thick ascending limb of Henle's loop. Pflugers Arch 415:449–460
Boron WF, Boulpaep EL (1989) The electrogenic Na/HCO_3 cotransporter. Kidney Int 36:392–402
Bourland WA, Day DK, Williamson HE (1977) The role of the kidney in the early nondiuretic action of furosemide to reduce elevated left atrial pressure in the hypervolemic dog. J Pharmacol Exp Ther 201:221–229
Braitsch R, Lohrmann E, Greger R (1990) Effect of probenecid on loop diuretic induced saluresis and diuresis. In: Puschett J (ed) Diuretics III, chemistry, pharmacology, and clinical applications. Elsevier, New York, pp 137–139
Brater DC (1990) Clinical pharmacology of loop diuretics. Klin Wochenschr 3:5–6
Brater DC (1991) Clinical pharmacology of loop diuretics. Drugs 41 [Suppl 3]:14–22
Brezis M, Rosen S, Silva P, Epstein FH (1984) Renal ischemia: a new perspective. Kidney Int 26:375–383
Brummet RE, Bendrick T, Himes D (1981) Comparative ototoxicity of bumetanide and furosemide when used in combination with kanamycin. J Clin Pharmacol 21:628–636
Burg MB (1982) Thick ascending limb of Henle's loop. Kidney Int 22:454–464
Burg MB, Bourdeau JE (1978) Function of the thick ascending limb. In: Vogel HG, Ullrich KJ (eds) New aspects of renal function. Elsevier, Amsterdam, pp 91–102
Burg MB, Green N (1973a) Effect of mersalyl on the thick ascending limb of Henle's loop. Kidney Int 4:245–251
Burg MB, Green N (1973b) Function of the thick ascending limb of Henle's loop. Am J Physiol 224(3):659–668
Burg MB, Stoner L (1976) Renal tubular chloride transport and the mode of action of some diuretics. Annu Rev Physiol 38:37–45
Burg MB, Stoner L, Cardinal J, Green N (1973) Furosemide effect on isolated perfused tubules. Am J Physiol 225:119–124
Burnham CE, Kidd J, Palfrey HC (1990) Expression of the $Na^+K^+2Cl^-$ cotransporter in Xenopus oocytes. Am J Physiol 259:F383–F388

Cantiello H, Copello J, Muller A, Mikulic L, Villami MF (1986) Effect of bumetanide on potassium transport and ionic composition of the arterial wall. Am J Physiol 251:F537–F546
Cornejo M, Guggino SE, Guggino WB (1987) Modification of Ca^{++}-activated K^{+} channels in cultured medullary thick ascending limb cells by N-bromoacetamide. J Membr Biol 99:147–155
Cornejo M, Guggino SE, Guggino WB (1989a) Ca^{++}-activated K^{+} channels from cultured renal medullary thick ascending limb cells: effects of pH. J Membr Biol 110:49–55
Cornejo M, Guggino SE, Sastre A, Guggino WB (1989b) Isomeric yohimbine alkaloids block calcium-activated K^{+} channels in medullary thick ascending limb cells of rabbit kidney. J Membr Biol 107:25–33
Costanzo LS, Windhager EE (1978) Calcium and sodium transport by the distal convoluted tubule of the rat. Am J Physiol 235:F492–F506
Costanzo LS, Windhager EE (1992) Renal regulation of calcium balance. In: Seldin DW, Giebisch G (eds) The kidney: physiology and pathophysiology. Raven, New York, pp 2375–2393
Daul A, Graben N, Bock KD (1987) Neuromyeloenzephalopathie nach hochdosierter Muzolimintherapie bei Dialysepatienten. Munch Med Wochenschr 129:542–543
Deetjen P (1966) Micropuncture studies on the site and mode of diuretic action of furosemide. Ann NY Acad Sci 139:408–415
Di Stefano A, Wittner M, Schlatter E, Lang HJ, Englert H, Greger R (1985) Diphenylamine-2-carboxylate, a blocker of the Cl^{-}-conductive pathway in Cl^{-}-transporting epithelia. Pflugers Arch 405 [Suppl 1]:S95–S100
Di Stefano A, Wangemann P, Friedrich T, Burckhardt G, Oekonomopoulos RR, Englert HC, Lang HJ, Greger R (1986) Photoaffinity labelling of the Na-2Cl-K carrier. Pflugers Arch 406:R59
Di Stefano A, Greger R, De Rouffignac C, Wittner M (1992) Active NaCl transport in the cortical thick ascending limb of Henle's loop of the mouse does not require the presence of bicarbonate. Pflugers Arch 420:290–296
Dikshit K, Vyden JK, Forrester JS, Chattjee K, Parkash R, Swan HJC (1973) Renal and extrarenal hemodynamic effects of furosemide in congestive heart failure after myocardial infarction. N Engl J Med 288:1087–1090
Dillingham MA, Schrier RW, Greger R (1993) Mechanisms of diuretic action. In: Schrier RW, Gottschalk CW (eds) Clinical disorders of fluid, electrolytes, and acid base. Little Brown, Boston, pp 2435–2452
Dykner T, Wester PO (1979) Ventricular extra systoles and intracellular electrolytes before and after potassium and magnesium infusions in hypokalemic patients on diuretic treatment. Am Heart J 97:12–18
Dykner T, Wester PO (1984) Intracellular magnesium loss after diuretic administration. Drugs 28 [Suppl 1]:161–166
Epstein M (1983) Renal sodium handling in cirrhosis. In: Epstein M (ed) The kidney in liver disease. Elsevier, New York, pp 25–53
Erne P, Bolli P, Bürgisser E, Bühler FR (1984) Correlation of platelet calcium with blood pressure. Effect of antihypertensive therapy. N Engl J Med 310:1084–1088
Faedda R, Satta A, Branca GF, Contu B, Bartoli E (1985) Demonstration of action of muzolimine on the serosal side of Henle's loop cells. Z Kardiol 74 [Suppl 2]:166–170
Fanelli GM, Bohn DL, Horbaty CA, Beyer KH, Scriabine A (1974) A new uricosuric-saluretic agent (6,7-dichloro-2-methyl-1-oxo-2-phenyl-5-indanyloxy) acetic acid (indanone): evaluation in the chimpanzee. Kidney Int 6:40A
Feit PW, Hoffmann EK, Schiodt M, Kristensen P, Jessen F, Dunham PB (1988) Purification of proteins of the Na/Cl cotransporter from membranes of Ehrlich ascites cells using a bumetanide-sepharose affinity column. J Membr Biol 103: 135–147

Fleckenstein A (1983) Calcium antagonism in heart and smooth muscle. In: Fleckenstein A (ed) Calcium antagonism. Wiley, New York, pp 209–285
Friedel HA, Buckley MMT (1991) Torasemide: a review of its pharmacological properties and therapeutic potential. Drugs 41:81–103
Friedman PA, Andreoli TE (1982) CO_2-stimulated NaCl absorption in the mouse renal cortical thick ascending limb of Henle. J Gen Physiol 80:683–711
Friedman PA, Gesek FA (1993) Calcium transport in renal epithelial cells. Am J Physiol 264:F181–F198
Geck P, Heinz E (1986) The Na-K-2Cl cotransport system. J Membr Biol 91:97–105
Geck P, Pietrzyk C, Burckhardt BC, Pfeiffer B, Heinz E (1980) Electrically silent cotransport of Na^+, K^+ and Cl^- in Ehrlich cells. Biochim Biophys Acta 600: 432–447
Good DW (1985) Sodium-dependent bicarbonate absorption by cortical thick ascending limb of rat kidney. Am J Physiol 248:F821–F829
Green TP, Mirkin BL (1980) Resistance of proteinuric rats to furosemide: urinary drug protein binding as a determinant of drug effect. Life Sci 26:623–630
Greger R (1981a) Chloride reabsorption in the rabbit cortical thick ascending limb of the loop the Henle. A sodium dependent process. Pflugers Arch 390:38–43
Greger R (1981b) Coupled transport of Na^+ and Cl^- in the thick ascending limb of Henle's loop of rabbit nephron. Scand Audiol 14:1–15
Greger R (1981c) Cation selectivity of the isolated perfused cortical thick ascending limb of Henle's loop of rabbit kidney. Pflugers Arch 390:30–37
Greger R (1985a) Wirkung von Schleifendiuretika auf zellulärer Ebene. Nieren Hochdr Krankh 14(6):217–220
Greger R (1985b) Ion transport mechanisms in thick ascending limb of Henle's loop of mammalian nephron. Physiol Rev 65:760–797
Greger R (1986) Chlorid-transportierende Epithelien. Physiol Akt 2:47–58
Greger R (1987) Pathophysiologie der ranalen Ischämie. Z Kardiol 76 S4:81–86
Greger R (1990) Possible sites of lithium transport in the nephron. Lithium in renal physiology. Kidney Int 37 [Suppl 28]:S26–S30
Greger R, Heidland A (1991) Action and clinical use of diuretics. In: Cameron S, Davison AM, Grünfeld JP (eds) Clinical nephrology. Oxford University Press, London, pp 197–224
Greger R, Schlatter E (1981) Presence of luminal K^+, a prerequisite for active NaCl transport in the thick ascending limb of Henle's loop of rabbit kidney. Pflugers Arch 392:92–94
Greger R, Schlatter E (1983a) Properties of the basolateral membrance of the cortical thick ascending limb of Henle's loop of rabbit kidney. A model for secondary active chloride transport. Pflugers Arch 396:325–334
Greger R, Schlatter E (1983b) Cellular mechanism of the action of loop diuretics on the thick ascending limb of Henle's loop. Klin Wochenschr 61:1019–1027
Greger R, Schlatter E (1983c) Properties of the lumen membrance of the cortical thick ascending limb of Henle's loop of rabbit kidney. Pflugers Arch 396:315–324
Greger R, Schlatter E (1984) Mechanism of NaCl secretion in rectal gland tubules of spiny dogfish (Squalus acanthias). II. Effects of inhibitors. Pflugers Arch 402: 364–375
Greger R, Wangemann P (1987) Loop diuretics. Renal Physiol 10:174–183
Greger R, Lang F, Deetjen P (1974) Urate handling by the rat kidney: IV. Reabsorption in the loops of Henle. Pflugers Arch 352:115–120
Greger R, Oberleithner H, Schlatter E, Cassola AC, Weidtke C (1983a) Chloride activity in cells of isolated perfused cortical thick ascending limbs of rabbit kidney. Pflugers Arch 399:29–34
Greger R, Schlatter E, Lang F (1983b) Evidence for electroneutral sodium chloride cotransport in the cortical thick ascending limb of Henle's loop of rabbit kidney. Pflugers Arch 396:308–314

Greger R, Weidtke C, Schlatter E, Wittner M, Gebler B (1984a) Potassium activity in cells of isolated perfused cortical thick ascending limbs of rabbit kidney. Pflugers Arch 401:52–57

Greger R, Wittner M, Schlatter E, Di Stefano A (1984b) $Na^+2Cl^-K^+$-cotransport in the thick ascending limb of Henle's loop and mechanism of action of loop diuretics. In Hoshi T (ed) Coupled transport in nephron. Mechanisms and pathophysiology. Miura Medical Research Foundation, Tokyo, pp 96–118

Greger R, Wangemann P, Wittner M, Di Stefano A, Lang HJ, Englert HC (1987) Blockers of active transport in the thick ascending limb of the loop of Henle. In: Andreucci VE, Dal Canton A (eds) Diuretics: basic, pharmacological, and clinical aspects. Nijhoff, Boston, pp 33–38

Greger R, Bleich M, Schlatter E (1991) Ion channel regulation in the thick ascending limb of the loop of Henle. Kidney Int 40:S33:S119–S124

Greven J, Heidenreich O (1978) Effects of ozolininone, a diuretic active metabolite of etozolin, on renal function. I. Clearance studies in dogs. Naunyn Schmiedebergs Arch Pharmacol 304:283–287

Griffing GT, Sindler BH, Aureccia SA, Melby JC (1983) Reversal of diuretic-induced secondary hyperaldosteronism and hypokalemia by enalapril (MK 421): a new angiotensin-converting enzyme inhibitor. Metabolism 32:711–716

Gross P (1988) Role of diuretics, hormonal derangements, and clinical setting of hyponatremia in medical patients. Klin Wochenschr 66:662–669

Guggino SE, Suarez-Isla BA, Guggino WB, Sacktor B (1985) Forskolin and antidiuretic hormone stimulate a Ca^{++}-activated K^+ channel in cultured kidney cells. Am J Physiol 249:F448–F455

Guggino SE, Guggino WB, Green N, Sacktor B (1987a) Ca^{2+}-activated K^+ channels in cultured medullary thick ascending limb cells. Am J Physiol 252:C121–C127

Guggino SE, Guggino WB, Green N, Sacktor B (1987b) Blocking agents of Ca^{2+}-activated K^+ channels in cultured medullary thick ascending limb cells. Am J Physiol 252:C128–C137

Haas M (1989) Properties and diversity of (Na-K-Cl) cotransporters. Annu Rev Physiol 51:443–457

Haas M, Forbush B (1987) Photolabeling of a 150-kDa ($Na^+K^+Cl^-$) cotransport protein from dog kidney with a bumetanide analogue. Am J Physiol 253:C243–C250

Haas JA, Larson MV, Marchand GR, Lang FC, Greger RF, Knox FG (1977) Phosphaturic effect of furosemide: role of PTH and carbonic anhydrase. Am J Physiol 232:F105–F110

Halperin ML, Kamel SK, Ethier JH, Stienbaugh BJ, Jungas RL (1992) Biochemistry and physiology of ammonium excretion. In: Seldin DW, Giebisch G (eds) The kidney: physiology and pathophysiology. Raven, New York, pp 2645–2679

HAPPHY (1987) Beta-blockers versus diuretics in hypertensive men: main results from the HAPPHY trial. J Hypertens 5:561–572

Häussinger D, Lang F (1991) Cell volume in the regulation of hepatic function: a mechanism for metabolic control. Biochim Biophys Acta 1071:331–350

Hebert SC, Andreoli TE (1984) Control of NaCl transport in the thick ascending limb. Am J Physiol 246:F745–F756

Heidland A, Wigand ME (1970) Einfluß hoher Furosemiddosen auf die Gehörfunktion bei Urämie. Klin Wochenschr 48:1052–1056

Heidland A, Hennemann H, Wigand ME (1972) Dosiswirkung und optimale Infusionsgeschwindigkeit hoher Furosemidgaben bei terminaler Niereninsuffizienz. In: Kluthe R (ed) Medikamentöse Therapie bei Nierenerkrankungen. Thieme, Stuttgart, pp 221–228

Hyams DE (1981) Amiloride: a review. In: Wood, Summerville (eds) Arrhythmias and myocardial infarction: the role of potassium. Royal Society of the medical in ternational congress symposium series. Academic Press, London, pp 65–73

Jacobs DB, Mookerja BK, Jang CY (1984) Furosemide inhibits glucose transport in isolated rat adipocytes via direct inactivation of carrier proteins. J Clin Invest 74:1679–1685

Johnston GD, Hiatt WR, Nies AS (1983) Factors modifying the early non diuretic effects of furosemide in man. The possible role of renal prostaglandins. Circ Res 53:630–635

Keller E, Hoppe-Seyler G, Mumm R, Schollmeyer P (1981) Influence of hepatic cirrhosis and end-stage renal disease on pharmacokinetics and pharmacodynamics of furosemide. Eur J Clin Pharmacol 20:27–33

Kikeri D, Sun A, Zeidel ML, Hebert SC (1989) Cell membranes impermeable to NH_3. Nature 339:478–480

Kinne R, Koenig B, Hannafin J, Kinne-Saffran E, Scott DM, Zierold K (1986) The use of membrane vesicles to study the NaCl/KCl cotransporter involved in active transepithelial chloride transport. Pflugers Arch 405 [Suppl 1]:S101–S105

Klinke R, Lahn W, Querfurth H, Scholtholt J (1981) Ototoxic side effects of diuretics. Almquist and Wiksell Periodical, Stockolm, pp 1–232

Knauf H, Mutschler E (1984) Pharmacodynamics and pharmacokinetics of xipamide in patients with normal and impaired kidney function. Eur J Clin Pharmacol 26:513–520

Knauf H, Mutschler E (1990) Saluretic effects of the loop diuretic torasemide in chronic renal failure. Interdependence of electrolyte excretion. Eur J Clin Pharmacol 39:337–343

Knauf H, Mutschler E (1991) Pharmacodynamic and kinetic considerations on diuretics as a basis for differential therapy. Klin Wochenschr 69:233–238

Knauf H, Mutschler E (1992a) Wirkprofil. In: Knauf H, Mutschler E (eds) Diuretika, Prinzipien der klinischen Anwendung. Urban and Schwarzenberg, Munich, pp 189–213

Knauf H, Mutschler E (1992b) Constant K^+/Na^+ excretion ratio during peak diuresis after piretanide but insignificant K^+ loss during 24 hours. Eur J Clin Pharmacol 43:23–27

Knauf H, Mutschler E (1993a) Piretanide. Marius, Carnforth, UK

Knauf H, Mutschler E (1993b) Low-dose segmental blockade of the nephron rather than high-dose diuretic monotherapy. Eur J Clin Pharmacol 44 [Suppl 1]:S63–S68

Knauf H, Mutschler E (1994) Functional state of the nephron and diuretic dose-response-rationale for low-dose combination therapy. Cardiology 84 [Suppl 2]: 18–26

Knauf H, Missmahl M, Schölmerich J, Gerok W, Mutschler E (1987) Altered kinetics of etozolin and its active metabolite ozolinone in hepatitis and hepatic cirrhosis with ascites. Drug Res 37:1385–1388

Knauf H, Wenk E, Schölmerich J, Gerok W, Leser HG, Mutschler E (1990) Prediction of diuretic mobilization of cirrhotic ascites by pretreatment fractional sodium excretion. Klin Wochenschr 68:545–557

Knauf H, Gerok W, Mutschler E (1992) Pharmakokinetik. In: Knauf H, Mutschler E (eds) Diuretika. Urban Schwarzenberg, Munich, pp 149–187

Knauf H, Cawello W, Schmidt G, Mutschler E (1994) The saluretic effect of the thiazide diuretic bemetizide in relation to the glomerular filtration rate. Eur J Clin Pharmacol 46:9–13

Knepper MA, Good DW, Burg MB (1985) Ammonia and bicarbonate transport by rat cortical collecting ducts perfused in vitro. Am J Physiol 249:F870–F877

Koenig B, Ricapito S, Kinne R (1983) Chloride transport in the thick ascending limb of Henle's loop; potassium dependence and stoichiometry of the NaCl cotransport system in plasma membrane vesicles. Pflugers Arch 399:173–179

Koomans HA, Boer WH, Dorhout Mees EJ (1989) Evaluation of lithium clearance as a marker of proximal sodium handling. Kidney Int 36:2–12

Lang F, Rehwald W (1992) Potassium channel in renal epithelial transport regulation. Physiol Rev 72:1–32

Lang F, Greger R, Deetjen P (1977) Effect of diuretics on uric acid metabolism and excretion. In: Siegenthaler W, Beckerhoff R, Vetter W (eds) Diuretics in research and clinics. Thieme, Stuttgart, pp 213–224

Lang F, Greger R, Knox FG, Oberleithner H (1981) Factors modulating the renal handling of phosphate. Renal Physiol 4:1–16

Laragh JM (1984) Modification of stepped care approach to antihypertensive therapy. Am J Med 777:78–86

Larson EB, Roach RC, Schoene RB, Hornbein TF (1982) Acute mountain sickness and acetazolamide: clinical efficacy and effect on ventilation. J Am Med Assoc 248:328–332

Lohrmann E, Nitschke RB, Nitschke R, Burhoff I, Masereel B, Pirotte B, Schlatter E, Delarge J, Lang HJ, Englert HC, Salomonsson M, Persson AEG, Eidelman O, Cabantchik ZI, Greger R (1991) The mode of action of diuretics. In: Hatano M (ed) Nephrology. Springer, Berlin Heidelberg New York, pp 1072–1081

Lohrmann E, Masereel B, Nitschke R, Pirotte B, Delarge J, Greger R (1992) Action of diuretics at the cellular level. Prog Pharmacol Clin Pharmacol 9:24–32

Lohrmann E, Burhoff I, Greger R (1993) Tubular effects of the loop diuretic torasemide. Cardiol 84 [Suppl 2]:135–142

Maher JJ, Schreiner GF (1965) Studies on ethacrynic acid in patients with refractory edema. Ann Intern Med 62:15

Marcus DC, Marcus NY, Greger R (1987) Sidedness of action of loop diuretics and ouabain on non sensory cells of utricle: a micro-Ussing chamber for inner ear tissues. Hear Res 30:55–64

McFarland KF, Carr AA (1977) Changes in fasting blood glucose sugar after hydrochlorothiazide and potassium supplementation. J Clin Pharmacol 17:13–17

Midecke M, Weisweiler P, Schwandt P, Holzgreve H (1987) Serum lipoproteins during antihypertensive therapy with beta-blockers and diuretics: a controlled long-term comparative trial. Clin Cardiol 10:94–98

MRFIT (1985) Multiple risk factor intervention trial research group. Baseline rest electrocardiographic abnormalities, antihypertensive treatment, and mortality in the multiple risk factor intervention trial. Am J Cardiol 55:1–55

Muschaweck R, Hajdú P (1964) Die salidiuretische Wirksamkeit der Chlor-N-(2-furylmethyl)-5-sulfamyl-anthranilsäure. Drug Res 14:44–47

Nishijma J (1984) Acute and chronic hemodynamic effects of the basic therapeutic regimen for congestive heart failure. Diuretics, low salt diet and bed rest. Jpn Heart J 25:571–585

Nitschke R, Schlatter E, Eidelman O, Lang HJ, Englert HC, Cabantchik ZI, Greger R (1989) Piretanide-dextran and piretanide-polyethylene glycol interact with high affinity with the Na^+ $2Cl^-$ K^+ cotransporter in the thick ascending limb of the loop of Henle. Pflugers Arch 413:559–561

Oberleithner H, Guggino W, Giebisch G (1983) The effect of furosemide on luminal sodium, chloride and potassium transport in the early distal tubule of Amphiuma kidney. Pflugers Arch 396:27–33

Palfrey HC, Alper SL, Greengard P (1980) cAMP-stimulated cation cotransport in avian erythrocytes: inhibition by "loop" diuretics. Am J Physiol 238:103–115

Paulais M, Teulon J (1990) cAMP-activated chloride channel in the basolateral membrane of the thick ascending limb of the mouse kidney. J Membr Biol 113:253–260

Quamme GA (1989) Control of magnesium transport in the thick ascending limb. Am J Physiol 256:F197–F210

Rajerison RM, Faure M, Morel F (1988a) Relationship between cell volume and cation content in thick ascending limb of rat kidney. Pflügers Archiv 412:497–502

Rajerison RM, Faure M, Morel F (1988b) Involvement of Na and Cl in ouabain-induced cell swelling in thick ascending limb of rat kidney. Pflügers Archiv 412:491–496

Rocha AS, Kokko JP (1973) Sodium chloride and water transport in the medullary thick ascending limb of Henle. J Clin Invest 52:612–623
Schlatter E (1993) Effects of various diuretics on membrane voltage of macula densa cells. Whole-cell patch-clamp experiments. Pflugers Arch 423:74–77
Schlatter E, Greger R, Weidtke C (1983) Effect of "high ceiling" diuretics on active salt transport in the cortical thick ascending limb of Henle's loop of rabbit kidney. Correlation of chemical structure and inhibitory potency. Pflugers Arch 396:210–217
Schlatter E, Greger R (1985) cAMP increases the basolateral Cl^- conductance in the isolated perfused medullary thick ascending limb of Henle's loop of the mouse. Pflugers Arch 405:367–376
Schlatter E, Salomonsson M, Persson AEG, Greger R (1989) Macula densa cells sense luminal NaCl concentration via the furosemide sensitive Na-2Cl-K cotransporter. Pflugers Arch 414:286–290
Schlatter E, Bleich M, Hirsch J, Greger R (1993) pH-sensitive K^+ channels in the distal nephron. Nephrol Dial Transplant 8:488–490
Schnermann J, Briggs JP (1991) Function of the juxtaglomerular apparatus: control of glomerular hemodynamics and renin secretion. In: Seldin DW, Giebisch G (eds) The kidney: physiology and pathophysiology. Raven, New York, pp 1249-- 1290
Schuurmans Stekhoen F, Bonting SL (1981) Transport adenosine triphosphatases: properties and functions. Physiol Rev 61:1–76
Skarfors ET, Lithell HO, Selinus I, Aberg H (1989) Do antihypertensive drugs precipitate diabetes in predisposed men? Br Med J 298:1147–1152
Spahn H, Knauf H, Mutschler E (1990) Pharmacokinetics of torasemide and its metabolites in healthy controls and in chronic renal failure. Eur J Clin Pharmacol 39:345–348
Stanton B, Giebisch G (1992) Renal potassium transport. In: Windhager EE (ed) Handbook of physiology: renal physiology. American Physiological Society, Rockville, pp 813–874
Suki WN, Eknoyan G (1992) Physiology of diuretic action. In: Giebisch G, Seldin DW (eds) The kidney. Raven, New York, pp 3629–3670
Suki WN, Yuin JJ, VonMinden M (1970) Acute treatment of hypercalcemia with furosemide. N Engl J Med 283:836–840
Thomsen K, Schou M, Steiness I, Hansen HE (1969) Lithium as an indicator of proximal sodium reabsorption. Pflugers Arch 308:180–184
Tilmann M, Kunzelmann K, Fröbe U, Cabantchik ZI, Lang HJ, Englert HC, Greger R (1991) Different types of blockers of the intermediate conductance outwardly rectifying chloride channel (ICOR) of epithelia. Pflugers Arch 418: 556–563
Ullrich KJ, Greger R (1985) Approaches to study tubule transport functions. In: Seldin DW, Giebisch G (eds) The kidney, physiology and pathophysiology. Raven, New York, pp 427–469
Ullrich KJ, Rumrich G, Klöss S (1989) Contraluminal organic anion and cation transport in the proximal renal tubule: V. Interaction with sulfamoyl- and phenoxy diuretics, and with β-lactam antibiotics. Kidney Int 36:78–88
Voith B, Spahn-Langguth H, Paliege, R, Knauf H, Mutschler E (1993a) Ethacrynic acid: determinations of pharmacokinetic and pharmacodynamic parameters. Fundam Clin Pharmacol 7:386
Voith B, Spahn-Langguth H, Paliege, R, Knauf H, Mutschler E (1993b) Ethacrynic acid: evidence for active metabolites. Naunyn Schmiedebergs Arch Pharmacol Suppl 349:594
Wang W, Giebisch G (1991) Dual modulation of renal ATP-sensitive K^+ channel by protein kinases A and C. Proc Natl Acad Sci USA 88:9722–9725
Wang W, White S, Geibel J, Giebisch G (1990) A potassium channel in the apical membrane of rabbit thick ascending limb of Henle's loop. Am J Physiol 258: F244–F253

Wangemann P, Wittner M, Di Stefano A, Englert HC, Lang HJ, Schlatter E, Greger R (1986) Cl^--channel blockers in the thick ascending limb of the loop of Henle. Structure activity relationship. Pflugers Arch 407 [Suppl 2]:S128–S141
Wangemann P, Braitsch R, Greger R (1987) The diuretic effect of muzolimine. Pflugers Arch 410:674–676
Weinberger MH (1985) Antihypertensive therapy and lipids: evidence, mechanisms, and implications. Arch Intern Med 145:1102–1105
Welsh MJ (1987) Electrolyte transport by airway epithelia. Physiol Rev 67:1143–1184
Wirz H (1956) Der osmotische Druck in den kortikalen Tubuli der Rattennieren. Helv Physiol Pharmacol Acta 14:353–362
Wittner M, Weidtke C, Schlatter E, Di Stefano A, Greger R (1984) Substrate utilization in the isolated perfused cortical thick ascending limb of rabbit nephron. Pflugers Arch 402:52–62
Wittner M, Di Stefano A, Schlatter E, Delarge J, Greger R (1986) Torasemide inhibits NaCl reabsorption in the thick ascending limb of the loop of Henle. Pflugers Arch 407:611–614
Wittner M, Di Stefano A, Wangemann P, Delarge J, Liegeois JF, Greger R (1987) Analogues of torasemide – structure function relationships. Experiments in the thick ascending limb of the loop of Henle of rabbit nephron. Pflugers Arch 408:54–62
Wittner M, Di Stefano A, Wangemann P, Nitschke R, Greger R, Bailly C, Amiel C, Roinel N, De Rouffignac C (1988) Differential effects of ADH on sodium, chloride, potassium, calcium and magnesium transport in cortical and medullary thick ascending limbs of mouse nephron. Pflugers Arch 412:516–523
Woltersdorf OW, deSolms SJ, Cragoe EJ (1978) The evolution of the (aryloxy)acetic acid diuretics. In: Cragoe EJ (ed) Diuretic agents. American Chemical Society, pp 190–232
Wright FS, Schnermann J (1974) Interference with feedback control of glomerular filtration rate by furosemide, triflocin and cyanide. J Clin Invest 53:1695–1708

CHAPTER 8

Thiazide Diuretics

H. VELÁZQUEZ, H. KNAUF, and E. MUTSCLER

A. Introduction

Thiazide diuretics have been in existence for more than 3 decades. They are benzothiadiazines or analogs and derivatives of the carbonic anhydrase-inhibiting sulfonamides. Their synthesis was spawned by a search for diuretics capable of promoting renal excretion of the chloride anion (instead of bicarbonate) with sodium. Thus, the "thiazide" effect is synonymous with the "saliuretic" effect as opposed simply to a natriuretic effect with bicarbonate being the predominant accompanying anion. One reason for the attractiveness of the use of these diuretics in humans was that they did not contain heavy metals and thus were less toxic than the mercury-containing diuretics in use at that time. Thiazides are currently used in the treatment of a wide variety of conditions including hypertension, edema, nephrolithiasis and diabetes insipidus.

Most thiazide diuretics in use today were synthesized during the late 1950s and early 1960s. Several excellent and extensive reviews examine the history, structure and chemistry of these agents in great detail (PETERS and ROCH-RAMEL 1969; BEYER and BAER 1961; SPRAGUE 1968; MAREN 1976; ALLEN 1983; LANT 1985). Thus, these topics will be summarized in brief and the reader referred to previous reviews. In contrast, more new information is emerging about the mechanism by which these agents inhibit salt transport and about the properties of thiazide-diuretic-sensitive receptors. A number of different cell types and cell lines have been found to respond to these agents. Thus, thiazide diuretics are becoming important tools in our efforts to understand how cells regulate sodium and chloride transport. Thus, the primary purpose of this chapter is to review recent advances in our understanding of the mechanism of action of the thiazide-type diuretics.

B. Chemical Structures

Thiazide diuretics (Fig. 1) evolved from the group of sulfonamide drugs known to inhibit the enzyme carbonic anhydrase. They contain unsubstituted R-SO_2NH_2 groups and for this reason act as mild carbonic anhydrase inhibitors (MAREN 1976). Two of the first thiazide-type diuretics synthesized were chlorothiazide and hydrochlorothiazide. Both are derived from

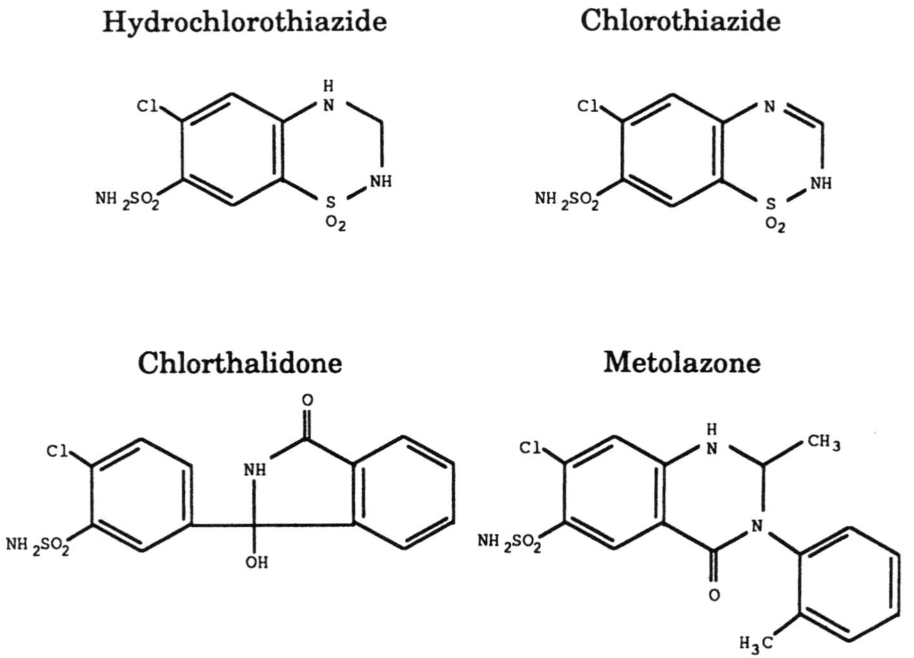

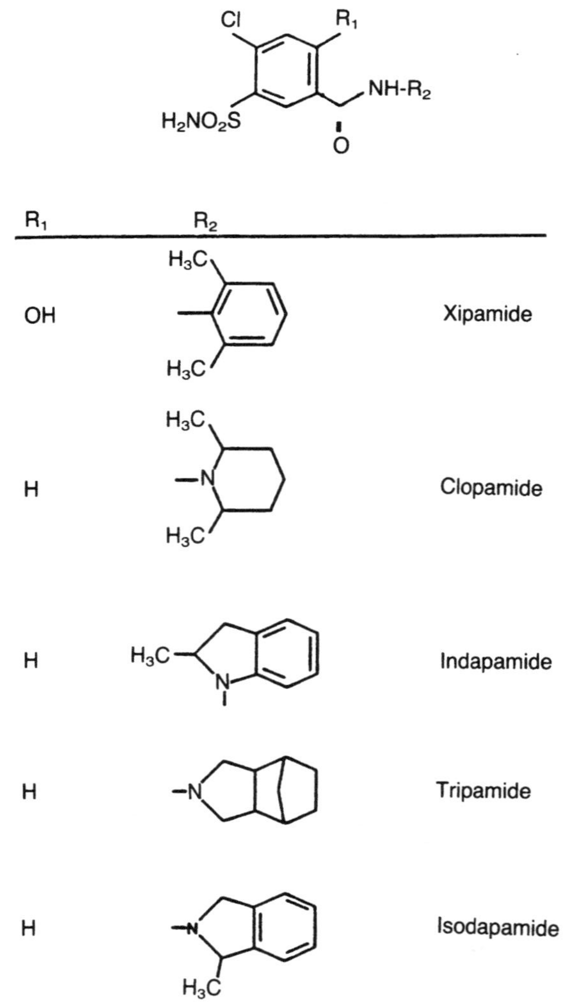

Fig. 1. Structures of several thiazide and thiazide-like diuretics: the benzothiadiazines hydrochlorothiazide (6-chloro-3,4-dihydro-2*H*-1,2,4-benzothiadiazine-7-sulfonamide-1,1-dioxide) and chlorothiazide (6-chloro-2*H*-1,2,4-benzothiadiazine-7-sulfonamide-1,1-dioxide); chlorthalidone (2-chloro-5-(1-hydroxy-3-oxo-1-isoindolinyl) benzene sulfonamide); and a quinazolinone derivative, metolazone (7-chloro-1,2,3,4-tetrahydro-2-methyl-3-(2-methylphenyl)-4-oxo-6-quinazoline sulfonamide). The benzamide derivatives xipamide, clopamide, indapamide, tripamide and isodapamide are also shown

chlorodisulfamoyl-aniline in a condensation reaction with formic acid or formaldehyde, respectively (PETERS and ROCH-RAMEL 1969; NOVELLO and SPRAGUE 1957; SPRAGUE 1958; DESTEVENS 1963; BEYER 1958). They were selected because of their ability to increase renal chloride and sodium excretion. Most of the benzothiadiazine diuretics that have been developed since then are derivatives of chlorothiazide and hydrochlorothiazide.

The basic structure of the thiazide-type diuretic is given in Fig. 1. All true thiazide diuretics contain a benzothiadiazine ring structure. The chlorine and sulfamoyl substitutions on the benzene ring are critical for an optimal thiazide effect. The long list of thiazide diuretics includes agents such as hydroflumethiazide, polythiazide, benzthiazide, cyclothiazide, trichlormethiazide, bendroflumethiazide, hydrochlorothiazide and chlorothiazide. An excellent detailed account of the history and development of these diuretics and a listing of the structures of 50 thiazides and related diuretics was published in the previous edition of this handbook (PETERS and ROCH-RAMEL 1969).

Other non-benzothiadiazine agents have been found to have a thiazide-like effect (increased renal chloride excretion with sodium). In contrast to the requirements for the substitutions on the benzene ring, the heterocyclic portion of the benzothiadiazine structure does not appear to be essential for this function. Thus, quinazolinone derivatives such as metolazone (SEELY and DIRKS 1977; SHETTY et al. 1970; BELAIR et al. 1969) (Fig. 1) and 4-chloro-3-sulfamido-benzoic acid derivatives such as chlorthalidone (STENGER et al. 1959) (Fig. 1), clopamide (FLÜCKIGER et al. 1963) and xipamide (HEMPELMANN et al. 1975, 1977) are also potent saliuretics. Numerous chemical modifications and structure-function correlations have been performed and are reviewed elsewhere (PETERS and ROCH-RAMEL 1969; MAREN 1976; BEYER and BAER 1961; SPRAGUE 1968; DESTEVENS 1963).

C. Pharmacokinetics

The literature on the pharmacokinetics of thiazide diuretics is extensive and cannot be reviewed here in detail. The reader is referred to Chap. 5 of this volume for a discussion of the metabolism of diuretics and to Chap. 6 for a discussion of renal excretion of diuretics. Previous reviews dealing with this area have been published (BEERMANN 1984; PETERS and ROCH-RAMEL 1969; WELLING 1986; LANT 1985).

Thiazide and thiazide-like diuretics are absorbed slowly and incompletely from the gastrointestinal tract. Total absorption of the drugs varies from only 10% to nearly 100% depending on the specific diuretic (Table 1). Bendroflumethiazide is absorbed nearly completely, whereas about 70% of hydrochlorothiazide, and less of chlorothiazide and chlorthalidone, is absorbed. Although the lipophilicity of the drug should favor its absorption, the absorption of most thiazides occurs in spite of large variations in their

Table 1. Pharmacokinetic data of thiazide diuretics (from Micromedex Computerized Clinical Information System 1994)

Type	Dose range (mg/day)	Usual dosage interval	Absorption (%)	Protein binding (%)	Volume of distribution (l/kg)	$t_{1/2}$ (h)	Duration of action (h)	Mode of elimination	Renal excretion (%/24 h)	Renal clearance (ml/min kg)
Bendroflume-thiazide	2.5–5	o.d.	~90		1.48	3	6–12	Renal		
Chlorothiazide	500–2000	o.d.-b.i.d.	~10[a]	95	0.2 ± 0.08	0.75–2 (1.5 ± 0.2[b])	6–12		92 ± 5[c]	4.5 ± 1.7
Chlorthalidone	12.5–100	o.d.-b.i.d.	65	75	3.9 ± 0.8	51–89 (44 ± 10[b])	24–72	Renal biliary	50–65	1.6 ± 0.3
Cyclothiazide	1–6	o.d.					18–24			
Hydrochloro-thiazide	12.5–100	o.d.	60–80	40 (58 ± 17[b])	3 (0.83 ± 0.31[b])	10–12 (2.5 ± 0.2[b])	6–12	Renal	50–70	4.9 ± 1.1
Indapamide	2.5–5	o.d.		71–79	24–60	15–25[d]	12–24	Renal biliary	5	
Metolazone	1.25–10	o.d.	56	95		8–14	12–24	Renal biliary	30–45	
Methyclothiazide	2.5–10	o.d.					>24			
Polythiazide	1–4	o.d.			4	25.7	24–48		25	
Trichlorme-thiazide	1–4	o.d.					12–24			239[e]
Xipamide	10–40	o.d.	~90	~98	21 ± 4	7 ± 1	12–24	Renal biliary	40	12 ± 2

[a] Not proportional to the dose administered.
[b] Contrasting data from those of Goodman Gilman (1990).
[c] Total renal excretion.
[d] Biphasic.
[e] (ml/min); 56 ml/min in patients with renal impairment.

degree of liposolubility. Administration of the diuretic with food can increase its absorption.

I. Protein Binding

Once absorbed, the drugs are largely protein bound (BEYER and BAER 1961). As far as their distribution in the body is concerned, chlorothiazide, e.g., is distributed in a volume nearly equal to body water and is concentrated in kidney, liver and small intestine of rats and mice when injected subcutaneously. The higher concentrations in liver and subsequent secretion back into the intestine via the bile contribute to the excretion of the diuretic via the gut. Bone and brain exhibit lower values than found in blood. Hydrochlorothiazide is accumulated in red blood cells (BEERMANN 1984; CHEN et al. 1992). Its high partitioning in the erythrocyte provides for an effective barrier for release during its passage through the kidney, liver and other organs (CHEN et al. 1992). Chlorthalidone in blood is bound nearly completely to carbonic anhydrase in red cells (BEERMANN 1984).

The magnitude of the volume of distribution and particularly the preferential binding to tissues and cellular components, respectively, serve to extend the half-life of the diuretic in the circulation. In this context, the lipophilicity and consequent tubular reabsorption also prolong the half-life.

II. Renal Excretion

The primary route of elimination of most thiazides is the kidney. They gain access to the luminal (or apical) side of the cell, their site of action, via two pathways: filtration at the glomerulus and secretion by the proximal tubule (BEYER and BAER 1975; GRANTHAM and CHONKO 1991; see also Chap. 6, this volume). Diuretic drugs are freely filtered at the glomerulus; however, since most are largely bound to either proteins or red blood cells, the concentration of the free form is very low and the quantity reaching the tubule lumen by this route is small. As discussed in Chap. 6, this volume, thiazides are taken up by the proximal tubule from the blood across the basolateral membrane and accumulated inside cells by an organic acid transport mechanism shared by *para*-aminohippurate and phenol red (competitively inhibited by probenecid). Subsequently, they diffuse across the apical membrane into the luminal fluid. Secretion into luminal fluid may be retarded by intracellular binding and compartmentalization of the drug. Secretion is inhibited by metabolic inhibitors such as cyanide or dinitrophenol, and by inhibition of the $(Na^{+}+K^{+})$ pump with ouabain.

Once in the tubular fluid, the diuretic is able to interact with its "receptor" on the apical membrane of the target cell. The high-affinity binding of the thiazides occurs primarily at the luminal membrane of the distal tubule cells. Fluid absorption by the upstream proximal tubule and loop of Henle, without significant reabsorption of the drug, raises the concentration of the free diuretic in tubular fluid to a value that is much higher than that of plasma or of glomerular filtrate and that is effective in inhibiting cellular ion transport mechanisms. As for hydrochlorothiazide, proximal tubular secretion exceeds glomerular filtration by about a factor of 4 (NIEMEYER et al. 1983).

The renal excretion of the active drug at concentrations almost 2 orders of magnitude higher in tubular fluid than in plasma characterizes these diuretics as "organotropic drugs."

Cumulative urinary excretion of the active diuretic, A_e, represents the fraction of administered dose which had access to the renal receptor. As will be shown below in patients with liver disease, the A_e determined in healthy controls does not represent the maximal fraction excreted via the kidney. The renal clearance of the drug, Cl_r, is given by the amount excreted into the urine, A_e, divided by the plasma concentration time curve, AUC. For diuretics strongly bound to protein, e.g., xipamide, the Cl_r related to the fraction not bound to protein gives a clearance of about 600 ml/min, equal to PAH clearance (KNAUF and MUTSCHLER 1984).

D. Pharmacodynamics

I. Thiazide-Sensitive Systems

At least two types of mechanisms of salt transport are known to be inhibited by thiazide diuretics: one in which both sodium and chloride are transported in the same direction electroneutrally by a single transporter (Na^+/Cl^- cotransport), and another in which chloride is exchanged for bicarbonate electroneutrally (Cl^-/HCO_3^- exchange). These mechanisms are found in different cell types in a wide range of tissues (kidney, bladder, gallbladder, intestine, smooth muscle) in a number of different species (rat, mouse, rabbit, fish, amphibian) and will be discussed below. The two types of mechanisms are involved in net absorption of salt and in the regulation of cell volume. Other effects of thiazides have been described for which the mechanism of action is not known.

1. Na^+/Cl^- Cotransport

Accumulating evidence from experiments in a variety of tissues and species suggests that thiazide diuretics block an Na^+/Cl^- cotransport mechanism. In the kidney, the major target of thiazides is the electroneutral, Na^+/Cl^- cotransporter in the distal nephron (ELLISON et al. 1987; VELÁZQUEZ et al. 1984; VELÁZQUEZ and WRIGHT 1986b; COSTANZO 1985; PLANELLES and ANAGNOSTOPOULOS 1992; SHIMIZU et al. 1988; TERADA and KNEPPER 1990). Na^+ and chloride transport by cells in this region is important to Na^+, chloride and volume homeostasis. Transport of K^+ by this mechanism does not appear to be involved and it is not inhibited by the high ceiling diuretics (e.g., bumetanide, an inhibitor of the $Na^+/2Cl^-/K^+$ cotransport mechanism in the thick ascending limb of the loop of Henle) (VELÁZQUEZ et al. 1992; VELÁZQUEZ and WRIGHT 1986b). A detailed discussion of thiazide-sensitive salt transport in the kidney is presented below.

In the bladder of the winter flounder Pseudopleuronectes americanus, a major component of Na^+ and chloride absorption (RENFRO 1978; DAWSON and ANDREW 1979) is inhibited by hydrochlorothiazide or by metolazone (DUFFEY and FRIZZELL 1984; STOKES 1984). The flounder bladder is a high-resistance epithelium and is well suited for electrophysiological and transepithelial tracer-flux assessment of ion transport characteristics. The effects of diuretics and ion substitutions were determined using sheets of the bladder mounted in modified Ussing chambers. Increasing mucosal chloride concentration stimulated mucosa-to-serosa Na^+ flux and increasing Na^+ concentration stimulated chloride efflux. These fluxes were independent of mucosal K^+ concentration. Na^+ and Cl^- fluxes were not sensitive to the loop diuretic bumetanide, suggesting that the transport mechanism was not the $Na^+/2Cl^-/K^+$ cotransporter. Na^+ and chloride fluxes were not affected by inhibitors of Na^+/H^+ exchange (amiloride) or Cl^-/HCO_3^- exchange (disulfonic stilbene, DIDS) or by the carbonic anhydrase inhibitor acetazolamide (STOKES 1984). These results argue against the presence of Na^+/H^+ exchange in parallel with Cl^-/HCO_3^- exchange as mediating net Na^+ and chloride absorption. Hydrochlorothiazide (10^{-4} mmol/l), however, decreased the net flux of Na^+ by 85% and of chloride by 65%. Together the results support the presence of an Na^+-Cl^- cotransport mechanism mediating Na^+ and chloride absorption in flounder bladder cells.

Rabbit gallbladder cells have at least three distinct apical mechanisms involved in Na^+ and chloride transport. One of these is thiazide-sensitive Na^+/Cl^- cotransport (CREMASCHI et al. 1987a,b, 1992; MEYER et al. 1990; CREMASCHI and PORTA 1992), which accounts for about 50% of net Na^+ and chloride absorption in this tissue. Most of the residual salt transport is accounted for by parallel Na^+/H^+ exchange and Cl^-/HCO_3^- exchange (double exchange). The Na^+/Cl^- cotransporter is not inhibited by removal of K^+ or by the addition of bumetanide, or disulfonic stilbenes (SITS). When hydrochlorothiazide is added in the absence of bicarbonate and CO_2 (to reduce the contribution to NaCl absorption by bicarbonate-dependent double exchange), virtually all Na^+ and chloride absorption by the rabbit gallbladder is blocked. After 15–30 min, in the continued presence of hydrochlorothiazide and absence of bicarbonate and CO_2, however, Na^+ and chloride absorption is reactivated (CREMASCHI et al. 1992). This new component of NaCl entry appears to be induced subsequent to cell shrinkage by the experimental maneuver as the cell attempts to restore the original cell volume. It is not thiazide sensitive but bumetanide sensitive and most likely represents $Na^+/2Cl^-/K^+$ cotransport.

Cardiac myocytes from rabbit ventricle regulate their volume in response to osmotic stress. Part of the response is postulated to occur secondary to Na^+ and chloride flux into the cell via an Na^+/Cl^- cotransport mechanism that is sensitive to chlorothiazide (DREWNOWSKA and BAUMGARTEN 1991). In addition, these cells exhibit a bumetanide-sensitive $Na^+/2Cl^-/K^+$ cotransport mechanism in parallel to the Na^+/Cl^- cotransport system. Single cells were isolated from the septum and volume measurements performed using optical techniques. The addition of the combination of chlorothiazide and bumetanide to the bathing solution altered the normal response to hypertonic and hypotonic media (DREWNOWSKA and BAUMGARTEN 1991). Furthermore, the steady state resting volume in an isotonic medium fell by approximately 15% compared to the control volume when chlorothiazide alone, bumetanide alone or both diuretics together were added to the bathing solution. Thus, thiazide-sensitive salt influx into heart ventricle cells may play a role in maintaining the normal cell volume, in addition to being involved in regulatory volume responses. The importance of this effect of thiazide diuretics on cardiac muscle in humans during diuretic therapy remains to be determined.

Not all Na^+/Cl^- cotransport systems are thiazide sensitive, however. In Ehrlich ascites cells, in rabbit medullary thick ascending limb cells and in bovine tracheal epithelial cells, an Na^+/Cl^- cotransport system has been identified that is inhibited by loop diuretics but does not require K^+ (ALVO et al. 1985; HOFFMANN et al. 1983; MUSCH and FIELD 1989). In the Ehrlich tumor cells, Na^+ and chloride enter the cell during the regulatory volume increase via an Na^+/Cl^- cotransporter and can be blocked by furosemide or bumetanide (HOFFMANN et al. 1983). In ouabain-treated renal thick ascending limb cells assayed in suspension, Na^+ and chloride influx were found to be interdependent (ALVO et al. 1985), suggesting the presence of an Na^+-Cl^- cotransport mechanism. Changing the medium osmolarity, however, was found to modify the nature of the response: in hypertonic medium, Na^+ and chloride uptake was now dependent on the K^+ concentration, an effect that was not present in isotonic media (EVELOFF and CALAMIA 1986). In the bovine tracheal cells, affinities for Na^+ (4 mmol/l) and chloride (54 mmol/l) of the transporter resemble that for the renal $Na^+/2Cl^-/K^+$ cotransporter (GREGER and SCHLATTER 1983a,b), although no dependence on K^+ could be demonstrated and the Hill coefficient for chloride was 1.0 (MUSCH and FIELD 1989). Based on these results, it is possible that thiazide-sensitive and loop diuretic-sensitive Na^+/Cl^- cotransporters are distinct transport proteins. Furthermore, it is not possible to determine whether Na^+/Cl^- cotransport and $Na^+/2Cl^-/K^+$ cotransport represent the same transporter with different affinities for K^+ depending on the intracellular ionic composition or whether they are distinct transport proteins (EVELOFF and WARNOCK 1987; HOFFMANN 1986).

Similarly, in the apical membrane of *Necturus* gallbladder (ERICSON and SPRING 1982), a bumetanide-sensitive component of Na^+/Cl^- cotransport has been identified. The administration of ouabain to Necturus gallbladder cells, mounted in a miniature Ussing chamber, leads to cell swelling caused by continued entry of salt across the apical membrane while active Na^+ extrusion is blocked. The salt and water influx across the apical membrane depends equally on the apical concentration of Na^+ and of chloride, the K_m for each ion being in the range between 20 and 25 mmol/l. These values are different from those reported for the $Na^+/2Cl^-/K^+$ cotransporter or the bovine tracheal Na^+/Cl^- cotransporter (see above). This ouabain-induced cell swelling can be blocked by bumetanide but not by bicarbonate removal or the transport inhibitors SITS and amiloride (ERICSON and SPRING 1982). The presence or absence of this mechanism in the apical membrane of these cells is determined by the level of cyclic AMP (GARVIN and SPRING 1992). Entry of Na^+ and chloride via Na^+/Cl^- cotransport predominates when cAMP levels are high. Without cAMP present, Na^+ and Cl^- entry via double exchange predominates.

2. Cl^-/HCO_3^- Exchange

In the amphibian renal distal tubule, Na^+ and Cl^- absorption is inhibited by thiazide diuretics (HANSEN et al. 1981; STANTON 1988; PLANELLES and ANAGNOSTOPOULOS 1992) and as mentioned above may be mediated by an Na^+/Cl^- cotransport mechanism. However, thiazides may also interfere with Cl^-/HCO_3^- exchange. Available evidence suggests the presence of two cell types (I and II) in the *Amphiuma* late distal tubule (STANTON et al. 1987). In the type II cells, the primary mechanism involved in NaCl absorption appears to be parallel apical Na^+/H^+ exchange and Cl^-/HCO_3^- exchange. Inhibitors of Na^+/H^+ exchange (amiloride) and of Cl^-/HCO_3^- exchange (DIDS) as well as hydrochlorothiazide and trichormethiazide appear to block chloride uptake (STANTON 1988) and decrease the cell chloride concentration. The cell membrane voltage hyperpolarizes in each case secondary to a decrease in cell chloride concentration (the basolateral membrane has a chloride conductance). In this system, thiazide inhibition of Cl^-/HCO_3^- exchange has been proposed as the mechanism by which Na^+ and Cl^- absorption by the distal tubule is decreased (STANTON 1988). These renal effects of thiazide diuretics will be discussed in detail below.

In the rabbit distal colon, a component of Cl^- transport is inhibited by thiazide diuretics (FERRIOLA et al. 1986). Sheets of tissue were mounted in a chamber to assess short circuit current and unidirectional chloride fluxes. Trichlormethiazide decreased net chloride absorption by 53%, bendroflumethiazide and hydrochlorothiazide had a smaller effect and *diazoxide*, a nondiuretic thiazide and K^+ channel opener, did not have an effect. Absorptive chloride flux was decreased without an effect on secretory unidirectional flux, and transport inhibition was not accompanied by an effect on the short circuit current. Chloride absorption proceeds even when Na^+ is removed from the bathing solutions. The results are consistent with thiazide-sensitive chloride absorption in the rabbit distal colon occurring via a Cl^-/HCO_3^- exchange mechanism. The effect is not likely to be secondary to inhibition of carbonic anhydrase since the potency of inhibition of transport by the diuretics tested is inversely correlated to the ability to inhibit carbonic anhydrase in vitro.

Thiazides have been tested in a number of systems to determine whether or not they inhibit Cl^-/HCO_3^- exchange. In most systems, Cl^-/HCO_3^- is not affected. In Madin-Darby canine kidney (MDCK) cell cysts grown in a collagen gel (TANNER et al. 1992), chlorothiazide does not affect basolateral Cl^-/HCO_3^- exchange, a process believed to be involved in fluid secretion.

3. Other Mechanisms

In some systems, thiazides inhibit electrogenic Na^+ transport (PENDLETON et al. 1968; SCHNIEDERS and LUDENS 1980; SULLIVAN et al. 1971; MARUMO et al. 1982; ERIKSSON and WISTRAND 1987). In the isolated toad (*Bufo marinus*) bladder preparation, bendroflumethiazide (1.6×10^{-4} mmol/l) when added to the serosal side of the bladder decreased the short circuit current, active Na^+ flux, and transepithelial voltage and conductance (PENDLETON et al. 1968). Hydrochlorothiazide (IC_{50} 6.4×10^{-4} mol/l) and trichlormethiazide (IC_{50} 3.1×10^{-4} mol/l), when added to both, serosal and mucosal sides, also decreased short circuit current (SCHNIEDERS and LUDENS 1980). This effect was not, however, mediated by direct inhibition of an Na^+ transport mechanism, but rather by an effect on cellular metabolism (SULLIVAN et al. 1971). Relatively high concentrations are generally needed and the effect is produced by exposure of the drug to the basolateral membrane through which the drug can enter the cell (PENDLETON et al. 1968). Bendroflumethiazide increases lactate production and inhibits oxygen consumption independent of the presence of Na^+ (McDOUGAL and SULLIVAN 1970).

In frog bladder, mucosal hydrochlorothiazide (5×10^{-4} mol/l) quickly and reversibly inhibited the short circuit current whereas serosal application of the drug had no effect (MARUMO et al. 1982). No effect could be observed on ADH-stimulated water transport by the bladder. Hydrochlorothiazide, diazoxide and other thiazides have been reported to inhibit phosphodiesterase activity in heart or aortic tissue (MOORE 1968; VULLIEMOZ et al. 1980). In the frog bladder, however, no evidence was obtained to support inhibition of phosphodiesterase by hydrochlorothiazide (MARUMO et al. 1982). In contrast, the activity of the (Na^+/K^+)-ATPase was inhibited by 64% by hydrochlorothiazide. The inhibition of short circuit current, thus, may occur as a direct consequence of inhibition of the Na^+ pump or by limiting apical Na^+ entry via an undefined mechanism, thus reducing Na^+ availability to the pump. The precise mechanism involved has not been resolved.

Chronic chlorothiazide administration may improve lung mechanics independent of any effect related to the diuresis (KAO et al. 1984; RUSH et al. 1990). In the isolated bronchial smooth muscle preparation, chlorothiazide in relatively high concentrations for plasma (10^{-4}–10^{-3} mol/l) potentiated the cholinergic contractile response to electrical stimulation (SCOTT et al. 1992). In this tissue, incubation with an endogenous acetylcholinesterase inhibitor (neostigmine, 10^{-5} mol/l) prevented further augmentation of contraction by chlorothiazide. The authors conclude that, taken together, their results are most consistent with an action of thiazides to inhibit acetylcholinesterase activity.

Metolazone (10^{-5} to 10^{-3} mol/l) has been reported to inhibit Na^+-coupled phosphate cotransport in rat renal brush border membranes (EGEL et al. 1985; KEMPSON et al. 1983). Neither acetazolamide nor chlorothiazide inhibited brush border membrane phosphate transport, suggesting that the effect of metolazone on phosphate transport was not related to inhibition of carbonic anhydrase (see below).

II. Thiazide Binding to Transporter Proteins

Important steps have been taken in the isolation and purification of the thiazide-sensitive Na^+/Cl^- cotransporter. The finding that metolazone and other thiazides displace peripheral-type benzodiazepine receptor ligands (RO5-4864, PK11195) in rat kidney cortex (BASILE et al. 1988; BEAUMONT et al. 1984; LUKEMAN and FANESTIL 1987) has led to the synthesis of a tritiated derivative of metolazone that has been utilized by a number of investigators to characterize the thiazide-diuretic "receptor" (BEAUMONT et al. 1988, 1989a,b; LUO et al. 1990; TRAN et al. 1990; ELLISON et al. 1991, 1993; SHETTY et al. 1970). The renal peripheral benzodiazepine "receptor" density is increased after both chronic furosemide and chlorothiazide infusion into rats (BASILE et al. 1988).

Autoradiography has been used in rat kidneys to localize this "receptor" specifically to short and sparsely distributed convoluted tubule profiles in the cortex (BEAUMONT et al. 1989a). These data suggest specific binding of thiazide diuretics primarily to the distal convoluted tubule. Higher resolution autoradiography suggested preferential apical localization of the labeling (BEAUMONT et al. 1989a).

In a preparation of rat kidney membranes, two types of binding sites for thiazides have been described, a low-affinity (K_D 289 nmol/l) and a high-affinity (K_D 4 nmol/l) binding site (BEAUMONT et al. 1988). The low-affinity site is blocked by 1 μmol/l acetazolamide and is found not only in kidney but in a number of other tissues in the rat (testis, lung, spleen, liver, brain, heart) (BEAUMONT et al. 1988). Thus, a portion of the thiazide binding may be to membrane-bound carbonic anhydrase. The high-affinity binding site was blocked effectively by other thiazide diuretics (methylclothiazide, polythiazide, bendroflumethiazide, benzthiazide), by metolazone (a quinazolinone diuretic with thiazide-like actions in the kidney) and by some derivatives of 3-sulfamoyl-4-chlorobenzoic acid (indapamide, chlorthalidone)

(Beaumont et al. 1988). These sites were found predominantly in kidney cortex. Some high-affinity binding was detectable at levels just above background in kidney outer medulla or inner medulla, testis, adrenal gland, lung, spleen, liver and pancreas.

Binding of ^{3}H-metolazone to the high-affinity site in kidney cortical membranes was measured in the presence of different anions and cations (Tran et al. 1990). The presence of Na^+ increased the affinity of the receptor for metolazone (concentration for half-maximal effect, $K_{1/2}$, is approximately 10 mmol/l); the presence of chloride decreased the affinity of the receptor for metolazone; the presence of Na^+ increased the affinity of the receptor for chloride. A tentative model has been proposed (Tran et al. 1990) depicting two binding sites on the thiazide-diuretic "receptor" that are accessible from the apical extracellular solution, one that accepts Na^+ and one that accepts either chloride or metolazone, but not both (see also Chap. 2, this volume). When Na^+ is bound, the affinity for chloride and metolazone is increased. If chloride binds, the "receptor" (or Na^+/Cl^- cotransport mechanism) translocates across the apical membrane and discharges both Na^+ and chloride into the cell. When a thiazide diuretic is present in luminal fluid, it competes effectively with chloride for binding to one of the sites, Na^+ and chloride translocation does not occur and transepithelial Na^+ and chloride absorption is inhibited.

The thiazide diuretic "receptor" has been solubilized and partially purified (Luo et al. 1990; Ellison et al. 1991). Optimal solubilization of the "receptor" from rat renal cortical homogenates was obtained using 0.5% Triton X-100 detergent (Luo et al. 1990). The properties of the solubilized "receptor" from rat kidney cortex are similar to those of the membrane-bound receptor: K_D for metolazone in this study was 11 nmol/l vs. 12 nmol/l (soluble vs. membrane bound); the rank order potency of metolazone and a number of thiazide diuretics to block ^{3}H-metolazone binding was similar (although not identical) to the rank order potency obtained for intact membranes (Beaumont et al. 1988; Luo et al. 1990) (see above); the ability of chloride to inhibit and of Na^+ to stimulate ^{3}H-metolazone binding was retained (Luo et al. 1990).

A high-affinity thiazide-diuretic receptor has also been identified in rabbit kidney (Ellison et al. 1991). It is restricted to renal distal tubules and present in greater amounts in apical membrane-enriched preparations. The affinity (K_D-2 nmol/l) is similar to that of the rat kidney high-affinity receptor. Optimal solubilization of the receptor from rabbit cortical homogenates was obtained using 6 mmol/l CHAPS as detergent. ^{3}H-Metolazone binding to the solubilized "receptor" is inhibited by chloride and enhanced by Na^+. A greater than 200-fold enrichment of the receptor was achieved by sequential Mg^{2+} precipitation and differential centrifugation, hydroxyapatite chromatography and size exclusion HPLC (Ellison et al. 1991). It is encouraging that the thiazide-diuretic "receptor" in both rats and rabbits retains so many of its characteristics when solubilized; this feature will no doubt aid in the purification, reconstitution and cloning endeavors.

Monoclonal antibodies generated to partially purified high-affinity thiazide diuretic "receptors" isolated from rabbit kidney have been used to try to identify immunohistochemically the localization of this "receptor" in the kidney (ELLISON et al. 1993). The hybridoma selected generated an antibody (JM5) that immunoprecipitated 80% of solubilized and ^{3}H-metolazone-labeled "receptors". Western blot analysis identified a 125-kDa protein in a solubilized renal cortical apical membrane preparation. The antibody stained the apical membrane of distal convoluted tubules and connecting tubules. Proximal tubules, glomeruli and all interstitial structures were completely negative. Faint staining was also detected in a subpopulation of thick ascending limb cells and in the apical membrane of principal cells of the collecting duct (ELLISON et al. 1993).

Taken together, the results available to date are consistent with the hypothesis that the high-affinity thiazide-diuretic "receptor" in the kidney cortex represents an Na^+-Cl^- cotransport protein of the apical membrane of distal tubule cells.

III. Cloning the Thiazide Diuretic Receptor

Work has begun in cloning the gene for the thiazide-sensitive Na^+/Cl^- cotransporter. Using the *Xenopus* oocyte expression system, a cDNA library from the winter flounder (*Pseudopleuronectes americanus*) bladder was screened and a clone isolated that expressed Na^+ tracer flux which could be inhibited by chloride removal and metolazone addition (GAMBA et al. 1993). Oocyte Na^+ uptake was saturable and depended on both external chloride and Na^+ concentrations with K_m values for Na^+ and Cl^- of 25 and 14 mmol/l, respectively. Hill coefficients of near unity for both Na^+ and chloride are consistent with a 1:1 stoichiometry. Polythiazide and metholazone were the most potent inhibitors of Na^+ uptake (K_i 3 μmol/l); cyclothiazide, hydrochlorothiazide and chlorothiazide were less effective; whereas furosemide, acetazolamide and EIPA were ineffective. The clone predicts a protein of 1023 amino acids and 12 putative membrane-spanning domains. The size of the in vitro translation product is similar to that of the native solubilized rabbit kidney thiazide-diuretic receptor (ELLISON et al. 1993). Although the affinity of thiazide diuretics for the cloned flounder bladder transporter is 3 orders of magnitude lower than that of renal membrane-bound or soluble protein, most of the other functional characteristics are consistent with this clone being a good candidate for the Na^+-Cl^- cotransporter.

IV. Renal Actions

Thiazide diuretics are used frequently in clinical practice to promote renal Na^+ and fluid loss and do so primarily by decreasing Na^+ and Cl^- absorption by the distal tubule and interfering with the ability of the nephron to dilute urine maximally (BERGER and WARNOCK 1986; EARLEY et al. 1961; BERLINER et al. 1966; CAFRUNY and ROSS 1962; SULLIVAN and PIRCH 1966). Thiazide diuretics comprise a major category of diuretics used when mild fluid loss is desired over a prolonged period of time (BERGER and WARNOCK 1986). Thiazides inhibit salt transport mechanisms of tubule cells, thus promoting increased fluid and Na^+ excretion accompanied primarily by chloride as the anion (KUNAU et al. 1975; BEYER and BAER 1975; BEYER 1958).

Thiazide diuretics have been used to increase renal salt loss in the treatment of hypertension for many years. Although they have been associated with a number of side effects, this mostly has been due to administration of doses that are too high. Thiazides are known to stimulate Ca^{2+} absorption by the nephron and are, therefore, important therapeutic agents in the treatment of hypercalciuria and nephrolithiasis (BERGER and WARNOCK 1986; COSTANZO and WINDHAGER 1978; COE 1980). Enhanced renal K^+ loss is often associated with thiazide therapy. This is not generally a desired effect, but it can be controlled effectively in patients being treated with these agents.

1. Proximal Effects

Although the primary site of action of thiazides is in the distal tubule, thiazides have been shown to affect fluid transport in proximal tubules (GRANTHAM 1973). Once taken up into the cell from the blood, the diuretic can inhibit cytoplasmic carbonic anhydrase and cellular phosphodiesterases. The doses generally administered are too low to elicit a large effect on phosphodiesterases in vivo (MOORE 1968). From the cytoplasm, the diuretic diffuses into the tubular fluid and interacts with luminal carbonic anhydrase or with specific luminal membrane transport mechanisms.

Micropuncture studies following acute intravenous administration of thiazides have demonstrated decreased fractional reabsorption of salt and water in the proximal tubule (OKUSA et al. 1989; KUNAU et al. 1975; FERNANDEZ and PUSCHETT 1973; ULLRICH et al. 1966). Net fluid absorption by isolated perfused proximal tubules is inhibited (GRANTHAM 1973). It is likely that this action of thiazides is through the inhibition of carbonic anhydrase. Thiazides vary in their ability to inhibit carbonic anhydrase (MAREN 1976; PETERS and ROCH-RAMEL 1969; BEYER and BAER 1975). Proximal tubule acid secretion is affected because these drugs are secreted into the lumen by proximal tubule cells, and have access to both cellular and luminal carbonic anhydrase. Most of the acid secreted by the proximal tubule occurs via an apical Na^+/H^+ exchanger.

Another effect of thiazides in the proximal tubule may be related to inhibition of anion exchange mechanisms. Proximal tubules absorb a large fraction of chloride in parallel with Na^+/H^+ exchange via mechanisms for

Cl^-/anion exchange (KARNISKI and ARONSON 1985; SCHILD et al. 1987; ALPERN 1987; WANG et al. 1992; WAREING and GREEN 1992). Because thiazides are thought to inhibit Cl^-/HCO_3^- exchange in a number of other tissues such as rabbit distal colon (FERRIOLA et al. 1986), red blood cells (COUSIN and MOTAI 1976) and *Amphiuma* distal tubules (STANTON 1988), an effect on the transporter in the proximal tubule could inhibit chloride absorption directly. Results from experiments in rat renal cortical brush border membrane vesicles (KARNISKI and ARONSON 1987) suggest that thiazides inhibit a mechanism that exchanges chloride for formate ($Cl^-/CHOO^-$ exchange). Primary inhibition of this mechanism would lead to an increased delivery of chloride, Na^+ and water to the downstream segments. By the time fluid reaches the distal site, however, absorption of sodium and chloride by the intervening thick ascending limb of Henle's loop compensates for the inhibitory effect of thiazides in the proximal tubule. Thus a large increase in diuresis cannot be attributed to a proximal effect.

2. Distal Effects

A major renal site of action of the thiazide diuretics is the distal convoluted tubule. The region of the mammalian nephron beginning after the cortical thick ascending limb of Henle's loop and extending to the confluence of this tubule with another is termed the distal tubule (see Chap. 1, this volume). This part of the nephron is composed of at least four cell types. The initial portion of the distal tubule (distal convoluted tubule, DCT) primarily contains distal convoluted tubule cells (CRAYEN and THOENES 1978; KAISSLING 1982). The next segment, the connecting tubule (CNT), primarily contains connecting tubule cells and some intercalated cells (CRAYEN and THOENES 1978; KAISSLING 1982) and is discussed separately below. The final portion is termed the initial collecting tubule (ICT) and resembles the cortical collecting duct in that it contains intercalated cells as well as principal cells (GRAYEN and THOENES 1978; KAISSLING 1982; Chap. 1, this volume).

It has been established for some time that chlorothiazide decreases distal tubule chloride absorption (KUNAU et al. 1975) and Na^+ absorption (COSTANZO and WINDHAGER 1978). It has also been shown that chlorothiazide decreases net transport of Na^+ and chloride by nearly equal amounts (ELLISON et al. 1987; VELÁZQUEZ and WRIGHT 1986b) and that the magnitude of the transepithelial voltage along the entire distal tubule is not affected when compared to control conditions. These data suggested that the mechanisms involved in Na^+ and Cl^- absorption might be electroneutral.

There is a transport mechanism for Na^+ and chloride in the rat distal tubule that is mutually dependent on both ions (VELÁZQUEZ et al. 1984). Na^+ absorption is decreased when lumen chloride concentration is reduced and chloride absorption is decreased when lumen Na^+ concentration is reduced with a $K_{1/2}$ of approximately 10 mmol/l for both Na^+ and chloride (VELÁZQUEZ et al. 1984). The mechanisms involved in Na^+ and chloride transport appear to be electroneutral since decreasing lumen chloride concentration reduced Na^+ absorption by more than 50% without affecting the transepithelial voltage (VELÁZQUEZ et al. 1982, 1984; ELLISON et al. 1985).

K^+ does not appear to be involved in the mechanism of Na^+ and chloride absorption by the distal convoluted tubule. In the studies in rats cited above (ELLISON et al. 1985, 1987; VELÁZQUEZ and WRIGHT 1986b; VELÁZQUEZ et al. 1984), large decreases in Na^+ and chloride transport were not associated with changes in K^+ transport. In the thick ascending limb K^+ binding to a transporter that binds $Na^+2Cl^-K^+$ is required before all three ions are translocated from lumen to cell (GREGER and SCHLATTER 1981). Loop diuretics such as furosemide and bumetanide are thought to inhibit specifically this $Na^+/2Cl^-/K^+$ cotransport mechanism (PALFREY et al. 1980; OSTERGAARD et al. 1972; Chap. 7, this volume) and decrease the rate of Na^+, chloride and K^+ absorption by the loop of Henle (VELÁZQUEZ and WRIGHT 1986b). In an effort to establish differences or similarities between the transporter in the distal tubule and the one present in the upstream segment, the thick ascending limb of Henle's loop, distal tubules were perfused with bumetanide, furosemide and chlorothiazide (VELÁZQUEZ and WRIGHT 1986b). While chlorothiazide decreased Na^+ and chloride transport dramatically in the distal tubule, bumetanide did not affect net Na^+ or chloride absorption (VELÁZQUEZ and WRIGHT 1986b). Interestingly, furosemide had a partial effect on Na^+ and chloride absorption; however, furosemide was not additive to a chlorothiazide effect, and chlorothiazide further decreased Na^+ and chloride absorption in the presence of furosemide. Thus, it is likely that the mechanism for Na^+ and Cl^- transport in the distal tubule differs from the one present in thick ascending limb cells: the apical Na^+-Cl^- cotransport mechanism in the rat distal tubule appears to be electroneutral with similar affinities for Na^+ and chloride, and does not involve K^+.

Evidence for the action of thiazides in the distal convoluted tubule segment of the distal tubule comes primarily from experiments performed in rats (ELLISON et al. 1987; COSTANZO 1984, 1985). When the DCT of rats was perfused separately from the rest of the distal tubule, thiazides either reduced (COSTANZO 1985) or abolished (ELLISON et al. 1987) net Na^+ and chloride transport. They had no effect on the initial collecting tubule when this segment was perfused separately from the rest of the distal tubule (ELLISON et al. 1987; COSTANZO 1985). The experiments do not rule out the possibility that thiazides are effective in the CNT segment of the rat distal tubule since in these studies some CNT cells may have been included in the perfused structure. In rats, however, it is not possible to study the connecting tubule directly and separately from the adjoining DCT and ICT segments.

Long-term administration of hydrochlorothiazide to control rats resulted in a decrease in (Na^+/K^+)-ATPase activity specifically in the DCT segment of the distal tubule, and an increase in activity in the downstream cortical collecting duct (GARG and NARANG 1987). The decreased activity in the DCT was attributed to a direct inhibition by thiazides of salt entry into these cells, whereas the increase in collecting duct enzyme activity may have been caused by the increased delivery of salt to this downstream site. Based on the available evidence in rats, it appears that the primary effect of thiazides is on distal convoluted tubule cells and that connecting tubule cells and initial collecting tubule cells are not the primary targets (see "Collecting Duct" below).

Thiazide-diuretic sensitive Na^+ and chloride absorption may be described as being directly coupled Na^+/Cl^- cotransport or as parallel Na^+/H^+ and Cl^-/HCO_3^- exchange. Either one or both of these alternatives for net Na^+ and chloride uptake could be present in the same membrane. Since thiazides abolish Na^+ and chloride absorption in the distal convoluted tubule of the rat (Ellison et al. 1987), it is likely that all of the Na^+ and chloride traverses a common pathway. It has not been possible to demonstrate an effect of high doses of amiloride (to inhibit Na^+/H^+ exchange) or of the stilbene DIDS (to inhibit Cl^-/HCO_3^- exchange) on Na^+ or chloride transport by the distal tubule of the rat (H. Velázquez, D.H. Ellison, M.D. Okusa and F.S. Wright, unpublished preliminary results). Thus, at least under the conditions of the experiment, perfusion with artificial tubular fluid or interstitial fluid, a major contribution to Na^+ and chloride absorption by a double exchange mechanism is unlikely but cannot be ruled out. It may be that in the rat most if not all Na^+ and chloride is absorbed via a mechanism that couples these two ions directly.

It has been possible to demonstrate that the rat distal convoluted tubule can absorb Na^+ and chloride absorption via an Na^+/H^+ and Cl^-/formate (or Cl^-/oxalate) double exchange mechanism when formate (0.5 mmol/l) is added to fluid used to perfuse distal tubules (Wang et al. 1993). Chloride absorption and volume absorption were not inhibited by DIDS or amiloride under control conditions consistent with our own unpublished results (see above). However, amiloride and DIDS do block the formate- and oxalate-dependent increases in transport. The fraction of chloride absorption stimulated by organic anions is not sensitive to chlorothiazide.

In the amphibian (*Amphiuma*) renal late distal tubule, a segment apparently homologous to the mammalian DCT, it has been shown (Stanton 1988) that apical application of both thiazides and amiloride decreases net Na^+ absorption by nearly 50%. The amphibian late distal tubule comprises two cell types (I and II), with the type II cell exhibiting thiazide-sensitive chloride transport. Because the basolateral membrane of these cells is conductive to chloride as well as K^+, changes in cell chloride would be expected to influence the membrane voltage. The basolateral membrane voltage hyperpolarized with apical hydrochlorothiazide (10^{-4} mol/l) (Stanton 1988; Hansen et al. 1981), DIDS (5×10^{-4} mol/l) and millimolar concentrations of amiloride (Stanton 1988). Cell chloride activity fell during luminal administration of hydrochlorothiazide and DIDS and correlated with basolateral membrane hyperpolarization (Stanton 1990). These results are consistent with the hypothesis that thiazides inhibit a mechanism for *Cl^-/HCO_3^- exchange*. It was concluded that at least part of the Na^+ and chloride absorption could be accounted for by the presence of parallel Na^+/H^+ and Cl^-/HCO_3^- exchange mechanisms in the luminal membrane although inhibition by thiazides of a mechanism coupling Na^+ and chloride directly in the same cells could not be excluded.

As will be discussed below, the rabbit DCT and the amphibian late distal type I and type II cells share an additional feature. The fractional resistance of the apical membrane is near unity (STANTON 1988; STANTON et al. 1987; VELÁZQUEZ and GREGER 1986). This limits conductive transport across the apical membrane and is consistent with electroneutral pathways for transepithelial salt absorption.

In another amphibian species (*Necturus*), the effects of hydrochlorothiazide on volume and ion transport were assessed using split drops in distal tubules (homologous to mammalian renal DCT cells) and using ion-selective microelectrodes (PLANELLES and ANAGNOSTOPOULOS 1992). The reabsorptive half-time of Na^+ was increased by chloride removal, and the reabsorptive half-time of chloride was increased by Na^+ removal. Neither K^+ removal nor addition of bumetanide or millimolar amiloride affected reabsorptive half-times. In contrast, increasing hydrochlorothiazide concentration increased reabsorptive half-times in a biphasic manner, suggesting a possible high-affinity ($0.5\,\mu mol/l$) and low-affinity ($115\,\mu mol/l$) site for this diuretic (PLANELLES and ANAGNOSTOPOULOS 1992). The present and previous (ANAGNOSTOPOULOS and PLANELLES 1987) results support the presence of a thiazide-inhibitable electroneutral Na^+-Cl^- cotransport mechanism in the *Necturus* distal tubule. It is of note that in this species as well as in the rat distal tubule (VELÁZQUEZ et al. 1984) and in the flounder bladder (STOKES 1984), furosemide by not bumetanide appears to inhibit partially the Na^+-Cl^- cotransporter. It remains to be investigated whether or not the furosemide effect is a nonspecific action on the thiazide-sensitive cotransport mechanism.

There is evidence supporting electroneutral Na^+/Cl^- cotransport in the DCT of other species as well. The rabbit DCT shares some of the properties of the rat DCT. In a series of preliminary studies (VELÁZQUEZ and GREGER 1985), rabbit DCT segments were dissected from slices of rabbit kidney. To permit positive identification of the DCT, the glomerulus and a portion of the thick ascending limb still attached was dissected. Subsequently, the thick ascending limb and glomerulus were cut off and the downstream DCT perfused. The transepithelial resistance was high ($\approx 100\,\Omega \cdot cm^2$) and the transepithelial voltage was near zero ($\approx -5\,mV$) (VELÁZQUEZ and GREGER 1985 and unpublished observations). At least 99% of the transcellular resistance resided in the apical membrane. This extremely low luminal membrane conductance limits the circular intraepithelial ionic current and, therefore, net ion absorption via conductive pathways. Na^+ and chloride absorption in this preparation of the rabbit DCT must occur via an electroneutral process.

The basolateral membrane is selective for K^+ ion (GREGER and VELÁZQUEZ 1987; VELÁZQUEZ and GREGER 1986; MOLONY and JACOBSON 1985). We have been unable to show the presence of a significant basolateral membrane chloride conductance directly by imposing rapid changes in bath chloride concentration, in both the absence and presence of Ba^{2+} (which blocks the dominant K^+ conductance and permits other conductances to manifest themselves). Although in one group of tubules a small hyperpolarization was measured with hydrochlorothiazide administration (VELÁZQUEZ and GREGER 1985), we have been unable to repeat this observation in subsequent measurements (H. VELÁZQUEZ, R. GREGER, and E. SCHLATTER, unpublished observations). Primarily because of the lack of conductive pathways in this tissue other than to K^+, changes in Na^+ and Cl^- transport by thiazides need not cause secondary changes in membrane voltage and, thus, other methods must be employed to confirm the presence of Na^+/Cl^- cotransport in the rabbit DCT.

Chloride taken up across the apical membrane via the thiazide-sensitive transporter must leave the cell across the basolateral membrane in order to accomplish transepithelial Na^+ and Cl^- absorption. Measurements of the K^+ activity in DCT cells of the rabbit indicate that the mechanism for chloride exit may be electroneutral and coupled to K^+. The electrochemical gradient for K^+ does not permit K^+ diffusion from cell to blood via its conductance (VELÁZQUEZ et al. 1988) since the equilibrium voltage for K^+, E_K, was not significantly different from the membrane voltage. An electroneutral K^+/Cl^- exit mechanism across the basolateral membrane may account for the extrusion of chloride taken up from the lumen.

In contrast to these observations, other investigators have reported that the rabbit DCT does not possess thiazide-sensitive Na^+ and Cl^- transport (SHIMIZU et al. 1988, 1989). Trichlormethiazide (10^{-4} mol/l) did not decrease lumen-to-bath chloride flux nor did it affect the lumen negative transepithelial voltage. The fractional resistance was not near unity (0.78), suggesting the presence of conductive pathways. The transepithelial voltage was depolarized by apical addition of amiloride (10^{-5} mol/l) (SHIMIZU et al. 1988; YOSHITOMI et al. 1989), and hyperpolarized by apical addition of Ba^{2+} (2 mmol/l) (YOSHITOMI et al. 1989). The presence of apical conductances to Na^+ and K^+ was further confirmed by measurement of apical membrane voltages during these same maneuvers using intracellular microelectrodes. In addition, the basolateral membrane possessed a small chloride conductance as well as a predominant K^+ conductance.

There is not a consensus regarding the cell model for the rabbit DCT. It is based primarily on the available evidence from isolated tubules perfused in vitro. Differences in rabbit strain, dissection technique and tubule identification are all possible contributors to this controversy and remain to be sorted out.

It has long been known that thiazides decrease renal Ca^{2+} excretion (LAMBERG and KUHLBACK 1959). The distal tubule plays a key role in this process. In the rat, thiazide diuretics stimulate net Ca^{2+} absorption by the distal convoluted tubule (COSTANZO and WINDHAGER 1978; COSTANZO 1985) concurrent with the inhibition of net Na^+ and Cl^- absorption. It was hypothesized that at least two cellular mechanisms could contribute to this effect (COSTANZO 1988; TAYLOR and WINDHAGER 1979; WALSER 1971). First, thiazides decrease Na^+ and Cl^- influx into the cell causing cell Na^+ and Cl^- activity to fall; the low intracellular Na^+ activity could increase the driving force for Na^+ flux from bath-to-cell via a basolateral Na^+/Ca^{2+} exchange mechanism which in turn would drive the secondary active extrusion of Ca^{2+} from the cell at a higher rate. Second, thiazides may hyperpolarize the apical membrane, either directly or via the lower intracellular chloride concentration, thus increasing the driving force for passive Ca^{2+} entry from the lumen into the cell.

It has been possible to test whether cell hyperpolarization plays a role in thiazide stimulation of Ca^{2+} transport by using a mouse kidney cell line derived from a thick ascending limb plus distal convoluted tubule preparation. Cells were originally isolated from collagenase digested mouse cortex by double antibody magnetic immunodissection using a Tamm-Horsfall antibody (PIZZONIA et al. 1991), which localizes to both the thick ascending limb and the distal convoluted tubule in this species. The cells in culture were infected with chimeric adenovirus 12/simian virus 40 construct, cloned and subsequently screened for the ability of thiazide diuretics to stimulate Ca^{2+} uptake (GESEK and FRIEDMAN 1992). Although the cells grow to confluence, they do not form tight junctions and do not develop a transepithelial voltage and resistance. In these cells, chlorothiazide (10^{-4} mol/l) blocked $^{22}Na^+$ uptake by 40% and $^{36}Cl^-$ by 50%, and stimulated $^{45}Ca^{2+}$ uptake by 45%, whereas bumetanide had no effect. These results are consistent with those observed in the rat distal tubule. Chlorothiazide hyperpolarized the membrane voltage of these cells (measured with the voltage-sensitive dye $DiOC_6(3)$) secondary to a fall in intracellular chloride. However, maneuvers that prevent a thiazide-induced change in membrane voltage [chloride channel blockade with nitrophenylpropylaminobenzoate (NPPB), removal of bath chloride] also prevented a change in calcium influx. Addition of the dihydropyridine nifedipine (10^{-5} mol/l) prevented thiazides from increasing Ca^{2+} influx. Thus, in this cell line, membrane hyperpolarization is required to stimulate calcium influx via Ca^{2+} channels.

Thiazides also stimulate Ca^{2+} absorption by the winter flounder bladder (ZIYADEH et al. 1987) and the turtle bladder (SABATINI and KURTZMAN 1988). In the winter flounder, hydrochlorothiazide induced net Ca^{2+} absorption from net secretion under control conditions, while hyperpolarizing the cell. This effect bears some resemblance to that seen in the mammalian distal tubule. In the turtle bladder, hydrochlorothiazide (10^{-3} mol/l) also stimulated mucosa-to-serosa Ca^{2+} flux but this effect occurred independent of changes in Na^+ flux and in the presence of basolateral ouabain to block the action of the Na^+ pump. The mechanism of action in the turtle bladder is unknown.

Thus, there is a consensus that in the distal convoluted tubule of the rat and of the mouse, and in the homologous segment of amphibian kidney, thiazide diuretics inhibit Na^+ and Cl^- transport. A controversy exists as to whether a thiazide-sensitive mechanism is present in the distal convoluted tubule of the rabbit. The reasons for the discrepancies are not apparent at present but it is possible that dissection techniques, identification criteria and differences in species strains all contribute.

Connecting Tubule. Available data for the connecting tubule are from the rabbit kidney. There is evidence for at least three separate mechanisms for Na^+ transport in the rabbit connecting tubule: one mediated by a channel and blocked by micromolar concentrations of amiloride, one mediated by an Na^+/H^+ exchange mechanism blocked by millimolar concentrations of amiloride and another pathway mediated by an *Na^+/Cl^-* cotransport mechanism and inhibited by thiazide diuretics (SHIMIZU et al. 1988; SHIMIZU and NAKAMURA 1992). Apical trichlormethiazide decreased net chloride absorption by 30% without an effect on the lumen negative transepithelial voltage. Unidirectional lumen-to-bath chloride flux was not affected by furosemide, whereas it was decreased by bicarbonate removal, luminal amiloride (10^{-3} mol/l) and luminal DIDS. Trichlormethiazide further inhibited chloride efflux in the presence of each of the above maneuvers, suggesting the presence of an Na^+/Cl^- cotransport mechanism distinct from either the $Na^+/2Cl^-/K^+$ cotransporter or a bicarbonate-dependent double exchange mechanism (SHIMIZU et al. 1988).

Na^+ flux across the apical membrane of the rabbit CNT was assessed by measuring the rate of cell swelling after inhibition of basolateral Na^+ extrusion via the (Na^+/K^+)-ATPase with ouabain (SHIMIZU and NAKAMURA 1992). Two Na^+ influx pathways contributed to the ouabain-induced cell swelling: one that could be blocked by amiloride and another inhibitable by trichlormethiazide. Cell swelling could be prevented by blocking both Na^+ influx pathways simultaneously and by chloride removal in the presence of amiloride. The results support the presence of a thiazide-sensitive chloride-dependent Na^+ transport mechanism in the apical membrane of the rabbit CNT.

Preliminary reports using in vitro primary culture of dissected rabbit connecting tubules on permeable supports do not provide evidence for a thiazide effect (VELÁZQUEZ et al. 1991). The monolayers absorbed Na^+ and secreted K^+. The lumen negative voltage was abolished by amiloride (10^{-5} mol/l), suggesting the presence of Na^+ channels and a pathway for conductive Na^+ absorption, and was hyperpolarized by apical Ba^{2+}. The monolayers did not absorb chloride, however, and neither Na^+ nor Cl^- transport was affected by hydrochlorothiazide. It may be that in culture these cells de-differentiate and do not express the thiazide-sensitive NaCl transporter that can be detected in the isolated perfused tubule preparation.

Another diuretic with uricosuric action: [6,7-dichloro-5-(*N*,*N*-dimethylsulfamoyl)-2,3-dihydrobenzofurancarboxylic acid], (S-8666), has its major site of action in the loop of Henle (SHIMIZU et al. 1991). However, in the CNT of the rabbit, it also inhibits lumen-to-bath chloride flux without affecting the transepithelial voltage and hyperpolarizes the basolateral membrane voltage. This latter effect resembles the effect of trichlormethiazide observed in this tubule segment (SHIMIZU et al. 1988).

In summary, in vivo data on the effect of thiazide diuretics on the connecting tubule are not available. In the rabbit tubule perfused in vitro there is evidence for the inhibition by thiazides of an Na^+/Cl^- cotransport mechanism. In vitro primary cultures of CNT explants, however, do not express thiazide-sensitive chloride transport.

Collecting Duct. In some species, thiazides act in the cortical collecting duct. This segment is not accessible from the surface of the kidney so that in vitro study using the isolated perfused tubule preparation is necessary. Most studies have been performed in rabbit kidney because of the relative ease of dissection of tubules. In this species, net Na^+ transport and the lumen negative transepithelial voltage are abolished by apical amiloride or basolateral ouabain (O'NEIL and SANSOM 1984; REIF et al. 1986; STONER et al. 1974; KOEPPEN et al. 1983; SCHLATTER and SCHAFER 1987). Furthermore, the application of basolateral ouabain in the presence of apical amiloride did not result in the cell depolarization that normally occurs when the apical Na^+ entry pathway is intact (SCHLATTER and SCHAFER 1987). These data suggest that most if not all of the Na^+ transport by the rabbit cortical collecting tubule occurs via an amiloride-sensitive conductive pathway and is mediated by the principal cell. The presence of an additional thiazide-sensitive pathway for Na^+ absorption in this preparation is unlikely.

In rats, however, the results from some experiments support the presence of electroneutral thiazide-sensitive Na^+ and Cl^- transport in the initial collecting tubule (COSTANZO 1985; ELLISON et al. 1987). This tissue is morphologically similar to the cortical collecting duct but is located upstream of the first confluence with another tubule to form the collecting duct. In one of two studies of the in vivo microperfused initial collecting tubule segment of the distal tubule (ELLISON et al. 1987), although no significant effect of chlorothiazide was found on Na^+ or Cl^- absorption, transport rates tended to be lower with the drug and an effect could not be ruled out. In the second study (COSTANZO 1985), a very small (6%) decrease of Na^+ absorption of borderline significance was found when chlorothiazide was perfused. These results do not argue strongly in favor or against the presence of a thiazide-sensitive mechanism present in the rat initial collecting tubule (cortical collecting tubule) in vivo.

Cortical collecting ducts were dissected out of DOC-treated rats and perfused in vitro, to assess the presence of a thiazide-sensitive component of Na^+ and Cl^- transport (TERADA and KNEPPER 1990; ROUCH et al. 1991).

Previous studies by one group of investigators (TOMITA et al. 1985, 1986; NONOGUCHI et al. 1989) had raised the possibility that a significant portion of Na^+ and Cl^- transport by the rat cortical collecting duct was electroneutral. Both bradykinin and atrial natriuretic peptide inhibited Na^+ and Cl^- absorption by approximately 50% without an effect on the lumen negative voltage (TOMITA et al. 1985, 1986; NONOGUCHI et al. 1989). Hydrochlorothiazide (10^{-4} mol/l) inhibited net Na^+ absorption and net Cl^- absorption by approximately 50% without an effect on transepithelial voltage (TERADA and KNEPPER 1990). This effect was observed both in the presence and absence of vasopressin. Amiloride alone also inhibited Na^+ absorption by approximately 50% and, together with hydrochlorothiazide, net Na^+ transport was inhibited by 82% (an effect significantly greater than that of amiloride alone). The results support the presence of two Na^+ transport pathways in parallel in the in vitro preparation of the rat cortical collecting duct, one conductive pathway inhibited by amiloride and the other an electroneutral pathway blocked by thiazide diuretics. Although the authors have not proven with the present data the presence of a directly coupled Na^+/Cl^- cotransport mechanism, they favor this mechanism over that of double exchange (TERADA and KNEPPER 1990) because this tissue lacks endogenous luminal carbonic anhydrase.

In contrast, however, another group investigating rat cortical collecting tubules was not able to show a thiazide effect. Hydrochlorothiazide (10^{-4} mol/l) had no effect on lumen-to-bath Na^+ flux or on the transepithelial voltage (ROUCH et al. 1991). Bradykinin and atrial natriuretic peptide were also without an effect. Amiloride and benzamil abolished net Na^+ absorption and the voltage. They have also shown that intracellular chloride activity in the principal cells was 17 mmol/l and was not altered by addition of hydrochlorothiazide, chloride channel blockers, disulfonic stilbenes or bicarbonate replacement (SCHLATTER et al. 1990). Cell chloride decreased gradually only upon bilateral chloride removal, suggesting that these cells do not contribute significantly to transepithelial chloride transport.

In the rat an effect of thiazides has been noted in the medullary portion of the collecting duct (WILSON et al. 1983, 1988). The collecting ducts were microcatheterized in retrograde fashion via the duct of Bellini in order to sample fluid at two sites, the border between the inner and outer medulla and the base of the papilla, and to assess transport by the intervening segment. Although chloride and water transport were reduced, Na^+ absorption was not. The mechanisms involved have not been resolved.

It is not clear to what extent a thiazide-sensitive mechanism plays a role in Na^+ and Cl^- absorption by the collecting duct. The controversy regarding the in vitro results in the cortical collecting duct has not yet been clarified but might be influenced by differences in diet, rat strain or the technique itself.

3. Effects on Renal Salt and Water Excretion

The administration of thiazides, within hours, leads to a sustained increase in renal excretion of Na^+, Cl^- and water (WALTER and SHIRLEY 1986). Although inhibition of salt and water absorption in the proximal tubule (see above), when thiazides are given acutely, can lead to an increase in the load delivered out of this tubule segment (KUNAU et al. 1975; FERNANDEZ and PUSCHETT 1973; ULLRICH et al. 1966), the loop of Henle is capable of buffering the increase to a large extent. The amount of fluid reaching the distal tubule is only increased by a relatively small amount (COSTANZO and WINDHAGER 1978; KUNAU et al. 1975; WALTER and SHIRLEY 1986). In free flow micropuncture studies, 10–14 days of hydrochlorothiazide administration to rats via osmotic minipump did not affect Na^+ and Cl^- delivery to the distal tubule (MORSING et al. 1991). Thus, in most circumstances, the increase in renal salt and water excretion observed with thiazides is primarily a function of the contribution of the distal tubule and not the proximal tubule or the loop of Henle although the collecting duct may also contribute (WILSON et al. 1983, 1988). When thiazides are administered long term, fractional absorption of water and Na^+ by the proximal tubule increases, presumably because of the volume contraction caused by the diuretic (WALTER and SHIRLEY 1986).

Both the structure and function of the distal tubule change during chronic increased Na^+ and Cl^- concentration in the lumen (ELLISON et al. 1989; KAISSLING and LEHIR 1982; KAISSLING et al. 1985; KAISSLING 1985). When the delivery of salt to the distal tubule of rats or rabbits was increased by increased salt in the diet or by infusion of furosemide by minipump, the DCT cells hypertrophied, the NaCl transport capacity increased and the density of ^3H-metolazone-binding sites increased (CHEN et al. 1990; BASILE et al. 1988). If hydrochlorothiazide was infused into rats for 10–14 days, the transport capacity for NaCl of the DCT was decreased by 47% (MORSING et al. 1991). In contrast, an increase in ^3H-metolazone binding in rat kidney cortex was measured after prior treatment with hydrochlorothiazide, indicating that there were a greater number of thiazide "receptors" (MORSING et al. 1991; CHEN et al. 1990) at a time when transport capacity was less. An increase in renal peripheral benzodiazepine "receptors" was also observed after 5 days of hydrochlorothiazide infusion (BASILE et al. 1988). Thus, transport function and the number of receptors does not correlate. It appears that more than one variable can regulate the activity and expression of the thiazide-sensitive "receptor." The possibilities include the concentration of Na^+ and Cl^- in luminal fluid and the occupancy of the receptor by either chloride or the diuretic drug.

The distal convoluted tubule is strategically located in the renal cortex just after the medullary diluting segment and just before the collecting duct which is responsible for fluid concentration. The DCT absorbs up to 10% of the filtered quantity of NaCl (HIERHOLZER 1985). Thiazide inhibition of Na^+

and Cl^- absorption in the relatively water impermeable DCT, therefore, decreases the diluting capacity of the kidney. It does not affect the concentrating ability of the kidney since the mechanisms of salt transport in the loop of Henle primarily responsible for the maintenance of the hypertonic medulla are unaffected by thiazides (SCHLATTER et al. 1983).

There are conditions, however, in which effects upstream to the distal tubule do contribute importantly to the effectiveness of thiazides. In certain clinical states of refractory fluid retention such as congestive heart failure or cirrhosis (EPSTEIN et al. 1977; OSTER et al. 1983; WOLLAM et al. 1981; MARONE et al. 1985), thiazides are often given together with loop diuretics because these agents alone are not effective enough in increasing renal sodium and fluid excretion (see also Chap. 13, this volume). Furosemide and thiazides together are much more effective. It does not appear that thiazides affect the pharmacokinetics of furosemide (MARONE et al. 1985). Recent results from microperfusion experiments in rats provide a possible explanation for this effect (ELLISON et al. 1987). DCT cells have a large reserve capacity to increase rates of Na^+ and chloride absorption when lumen concentrations are increased. In the rat distal tubule, rates of Na^+ and Cl^- absorption increased linearly with increases in the luminal Na^+ and chloride concentration and did not show signs of saturation (ELLISON et al. 1987). In the clinical settings in which combined therapy would be used, the glomerular filtration rate is usually low and fractional proximal absorption rate high. Administration of furosemide alone would be expected to produce a relatively small increase in tubule fluid flow rate because tubule volume flow rate is low initially. However, a large increase in Na^+ and chloride concentration in the distal tubule would occur secondary to inhibition of $Na^+2Cl^-K^+$ cotransport in the thick ascending limb. Since the distal convoluted tubule, the segment immediately downstream from the loop segment, has such a large reserve capacity to increase its rate of Na^+ and Cl^- absorption in response to an increase in lumen NaCl concentration, it promptly absorbs the salt delivered to it from the loop of Henle. To be able to excrete effectively the fraction of salt absorption inhibited by furosemide in the loop of Henle, the mechanisms for salt absorption in the downstream DCT must also be inhibited (ELLISON et al. 1987). This can be accomplished by giving a thiazide diuretic together with a loop diuretic. In addition, thiazides could enhance the diuretic effect of furosemide because, under these circumstances, a proximal inhibitory effect of thiazides such as that seen with acute administration can no longer be buffered as effectively as when the loop of Henle transport mechanisms are operating.

During ischemia rapid yet reversible downregulation of thiazide-diuretic "receptors" has been demonstrated (BEAUMONT et al. 1989b). Ten minutes of acute ischemia of rat kidneys caused a 90% inhibition of ^{3}H-metolazone binding measured in a renal membrane preparation. Reperfusion for 10 min after clamping the renal artery for 10 min increased "receptor" density to 40% of control levels. It is not possible to identify at present the mechanisms involved in this rapid and reversible change in transporter activity. If most of the salt entry into the distal tubule cell occurs via the Na^+/Cl^- cotransport mechanism, its downregulation may serve to conserve ATP during ischemia (Beaumont et al. 1989b).

Chlorouresis achieved by giving thiazide diuretics can be blunted indomethacin (Kirchner et al. 1987; Scriabine et al. 1980; Cooling and Sim 1978). Thiazide diuretics have been reported to increase urinary excretion of prostaglandins; indomethacin administration reduces prostaglandin excretion and also the rate of chloride excretion. Since renal hemodynamics or proximal tubule chloride absorption were not affected, it was suggested that increased absorption of chloride by the loop of Henle during indomethacin administration limited the delivery of chloride to the site of action of chlorothiazide in the distal tubule (Kirchner et al. 1987).

4. Effects on Renal K^+ Excretion

K^+ excretion is generally stimulated when thiazides are administered (Costanzo and Windhager 1978; Walter and Shirley 1986; Velázquez and Wright 1986a). This effect is, however, not a direct effect of the diuretic agent on a K^+ transport mechanism but rather a flow-dependent process (Wright 1982). When distal tubules are perfused directly in vivo with artificial solutions that attempt to maintain the initial ionic composition, osmolarity of luminal fluid and rates of net fluid transport constant, thiazides do not affect K^+ secretion (Velázquez and Wright 1986b). It appears that the inhibition of Na^+ and chloride absorption upstream to the secretory site for K^+ results in a slightly elevated delivery of K^+ to the distal tubule and an increased lumen fluid flow rate (Costanzo and Windhager 1978). It has been shown that an increase in fluid flow rate by itself can stimulate K^+ secretion by the distal tubule (Good and Wright 1979). Thus, when ion transport by the DCT is inhibited, luminal fluid becomes less hypotonic and less fluid is absorbed by the downstream connecting tubule and initial collecting tubule, resulting in an increased flow rate at the site of K^+ secretion. Taken together, these data are consistent with the notion that thiazides do not have a direct effect on a K^+ transport mechanism but act via a secondary increase in lumen flow rate.

A second mechanism by which thiazides may increase distal K^+ secretion is by affecting the lumen Ca^{2+} concentration. In microperfusion experiments performed in the rat distal tubule, the secretory K^+ flux was stimulated when lumen Ca^{2+} concentration was progressively lowered below 1.5 mmol/l (Okusa et al. 1990). Thus, thiazide-induced stimulation of Ca^{2+} absorption would lower the luminal Ca^{2+} concentration and might be expected to stimulate net K^+ secretion secondarily. This hypothesis remains to be tested.

Inhibition of K^+ secretion by thiazide diuretics under specific circumstances has also been demonstrated (Velázquez et al. 1992). Distal tubule

K^+ secretion is greatly stimulated by lowering the chloride concentration (VELÁZQUEZ et al. 1982; ELLISON et al. 1985) and by raising the lumen Na^+ concentration simultaneously (VELÁZQUEZ et al. 1987, 1992). In this stimulated state, with high lumen Na^+ concentration and low lumen chloride concentration, the addition of chlorothiazide decreases K^+ secretion (VELÁZQUEZ et al. 1992). Tubule fluid flow rate, osmolarity and net volume absorption were similar under both control and experimental conditions; thus changes in flow rate could not account for the differences in net K^+ secretion. We believe that this effect of thiazides is to decrease K^+ secretion by DCT cells secondary to inhibition of Na^+ and Cl^- transport in the same cells. Recent evidence suggests that K^+ secretion by the DCT does occur and can be regulated (SCHNERMANN et al. 1987; VELÁZQUEZ et al. 1987). It has been postulated (VELÁZQUEZ et al. 1992) that a secretory K^+-Cl^- cotransport mechanism is located in the luminal membrane in parallel with the thiazide-sensitive Na^+-Cl^- cotransport mechanism (VELÁZQUEZ et al. 1987). To explain the decrease in K^+ secretion by thiazides, we hypothesize (VELÁZQUEZ et al. 1992) that in the DCT cells inhibition of chloride uptake via Na^+-Cl^- cotransport lowers cell chloride concentration, thus decreasing the driving force for secretory Cl^- and for K^+ movement from cell to lumen.

5. Effects on Renal Ca^{2+} Excretion

Under normal conditions approximately 95% of the ultrafiltered Ca^{2+} is reabsorbed. The proximal tubule is responsible for reabsorbing the bulk of filtered Ca^{2+}, while the loop of Henle and the distal tubule reabsorb smaller quantities (COSTANZO and WINDHAGER 1978; LASSITER et al. 1963). The distal tubule is responsible for the fine regulation of Ca^{2+} excretion since the fraction of the filtered load of calcium reabsorbed in this segment is influenced by circulating hormones, volume status, acid base status and diuretics.

Chronic thiazide administration decreases urinary Ca^{2+} excretion which provides the basis for its use in states of hypercalciuria and stone formation (COE and KAVALACH 1974; COE 1981; YENDT and COHANIM 1978; KRAUSE et al. 1989). There are several mechanisms by which thiazides decrease renal Ca^{2+} excretion (see above). Ca^{2+} absorption by the proximal tubule is stimulated secondary to the volume contraction caused by this agent (AGUS et al. 1977; POUJEOL et al. 1976). Ca^{2+} absorption by the distal tubule is stimulated secondary to changes in intracellular ionic composition that affect membrane voltage or driving forces for Ca^{2+} extrusion.

E. Pharmacokinetics in Disease States

Of the modes of elimination – biotransformation, biliary excretion, renal excretion – it is the latter route which allows specific interaction of the drug with the "receptor". Any portion of a diuretic which undergoes biotransformation to inactive metabolites or biliary excretion cannot exert a saluretic effect. The functional status of those organs which govern the elimination kinetics of drugs may therefore be expected to determine the pharmacodynamics of these substances.

I. Chronic Renal Failure

In progressive renal failure the reduction in the number of intact nephrons can be deduced from the reduction in GFR. Homeostasis of the body and balance between salt and water intake and output is maintained by distal tubular salt "rejection," yielding an increase in fractional sodium excretion, FE_{Na}. Every time the GFR is halved, FE_{Na} is doubled ("magnification phenomenon," BRICKER et al. 1978) up to the limit of about 25% due to the transport capacity of the kidney. Given a constant dose of a diuretic the saluretic effect of the drug should be proportional to its renal clearance. By making use of an i.v. formulation of xipamide it has been shown that the total plasma clearance ($dose_{i.v.}/AUC$) was decreased in parallel to the reduction in renal clearance (Fig. 2). The nonrenal clearance given by the intercept with the ordinate at zero GFR remained constant and did not compensate for the decrease in renal clearance (KNAUF and MUTSCHLER 1984). For hydrochlorothiazide (HCTZ), which is much less bound to protein

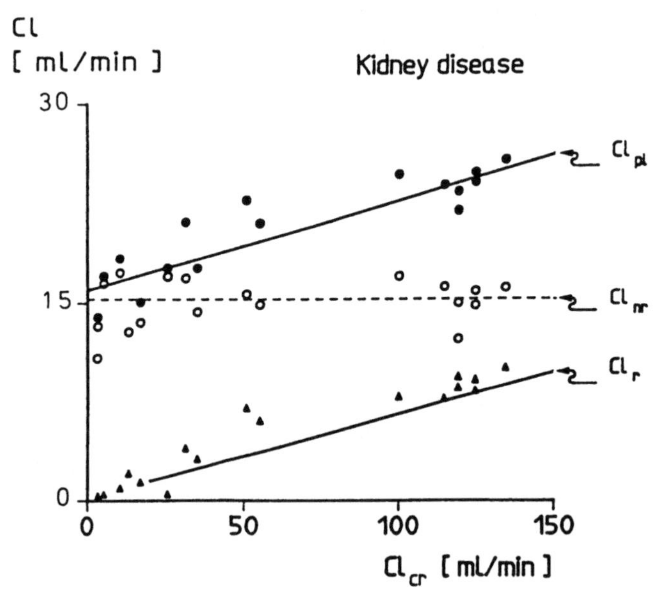

Fig. 2. Correlation between the total plasma clearance (Cl_{pl}) (●), nonrenal clearance (Cl_{nr}) (○), renal clearance (Cl_r) (▲) and creatinine clearance (Cl_{cr}) of test subjects after a single dose of 40 mg xipamide. (Data from KNAUF and MUTSCHLER 1984)

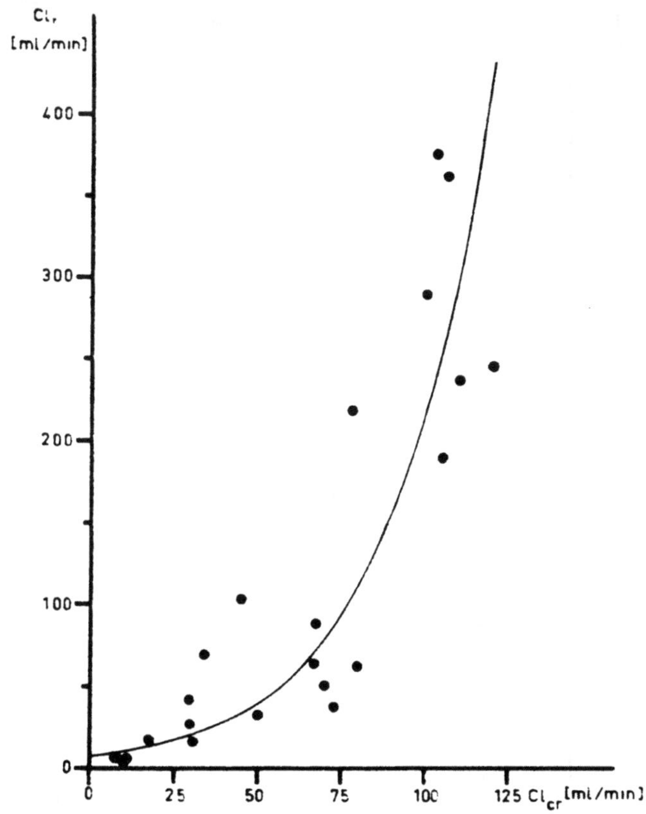

Fig. 3. Relationship between renal clearance (Cl_r) of hydrochlorothiazide and endogenous creatinine clearance (Cl_{cr}) in patients with different degrees of renal failure. (Data from NIEMEYER et al. 1983)

than xipamide, the relationship between patient renal drug clearance and GFR is best described by a nonlinear regression curve (Fig. 3) (NIEMEYER et al. 1983). If a diuretic such as xipamide is almost totally bound to protein, in contrast to HCTZ (Table 1), the renal clearance related to the fraction not bound to protein may approach PAH clearance (KNAUF and MUTSCHLER 1984).

As shown by ANDERSON et al. (1961) and BEERMANN and GROSCHINSKY-GRIND (1977), at least 95% of an intravenous dose of HCTZ is excreted unchanged by the kidney. It can therefore be expected that on long-term treatment HCTZ would accumulate if it were administered to patients with chronic renal failure. Indeed, the half-life of elimination was shown to increase from an average of 6.4 h in normal subjects to 21 h in patients whose GFR was below 30 ml/min (NIEMEYER et al. 1983).

The relationship between the elimination rate constant k_{el} and the GFR of the patients is shown in Fig. 4. The intercept of the regression line on the

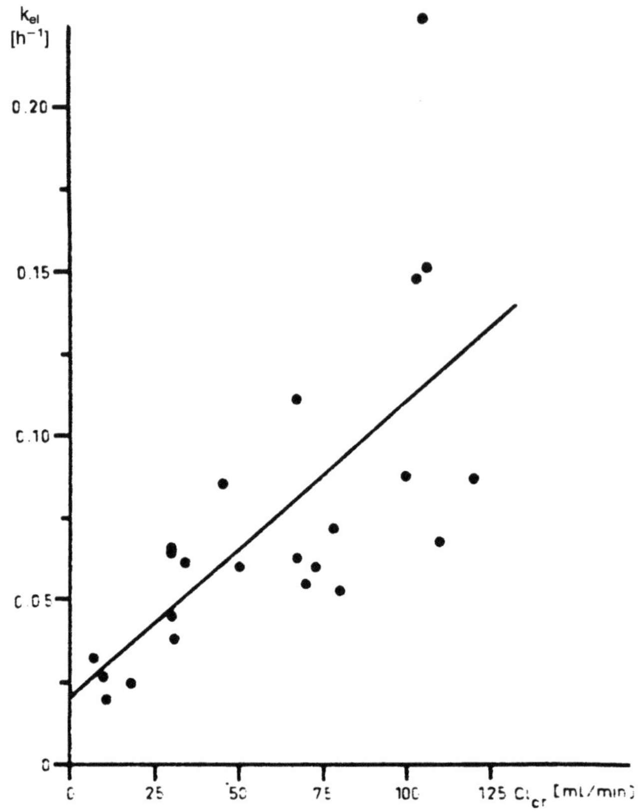

Fig. 4. Relationship between elimination rate constant (k_{el}) of HCTZ and endogenous creatinine clearance (Cl_{cr}). (Data from NIEMEYER et al. 1983)

ordinate yields a nonrenal elimination constant of 0.02/h corresponding to an elimination half-life of 34 h at zero GFR.

In contrast to HCTZ, for intravenously given xipamide recovery in urine is only about 40% (KNAUF and MUTSCHLER 1984). A greater fraction of xipamide therefore has to be eliminated by biotransformation and/or hepatobiliary excretion. This fraction may be obtained from a plot similar to that shown in Fig. 4 for HCTZ. A value of 0.075/h was thus determined for xipamide, corresponding to an elimination half-life of 9 h in end-stage renal disease.

II. Liver Disease

It has been shown for antikaliuretic agents (ANTONIN et al. 1982; MUTSCHLER et al. 1983; SPAHN et al. 1987) and for the lipid-lowering drug gemfibrocil (KNAUF et al. 1990a) that in liver disease nonrenal drug clearance is commonly reduced. This is primarily a result of an impairment of hepatobiliary drug excretion. Biotransformation is only reduced in end-stage liver disease. The reduction of nonrenal clearance can lead to an increase in urinary recovery, A_e, of both the parent drug and its metabolites. For thiazides this was first shown for xipamide (KNAUF et al. 1990b). In patients with liver cirrhosis, urinary recovery of parent xipamide rose from 30% to 60% and that of xipamide glucuronide from 10% to 20%, yielding an A_e of the parent drug and the metabolite of about 80% (Fig. 5).

Interestingly, of the varying parameters of liver disease there was only a positive correlation between A_e of the drug and the direct bilirubin concentration of the patients (Fig. 6). As the area under the plasma level time curve (AUC) increased in proportion to A_e, the renal clearance remained practically constant in the patients with liver disease (Fig. 7).

Obviously, conjugation still occurs in liver cirrhosis but the transfer of the parent drug and the metabolite to the biliary canaliculus is impaired in this state. It is generally assumed that drugs are transported across the sinusoidal and canalicular membranes of the hepatocyte by carrier-mediated transport systems that correspond to those described in the small intestine

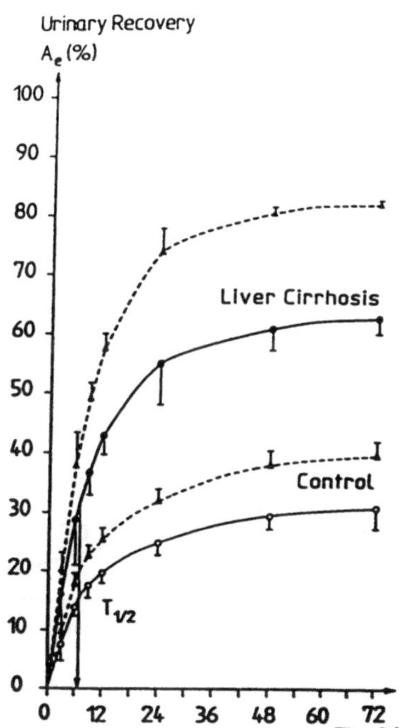

Fig. 5. Urinary recovery (percentage) of the 40-mg oral dose of xipamide given to patients with cirrhosis and to healthy control subjects. *Circles* represent parent xipamide; *triangles* represent the sum of parent xipamide and xipamide glucuronide. (Data from KNAUF et al. 1990b)

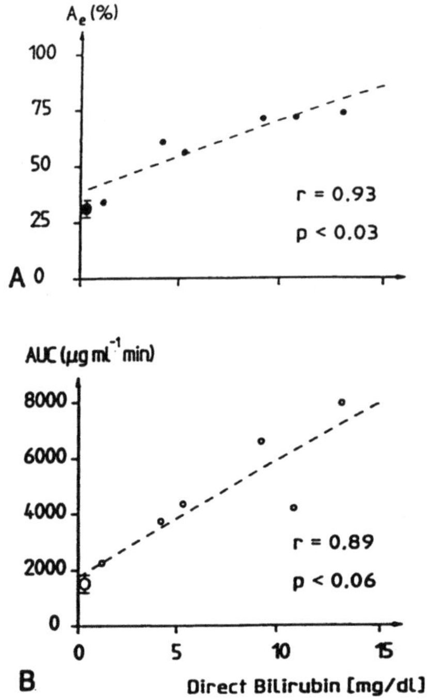

Fig. 6A,B. Correlation between urinary recovery, A_e (**A**), and area under the plasma concentration time curve. AUC (**B**), and the direct plasma bilirubin of the patients. *Large symbols with bars* represent the control subjects (mean ± SD; $n = 10$). (Data from Knauf et al. 1990b)

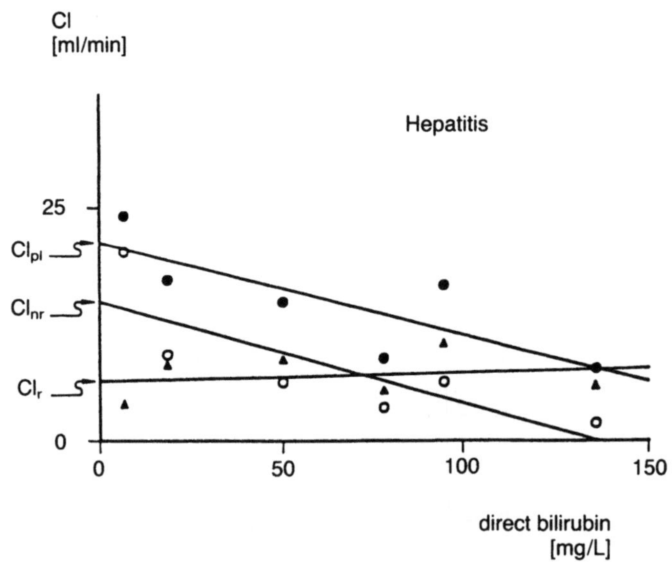

Fig. 7. Clearance pattern of i.v. administered xipamide in relation to liver function, quantified by the direct bilirubin of the patients with acute hepatitis. Total plasma clearance (Cl_{pl}) = renal clearance (Cl_r) + nonrenal clearance (Cl_{nr}). (Data from Knauf and Mutschler 1991)

and the renal proximal tubule (KNAUF et al. 1992). The present findings suggest that in liver dysfunction the carrier mechanisms operating the transfer to the bile duct seem to be primarily impaired. This assumption easily explains both jaundice and the increase in AUC and A_e of the parent drug and its glucuronide. Consequently, there is a shift from biliary excretion toward renal excretion. Although in cirrhotic patients the Cl_r of xipamide is not increased to compensate for the drastic reduction in Cl_{nr}, greater amounts of the drug reach the urine.

The change in the route of diuretic excretion with its consequent accumulation can be clinically relevant. The Cl_{nr} of xipamide in healthy control subjects represents that part of the drug which is lost for the saluretic effect. In liver disease, total plasma clearance may be taken over by Cl_r. These findings in liver disease indicate possible potentiation of the diuretic effects in patients with cholestatic conditions.

F. Saluretic Effects of Thiazides

I. Effects in Healthy Controls

The effects of a single oral dose of 25 mg HCTZ on urinary electrolyte excretion are illustrated in Fig. 8 and are compared with those induced by other classes of diuretics. The HCTZ-induced natriuresis lasts for about 12 h and is associated with significant kaliuresis. Calciuresis is very poor in contrast to that of loop diuretics. Several studies with different diuretics have been undertaken to determine the saluretic efficacy of the drugs (KNAUF and MUTSCHLER 1984, 1989, 1991, 1993, 1994, 1995; KNAUF et al. 1990b, 1994).

In these studies the diuretic-induced excretion of electrolytes was expressed as the increase in excretion compared with the control state in each patient and related to the duration of action of the diuretic. The latter is given by the period during which natriuresis exceeds the pretreatment (control) value. A fixed relationship between the diuretic-induced increase in various electrolytes was established for the three classes of diuretics (KNAUF and MUTSCHLER 1990) given in Table 2. For example, after administration of hydrochlorothiazide, the urinary excretion ratio K^+/Na^+ was 0.20 and after administration of piretanide it was 0.13. Antikaliuretics "save" K^+

Table 2. Ratios of diuretic-induced excretion of electrolytes during peak diuresis

	K^+/Na^+	Ca^{2+}/Na^+	Mg^{2+}/K^+
Loop diuretics, e.g., furosemide	0.12	0.02	0.15
Thiazides, e.g., hydrochlorothiazide	0.20	0.008	0.14
Antikaliuretis, e.g., amiloride	−0.21	0.004	~0.1

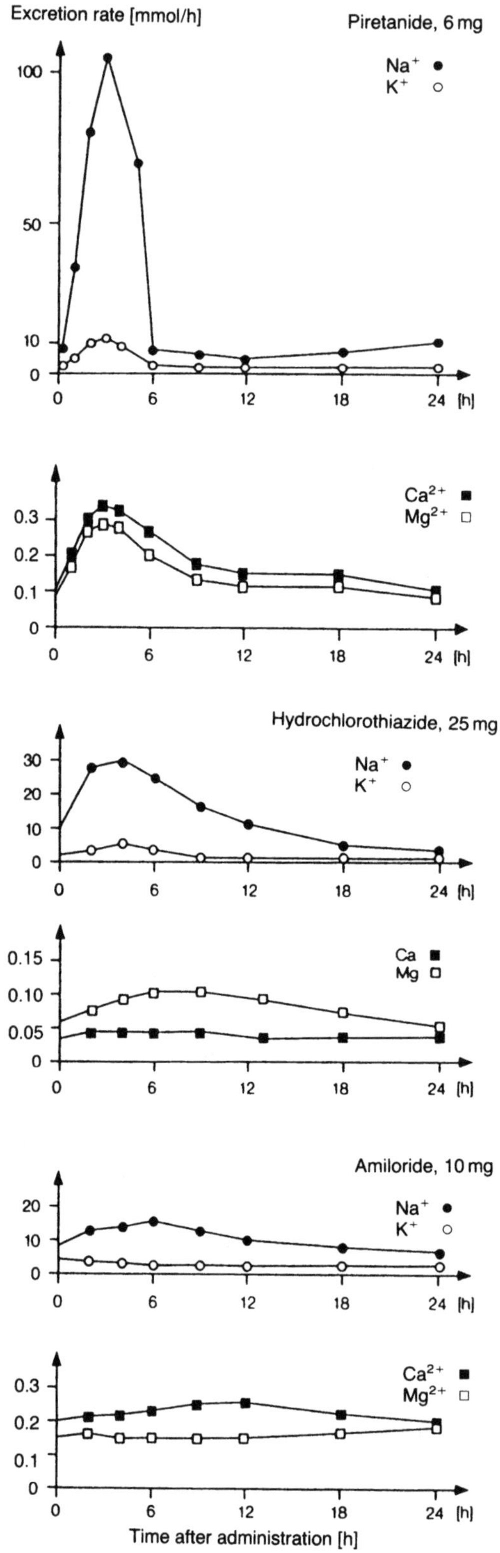

Fig. 8. Time course and pattern of electrolyte excretion following the administration of three classes of diuretics. (Data from KNAUF and MUTSCHLER 1992b)

and Mg^{2+} compared with the control state. The diuretic-induced excretory patterns for other electrolytes, such as Ca^{2+} and Mg^{2+}, also follow similar fixed relationships. As will be outlined below, these fixed relationships are independent of the glomerular filtration rate. Obviously, the thiazide diuretics lead to greater excretion of Na^+ due to their longer duration of action compared with diuretics acting in the loop of Henle. However, consideration of the maximal Na^+ excretion as a percentage of the amount filtered (fractional excretion of sodium, FE_{Na}) demonstrates that loop diuretics give the greatest amount of filtrate that can be achieved (see also below).

II. Responses in Renal Failure

As shown above (Fig. 2), the renal clearance of diuretics is reduced in direct proportion to the decrease in glomerular filtration rate. For this reason, with diuretics whose natriuretic activity is directly dependent on renal clearance of the drug, the excretion of electrolytes also decreases in direct proportion to the decrease in glomerular filtration rate. Moreover, the fixed relationship between the increase in urinary excretion of the various electrolytes (see above) is maintained in renal failure, i.e., it is independent of the number of functioning nephrons.

Recent studies (KNAUF and MUTSCHLER 1995) in patients with chronic renal failure show that HCTZ does induce a saluretic effect even in end-stage renal disease, presenting the linear relationship for electrolyte excretion and GFR (Fig. 9) known for loop diuretics (KNAUF and MUTSCHLER 1990, 1992). Thus, there is no "cutoff" at the GFR <30 ml/min as claimed earlier (WILCOX 1991).

The same findings were reported earlier for other thiazides such as xipamide (KNAUF and MUTSCHLER 1984) and bemetizide (KNAUF et al. 1994) studied in advanced renal failure. Obviously, nephron adaptation in renal failure maintains the ion transport characteristics of the early distal tubule, the site of action of thiazides.

III. Coadministration with Loop Diuretics in Renal Failure

The maintenance of saluretic effectiveness of thiazides in renal failure provides a rational basis for combining thiazides with loop diuretics to increase their efficacy. Indeed, long before the above-cited (Fig. 9) pharmacodynamic properties of thiazides were known, it was reported (WOLLAM et al. 1982) that the diuretic potency of combined hydrochlorothiazide and furosemide was greater than increasing the doses of loop diuretics alone. In recent randomized single-blind placebo-controlled crossover trials in patients whose GFR varied from 140 to 5 ml/min (KNAUF and MUTSCHLER 1993, 1994, 1995), it was shown that across the whole range of renal impairment studied the excretory dose responses of both HCTZ (or xipamide) and

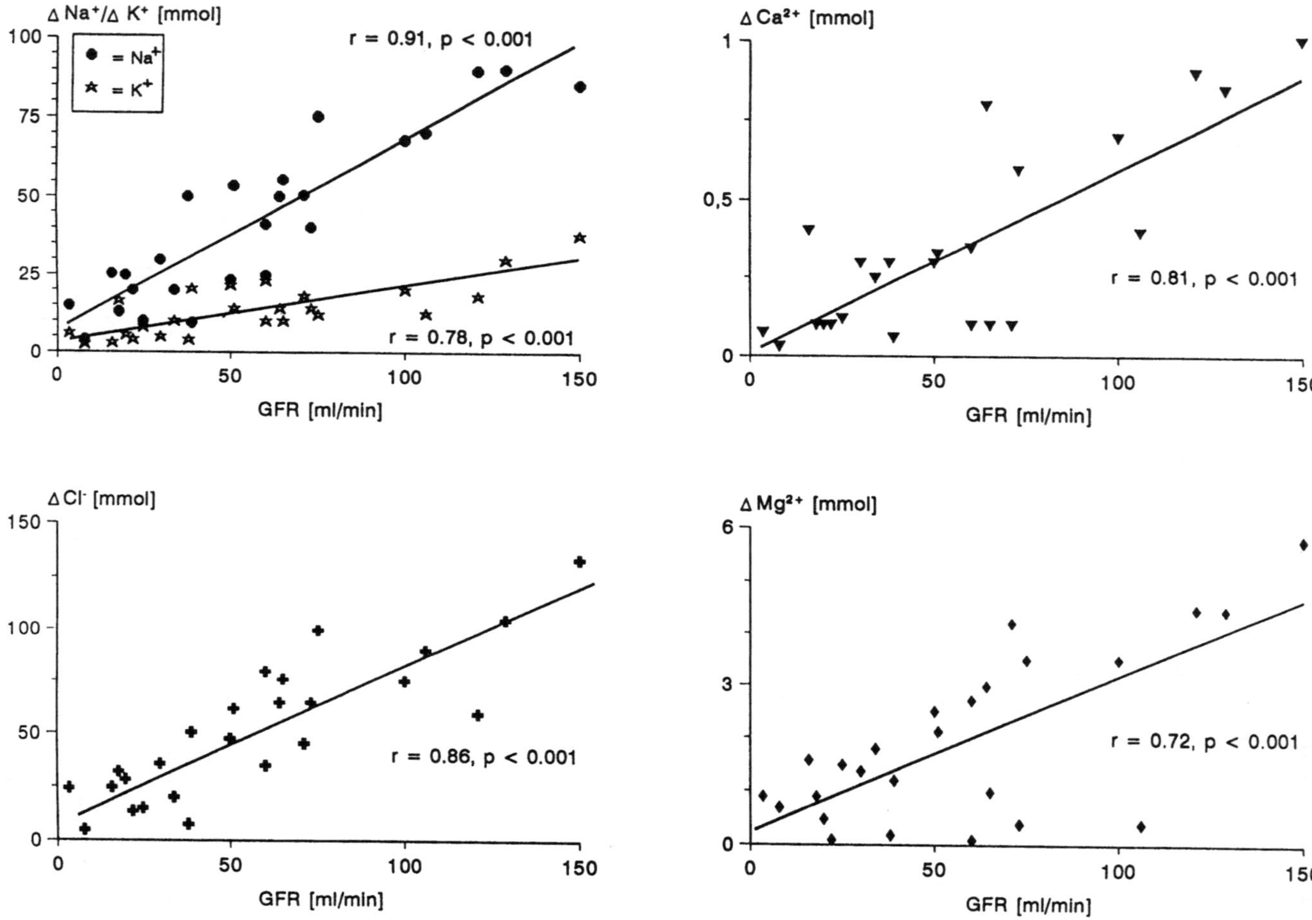

Fig. 9. Correlation between the HCTZ-induced net excretion of electrolytes during τ and the GFR of the subjects. (Data from KNAUF and MUTSCHLER 1995)

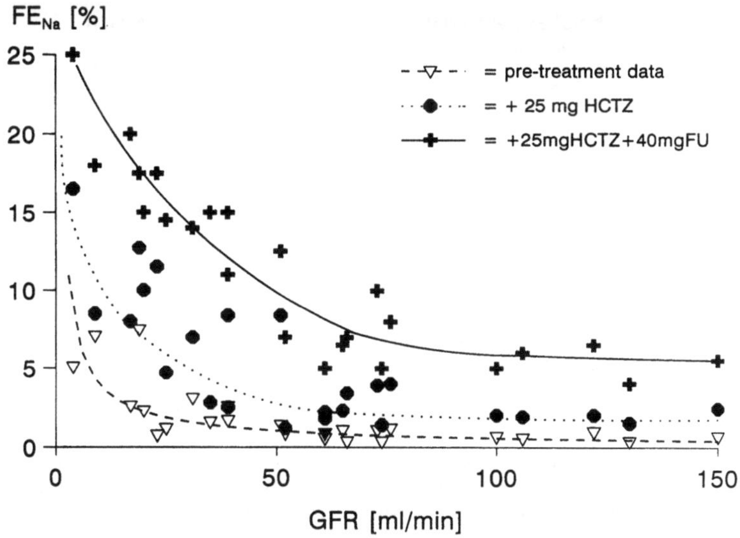

Fig. 10. Fractional sodium excretion of the subjects as a function of the GFR before and after administration of 25 mg HCTZ or coadministration of 25 mg HCTZ and 40 mg FU. The normal hyperbolic function ($FE_{Na} \times GFR = const.$) is given according to Slatopolsky et al. (Data from KNAUF and MUTSCHLER 1995)

furosemide monotherapy were relatively flat. In contrast, their coadministration resulted in a substantial and significant increase in saluresis which was independent of the degree of renal impairment (Fig. 10). Thus, the coadministration of low doses of diuretics acting at different sites in the nephron is more effective in patients with renal failure than increasing the dose of a diuretic acting at a single nephron site (HEIDLAND et al. 1985; RUDY et al. 1991). In accordance with these findings patients on hemo- and peritoneal dialysis presenting a rest diuresis also profited from low-dose coadministration of loop diuretics with thiazides compared with "conventional" high-dose monotherapy with loop diuretics (BOESKEN et al. 1995).

A number of mechanisms are probably involved in the enhanced diuretic activity that results when thiazides are combined with those exerting their activity in the loop of Henle. Impairment of Na^+ absorption in the loop of Henle results in a greater amount of Na^+ being delivered to the distal tubule. As the transport mechanisms in the distal nephron segment are unsaturated this results in partial reabsorption of the Na^+ rejected in the loop of Henle (SIGURD et al. 1975; ELLISON 1991). Moreover, chronically increased Na^+ delivery (KAISSLING and STANTON 1988), increased basolateral infoldings with increased $(Na^+ + K^+)$-ATPase activity (LE HIR et al. 1982), increased numbers of thiazide-sensitive NaCl transporters (SCHERZER et al. 1987) and increased transcellular NaCl transport capacity (STANTON and KAISSLING 1988) have been reported. Coadministration of thiazides with

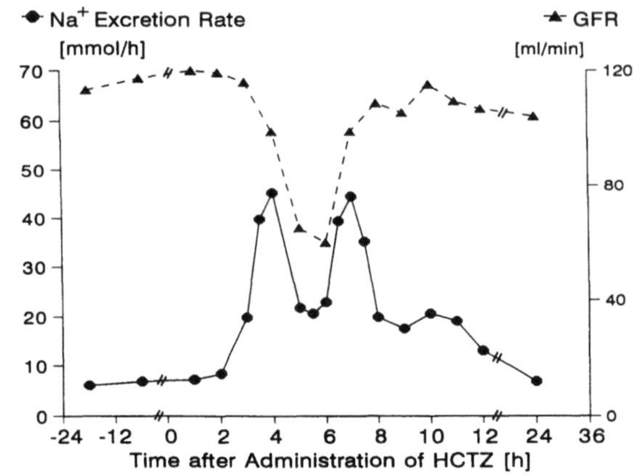

Fig. 11. Time course of Na^+ excretion rate following a single oral dose of 25 mg HCTZ in a healthy subject. *Right ordinate* gives the GFR of the subject obtained during the experiment via the creatinine clearance. (Data from KNAUF and MUTSCHLER 1995)

loop diuretics thus prevents these adaptive processes and maintains the diuretic efficiency of the loop diuretics (KAISSLING et al. 1985).

A further mechanism relates to the tubulo-glomerular feedback system. This feedback mechanism is inhibited by loop diuretics so that adding a loop diuretic enhances the diuretic activity of a thiazide by inhibiting its self-limiting effectiveness. Thiazide diuretics are known to reduce reversibly the glomerular filtration rate (VILLAREAL et al. 1966). In recent studies with bemetizide (KNAUF et al. 1994) and hydrochlorothiazide (KNAUF and MUTSCHLER 1995) it was shown that administration of these thiazides was followed by double-peak natriuresis paralleled by a reversible reduction in GFR (Fig. 11). The abrupt reduction in GFR was observed only if the Na^+ excretion rate exceeded the critical threshold value of approximately 40 mmol/h in normal subjects and about 15 mmol/h in patients with advanced renal failure. This dose-dependent effect is probably a physiological regulatory mechanism which prevents rapid and excessive volume depletion and which is a self-limiting homeostatic mechanism specific to the thiazide group of diuretics and which is not seen with loop diuretics (KNAUF and MUTSCHLER 1990). These latter drugs specifically block the $Na^+2Cl^-K^+$ cotransport system present in the macula densa cells (SCHLATTER et al. 1990) and interrupt the physiological tubulo-glomerular feed back loop which may result in precipitous diuresis and excessive volume loss. This problem does not exist in the case of thiazide administration.

IV. Coadministration with Other Diuretics in Edematous States with Normal Kidney Function

Low-dose combination therapy of thiazides with loop diuretics has been demonstrated to be superior to high-dose monotherapy for other states of poor or insufficient diuretic response to loop diuretics ("diuretic resistance") in patients with normal intrinsic renal function such as congestive heart failure (OLESEN and SIGURD 1971; GUNSTONE et al. 1971; SIGURD et al. 1975) or hepatic ascites (EPSTEIN et al. 1977; OSTER et al. 1983). This is not only true for the frequently used metolazone, but also for other thiazides such as hydrochlorothiazide, bendroflumethiazide and quinethazone. The mechanisms for the observed synergistic effects have not yet been definitely established in these disease states. Besides the blockade of different tubular segments, thereby avoiding partial attenuation of the diuretic response to monotherapy (see above), a further argument for the enhanced effectiveness of combination therapy relates to the pharmacokinetic characteristics of the drugs used (MARONE et al. 1985). Thiazides have a long duration of action. In contrast, loop diuretics such as furosemide, with a short half-life, induce a rebound natriuresis following their peak diuretic activity, which may negate a beneficial long-term effect of loop diuretics on sodium balance (ELLISON 1991; LOON et al. 1989).

The pathogenesis of Na^+ and water retention in edematous states such as congestive heart failure, liver cirrhosis and nephrotic syndrome has been analyzed by SCHRIER (1988) and SCHRIER et al. (1988). These diseases appear to have as a hallmark of their clinical syndrome disturbances of volume regulation. The effective arterial blood volume may be reduced

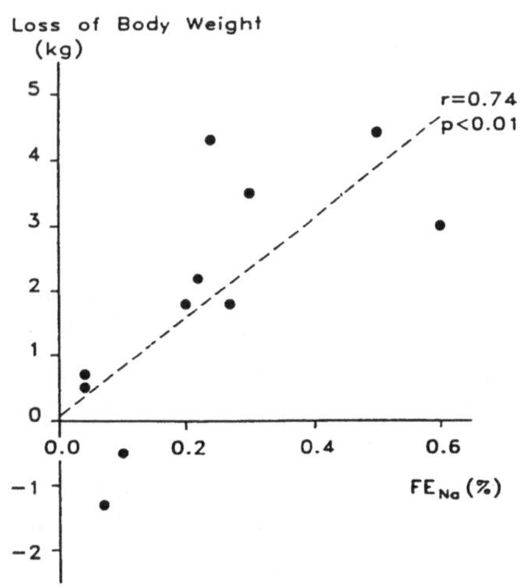

Fig. 12. Relationship between xipamide-induced loss of body weight and pretreatment fractional Na^+ excretion (Data from KNAUF et al. 1990c)

("underfilling") by low cardiac output (heart failure), by peripheral arterial vasodilation (liver cirrhosis) or by diminished plasma oncotic pressure (nephrotic syndrome). The kidney responds to vascular underfilling by proximal tubular hyperreabsorption (Skorecki and Brenner 1982) followed by a reduced Na^+ load delivered to the loop of Henle and the distal tubule, the sites of diuretic action. The resulting low fractional Na^+ excretion can be used as a predictor of diuretic resistance (Fig. 12) (Knauf et al. 1990c). The effectiveness of acetazolamide coadministered with thiazides corroborates increased proximal-tubular Na^+ reabsorption as the major alteration in such disease states (Knauf and Mutschler 1994).

Despite the positive aspects of these regimens, there are disturbing observations regarding adverse effects that may result from massive diuresis with electrolyte losses and circulatory collapse (Bamford 1981; Black et al. 1978; Asscher 1974; Allen et al. 1981; Gunstone et al. 1971).

G. Diuretics in Nonedematous States

I. Hypertension

Over the past 30 years, thiazide diuretics have become a mainstay of antihypertensive therapy and their efficacy has been documented in many trials. (For reviews see Hansson and Dahlöf 1990; Flack and Grimm 1989; Kaplan 1990.) Whereas the benefits of diuretic treatment on total mortality and stroke have been widely accepted, the effects of diuretics on coronary heart disease have been a matter of debate. Recent clinical trials have added important new data confirming the beneficial effects of diuretics in reducing morbidity and mortality in the treatment of elderly hypertensive patients (for review see Moser 1994). In addition a significant decrease in CHD events has been reported (Table 3). Furthermore, long-term treatment with diuretics has resulted in regression of left ventricular hypertrophy, prevention of congestive heart failure and slowing of the progression of renal deterioration. On the other hand, there has been little evidence that the use of diuretics produces clinically significant changes in serum lipids or serum glucose levels (Moser 1994). Potential adverse effects and complications which have been reported to develop in the treatment of hypertension (Birkenhäger 1990; Moser 1990; Dollery 1989; Fröhlich 1987; Lant 1985) have been in part due to the use of doses that were too high (Leonetti et al. 1989). An effective antihypertensive effect can be achieved in most patients with a low dose (12.5 mg HCTZ/day) (Birkenhäger 1990; Carney et al. 1976; Russell et al. 1981). The recommended maximum dose is 50 mg/day (Schnaper et al. 1989). Most complications have occurred when doses in the range of up to 200 mg/day were used (Birkenhäger 1990; Veterans Administration Cooperative Study Group on Antihypertensive Agents 1982; Leonetti et al. 1989).

Table 3. Antihypertensive drug trials: risk reductions for coronary heart disease in diuretic-based studies (reproduced with permission from Moser 1994)[a]

	Active			Control			Risk reduction (%)	
	Fatal	Total	Number of subjects	Fatal	Total	Number of subjects	Fatal	Total
Fourteen previous trials	316	671	18487	356	771	18407	11 (NS) ($P = 0.12$)	14 ($P = 0.007$)
SHEP[b]	59	104	2365	74	142	2371	21 (NS) ($P = 0.19$)	28 ($P = 0.01$)
STOP Hypertension[c]	10	31	812	20	32	815	49 ($P = 0.07$)	3 (NS) ($P = 0.91$)
MRC older adults[d]								
Diureties + β-blockers	85	128	2183	110	159	2213	22 (NS) ($P = 0.08$)	19 (NS) ($P = 0.08$)
Seventeen trials[e]	470	934	23847	560	1104	23806	16 ($P = 0.006$)	16 ($P = 0.0001$)
	33	48	1081	47[f]	70[f]	–	30 ($P = 0.006$)	32 ($P = 0.007$)

[a] 30%–50% of subjects had an additional antihypertensive agent added during some phase of the study.
[b] Systolic Hypertension in the Elderly Program.
[c] Swedish Trial in Old Patients with Hypertension; includes all subjects on β-blockers or diuretics.
[d] Subjects treated with diuretics only.
[e] Also includes patients initially randomized to β-blockers in MRC older adults and STOP Hypertension studies.
[f] Estimated from available data.

As the primary therapy of hypertension, diuretics lower blood pressure in about half the patients with essential hypertension (for review see Hansson and Dahlöf 1990). This response rate is similar to, or better than, that found with other monotherapies. The variability of responders depends on several factors including the demographics of the patient population studied, diet and humoral/hormonal profiles (salt sensitivity). Other hypertensive agents gain efficacy when added to diuretic therapy (e.g., ACE inhibitors).

The question as to how diuretics lower high blood pressure has so far not yet been definitely answered. Relatively little progress has been made since Tobian's review of this subject over 20 years ago. Two possible mechanisms of action of the mostly used thiazide diuretics must be considered: (1) The hypotensive effect is a direct or indirect consequence of sodium (chloride) depletion and/or (2) diuretics act by direct or indirect vascular effects.

1. Many studies have shown that the hypotensive response to diuretics is reflected by acute and then chronic hemodynamic correlates as depicted in Fig. 13 (Dollery et al. 1959; Wilson and Freis 1959; Frohlich et al. 1960; Shah et al. 1978; van Brummelen et al. 1980). With chronic diuretic administration lowered blood volume returns towards or to normal, and peripheral resistance falls to below pretreatment values. A small degree of volume depletion relative to salt intake is sustained. The reduction of peripheral resistance may be taken to indicate (Pecker 1990) that a "near-normal" absolute level of blood volume is inappropriately low for the state of the vascular tree. Under these conditions the reactivity of the resistance vessels to pressor agents (or pressor response) is reduced (Weidmann 1980). Upon cessation of diuretics both body weight (1–2 kg) and plasma volume increase (Wilson and Freis 1959).

Ingestion of large amounts of salt can prevent or reverse the hypotensive action of diuretics (Freis et al. 1958; Winer 1961). On the other hand, the hemodynamic and extracellular volume effects of 150 mg chlorothiazide daily have been found to be quite similar to those of draconian salt restriction (less than 10 mmol/day) (Tobian 1967; Murphy 1950). This holds true for the noncommonly used thiazide doses when compared with hemodynamic, biochemical and hormonal profiles in patients on low-salt diets (Laragh and Pecker 1983). The antihypertensive effects of diuretics and dietary salt restriction would therefore appear to have the same basis.

2. In several experimental studies it was shown that diuretics inhibit ion transporters in vascular smooth muscle cell (Haddy et al. 1985; Cantiello et al. 1986). The resulting decrease in intracellular sodium and calcium concentration was supposed to reduce smooth muscle contractility. However, the concentrations at which these effects occur are usually above those found with clinical use of these drugs. Furthermore, peripheral vascular resistance rises acutely with thiazide administration (Fig. 13), and diuretics do not lower blood pressure in hypertensive patients on chronic hemodialysis

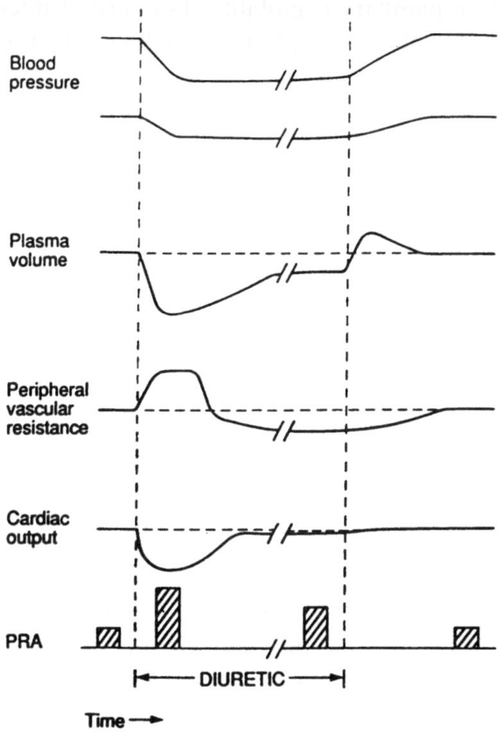

Fig. 13. Diagram of hemodynamic correlates of the antihypertensive effects of diuretic therapy (from PECKER 1990)

(BENNET et al. 1977; ORGISON 1962) or in nephrectomized animals. These findings underline a key role for the kidney in the diuretic action.

Recent studies on long-term use of thiazides attribute the sustained decrease in vascular resistance (BIRKENHÄGER 1990; VAN BRUMMELEN et al. 1979, FRÖHLICH 1987; AMBROSIONI et al. 1987) to the prostacyclin-induced fall in resistance of the renal vascular tree. The hydrochlorothiazide-induced relaxation of human, guinea pig and rat small arteries was discussed to involve Ca^{2+}-activated K^+ channels in the vasorelaxant effect of thiazides (CALDER et al. 1992).

II. Diabetes Insipidus

Thiazides can produce an antidiuretic effect and thus achieve a paradoxic 50% reduction in urine flow in patients with central or nephrogenic diabetes insipidus. This effect is related to two mechanisms of action (WILCOX 1991):

1. Thiazide-induced volume contraction or ECV depletion, respectively, reduces GFR and enhances Na^+ reabsorption by the proximal tubule, thereby curtailing the delivery of filtrate to the diluting segment.

2. Thiazides increase papillary osmolality. For further details of the clinical use of thiazides in diabetes insipidus, see Chap. 13, this volume.

III. Nephrolithiasis

Thiazides are known to stimulate tubular reabsorption of Ca^{2+} by a parathormone-independent mechanism. Thus, they produce a sustained reduction in renal Ca^{2+} excretion which is accompanied by a small rise in serum Ca^{2+} concentration. Therefore, thiazides are used in patients prone to stone formation as they reduce the underlying hypercalciuria. In contrast to loop diuretics they do not cause osteoporosis to worsen. For further details of their clinical uses, see Chap. 13, this volume.

H. Side Effects of Diuretic Therapy

Although there are a number of side effects associated with the use of thiazide diuretics, these are generally of limited severity and clinical importance and can be controlled effectively in a clinical setting (MOSER 1989a; LEONETTI et al. 1989; FIELD and LAWRENCE 1986). Adverse effects may be (a) the direct consequence of the renal action of the diuretics such as orthostatic symptoms due to volume depletion, hypokalemia, Mg^{2+} depletion, hyponatremia and hyperuricemia; (b) secondary effects such as glucose intolerance and hyperlipidemia; or (c) idiosyncratic drug reaction.

I. Hypokalemia

Hypokalemia [plasma (K^+) <3.5 mmol/l] induced by thiazides has been well documented and is more prominent when high doses are administered (KAU 1988; HOLLAND et al. 1981; PAPADEMETRIOU et al. 1988; NORDREHAUG and VON DER LIPPE 1983; DYCKNER et al. 1975; HOLLIFIELD 1986). It is controversial whether or not hypokalemia increases the likelihood of ventricular arrhythmias or myocardial infarction (MOSER 1990; PAPADEMETRIOU et al. 1983; PAPADEMETRIOU et al. 1988; MADIAS et al. 1984; ANDERSSON et al. 1991). Low blood K^+ concentrations generally can be corrected by dietary supplementation or by concurrent administration of K^+-sparing diuretics and thiazides (EUROPEAN WORKING PARTY ON HIGH BLOOD PRESSURE IN THE ELDERLY 1985; BLACK 1991; SCHNAPER et al. 1989; MOSER 1989a).

II. Mg^{2+} Depletion

Reduction of plasma levels and whole body Mg^{2+} content can occur during long-term thiazide use and is generally associated with hypokalemia. Coadministration of a K^+-sparing diuretic has been used to prevent both hypomagnesemia and hypokalemia (HOLLIFIELD 1986, 1989).

III. Hyponatremia

In the absence of ADH the osmolality of the tubule fluid delivered from the loop of Henle is reduced further by reabsorption of Na^+Cl^- from the early distal convoluted tubule. Therefore, thiazides impair maximal urinary dilution as they inhibit the generation of "free water" (WILCOX 1991). Unlike loop diuretics, thiazides do not block ion transport in the thick ascending limb of Henle's loop, which is required for elaboration of a hypertonic medullary interstitium. Therefore, thiazides do not impair the urinary concentrating mechanism. The combination of increased Na^+Cl^- excretion and impaired dilution but intact concentration ability predisposes patients to the development of hyponatremia during long-term thiazide therapy. Hyponatremia can be severe, and can be reproduced in susceptible individuals with single-dose rechallenge (FRIEDMAN et al. 1989). Elderly patients may be more susceptible to this disorder (FIELD and LAWRENCE 1986; FICHMANN et al. 1971; DYCKNER and WESTER 1981; STRYKERS et al. 1984).

IV. Hyperuricemia

Hyperuricemia is a common complication of thiazide diuretic therapy and is – as is hypokalemia and Mg^{2+} wasting – dose related. It is observed in about 30% of untreated mild hypertensives (MESSERLI et al. 1980) and serum uric acid rises by about 10 mg/l on thiazide treatment. For most patients this is clinically irrelevant. Gout may occur rarely (MOSER 1990; HYPERTENSION DETECTION AND FOLLOW-UP PROGRAM COOPERATIVE GROUP 1979; BLACK 1991; CARLSON et al. 1990).

V. Hyperglycemia

Thiazide use in hypertensives appears to be related to impairment of insulin release by glucose and increased peripheral resistance to insulin (BLACK 1990; MURPHY et al. 1982; POLLARE et al. 1989; GIFFORD 1986; ANDERSSON et al. 1991). The diuretic drug itself does not appear to have a direct effect on insulin secretion, but may act through low blood K^+ concentrations secondary to increased renal K^+ excretion (ANDERSSON et al. 1991; HELDERMAN et al. 1983).

VI. Hyperlipidemia

Thiazides have also been reported to cause an increase in serum cholesterol, triglycerides and low-density lipoprotein levels (HUNNINGHAKE 1991; MOSER 1990; ACES 1983; MOSER 1989b; GRIMM et al. 1981; SCHOENFELD AND GOLDBERGER 1964; AMES and HILL 1976; GRIMM 1991). The duration of the therapy may play a role in determining the degree of this imbalance (GIFFORD 1986; FREIS 1990; BALLANTYNE 1990; MOSER 1989a). This may represent an increased cardiovascular risk (BLACK 1991; SEED et al. 1990; GRIMM 1991) yet has recently been questioned (MOSER 1994). Dietary adjustments may further diminish any adverse effects of thiazides on plasma lipid levels.

VII. Allergy

Binding of thiazide diuretics (and other sulfonamides or furosemide) molecules to circulating proteins can subsequently elicit an immune response and cause an allergic reaction (SULLIVAN 1991). Patients who have experienced one allergic reaction to an antibacterial drug have a tenfold increased risk of reacting to other structurally unrelated antimicrobial drugs (SULLIVAN et al. 1989). In this context hydrochlorothiazide has been reported to be linked to acute allergic interstitial pneumonitis (noncardiogenic pulmonary edema syndrome) (BIRON et al. 1991; KAVARU et al. 1990; COOPER and MATTHAY 1987; WHITE and WARD 1985; STEINBERG 1968; BEAUDRY and LAPLANTE 1973; WEDDINGTON et al. 1973; DORN and WALKER 1981; FARRELL and SCHILLACI 1976). Although extremely rare with an incidence of less than 1%, this reaction is potentially fatal. It develops rapidly and is apparent within 1 h of administration of the drug. Rarely reported clinical conditions associated with thiazide therapy include acute pancreatitis, interstitial nephritis, vasculitis, thrombocytopenia, leukopenia and hemolytic anemia.

VIII. Erectile Dysfunction

Thiazides, like other blood pressure lowering drugs, can be associated with impotence as observed in long-term trials (WEISS 1991; GREENBERG et al. 1984; SMITH and TALBERT 1986; BEELEY 1984; REYES et al. 1983). Less common diuretic-related side effects include dizziness, lethargy, dry mouth, palpitations, orthostatic hypertension, headache, skin rash, nausea and muscle cramps. Elderly hypertensive patients have not appeared to suffer more side effects than younger patients (SYSTOLIC HYPERTENSION IN THE ELDERLY PROGRAM 1985; HYPERTENSION DETECTION AND FOLLOW-UP PROGRAM COOPERATIVE GROUP 1979; EUROPEAN WORKING PARTY ON HIGH BLOOD PRESSURE IN THE ELDERLY 1985).

I. Drug Combinations

Thiazide diuretics are often used in conjunction with other agents in combination therapy (BLACK 1990; LARDINOIS and NEUMAN 1988). When used in combination with β-blockers, glucose intolerance appears to be worse than with either agent alone (BLACK 1991; BENGTSSON et al. 1984; HELGELAND et al. 1984; DORNHURST et al. 1985; GREENBERG et al. 1984). The combination with ACE inhibitors appears to reduce the degree of glucose intolerance and of hypokalemia, thus providing a favorable metabolic effect (WEINBERGER 1983; HERRERA-ACOSTA et al. 1985; COSTA et al. 1988; AMBROSIONI et al. 1987). The combination may have a possible synergistic effect in reducing diastolic blood pressure (DE QUATTRO 1991). When used in combination with loop diuretics thiazides are more effective than loop diuretics alone in mobilizing edema (see Sects. F.III, F.IV).

References

Aces RP (1983) Metabolic disturbances increasing risk of coronary heart disease during diuretic-based antihypertensive therapy: lipid alterations and glucose intolerance. Am Heart J 106:1207–1214

Agus ZS, Chiu PJS, Goldberg M (1977) Regulation of urinary calcium excretion in the rat. Am J Physiol 232:F545–F549

Allen RC (1983) Sulfonamide diuretics. In: Cragoe EJ Jr (ed) Diuretics. Chemistry, pharmacology, and medicine. Wiley, New York, pp 49–200

Allen JM, Hind CRK, McMichael HB (1981) Synergistic action of metolazone with "loop" diuretics (letter). Br Med J 282:1873

Alpern J (1987) Apical membrane chloride/base exchange in the rat proximal convoluted tubule. J Clin Invest 79:1026–1030

Alvo M, Calamia J, Eveloff J (1985) Lack of potassium effect on Na-Cl cotransport in the medullary thick ascending limb. Am J Physiol 249:F34–F39

Ambrosioni E, Borghi C, Costa FV (1987) Captopril and hydrochlorothiazide: rationale for their combination. Br J Clin Pharmacol 23 [Suppl 1]:43S–50S

Anagnostopoulos T, Planelles G (1987) Cell and luminal activities of chloride, potassium, sodium and protons in the late distal tubule of Necturus kidney. J Physiol (Lond) 393:73–89

Anderson KV, Brettell HR, Aikawa JK (1961) ^{14}C-labeled hydrochlorothiazide in human beings. Arch Intern Med 107:736–742

Andersson OK, Gudbrandsson T, Jamerson K (1991) Metabolic adverse effects of thiazide diuretics: the importance of normokalaemia. J Intern Med 229:89–96

Antonin KH, Antonin E, Möhrke W, Mutschler E, Völger KD (1982) Veränderungen der Pharmakokinetik von Triamteren und Hydrochlorothiazid nach ein- und mehrmaliger Gabe an Patienten mit akuten und chronischen Lebererkrankungen. Therapiewoche 32:3905–3917

Asscher AW (1974) Treatment of frusemide resistant oedema with metolazone. Clin Trials J (London) 4:134–139

Ballantyne D (1990) Long-term effects of antihypertensives on blood lipids. J Hum Hypertens 4 [Suppl 2]:35–37

Bamford JM (1981) Synergistic action of metolazone with "loop" diuretics (letter). Br Med J 283:618

Basile AS, Lueddens HW, Skolnick P (1988) Regulation of renal peripheral benzodiazepine receptors by anion transport inhibitors. Life Sci 42:715–726

Beaudry C, Laplante L (1973) Severe allergic pneumonitis from hydrochlorothiazide. Ann Intern Med 78:251–253

Beaumont K, Healy DP, Fanestil DD (1984) Autoradiographic localization of benzodiazepine receptors in the rat kidney. Am J Physiol 247:F718–F724

Beaumont K, Vaughn DA, Fanestil DD (1988) Thiazide diuretic drug receptors in rat kidney: identification with [^{3}H]metolazone. Proc Natl Acad Sci USA 85: 2311–2314

Beaumont K, Vaughn DA, Healy DP (1989a) Thiazide diuretic receptors: autoradiographic localization in rat kidney with [^{3}H]metolazone. J Pharmacol Exp Ther 250:414–419

Beaumont K, Vaughn DA, Maciejewski AR, Fanestil DD (1989b) Reversible downregulation of thiazide diuretic receptors by acute renal ischemia. Am J Physiol 256:F329–F334

Beeley L (1984) Drug-induced sexual dysfunction and infertility. Adverse Drug React Acute Poisoning Rev 3:23–42

Beermann B (1984) Aspects on pharmacokinetics of some diuretics. Acta Pharmacol Toxicol 54 [Suppl 1]:17–29

Beermann B, Groschinsky-Grind M (1977) Pharmacokinetics of hydrochlorothiazide in man. Eur J Clin Pharmacol 12:297–303

Beermann B, Groschinsky-Grind M (1978) Antihypertensive effect of various doses of hydrochlorothiazide and its relation to plasma level of the drug. Eur J Clin Pharmacol 13:195–201

Beermann B, Groschinsky-Grind M (1978) Gastrointestinal absorption of hydrochlorothiazide enhanced by concomitant intake of food. Eur J Clin Pharmacol 13:125–128

Belair E, Kaiser F, Van Deuberg B, Borelli A, Lawler R, Panasevich R, Yelnosky J (1969) Pharmacology of SR 720–22. Arch Int Pharmacodyn Ther 177:71–70

Bengtsson C, Blohme G, Lapidus L, Lindquist O, Lundgren H, Nystrom E, Petersen K, Sigurdsson JA (1984) Do antihypertensive drugs precipitate diabetes? Br Med J 289:1495–1497

Bennett WM, McDonald WJ, Kuehnel E, Hartnett MN, Porter GA (1977) Do diuretics have antihypertensive properties independent of natriuresis? Clin Pharmacol Ther 22:499–504

Berger BE, Warnock DG (1986) Clinical uses and mechanisms of action of diuretic agents. In: Brenner BM, Rector FC (eds) The kidney. Saunders, Philadelphia, pp 433–455

Berliner RW, Dirks JH, Cirksena WJ (1966) Action of diuretics in dogs studied by micropuncture. Ann NY Acad Sci 139:424–432

Beyer KH, Baer JE (1975) The site and mode of action of some sulfonamide-derived diuretics. Med Clin North Am 59:735–750

Beyer KH, (1958) The mechanism of action of chlorothiazide. Ann NY Acad Sci 71:363–379

Beyer KH, Baer JE (1961) Physiological basis for the action of newer diuretic agents. Pharmacol Rev 13:517–562

Birkenhäger WH (1990) Diuretics and blood pressure reduction: physiologic aspects. J Hum Hypertens 8:S3–S7

Biron P, Dessureault J, Napke E (1991) Acute allergic interstitial pneumonitis induced by hydrochlorothiazide (published erratum appears in Can Med Assoc J 1991 Sep 1;145(5):391). Can Med Assoc J 145:28–34

Black HR (1990) The coronary artery disease paradox: the role of hyperinsulinemia and insulin resistance-implications for therapy. J Cardiovasc Pharmacol 15: 26–S38

Black HR (1991) Metabolic considerations in the choice of therapy for the patient with hypertension. Am Heart J 121:707–715

Black WD, Shiner PT, Roman J (1978) Severe electrolyte disturbances associated with metolazone and furosemide. South Med J 71:381–385

Boesken WD, Hirth W, Strupp R, Intemann H, Oser B, Stauß H, Knauf H (1995) Low-dose diuretic combination therapy in patients on chronic dialysis. Kidney Int (Submitted for publication)

Bricker NS, Fine LG, Kaplan MA, Epstein M, Bourgoignie JJ, Lich A (1978) "Magnification phenomenon" in chronic renal disease. N Engl J Med 299: 1287–1293

Cafruny EJ, Ross C (1962) Involvement of the distal tubule in diuresis produced by benzothiadiazines. J Pharmacol Exp Ther 137:324

Calder JA, Schachter M, Sever PS (1992) Direct vascular actions of hydrochlorothiazide and indapamide in isolated small vessels. Eur J Pharmacol 220: 19–26

Cantiello H, Copello J, Muller A, Mikulic L, Villami MF (1986) Effect of bumetanide on potassium transport and ionic composition of the arterial wall. Am J Physiol 251:F537–F546

Carlson JE, Kober L, Torp-Pedersen C, Johansen P, Moser M (1990) Relation between dose of bendrofluazide, antihypertensive effect, and adverse biochemical effects. Br Med J 300:975–978

Carney S, Gillies AI, Morgan T (1976) Optimal dose of a thiazide diuretic. Med J Aust 2:692–693

Chen T-M, Abdelhameed MH, Chiou WL (1992) Erythrocytes as a total barrier for renal excretion of hydrochlorothiazide: slow influx and efflux across erythrocyte membranes. J Pharm Sci 81:212–218

Chen ZF, Vaughn DA, Beaumont K, Fanestil DD (1990) Effects of diuretic treatment and of dietary sodium on renal binding of ^{3}H-metolazone. J Am Soc Nephrol 1:91–98

Coe FL (1980) Clinical stone disease. In: Brenner BM (ed) Nephrolithiasis: contemporary issues in nephrology. Churchill Livingstone, New York

Coe FL (1981) Prevention of kidney stones. Am J Med 71:514

Coe FL, Kavalach AG (1974) Hypercalciuria and hyperuricosuria in patients with calcium nephrolithiasis. N Engl J Med 291:1344

Cooling MJ, Sim MF (1978) Effects of prostaglandin synthase inhibition on natriuresis induced by diuretics and sodium loading in the rat. Br J Pharmacol 64: 439P–440P

Cooper JA Jr, Matthay RA (1987) Drug-induced pulmonary disease. Dis A Month 33:61–120

Costa FV, Borghi C, Mussi A, Ambrosioni E (1988) Hypolipidemic effects of long-term antihypertensive treatment with captopril. A prospective study. Am J Med 84:159–161

Costanzo LS (1984) Comparison of calcium and sodium transport in early and late rat distal tubules: effect of amiloride. Am J Physiol 246:F937–F945

Costanzo LS (1985) Localization of diuretic action in microperfused rat distal tubules: Ca and Na transport. Am J Physiol 248:F527–F535

Costanzo LS (1988) Mechanism of action of thiazide diuretics. Semin Nephrol 8:234–241

Costanzo LS, Windhager EE (1978) Calcium and sodium transport by the distal convoluted tubule of the rat. Am J Physiol 235:F492–F506

Cousin JL, Motai R (1976) The role of carbonic anhydrase inhibitors on anion permeability into ox red blood cells. J Physiol (Lond) 256:61–80

Crayen ML, Thoenes W (1978) Architecture and cell structures in the distal nephron of the rat kidney. Eur J Cell Biol 17:197–211

Cremaschi D, Porta C (1992) Sodium salt neutral entry at the apical membrane of the gallbladder epithelium: comparing different species. Comp Biochem Physiol [A] 103A:619–633

Cremaschi D, Meyer G, Bottá G, Rossetti C (1987a) The nature of the neutral Na^{+}-Cl^{-}-coupled entry at the apical membrane of rabbit gallbladder epithelium: II. Na^{+}-Cl^{-}-symport is independent of K. J Membr Biol 95:219–228

Cremaschi D, Meyer G, Rossetti C, Bottá G, Palestini P (1987b) The nature of the neutral Na^{+}-Cl^{-}-coupled entry at the apical membrane of rabbit gallbladder epithelium: I. Na^{+}/H^{+}, Cl^{-}/HCO_3^{-} double exchange and Na^{+}-Cl^{-} symport. J Membr Biol 95:209–218

Cremaschi D, Porta C, Bottà G, Meyer G (1992) Nature of the neutral Na^{+}-Cl^{-} coupled entry at the apical membrane of rabbit gallbladder epithelium: IV. Na^{+}/H^{+}, Cl^{-}/HCO_3^{-} double exchange, hydrochlorothiazide-sensitive Na^{+}-Cl^{-} symport and Na^{+}-K^{+}-$2Cl^{-}$ cotransport are all involved. J Membr Biol 129: 221–235

Dawson DC, Andrew D (1979) Differential inhibition of NaCl absorption and short-circuit current in the urinary bladder of the winter flounder, Pseudopleuronectes amer. Bull Mt Desert Isl 19:46–49

Dequattro V (1991) Comparison of benazepril and other antihypertensive agents alone and in combination with the diuretic hydrochlorothiazide. Clin Cardiol 14:28–32
DeStevens G (1963) Diuretics. Chemistry and pharmacology. Academic, New York
Dollery CT (1989) Diuretic agents and beta-blockers in the treatment of hypertension. Hypertension 13:I62–I65
Dollery CT, Harington M, Kaufmann G (1959) The mode of action of chlorothiazide in hypertension: with special reference to potentiation of ganglion-blocking drugs. Lancet I:1215–1218
Dorn MR, Walker BK (1981) Non-cardiogenic pulmonary edema associated with hydrochlorothiazide therapy. Chest 79:482–483
Dornhurst A, Powell SH, Pensky J (1985) Aggravation by propranolol of hyperglycemic effect of hydrochlorothiazide in type II diabetics without alteration of insulin secretion. Lancet I:123–126
Drewnowska K, Baumgarten CM (1991) Regulation of cellular volume in rabbit ventricular myocytes: bumetanide, chlorothiazide, and ouabain. Am J Physiol Cell Physiol 260:C122–C131
Duffey ME, Frizzell RA (1984) Flounder urinary bladder: mechanism of inhibition by hydrochlorothiazide (HCT). Fed Proc 43:444
Dyckner T, Wester PO (1981) Effects of magnesium infusions in diuretic induced hyponatremia. Lancet 1:585–586
Dyckner T, Helmers C, Lundman T, Wester PO (1975) Initial serum potassium level in relation to early complications and prognosis in patients with acute myocardial infarction. Acta Med Scand 197:207–210
Earley LE, Kahn M, Orloff J (1961) The effects of infusions of chlorothiazide on urinary dilution and concentration in the dog. J Clin Invest 40:857
Egel J, Pfanstiel J, Puschett JB (1985) Diuretic effects on renal brush border membrane transport and metabolism. Life Sci 37:1675–1681
Ellison DH (1991) The physiological basis of diuretic synergism: its role in treating diuretic resistance. Ann Int Med 114:886–894
Ellison DH, Velázquez H, Wright FS (1985) Stimulation of distal potassium secretion by low lumen chloride in the presence of barium. Am J Physiol 248:F638–F649
Ellison DH, Velázquez H, Wright FS (1987) Thiazide sensitive sodium chloride cotransport in the early distal tubule. Am J Physiol 253:F546–F554
Ellison DH, Velázquez H, Wright FS (1989) Adaptation of the distal convoluted tubule of the rat. Structural and functional effects of dietary salt intake and chronic diuretic infusion. J Clin Invest 83:113–126
Ellison DH, Morrisey J, Desir GV (1991) Solubilization and partial purification of the thiazide diuretic receptor from rabbit renal cortex. Biochim Biophys Acta Bio Membr 1069:241–249
Ellison DH, Biemesderfer D, Morrisey J, Lauring J, Desir GV (1993) Immunocytochemical characterization of the high-affinity thiazide diuretic receptor in rabbit renal cortex. Am J Physiol Renal Fluid Electrolyte Physiol 264:F141–F148
Epstein M, Lepp BA, Hoffman DS, Levinson R (1977) Potentiation of furosemide by metolazone in refractory edema. Curr Ther Res 21:656–667
Ericson A-C, Spring KR (1982) Coupled NaCl entry into Necturus gallbladder epithelial cells. Am J Physiol 243:C140–C145
Eriksson Ö, Wistrand PJ (1987) A search for a model tissue for studying effects of thiazide diuretics. Acta Physiol Scand 129:171–179
European Working Party on High Blood Pressure in the Elderly (1985) Mortality and morbidity results from the European working party on high blood pressure in the elderly. Lancet I:1349–1354
Eveloff J, Calamia J (1986) Effect of osmolarity on cation fluxes in medullary thick ascending limb cells. Am J Physiol 250:F176–F180
Eveloff JL, Warnock DG (1987) Activation of ion transport systems during cell volume regulation. Am J Physiol 252:F1–F10

Fernandez PC, Puschett JB (1973) Proximal tubular actions of metolazone and chlorothiazide. Am J Physiol 225:954–961

Ferriola PC, Acara MA, Duffey ME (1986) Thiazide diuretics inhibit chloride absorption by rabbit distal colon. J Pharmacol Exp Ther 238:912–915

Fichman MP, Vorherr H, Kleeman CR, Telfer N (1971) Diuretic-induced hyponatremia. Ann Intern Med 75:853–863

Field MJ, Lawrence JR (1986) Complications of thiazide diuretic therapy: an update. Med J Aust 144:641–644

Flack JM, Grimm RH (1989) Diuretics. In: Kaplan NM, Brenner BM, Laragh JH (eds) New therapeutic strategies in hypertension Raven, New York, pp 17–32

Flückiger VE, Schalch W, Taeschler M (1963) Das neue Salureticum Brinaldix (DT-327). Schweiz Med Wochenschr 93:1232–1237

Freis ED (1990) The cardiotoxicity of thiazide diuretics: review of the evidence. J Hypertens 8:S23–S32

Freis ED, Wanko A, Wilson IM, Parish AE (1958) Chlorothiazide in hypertensive and normotensive patients. Ann NY Acad Sci 71:450–455

Friedman E, Shadel M, Halkin H, Farfel Z (1989) Thiazide-induced hyponatremia. Reproducibility by single dose rechallenge and an analysis of pathogenesis. Ann Intern Med 110:24–30

Fröhlich ED (1987) Diuretics in hypertension. J Hypertens [Suppl] 5:S43–S49

Fröhlich ED, Schnaper HW, Wilson IM, Freis ED (1960) Hemody namic alterations in hypertensive patients due to chlorothiazide. N Engl J Med 262:1261–1263

Gamba G, Saltzberg SN, Lombardi M, Miyanoshita A, Lytton J, Hediger M, Brenner B, Hebert SC (1993) Primary structure and functional expression of a cDNA encoding the thiazide-sensitive sodium-chloride cotransporter. Proc Natl Acad Sci USA 90:2749–2753

Garg LC, Narang N (1987) Effects of hydrochlorothiazide on Na-K-ATPase activity along the rat nephron. Kidney Int 31:918–922

Garvin JL, Spring KR (1992) Regulation of apical membrane ion transport in Necturus gallbladder. Am J Physiol Cell Physiol 263:C187–C193

Gesek FA, Friedman PA (1992) Mechanism of calcium transport stimulated by chlorothiazide in mouse distal convoluted tubule cells. J Clin Invest 90:429–438

Gifford RW (1986) Role of diuretics in treatment of essential hypertension. Am J Cardiol 58:15A–17A

Good DW, Wright FS (1979) Luminal influences on potassium secretion: sodium concentration and fluid flow rate. Am J Physiol 236:F192–F205

Goodman Gilman (1990) The pharmacological basis of therapeutics, 8th edn. MacMillan, New York

Grantham J (1973) Sodium transport in isolated renal tubules. In: Lant EF, Wilson GM (eds) Modern diuretic therapy in the treatment of cardiovascular and renal disease. Exerpta Medica Foundation, Amsterdam, pp 220–228

Grantham JJ, Chonko AM (1991) Renal handling of organic anions and cations; metabolism and excretion of uric acid. In: Brenner BM, Rector FC (eds) The kidney. Saunders, Philadelphia, pp 483–509

Greenberg G, Brennan PJ, Miall WE (1984) Effects of diuretic and beta-blocker therapy in the medical research council trial. Am J Med 76:45–51

Greger R, Schlatter E (1981) Presence of luminal K^+, a prerequisite for active NaCl transport in the cortical thick ascending limb of Henle's loop of rabbit kidney. Pflugers Arch 392:92–94

Greger R, Schlatter E (1983a) Properties of the basolateral membrane of the cortical thick ascending limb of Henle's loop of rabbit kidney. Pflugers Arch 396: 325–334

Greger R, Schlatter E (1983b) Properties of the lumen membrane of the cortical thick ascending limb of Henle's loop of rabbit kidney. Pflugers Arch 396: 315–324

Greger R, Velázquez H (1987) The cortical thick ascending limb and early distal convoluted tubule in the urinary concentrating mechanism. Kidney Int 31: 590–596

Grimm RH Jr (1991) Antihypertensive therapy: taking lipids into consideration. Am Heart J 122:910–918

Grimm RH Jr, Leon AS, Hunninghake DB, Lenz K, Hannan P, Blackburn H (1981) Effects of thiazide diuretics on plasma lipids and lipoproteins in mildly hypertensive patients: a double-blind controlled trial. Ann Intern Med 94:7–11

Gunstone RF, Wing AJ, Shani HGP, Njemo D, Sabuka EMW (1971) Clinical experience with metolazone in fifty-two African patients: synergy with frusemide. Postgrad Med J 47:789–793

Haddy FJ, Pamnani MB, Swindall BT, Johnston J, Cragoe EJ (1985) Sodium channel blockers are vasodilators as well as natriuretic and diuretic agents. Hypertension 7 [Suppl I]:121–126

Hansen LL, Schilling AR, Wiederholt M (1981) Effect of calcium, furosemide and chlorothiazide on net volume reabsorption and basolateral membrane potential of the distal tubule. Pflugers Arch 389:121–126

Hansson L, Dahlöf B (1990) What are we really achieving with long-term antihypertensive drug therapy? In: Laragh JH, Brenner BM (eds) Hypertension, vol 2. Raven, New York, pp 2131–2142

Heidland A, Teschner M, Götz R, Heidbrender E (1985) Indications for high-dose furosemide treatment – developing a concept of therapy and present range of indications. Nieren- und Hochdruckkrankheiten 14:208–217

Helderman JH, Elahi D, Andersen DK, Raizes GS, Tobin JD, Shoken D, Andres R (1983) Prevention of the glucose intolerance of thiazide diuretics by maintenance of body potassium. Diabetes 32:106–111

Helgeland A, Leren P, Foss OP, Hjermann I, Lund-Larson PG (1984) Serum glucose levels during long term observation of treated and untreated men with mild hypertension: the Oslo Study. Am J Med 76:802–805

Hempelmann FW (1975) Die Proteinbindung von Xipamid (4-Chlor-5-Sulfamoyl-2′-6′-salicyloxylidid). Arzneimittel forsch (Drug Res) 25:258–259

Hempelmann FW, Dieker P (1977) Untersuchungen mit Xipamid (4-Chlor-5-sulfamoyl-2′,6′-salicycloxylidid), Teil II: Pharmakokinetik beim Menschen. Arzneimittelforsch (Drug Res) 27:2143–2151

Herrera-Acosta J, Perez-Grovas H, Fernandez M, Arriaga J (1985) Enalapril in essential hypertension. Drugs 30:35–46

Hierholzer K (1985) Sodium reabsorption in the distal tubular system. In: Seldin DW, Giebisch G (eds) The kidney: physiology and pathophysiology. Raven, New York

Hoffmann EK (1986) Anion transport systems in plasma membrane of vertebrate cells. Biochim Biophys. Acta 864:1–31

Hoffmann EK, Sjoholm C, Simonsen LO (1983) Na, Cl cotransport in Ehrlich ascites tumor cells activated during volume regulation (regulatory volume increase). J Membr Biol 76:269–280

Holland OB, Nixon JV, Kuhnert L (1981) Diuretic-induced ventricular ectopic activity. Am J Med 70:762–768

Hollifield JW (1986) Thiazide treatment of hypertension. Effects of thiazide diuretics on serum potassium, magnesium, and ventricular ectopy. Am J Med 80:8–12

Hollifield JW (1989) Electrolyte disarray and cardiovascular disease. Am J Cardiol 63:21B–26B

Hunninghake DB (1991) Effects of celiprolol and other antihypertensive agents on serum lipids and lipoproteins. Am Heart J 121:696–701

Hypertension Detection and Follow-up Program Cooperative Group (1979) Five-year findings of the hypertension detection and follow-up program: II. Mortality by race, sex and age. JAMA 242:2572–2577

Kaissling B (1982) Structural aspects of adaptive changes in renal electrolyte excretion. Am J Physiol 243:F211–F226

Kaissling B (1985) Structural adaptation to altered electrolyte metabolism by cortical distal segments. Fed Proc 44:2710–2716

Kaissling B, LeHir M (1982) Distal tubular segments of the rabbit kidney after adaptation to altered Na- and K-intake. Cell Tissue Res 224:469–492

Kaissling B, Stanton BA (1988) Adaption of distal tubule and collecting duct to increased sodium delivery. I. Ultrastructure. Am J Physiol 255:F1258–F1268

Kaissling B, Bachmann S, Kriz W (1985) Structural adaptation of the distal convoluted tubule to prolonged furosemide treatment. Am J Physiol 248:F374–F381

Kao LC, Warburton D, Cheng MH Cedeno C, Platzker ACG, Keens TG (1984) Effect of oral diuretics on pulmonary mechanics in infants with chronic bronchopulmonary dysplasia: results of a double blind crossover sequential trial. Pediatric 74:37–44

Kaplan NM (1990) Clinical hypertension, 5 th edn. Williams & Wilkins, Baltimore

Karniski LP, Aronson PS (1985) Chloride/formate exchange with formic acid recycling: a mechanism of active chloride transport across epithelial membranes. Proc Natl Acad Sci USA 82:6362–6365

Karniski LP, Aronson PS (1987) Anion exchange pathways for chloride transport in rat renal microvillus membranes. Am J Physiol 253:F513–F521

Kau ST (1988) Diuretics: what we have and what we need. In: Reyes AJ, Leary WP (eds) Progress in pharmacology: clinical pharmacology and therapeutic uses of diuretics, vol 6. Fischer, Stuttgart, pp 1–63

Kavaru MS, Ahmad M, Amirthalingam KN (1990) Hydrochlorothiazide-induced acute pulmonary edema. Cleve Clin J Med 57:181–184

Kempson SA, Kowalski JC, Puschett JB (1983) Direct effect of metolazone on sodium-dependent transport across the renal brush border membrane. J Lab Clin Med 101:308–316

Kirchner KA, Brandon S, Mueller RA, Smith MJ, Bower JD (1987) Mechanism of attenuated hydrochlorothiazide response during indomethacin administration. Kidney Int 31:1097–1103

Knauf H, Mutschler E (1984) Pharmacodynamics and pharmacokinetics of xipamide in patients with normal and impaired kidney function. Eur J Clin Pharmacol 26:687–693

Knauf H, Mutschler E (1989) The Na load in the nephron segment determines the ceiling quality of a diuretic. In: Puschett JB, Greenberg A (eds). Diuretics III, chemistry, pharmacology, and clinical applications. Elsevier, Amsterdam, pp 359–362

Knauf H, Mutschler E (1990) Saluretic effects of the loop diuretic torasemide in chronic renal failure. Interdependence of electrolyte excretion. Eur J Clin Pharmacol 39:337–343

Knauf H, Mutschler E (1991) Pharmacodynamic and kinetic considerations on diuretics as a basis of differential therapy. Klin Wschr 69:239–250

Knauf H, Muschler E (1992a) Constant K^+/Na^+ excretion ratio by piretanide during peak diuresis, but insignificant K^+ loss during 24 hours. Eur J Clin Pharmacol 43:23–27

Knauf H, Mutschler E (1992b) Wirkprofil. In: Knauf H, Mutschler E (eds) Diuretika, 2nd edn. Urban & Schwarzenberg, Munich, pp 189–213

Knauf H, Mutschler E (1993a) Low-dose segmental blockade of the nephron rather than high-dose diuretic monotherapy. Eur J Clin Pharmacol 44 [Suppl 1]:63–68

Knauf H, Mutschler E (1993b) K^+ loss induced by diuretics. In: Puschett JB (ed) Diuretics IV, Elsevier, Amsterdam, pp 253–256

Knauf H, Mutschler E (1994) Functional state of the nephron and diuretic dose response – rationale for low-dose combination therapy. Cardiology 84 [Suppl 2]:18–26

Knauf H, Mutschler E (1995) Effectiveness of hydrochlorothiazide and furosemide alone and in combination in chronic renal failure. J Cardiovasc Pharmacol (in press)

Knauf H, Kölle EU, Mutschler E (1990a) Gemfibrozil absorption and elimination in kidney and liver disease. Klin Wochenschr 68:692–698
Knauf H, Gerok W, Mutschler E, Schölmerich J, Spahn H, Wietholtz H (1990b) Disposition of xipamide in liver cirrhosis. Clin Pharm Ther 48:628–632
Knauf H, Wenk E, Schölmerich J, Goerk KJ, Gerok W, Leser HG, Mutschler E (1990c) Prediction of diuretic mobilization of cirrhotic ascites by pretreatment fractional sodium excretion. Klin Wochenschr 68:545–551
Knauf H, Gerok W, Mutschler E (1992) Pharmakokinetik. In: Knauf H, Mutscher E (eds) Diuretika, 2nd edn. Urban & Schwarzenberg, Munich, pp 149–187
Knauf H, Cawello W, Schmidt G, Mutschler E (1994) The saluretic effect of the thiazide-diuretic bemetizide in relation to the glomerular filtration rate. Eur J Clin Pharmacol 46:9–13
Kunau RT Jr, Weller DR, Webb HL (1975) Clarification of the site of action of chlorothiazide in the rat nephron. J Clin Invest 56:401–407
Lamberg B-A, Kuhlback B (1959) Effect of chlorothiazide and hydrochlorothiazide on the excretion of calcium in urine. Scand J Clin Lab Invest 11:351–357
Lant A (1985) Diuretics. Clinical pharmacology and therapeutic use, part I. Drugs 29:57–87
Laragh JH, Pecker MS (1983) Dietary sodium and hypertension. Ann Intern Med 77:1297–1305
Lardinois CK, Neuman SL (1988) The effects of antihypertensive agents on serum lipids and lipoproteins. Arch Int Med 148:1280–1288
Lassiter WC, Gottschalk CW, Mylle M (1963) Micropuncture study of renal tubular reabsorption of calcium in normal rodents. Am J Physiol 204:771–775
Le Hir M, Kaissling B, Dubach UC (1982) Distal tubular segments in the rabbit kidney after adaptation to altered Na- and K-intake. Changes in Na^+-K^+-ATPase activity. Cell Tissue Res 224:493–504
Leonetti G, Terzoli L, Bragato R (1989) Advantages and limitations of diuretic therapy in essential hypertension. Am J Hypertens 2:82S–85S
Loon NR, Wilcox CS, Unwin RJ (1989) Mechanism of impaired natriuretic response to furosemide during prolonged therapy. Kidney Int 36:682–689
Lukeman DS, Fanestil DD (1987) Interactions of diuretics with the peripheral-type benzodiazepine receptor in rat kidney. J Pharmacol Exp Ther 241:950–955
Luo H, Beaumont K, Vaughn DA, Fanestil DD (1990) Solubilization of thiazide diuretic receptors from rat kidney membranes. Biochim Biophys Acta Mol Cell Res 1052:119–122
Madias JR, Madias NE, Gavras HP (1984) Nonarrhythmogenicity of diuretic induced hypokalemia: its evidence in patients with uncomplicated hypertension. Arch Int Med 144:2171–2176
Maren TH (1976) Relations between structure and biological activity of sulfonamides. Annu Rev Pharmacol 16:309–327
Marone C, Muggli F, Lahn W, Frey FJ (1985) Pharmacokinetic and pharmacodynamic interaction between furosemide and metolazone in man. Eur J Clin Invest 15:253–257
Marumo F, Mishina T, Shimada H (1982) Effects of diazoxide and hydrochlorothiazide on water permeability and sodium transport in the frog bladder. Pharmacology 24:175–180
McDougal B, Sullivan LP (1970) Studies of the effect of bendroflumethiazide on sodium transport and metabolism in the toad bladder. J Pharmacol Exp Ther 172:203–210
Messerli FH, Frohlich ED, Dreslinski GR, Suarez DH, Aristimuno GE (1980) Serum uric acid in essential hypertension: an indicator of renal vascular involvement. Ann Int Med 93:817–821
Meyer G, Bottà G, Rossetti C, Cremaschi D (1990) The nature of the neutral Na^+-Cl^- coupled entry at the apical membrane of rabbit gallbladder epithelium: III. Analysis of transports on membrane vesicles. J Membr Biol 118:107–120

Micromedex Computerized Clinical Information System (1994) vol 80, Micromedex, Denver
Molony DA, Jacobson HR (1985) Comparison of distal convoluted tubule with cortical collecting tubule via electrophysiological methods – evidence for distal nephron heterogeneity. Kidney Int 27:317
Moore PF (1968) The effects of diazoxide and benzothiadiazine diuretics upon phosphodiesterase. Ann NY Acad Sci 150:256–260
Morsing P, Velázquez H, Wright FS, Ellison, DH (1991) Adaptation of distal convoluted tubule of rats. II. Effects of chronic thiazide infusion. Am J Physiol 261:F137–F143
Moser M (1989a) Relative efficacy of, and some adverse reactions to, different antihypertensive regimens. Am J Cardiol 63:2B–7B
Moser M (1989b) Lipid abnormalities and diuretics. Am Fam Phys 40:213–220
Moser M (1990) Antihypertensive medications: relative effectiveness and adverse reactions. J Hypertens 8:S9–S16
Moser M (1994) Effect of diuretics on morbidity and mortality in the treatment of hypertension. Cardiology 84 [Suppl 2]:27–35
Murphy RJF (1950) The effect of "rice diet" on plasma volume and extracellular fluid space in hypertensive subjects. J Clin Invest 29:912–917
Murphy MB, Lewis PJ, Kohner E, Schumer B, Dollery CT (1982) Glucose intolerance in hypertensive patients treated with diuretics, a fourteen-year follow-up. Lancet 2:1293–1295
Musch MW, Field M (1989) K-independent Na-Cl cotransport in bovine tracheal epithelial cells. Am J Physiol 256:C658–C665
Mutschler E, Gilfrich HJ, Knauf H, Möhrke W, Völger KD (1983) Pharmakokinetik von Triamteren bei Probanden und Patienten mit Leber- und Nierenfunktionsstörungen. Klin Wochenschr 61:683–691
Niemeyer C, Hasenfuß G, Wais U, Knauf H, Schäfer-Korting M, Mutschler E (1983) Pharmacokinetics of hydrochlorothiazide in relation to renal function. Eur J Clin Pharmacol 24:661–665
Nonoguchi H, Sands JM, Knepper MA (1989) ANF inhibits NaCl and fluid absorption in cortical collecting duct of rat kidney. Am J Physiol 256:F179–F186
Nordrehaug JE, Von Der Lippe G (1983) Hypokalaemia and ventricular fibrillation in acute myocardial infarction. Br Heart J 50:525–529
Novello FC, Sprague JM (1957) Benzothiadiazine dioxides as novel diuretics. J Am Chem Soc 79:2028–2029
Olesen KH, Sigurd B (1971) The supra-additive natriuretic effect of addition of quinethazone or bendroflumethiazide during long-term treatment with furosemide and spironolactone: permutation trial tests in patients with congestive heart failure. Acta Med Scand 190:233–240
Okusa MD, Persson AEG, Wright FS (1989) Chlorothiazide effect on feedback-mediated control of glomerular filtration rate. Am J Physiol 257:F137–F144
Okusa MD, Velázquez H, Ellison DH, Wright FS (1990) Luminal calcium regulates potassium transport by the renal distal tubule. Am J Physiol 258:F423–F428
Orgison JL (1962) Failure of chlorothiazide to influence tissue electrolytes in hypertensive and non-hypertensive nephrectomized dogs. Proc Soc Exp Biol Med 110:161–164
O'Neil RG, Sansom SC (1984) Characterization of apical cell membrane Na^+ and K^+ conductances of cortical collecting duct using microelectrode techniques. Am J Physiol 247:F14–F24
Oster JR, Epstein M, Smoler S (1983) Combined therapy with thiazide-type and loop diuretic agents for resistant sodium retention. Ann Intern Med 99:405–406
Ostergaard EH, Magnussen MP, Nielsen CK, Eilertsen E, Frey H-H (1972) Pharmacological properties of 3-*n*-butylamino-4-phenoxy-5-sulfamylbenzoic acid (bumetanide), a new potent diuretic. Drug Res 22:66–72

Palfrey MC, Feit PW, Greengard P (1980) cAMP-stimulated cation cotransport in avian erythrocytes: inhibition by "loop" diuretics. Am J Physiol 238:C139–C148
Papademetriou V, Fletcher R, Khatri IM, Freis ED (1983) Diuretic-induced hypokalemia in uncomplicated systemic hypertension: effect of plasma potassium correction in cardiac arrhythmias. Am J Cardiol 52:1017–1022
Papademetriou V, Burris JF, Notargiacomo A, Fletcher RD, Freis ED (1988) Thiazide therapy is not a cause of arrhythmia in patients with systemic hypertension. Arch Int Med 148:1272–1276
Pecker MS (1990) Pathophysiologic effects and strategies for long-term diuretic treatment of hypertension. In: Laragh JH, Brenner BM (eds) Hypertension, vol 2. Raven, New York, pp 2143–2167
Pendleton RG, Sullivan LP, Tucker JM, Stephenson RE (1968) The effect of a benzothiadiazide on the isolated toad bladder. J Pharmacol Exp Ther 164: 348–361
Peters G, Roch-Ramel F (1969) Thiazide diuretics and related drugs. In: Herken H (ed) Diuretics. Springer, Berlin Heidelberg New York, pp 257–385 (Handbook of experimental pharmacology, vol 24)
Pizzonia JH, Gesek FA, Kennedy SM, Coutermarsh BA, Bacskai BJ, Friedman PA (1991) Immunomagnetic separation, primary culture, and characterization of cortical thick ascending limb plus distal convoluted tubule cells from mouse kidney. In Vitro Cell Dev Biol 27A:409–416
Planelles G, Anagnostopoulos T (1992) Thiazide-sensitive Na-Cl cotransport mediates NaCl absorption in amphibian distal tubule. Pflugers Arch 421:307–313
Pollare T, Lithell H, Berne C (1989) A comparison of the effects of hydrochlorothiazide and captopril on glucose and lipid metabolism in patients with hypertension. N Engl J Med 321:868–873
Poujeol P, Chabardes D, Rionel N, Derouffignac C (1976) Influence of extracellular fluid volume expansion on magnesium, calcium and phosphate handling along the rat nephron. Pflugers Arch 365:203–211
Reif MC, Troutman SL, Schafer JA (1986) Sodium transport by rat cortical collecting tubule. J Clin Invest 77:1291–1298
Renfro JL (1978) Interdependence of active Na^+ and Cl^- transport by the isolated urinary bladder of the teleost, Pseudopleuronectes americanus. J Exp Zool 199:383–390
Reyes AJ, Olhaberry JV, Leary WP, Lockett CJ, Van Der Byl K (1983) Urinary zinc excretion, diuretics, zinc deficiency and some side effects of diuretics. S Afr Med J 64:936–941
Rouch AJ, Chen L, Troutman SL, Schafer JA (1991) Na^+ transport in isolated rat CCD: effects of bradykinin, ANP, clonidine, and hydrochlorothiazide. Am J Physiol 260:F86–F95
Rudy DW, Voelker JR, Green PK, Esparza FA, Brater DC (1991) Loop diuretics for chronic renal insufficiency: a continuous infusion is more potent than bolus therapy. Ann Intern Med 115:360–366
Rush MG, Engelhardt B, Parker RA, Hazinski TA (1990) Double blind, placebo-controlled trial of alternate-day furosemide therapy in infants with chronic bronchopulmonary dysplasia. J Pediatr 117:112–118
Russell JG, Mayhew SR, Humphries IS (1981) Chlorthalidone in mild hpypertension: dose-response relationship. Eur J Clin Pharmacol 20:407–411
Sabatini S, Kurtzman NA (1988) Thiazides stimulate calcium absorption in the turtle bladder. Kidney Int 33:954–958
Scherzer P, Wald H, Popovizer M (1987) Enhanced glomerular filtration and Na^+-K^+-ATPase with furosemide administration. Am J Physiol 252:F910–F915
Schild L, Giebisch GH, Karniski LP, Aronson PS (1987) Effect of formate on volume reabsorption in the rabbit proximal tubule. J Clin Invest 79:32–38
Schlatter E, Schafer JA (1987) Electrophysiological studies in principal cells of rat cortical collecting tubules: ADH increases the apical membrane Na^+ conductance. Pflugers Arch 409:81–92

Schlatter E, Greger R, Weidtke C (1983) Effect of "high ceiling" diuretics on active salt transport in the cortical thick ascending limb of Henle's loop of rabbit kidney. Correlation of chemical structure and inhibitory potency. Pflugers Arch 396:210–217
Schlatter E, Greger R, Schafer JA (1990) Principal cells of cortical collecting ducts of the rat are not a route of transepithelial Cl^- transport. Pflugers Arch 417: 317–323
Schlatter E, Salomonsson M, Persson AEG, Greger R (1990) Macula densa cells reabsorb NaCl via furosemide sensitive Na^+-K^+-$2Cl^-$ cotransport. In: Puschett JB, Greenberg A (eds) Diuretics III; chemistry, pharmacology, and clinical applications. Elsevier, Amsterdam, pp 756–758
Schnaper HW, Freis ED, Friedman RG, Garland WT, Hall WD, Hollifield J, Jain AK, Jenkins P, Marks A, Mcmahon FG, Sambol NC, Williams RL, Winer N (1989) Potassium restoration in hypertensive patients made hypokalemic by hydrochlorothiazide. Arch Intern Med 149:2677–2681
Schnermann J, Steipe B, Briggs JP (1987) In situ studies of distal convoluted tubule in rat. II. K^+ secretion. Am J Physiol 252:F970–F976
Schnieders JR, Ludens JH (1980) Comparison of effects of standard diuretics and indanone in isolated toad cornea and bladder. Am J Physiol 238:R70–R75
Schoenfeld MR, Gold Berger E (1964) Hypercholesterolemia induced by thiazides: a pilot study. Curr Ther Res 6:180–184
Schrier RW (1988) Pathogenesis of sodium and water retention in high and low output cardiac failure, cirrhosis, nephrotic syndrome, and pregnancy. N Engl J Med 319:1065–1072
Schrier RW, Arroyo V, Bernardi M, Epstein M, Henriksen JH, Rodes J (1988) Peripheral arterial vasodilation hypothesis: a proposal for the initiation of renal sodium and water retention in cirrhosis. Hepatology 8:1151–1158
Scott JY, Brunner SL, Tanaka DT (1992) The effect of chlorothiazide on neurally mediated contraction of rabbit bronchial smooth muscle. Am Rev Respir Dis 145:75–79
Scriabine A, Watson LS, Fanelli GM, Shum WK, Blaine EH, Russo Hfand Bokider NR (1980) Studies on the interaction of indomethacin with various diuretics. In: Scriabine A, Lifter AM, Kuehl FA (eds) Prostaglandins in cardiovascular and renal function. Spectrum Publications, Holliswood, New York, pp 471–483
Seely JF, Dirks JH (1977) Site of action of diuretic drugs. Kidney Int 11:1
Shah S, Khatri I, Freis ED (1978) Mechanism of antihypertensive effect of thiazide diuretics. Am Heart J 95:611–618
Shetty BV, Campanella LA, Thomas TL, Fedorchuk M, Davidson TA, Michelson L, Volz H, Zimmermann SE, Belair EJ, Truant A (1970) Synthesis and activity of some 3-aryl- and 3-aralkyl-1,2,3,4-tetrahydro-4-oxo-6-quinazolinesulfonamides. J Med Chem 13:886–895
Shimizu T, Nakamura M (1992) Ouabain-induced cell swelling in rabbit connecting tubule: evidence for thiazide-sensitive Na^+-Cl^- cotransport. Pflugers Arch 421:314–321
Shimizu T, Yoshitomi K, Nakamura M, Imai M (1988) Site and mechanism of action of trichlormethiazide in rabbit distal nephron segments perfused in vitro. J Clin Invest 82:721–730
Shimizu T, Yoshitomi K, Taniguchi J, Imai M (1989) Effect of high NaCl intake on Na^+ and K^+ transport in the rabbit distal convoluted tubule. Pflugers Arch 414:500–508
Shimizu T, Nakamura M, Imai M (1991) Effect of S-8666 on Cl transport in the rabbit connecting tubule perfused in vitro. Tohoku J Exp Med 164:293–298
Sigurd B, Olesen KH, Wennevold A (1975) The supra-additive natriuretic effect addition of bendroflumethiazide and bumetanide in congestive heart failure. Am Heart J 89:163–170
Skorecki KL, Brenner BM (1982) Body fluid homeostasis in congestive heart failure and cirrhosis with ascites. Am J Med 72:323–338

Smith PJ, Talbert RL (1986) Sexual dysfunction with antihypertensive and antipsychotic agents. Clin Pharm 5:373–384
Slatopolsky E, Elken I, Weerts C, Bricker NS (1968) Studies on the characteristics of the control system governing sodium excretion in uremic man. J Clin Invest 47:521–530
Spahn H, Reuter K, Mutschler E, Gerok W, Knauf H (1987) Amiloride pharmacokinetics in renal and hepatic disease. Eur J Clin Pharmacol 33:493–498
Sprague JM (1958) The chemistry of diuretics. Ann NY Acad Sci 71:328–343
Sprague JM (1968) Diuretics. In: Rabinowitz JL, Myerson RM (eds) Topics in medicinal chemistry, vol 2. Wiley, New York, pp 1–63
Stanton BA (1988) Electroneutral NaCl transport by distal tubule: evidence for Na^+/H^+-Cl^-/HCO_3^- exchange. Am J Physiol 254:F80–F86
Stanton BA (1990) Cellular actions of thiazide diuretics in the distal tubule. J Am Soc Nephrol 1:832–836
Stanton BA, Kaissling B (1988) Adaptation of distal and collecting duct to increased sodium delivery. II Na^+ and K^+ transport. Am J Physiol 255:F1269–F1275
Stanton BA, Omerovic A, Koeppen BM, Giebisch GH (1987) Electroneutral H^+ secretion in distal tubule of Amphiuma. Am J Physiol 252:F691–F699
Steinberg AD (1968) Pulmonary edema following ingestion of hydrochlorothiazide. JAMA 204:825–827
Stenger EG, Wirz H, Pulver R (1959) Hygroton (G 33182) ein neues Salidiuretikum mit protrahierter Wirkung. Schweiz Med Wochenschr 89:1126–1130
Stokes JB (1984) Sodium chloride absorption by the urinary bladder of the winter flounder: a thiazide-sensitive electrically neutral transport system. J Clin Invest 74:7–16
Stoner LC, Burg MB, Orloff J (1974) Ion transport in cortical collecting tubule; effect of amiloride. Am J Physiol 227:453–459
Strykers PH, Stern RS, Morse BM (1984) Hyponatremia induced by a combination of amiloride and hydrochlorothiazide. JAMA 252:389
Sullivan LP, Pirch JH (1966) Effect of bendroflumethiazide on distal nephron transport of sodium, potassium and chloride. J Pharmacol Exp Ther 151:168
Sullivan LP, Tucker JM, Scherbenske MJ (1971) Effect of furosemide on sodium transport and metabolism in toad bladder. Am J Physiol 220:1316–1324
Sullivan TJ (1991) Cross-reactions among furosemide, hydrochlorothiazide, and sulfonamides. JAMA 265:120–121
Sullivan TJ, Ong RC, Gilliam LK (1989) Studies of the multiple drug allergy syndrome. J Allergy Clin Immunol 83:270
Systolic Hypertension in the Elderly Program (1985) Antihypertensive efficacy of chlorthalidone. Am J Cardiol 56:913–920
Tanner GA, Maxwell WR, McAteer JA (1992) Fluid transport in a cultured cell model of kidney epithelial cyst enlargement. J Am Soc Nephrol 2:1208–1218
Taylor A, Windhager EE (1979) Possible role of cytosolic calcium and Na^+-Ca^+ exchange in regulation of transepithelial sodium transport. Am J Physiol 236: F505–F512
Terada Y, Knepper MA (1990) Thiazide-sensitive NaCl absorption in rat cortical collecting duct. Am J Physiol 259:F519–F528
Tobian L (1967) Why do diuretics lower blood pressure in essential hypertension? Annu Rev Pharmacol 7:399–408
Tomita K, Pisano JJ, Knepper MA (1985) Control of sodium and potassium transport in the cortical collecting duct of the rat. J Clin Invest 76:132–136
Tomita K, Pisano JJ, Burg MB, Knepper MA (1986) Effects of vasopressin and bradykinin on anion transport by the rat cortical collecting duct. J Clin Invest 77:136–141
Tran JM, Farrell MA, Fanestil DD (1990) Effect of ions on binding of the thiazide-type diuretic metolazone to kidney membrane. Am J Physiol 258:F908–F915

Ullrich KJ, Baumann K, Löschke K, Rumrich G, Stolte H (1966) Micropuncture experiments with saluretic sulfonamides. Ann NY Acad Sci 139:416–423
van Brummelen P, Man in't Veld AJ, Schalekamp MADH (1980) Hemodynamic changes during long-term thiazide treatment of essential hypertension in responders and nonresponders. Clin Pharmacol Ther 27:328–336
Velázquez H, Greger R (1985) K^+ and Cl^- permeabilities in cells of the rabbit early distal convoluted tubule. Kidney Int 27:322
Velázquez H, Greger R (1986) Influences on basolateral membrane K^+ conductance of cells of early distal convoluted tubule. Kidney Int 29:409
Velázquez H, Wright FS (1986a) Control by drugs of renal potassium handling. Annu Rev Pharmacol Toxicol 26:293–309
Velázquez H, Wright FS (1986b) Effects of diuretic drugs on Na^+, Cl^-, and K^+ transport by rat renal distal tubule. Am J Physiol 250:F1013–F1023
Velázquez H, Wright FS, Good DW (1982) Luminal influences on potassium secretion: chloride replacement with sulfate. Am J Physiol 242:F46–F55
Velázquez H, Good DW, Wright FS (1984) Mutual dependence of sodium and chloride absorption by renal distal tubule. Am J Physiol 247:F904–F911
Velázquez H, Ellison DH, Wright FS (1987) Chloride dependent potassium secretion in early and late renal distal tubules. Am J Physiol 253:F555–F562
Velázquez H, Ellison DH, Wright FS (1988) Multiple pathways for potassium transport across basolateral membrane of rabbit distal convoluted tubule cell. Kidney Int 33:429
Velázquez H, Ellison DH, Wright FS (1991) Similarity of rabbit CNT and DCT cells in culture. J Am Soc Nephrol 2:754
Velázquez H, Ellison DH, Wright FS (1992) Luminal influences on potassium secretion: chloride, sodium and thiazide diuretics. Am J Physiol 262:F1076–F1082
Veterans Administration Cooperative Study Group on Antihypertensive Agents (1982) Comparison of propranolol and hydrochlorothiazide for the initial treatment of hypertension. II. Results of long-term therapy. JAMA 248:2004–2011
Villareal H, Revollo A, Exaire JE, Larondo F (1966) Effects of chlorothiazide on renal hemodynamics. Circulation 26:409–412
Vulliemoz Y, Verosky M, Triner L (1980) Effect of benzothiadiazine derivatives on cyclic nucleotide phosphodiesterase and on the tension of the aortic strip. Blood Vessels 17:91–103
Walser M (1971) Calcium-sodium interdependence in renal transport. In: Fischer JW (ed) Renal pharmacology. Appleton-Century-Crofts, New York, pp 21–41
Walter SJ, Shirley DG (1986) The effect of chronic hydrochlorothiazide administration on renal function in the rat. Clin Sci 70:379–387
Wang T, Giebisch G, Aronson PS (1992) Effects of formate and oxalate on volume absorption in rat proximal tubule. Am J Physiol Renal Fluid Electrolyte Physiol 263:F37–F42
Wang T, Agulian SK, Giebisch G, Aronson PS (1993) Effects of formate and oxalate on chloride absorption in rat distal tubule. Am J Physiol 264:F730–F736
Wareing M, Green R (1992) Effect of changes in oxalate concentration on fluid reabsorption in the proximal convoluted tubule of the anaesthetized rat. J Physiol (Lond) 452:80P
Weddington WW, Mulroy MF, Sandri SR (1973) Pneumonitis and hydrochlorothiazide. Ann Intern Med 79:283
Weidmann P (1980) Recent pathogenic aspects in essential hypertension and hypertension associated with diabetes mellitus. Klin Wschr 58:1971
Weinberger MH (1983) Influence of an angiotensin-converting enzyme inhibitor on diuretic-induced metabolic effects in hypertension. Hypertension 5:132–138
Weiss RJ (1991) Effects of antihypertensive agents on sexual function. Am Fam Phys 44:2075–2082

Welling PG (1986) Pharmacokinetics of the thiazide diuretics. Biopharm Drug Dispos 7:501–535
White JP, Ward MJ (1985) Drug-induced adverse pulmonary reactions. Adverse Drug React Acute Poisoning Rev 4:183–211
Wilcox CS (1991) Diuretics. In: Brenner BM, Rector FC (eds) The kidney. WB Saunders, Philadelphia, pp 2123–2147
Wilson IM, Freis IM (1959) Relationship between plasma and extracellular fluid volume depletion and the antihypertensive effect of chlorothiazide. Circulation 20:1028–1036
Wilson DR, Honrath U, Sonnenberg H (1983) Thiazide diuretic effect on medullary collecting duct function in the rat. Kidney Int 23:711–716
Wilson DR, Honrath U, Sonnenberg H (1988) Interaction of amiloride and hydrochlorothiazide with atrial natriuretic factor in the medullary collecting duct. Can J Physiol Pharmacol 66:648–654
Winer BM (1961) The antihypertensive mechanism of salt depletion induced by hydrochlorothiazide. Circulation 24:788–796
Wollam GL, Tarazi RC, Bravo EL, Dustan HP (1981) Diuretic potency of combined hydrochlorothiazide and furosemide therapy in patients with azotemia. Am J Med 72:929–936
Wollam GL, Tarazi RC, Bravo EL, Dusten HP (1982) Diuretic potency of combined hydrochlorothiazide and furosemide therapy in patients with azotemia. Am J Med 72:929–938
Wright FS (1982) Flow dependent transport processes: filtration, absorption, secretion. Am J Physiol 243:F1–F11
Yendt ER, Cohanim M (1978) Prevention of calcium stones with thiazides. Kidney Int 13:397
Yoshitomi K, Shimizu T, Taniguchi J. Imai M (1989) Electrophysiological characterization of rabbit distal convoluted tubule cell. Pflugers Arch 414:457–463
Ziyadeh FN, Kelepouris E, Agus ZS (1987) Thiazides stimulate calcium absorption in urinary bladder of winter flounder. Biochim Biophys Acta 897:52–56

CHAPTER 9

Potassium-Retaining Diuretics: Aldosterone Antagonists

H. ENDOU and M. HOSOYAMADA

A. Chemical Structure and Properties, Structure-Activity Relationships

I. Introduction

Aldosterone antagonists are compounds which decrease the biological responses induced by aldosterone. The main effect of aldosterone is the enhancement of urinary sodium absorption and potassium excretion. Thus, aldosterone antagonists increase urinary sodium excretion and decrease potassium excretion (HERKEN 1969).

SALA and LUETSCHER (1954) found that cortisone inhibited the antinatriuretic effect of injected aldosterone in adrenalectomized rats. LANDAU et al. (1955) reported that progesterone blocked the sodium-retaining action of aldosterone in a patient with Addison's disease. Although progesterone possesses only 25% of the affinity of aldosterone for the mineralocorticoid receptor, it competitively inhibits the action of the native mineralocorticoid on sodium retention, invoking the renin-angiotensin system. In contrast, synthetic progestagens do not possess hypotensive action and exhibit negligible affinity for the mineralocorticoid receptor. Nevertheless, R-5020, the ideal synthetic progestagen, remains an important tool for the molecular differentiation of the mineralocorticoid receptor. However, none of these steroids is clinically relevant due to their strong progestomimetic action.

CELLA and KAGAWA (1957) studied the urinary sodium/potassium concentration (Na^+/K^+) ratio in 4-h urine samples from adrenalectomized rats as a screening test for mineralocorticoid and antimineralocorticoid activity (Kagawa's test), and demonstrated that some steroidal derivatives act by direct competition with aldosterone at the target cell level. These aldosterone antagonists have been called spirolactones and synthesized from Searle laboratories as SC compounds.

Spironolactone (SC9420), one of the spirolactones, has been used in antimineralocorticoid therapy since 1960, and has kept its value as an orally active drug. But its endocrine side effects, gynecomastia and impotence observed in man, make it difficult to use in long-term therapy. Thus, the search for new aldosterone antagonists has continued over 30 years, and the newly discovered mineralocorticoid receptor antagonists are categorized into the following groups (Fig. 1):

1. Modifications in 17α side chain: SC compounds
2. Alkylation in the 7α-position of the spiroether: RU compounds
3. 15β,16β-methylene spirolactones: ZK compounds
4. Other recently synthesized compounds

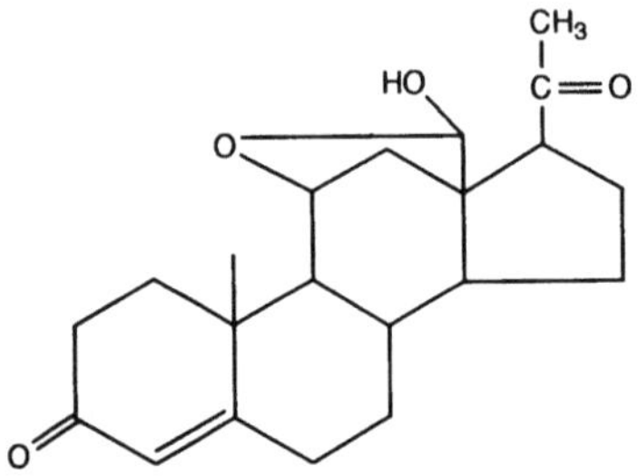

Aldosterone

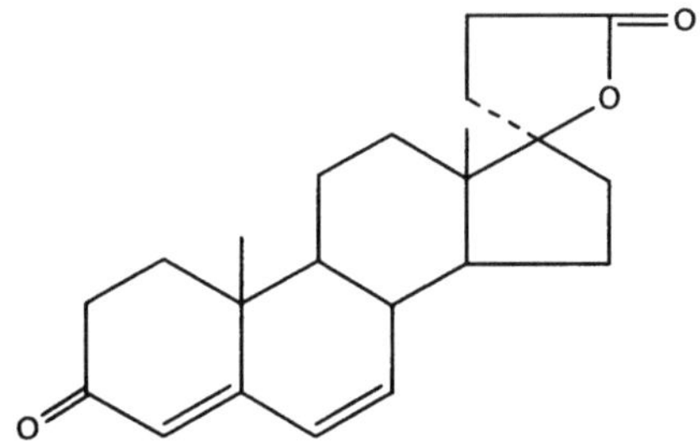

Canrenone (SC9376)

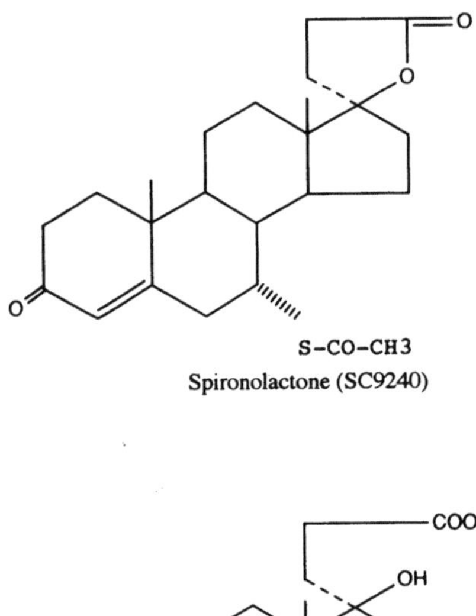

Spironolactone (SC9240)

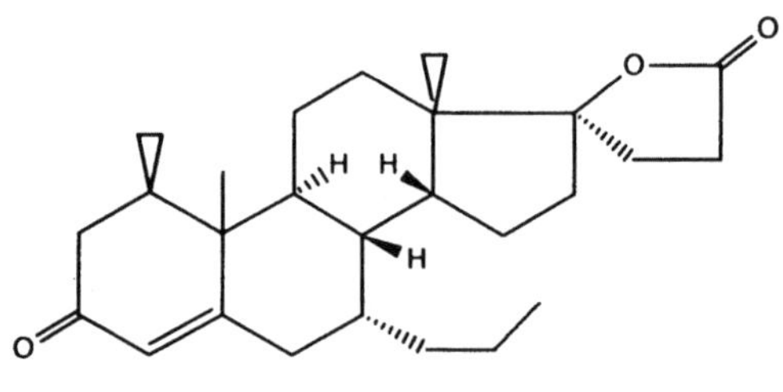

RU26752

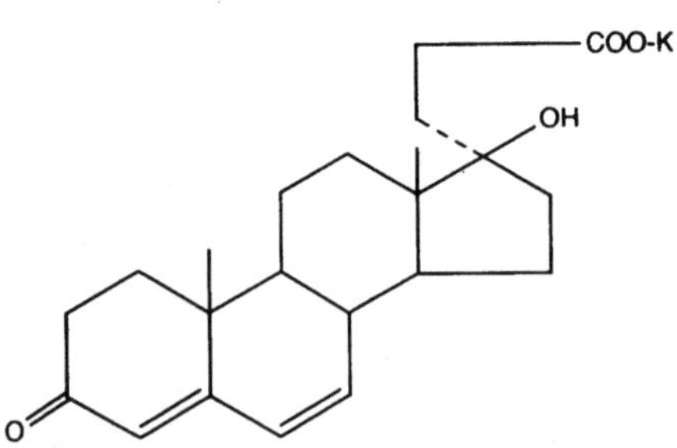

Potassium canrenoate (SC14226)

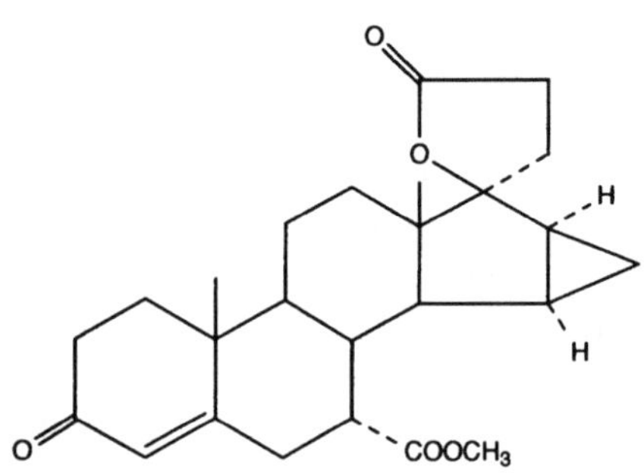

ZK91587

Fig. 1. Structural formulas of antimineralocorticoids in relation to the natural hormone aldosterone

II. Chemical Structure and Properties

1. Modifications of 17α Side Chain: SC Compounds

Adrenocorticosteroids have four rings in their structure – A, B, C, and D – which are not in a flat plane (Fig. 2). The planarity of the valence angles about the double bond between C-4 and C-5 prevents the chair form of ring A from being an energetically favorable conformational state, and is responsible for ring A being in a half-chair conformation. This 4,5 double bond and the 3-ketone are both necessary for typical adrenocorticosteroid activity. With respect to the mineralocorticoid receptor, it has been proposed that the binding of a steroid to the receptor is the result of a tight association between the receptor and the steroidal ring A (DUAX and GRIFFIN 1989).

Opening of the 17-lactone ring of spironolactone, leading to canrenone or prorenone, dramatically diminished their affinities for mineralocorticoid receptor in vitro; yet the antimineralocorticoid activity of the 17α-hydroxypropyl or the 17β-hydroxyl derivatives remained unimpaired in vivo (SAKAUYE and FELDMAN 1976). However, shortening or lengthening of the 17α side chain resulted in concurrent loss of in vivo and in vitro activities. PETERFALVI et al. (1980) demonstrated the critical importance of the presence of the lactone ring for the binding of spirolactones to the receptors with methylated canrenoate, an inactive canrenoate analog in the kidney (Fig. 3). Introduction of a hydrogen bond acceptor, such as an oxygen atom, in the 17β-position of ring D in the spiroether retained both the affinity for the mineralocorticoid receptor in vitro and antimineralocorticoid activity in vivo (NEDELEC et al. 1986).

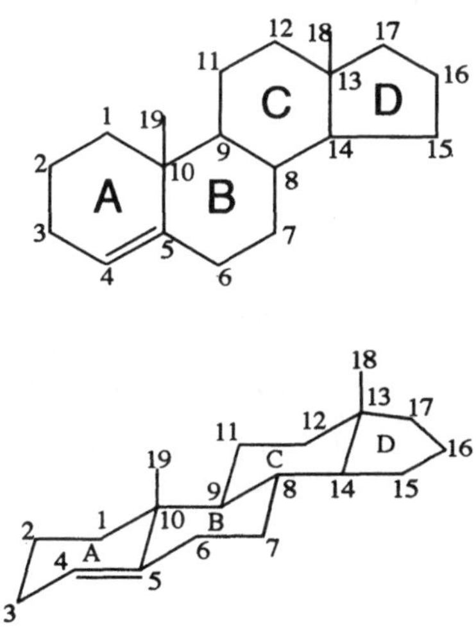

Fig. 2. Basic structures of adrenocorticosteroids

STEROIDS	a. ACTITITY IN VIVO	b. BINDING IN VITRO
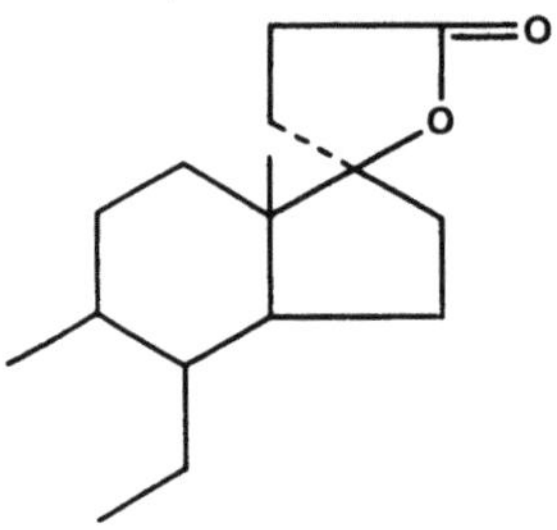 Spironolactone	[5 mg/kg] 96%	24%
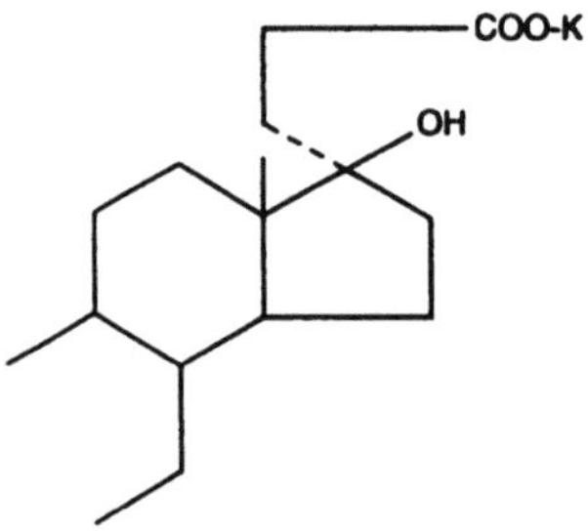 Potassium Canrenoate	[10 mg/kg] 63%	0.06%
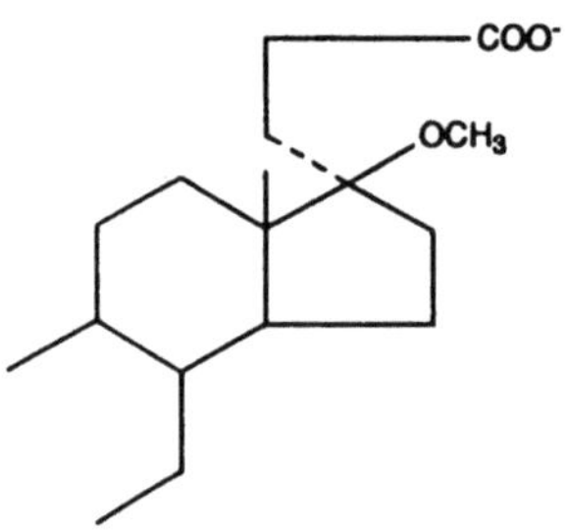 17-O-Methyl 5.6 dihydrocanrenoic acid	[20 mg/kg] 0%	0%

Fig. 3. Comparison of in vivo activity and in vitro binding of three spironolactones: as percentage competitive potency of aldosterone taken as reference and as percentage increase in sodium-potassium ratio compared with the aldosterone-treated reference group. The 100% activity corresponds to the concentration ratio of unlabeled to ^{3}H-aldosterone needed to obtain 50% inhibition of binding to mineralocorticoid sites. (Data are from PETERFALVI et al. 1980)

2. Structural Modification of Ring B: RU26752 and RU28318

Derivatives of spiroether alkylated in the 7α-position have been synthesized by Roussel-Uclaf. 7α-Propylspirolactone (RU26752) and its corresponding salt (RU28318), which needs to be transformed in vivo to the lactone in RU 26752, appeared most promising in retaining mineralocorticoid receptor-mediated antagonist activity and yet largely devoid of the afore-mentioned side effects due to negligible affinity for the androgen and progesterone receptors.

3. Structural Modification in Ring D: Mespirenone (ZK94679) and ZK91587

In the series of 15β,16β-methylene spirolactones synthesized at Schering Laboratories, ZK94679 (Mespirenone) and ZK91587 (dethiolated Mespirenone) possessed two to four times the antimineralocorticoid activity of the parent compound, possibly due to increased affinity for the mineralocorticoid receptor, and negligible affinity for androgen and progesterone receptors (LOSERT et al. 1986).

Both RU26752 and ZK91587 have been shown to reverse a number of parameters of aldosterone-induced hypertension in the rat (KALIMI et al. 1990; OPOKU et al. 1991) and have been invaluable in the molecular differentiation of the mineralocorticoid receptor.

4. Recent Structural Modifications

A number of 19-nor analogues were tested by YAMAKAWA et al. (1987) for binding to mineralocorticoid receptors and it was found that their affinity was directly related to the flatness of ring A, but no information is available as to the biological activity.

11β,18-Epoxypregnane (derivative 31) was the most promising in the series synthesized at Shionogi Laboratories due to its low affinity for receptors other than the mineralocorticoid receptor, although it possesses some aldosterone agonistic activity (KAMATA et al. 1985).

III. Steroidogenesis Inhibitors and Secretion Inhibitors

Steroidogenesis inhibitors and secretion inhibitors have also widely been reported to block endogenous aldosterone effects. Spironolactone and other aldosterone antagonists can inhibit aldosterone production in adrenal tissue at a dose higher than the therapeutic dose.

As far as aldosterone secretion inhibitors are concerned, animal experiments have shown that atrial natriuretic peptide (GANGULY 1992), calcitonin gene-related peptide (MURAKAMI et al. 1989), and a novel neuropeptide in bovine chromaffin cells (NGUYEN et al. 1989) inhibited angiotensin II- or ACTH-stimulated aldosterone secretion.

B. Pharmacodynamics

I. Renal Effects

1. Increase in Urinary Sodium-Potassium Ratio

Because the renal action of aldosterone is the enhancement of reabsorption of Na^+ and the secretion of K^+ and H^+, aldosterone antagonists have an effect on Na^+ excretion and K^+ retention. Thus, they have been and are used as potassium-sparing diuretics. On the other hand, loop diuretics or thiazides increase Na^+ delivery to the distal nephron segment and delivered Na^+ enhances K^+ secretion in the collecting duct (HORISBERGER and GIEBISCH 1987).

The increase of the Na^+/K^+ ratio in 4-h urine samples from adrenalectomized rats has been used as a screening test for antimineralocorticoid activity (Kagawa's test). Rats are adrenalectomized a few days before the experiment and fludrocortisone substituted until the experiment. Na^+/K^+ ratios are studied in three experimental groups: the control group is infused with NaCl solution without aldosterone, the second group is infused with NaCl solution with 1 μg aldosterone/kg body weight/h, and the test group rats are infused with NaCl solution with 1 μg aldosterone/kg body weight per hour plus the test compound. The urinary Na^+/K^+ ratio of the second group decreases with respect to the control group, and the antimineralocorticoid activity of the test compound is shown as an increase in the ratio in the test group.

For most of the aldosterone antagonists, several milligrams compound/kg body weight per hour are required to antagonize 1 μg aldosterone/kg body weight per hour in vivo, i.e., aldosterone antagonists require 1000 times higher concentrations to antagonize aldosterone. However, they have K_d values for the mineralocorticoid receptor in vitro close to those of the mineralocorticoids.

2. Target Nephron Segments: CCT and OMCT, and Other Segments

The principal and best-documented renal target sites for the steroid aldosterone are the cortical and outer medullary collecting tubules (CCT and OMCT) (GROSS et al. 1975; SCHWARTZ and BURG 1978; STONE et al. 1983). In experiments with radio-ligand (DOUCET and KATZ 1981; FARMAN et al. 1982; BONVALET 1987), or with polyclonal antibodies (MIRSHAHI et al. 1992), the same segments were shown as target sites. At the CCT, the primary role of mineralocorticoids is to enhance Na^+ reabsorption and K^+ secretion, while in the OMCT, aldosterone stimulates H^+ secretion. There are three types of cell in these segments: principal cells, intercalated A cells, and intercalated B cells. Principal cells reabsorb Na^+ through apical sodium channels and basolateral $(Na^+{+}K^+)$-ATPase. Protons are excreted by intercalated B cells which have H^+-ATPase in the luminal membrane.

There is also indirect evidence to suggest that the medullary thick ascending limb of Henle (mTALH) (STANTON 1986; WORK and JAMISON 1987), the distal convoluted tubule (DCT) (FARMAN et al. 1982; ELLISON et al. 1989), the connecting tubule (CNT) (GARG et al. 1981) and the papillary collecting duct (PCD) (HIGASHIHARA et al. 1984) may also be aldosterone sensitive, although the exact role played by mineralocorticoids in these segments is less clear.

3. Intracellular Mechanism of Aldosterone Antagonists

In earlier studies it was shown that the natural hormone aldosterone binds to the renal mineralocorticoid receptor (MCR), the complex is translocated to the nucleus, and these receptors are competitively occupied by spirolactones. MARVER et al. (1974) first demonstrated that in the rat kidney a ^{3}H-spirolactone (SC26304) binds to cytoplasmic proteins, which are probably mineralocorticoid receptors (MCRs). CLAIRE et al. (1979) and ROSSIER et al. (1983) reported specific binding of ^{3}H-prorenone to rat kidney and toad urinary bladder, respectively. On the other hand, these investigators detected no translocation of antagonist receptor complexes into the nucleus. But in an autoradiographic study specific nuclear spironolactone labeling in rabbit CCT was found (BONVALET et al. 1991). It was, therefore, of interest to determine the steps responsible for the antagonist activity of spirolactones in the cascade of events from steroid binding to transcriptional activation.

In later studies, spirolactone was found to occupy sites distinct from aldosterone-specific receptors (LUZZANI and GLASSER 1984), still different from moieties specific to the progestin R5020 (AGARWAL and PAILLARD 1979).

In addition, the availability of chemically defined derivatives of spirolactone permitted precise molecular cartography of MCR. The 7α-propyl derivatives RU26752 interacted with two molecular species of MCR, one of which did not bind aldosterone at all. The affinity for both sites increased with the introduction of a methoxycarbonyl residue in the derivative ZK91587. No antagonist exhibited affinity for blood serum transcortin, and/or receptors in organs like the lung and the liver which are not mineralocorticoid targets. Both of these derivatives rendered the MCR unstable at 35°C, but important differences were observed in the activation of the MCR as evidenced by binding to DNA-cellulose, DEAE-cellulose, and during sucrose gradient analysis. It appeared that ZK91587 may act by inhibiting MCR activation but that RU26752 interferes with a step beyond activation (AGARWEL and KALIMI 1988).

Some organ-specific differences were also noted when heart was used instead of rat kidney (AGARWEL and KALIMI 1989), suggesting molecular heterogeneity that appeared steroid, organ, and species dependent, in contrast to the classical notion of a homogeneous, unitary receptor protein. However, multiple binding sites on the receptor protein could also give rise to the impression of heterogeneity (CHANG et al. 1989).

It has been well established that free steroid receptors are recovered in the cytosol of target cells associated with the 90-kDa heat shock protein (HSP90) (Baulieu 1987; Pratt 1990). These hetero-oligomeric complexes, which sedimented as 8–9S entities, are unable to interact with specific DNA sequences. Dissociation of HSP90-receptor complexes leads to a 4S form of the receptor capable of DNA binding and transcriptional regulation.

Synthetic glucocorticoid antagonist, RU486, keeps the glucocorticoid receptor in its 8S untransformed state. Aldosterone antagonists, however, facilitate HSP90 release from the steroid-binding subunit of the mineralocorticoid receptor. The inactivation of mineralocorticoid receptors might be related to the attenuated HSP90-MCR interaction induced by antimineralocorticoids leaving the receptor unprotected in the absence of HSP90 and susceptible to protease actions. Rafestin-Oblin et al. (1992) demonstrated that aldosterone antagonists dissociate much more rapidly from both rat and human mineralocorticoid receptors than does aldosterone, despite the similarity of their affinity constants.

II. Extrarenal Effects

1. Tissue Distribution of Type I Receptors

Lombes et al. (1990) have developed an immunohistochemical technique for the localization of mineralcorticoid receptors using an anti-idiopathic antibody, and detected mineralocorticoid receptors in the tunica media and adventitia of arteries (aorta and mesenteric vessels), colon, ureter, bladder and salivary grands of the rabbit, in addition to the kidney. In the brain, Grillo et al. (1990) studied the binding of [^{3}H]ZK91587, a specific mineralocorticoid receptor (type I) ligand, which binds to receptors located in the CA1 and CA3 regions of hippocampus, some amygdaloid nuclei and lateral septum.

2. Cross-reactivity with the Glucocorticoid Receptors

At first, it seems possible that aldosterone antagonists have extrarenal effects associated with these extrarenal mineralocorticoid receptors. But the absence of 11β-hydroxysteroid dehydrogenase and corticosteroid-binding globulin (CBG, transcortin), which protect the mineralocorticoid receptor in organs such as the kidney and parotid glands from excessive stimulation by the naturally occurring glucocorticoids, reduces the specificity of aldosterone to the mineralocorticoid receptor of target organs. This implies that aldosterone antagonists can reduce the type I receptor mediated glucocorticoid effect in target organs.

3. Epithelia

Epithelia, such as colon, sweat gland ducts, bile canalicula (Omland and Mathisen 1991) and endolymphatic sac of the ear (Mori et al. 1991) are also target organs of aldosterone. Interestingly, dexamethasone and RU28362, a pure type II receptor agonist, also have an aldosterone-like effect in the colon. Coadministration of spironolactone inhibited the effects of aldosterone on sodium and potassium transport, but in dexamethasone-treated animals spironolactone resulted in a pattern of response similar to that found in RU28362-treated animals (Binder et al. 1989).

4. Cardiovascular System

In nonepithelial tissues, mineralocorticoid receptors may be associated with functions other than sodium and potassium transport. In the cardiovascular system, spironolactone is effective in lowering blood pressure despite normal levels of aldosterone and without necessarily producing significant changes in sodium and potassium urinary excretion. Hence spironolactone has direct vasodilatory activity, which has been shown in clinical studies (Clement 1990; Lagrue et al. 1990), and in in vitro studies with portal vein strips. Thereby, spironolactone acts primarily on the plasma membrane by depressing an inward current through slow calcium channels, and may also depress contractions dependent on the release of calcium from the sarcoplasmic reticulum (Dacquet et al. 1987). Using the patch-clamp method, it was shown directly that spironolactone acts on slow calcium channels in a manner similar to that of calcium blockers (Mironneau 1990).

As the chemical structure of spironolactone and canrenone is similar to that of ouabain, they may also inhibit ($Na^{+}+K^{+}$)-ATPase. Some studies have shown the relationship of the hypotensive activity of aldosterone antagonists and their ouabain-like effect (De Mendonca et al. 1992).

In addition, myocardial fibrosis in experimental models of arterial hypertension in rats was prevented by administration of low doses of spironolactone (Brilla et al. 1993). This is an important finding because activation of the renin-angiotensin-aldosterone system in arterial hypertension can lead to remodeling of the myocardial collagen network with progressive collagen accumulation in the cardiac interstitium. This is one determinant of diastolic dysfunction and pathologic hypertrophy.

5. Central Nervous System

In the brain, corticosterone binds to both type I and type II receptors but with a difference in affinity of one order of magnitude. One of the consequences of the very high affinity of type I receptor in limbic neurons is that these receptors are more than 80% occupied by the endogenous hormone even under basal resting conditions. On the other hand, the occupancy of glucocorticoid receptors varies in parallel with changes in the plasma corticosterone level during circadian variations or following stress. Some biochemical and behavioral observations support the concept that tonic influence on brain functions is mediated through type I receptors at low basal corticosterone levels, and phasic responses to episodically elevated corticosterone are regulated through type II receptors (DE KLOET and REAL 1987).

The intraventricularly administered antimineralocorticoid RU28318, a pure type I receptor antagonist, as already mentioned above, elevates the basal plasma corticosterone level and enhances adrenocortical secretion following stress, but the antiglucocorticoid RU38486 pure type II receptor antagonist has no effect on the basal plasma corticosterone level in the morning; however, it interferes with a glucocorticoid negative feedback following stress (RATKA et al. 1989). It has also been reported that the intracerebroventricular infusion of RU28318, at doses that are ineffective when administered subcutaneously, inhibits the development of hypertension produced by subcutaneous infusion of aldosterone or deoxycorticosterone in normotensive rats (GOMEZ SANCHEZ et al. 1990) and Dahl S/JR genetically hypertensive rats (GOMEZ SANCHEZ et al. 1992).

6. Steroidogenesis Inhibition

Spironolactone inhibits adrenal and testicular cytochrome P450 (MENARD et al. 1979) and the terminal oxidases for several steroidogenic enzymes (HALL 1985). Mespirenone, the $\delta 1,2-15\beta,16\beta$-methylene derivate of spironolactone, also inhibits the production of aldosterone whereas the production of corticosterone is significantly elevated at 0.1 mmol/l.

The direct effects of spironolactone and 7α-thiospironolactone, a putative intermediate in the activation pathway, on the adrenal cortex, causing a decrease in ACTH-stimulated cortisol production, result from the selective inhibition of 17α-hydroxylation. Since 17α-hydroxylase activity is apparently required for the activation of both compounds, suicide inhibition of the enzyme may be the mechanism of action (ROURKE et al. 1991). Mespirenone blocks not only the 18-hydroxylase in the main pathway of aldosterone synthesis, but also interferes with an alternative pathway of aldosterone biosynthesis, i.e., that of 18- and 21-hydroxylation (WEINDEL et al. 1991).

Furthermore, it has been observed that spironolactone is an effective inducer of several hepatic phase I enzymes of drug metabolism (FELLER and GERALD 1971; STRIPP et al. 1971), and it also induces hepatic bilirubin UDP-glucuronyltransferase (SOLYMOSS and ZSIGMOND 1973). In the female but not in the male rat, *p*-nitrophenol UDP-glucuronyltransferase is induced by spironolactone in the intestine and liver (CATANIA et al. 1990).

7. Antiandrogen Effects

Some aldosterone antagonists have an antiandrogen effect. Thus spironolactone produces antiacne effects and has recently been shown to inhibit 5α-dihydrotestosterone (5α-DHT) receptors in human sebaceous glands (AKAMATSU et al. 1993).

C. Pharmacokinetics

I. Absorption

About 70% of orally administered spironolactone is absorbed. It has been reported that oral administration of a single dose of 200 mg spironolactone to healthy men showed a peak serum concentration of the unchanged drug at 1 h (OVERDIEK et al. 1985; DOIGNON et al. 1990; VARIN et al. 1992). This finding is in contrast to the widely accepted notion that spironolactone is metabolized too rapidly to be detected in serum after oral administration (ABSHAGEN et al. 1976). This discrepancy is based on differences in analytical methods of spironolactone and its metabolites. The former results were obtained by a specific and sensitive high-performance liquid chromatography (HPLC) method, whereas the later data were obtained by the fluorimetric method, in which not only spironolactone but also its metabolites such as canrenone, 7α-thiomethylspironolactone and 6β-hydroxy-7α-thiomethylspironolactone can be simultaneously measured.

Another active potassium agent canrenoate, when given orally, can be absorbed in a similar fashion and bypasses the liver unchanged (SADEE et al. 1973).

II. Plasma Concentrations

Specific and sensitive HPLC has made it possible to analyze steady-state serum concentrations of spironolactone and its metabolites.

According to GARDINER et al. (1989), 12 healthy males received 100 mg spironolactone, once daily, for 15 days. Peak serum concentrations of spironolactone were 72 $\pm$ 45 μg/l (mean $\pm$ SD) on day 1, and 80 $\pm$ 20 μg/l on day 15. The area under the curve (AUC) (0–24) values on day 15 were 231 $\pm$ 50 μg/l/h, and the post-steady-state elimination half-life was 1.4 $\pm$ 0.5 h. A similar result (C_{max}, 185 $\pm$ 51 μg/l; $t_{1/2}$, 1.3 $\pm$ 0.3 h; AUC (0–24), 473 $\pm$ 149 μg/l) was reported by OVERDIEK et al. (1985).

These results indicate that unmetabolized spironolactone is present in a reasonable quantity for a meaningful period.

III. Metabolism

1. Spironolactone and Canrenoate

Large amounts of absorbed spironolactone are metabolized in the liver during the first passage, and there is a hepatointestinal recirculation of spironolactone. Figure 4 shows a possible metabolic pathway of spironolactone (VARIN et al. 1992). In human subjects, the main metabolite of spironolactone is 7α-thiomethylspironolactone (II in Fig. 4) through 7α-thiospironolactone (I) (OVERDIEK et al. 1985; VARIN et al. 1992). The AUC

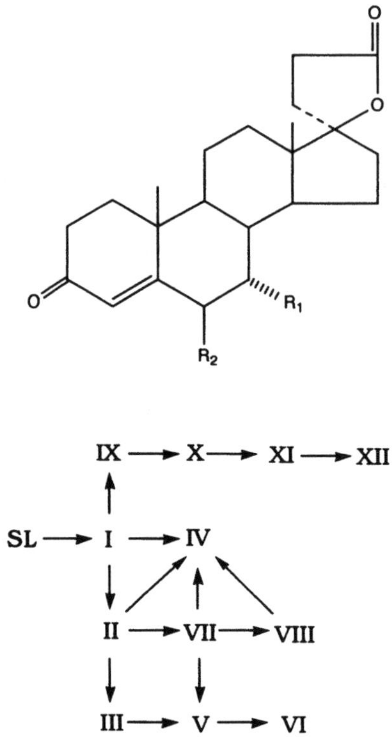

	R1	R2
SL:	$SCOCH_3$	H
I:	SH	H
II:	SCH_3	H
III:	SCH_3	OH
IV*:	H	H
V:	$SOCH_3$	OH
VI:	SO_2CH_3	OH
VII:	$SOCH_3$	H
VIII:	SO_2CH_3	H
IX:	S	H
X:	SOH	H
XI:	SO_2H	H
XII:	SO_3H	H

(* : 6-ene)

Fig. 4. Proposed pathways for the biotransformation of spirolactone (*SL*) (VARIN et al. 1992). Abbreviations: *I* = 7α-thio-SL; *II* = 7α-thiomethyl-SL; *III* = 6β-hydroxy-7α-thiomethyl-SL; *IV* = canrenone; *V* = 6β-hydroxy-7α-methylsulfonyl-SL; *VI* = 6β-hydroxy-7α-methylsulfone-SL; *VII* = 7α-methylsulfonyl-SL; *VIII* = 7α-methylsulfonyl-SL; *IX* = 7α-thio-SL; *X* = 7α-sulfonyl-SL; *XI* = 7α-sulfone-SL; *SII* = 7α-sulfone-SL. Canrenone may be further hydrolyzed (to canrenoate), reduced (to di-, tetra- and hexahydro- derivatives) or hydroxylated

(0–24) values of spironolactone and its metabolites after oral administration for 15 days were 231 ± 50 (μg/l/h, mean ± SD) for spironolactone (SL in Fig. 4), 2173 ± 312 h for canrenone (IV), 2804 ± 777 h for 7α-thiomethylspironolactone (II) and 1727 ± 367 h for 6β-hydroxy-7α-thiomethylspironolactone (III). Their post-steady-state half-life values were 1.4 ± 0.5 h, 16.5 ± 6.3 h, 13.8 ± 6.4 h, and 15.0 ± 4.0 h, respectively (GARDINER et al. 1989).

In rats, approximately 18% of an in vivo dose of spironolactone is metabolized to canrenone and related compounds, and in vitro 20%–30% of spironolactone is dithioacetylated to canrenone and its metabolites by liver 900 g supernatant (S9) (COOK et al. 1988).

Canrenoate is metabolized by rat hepatic S9 to canrenoic acid and 6α,7α- and 6β,7β-epoxy canrenone. The β-epoxide is further metabolized to its 3α- and 3β-hydroxy derivatives and its glutathione conjugate (Cook et al. 1988).

2. Spironolactone and Cytochrome P450 Destruction

Administration of spironolactone (SL) to experimental animals decreases the adrenal and testicular microsomal content of cytochrome P450, the terminal oxidase for several important steroid-producing enzymes, including the 17α- and 21-hydroxylases. A metabolite, 7α-thiol-SL, is an obligatory intermediate in the action of SL on adrenal monooxygenases but requires further metabolism for its toxicity. Hepatic cytochrome P450s are unaffected by SL, because 7α-thiol-SL is not metabolized by the liver (Sherry et al. 1986). Cook et al. (1988) hypothesized that spironolactone and its S-containing metabolites specifically inhibit an isoenzyme of cytochrome P450, or that spironolactone is a preferred substrate over canrenoate and/or canrenone for the metabolizing enzymes.

IV. Excretion

Although the exact ratios of metabolites of antialdosterones between urinary and biliary excretion have not been established, Sadee et al. reported in 1973 that 14%–24% of a given dose of spironolactone, canrenone or canrenoate is excreted into the urine within 5 days as measured by fluorescence assays, and 33% of an oral dose of potassium ^{3}H-canrenoate within 4 days as measured by radioactivity analysis.

D. Therapeutic Use (Indications, Dosage, Contraindications)

I. Indications

Because of their clinical usefulness, aldosterone antagonists have been searched for and investigated exhaustively, but only two drugs, spironolactone and potassium canrenoate, have been and are used clinically as antimineralocorticoid agents. Although both RU26752 and ZK91587 have potent antimineralocorticoid activity in vitro, both compounds have been abandoned as potential drugs for the treatment of hypertension. RU28318 entered clinical trials but its antimineralocorticoid effect in humans did not match its experimentally demonstrated potency in the rat. Spirorenone was severalfold more active than spironolactone, but it has been discontinued as a drug due to its progestational side effects (Agarwal and Lazar 1991).

The therapeutic use of spironolactone is based on two main effects. Firstly, it has the renal effects of increasing sodium and water excretion and reducing potassium excretion. Thus, it is used in the treatment of secondary hyperaldosteronism, such as refractory edema associated with congestive heart failure, ascites in cirrhosis of the liver, or in nephrotic syndrome. In the treatment of hypertension, spironolactone is not recommended as a drug of first choice. However, in the diagnosis and treatment of primary aldosteronism, spironolactone is used in a dose of 400 mg/day as an antihypertensive agent.

Secondly, spironolactone has antagonistic effects by inhibiting 5α-dihydrotestosterone (5α-DHT) receptors. Furthermore, because the antiandrogen drug cyproterone acetate is not readily available in some countries, spironolactone has been used in antiandrogen therapy: in the treatment of acne, hirsutism, premenstral syndrome and hormone therapy of prostatic carcinoma.

1. Congestive Heart Failure

WEBER and VILLARREAL (1993) recommend that anti-aldosterone therapy in patients with secondary hyperaldosteronism due to myocardial failure must do the following: reduce or preferably normalize plasma aldosterone levels by limiting synthesis, antagonize the renal and systemic effects of aldosterone at its receptor sites, and eliminate or minimize the multiple stimuli to aldosterone secretion. Some prospective open studies and other anecdotal reports have suggested that a combination of spironolactone and an angiotensin-converting enzyme inhibitor results in clinical improvement without serious hyperkalemia in heart failure patients who have normal renal function and cannot tolerate a further increase in the dose of angiotensin-converting enzyme inhibitor because of hypotension (ZANNAD 1993; DAHLSTROM and KARLSSON 1993).

2. Liver Cirrhosis

Only 10% of ascitic patients respond to dietary sodium restriction and bed rest. Most require pharmacotherapy consisting of spironolactone, which increases the proportion of responding patients to 65%, and loop diuretics, which may produce clinical improvement in an additional 20% (GERBES 1993).

Traditionally, spironolactone is used in doses of 25–50 mg four times daily, but ROCCO and WARE (1986) have shown that higher doses of 300–600 mg (and occasionally 800–1000 mg)/day are safe and effective.

3. Nephrotic Syndrome

Nephrotic patients, who have a normal glomerular-filtration rate, can show abnormal sodium retention, and they do not have high plasma renin activity and high plasma aldosterone because of their hypervolemic state. Angiotensin-converting-enzyme inhibitors failed to increase sodium excretion (BROWN et al. 1984). The aldosterone antagonist spironolactone, however, increased sodium excretion in sodium-loaded nephrotic patients but not in normal subjects (SHAPIRO et al. 1990).

4. Hypertension

In essential hypertension, spironolactone has been thought to be of no use as an antihypertensive agent because of its side effects (FERGUSON et al. 1977). But as mentioned above, spironolactone may have a vasodilating effect or an antihypertensive effect via its action in the central nervous system. Today many combination therapies, such as the combination with thiazide (DUEYMES 1990) or with angiotensin-converting enzyme inhibitors (SCHOHN et al. 1990), have been attempted.

5. Endocrine Disorders

a) *Primary Hyperaldosteronism*

In eight patients with primary aldosteronism due to adenomas, plasma aldosterone concentrations were not altered by spironolactone 75–225 mg daily. However, significant increases in plasma deoxycorticosterone were observed in the 3rd week, and increases in plasma corticosterone and cortisol together with suppressed plasma renin activity in the 2nd week. Low levels of serum potassium returned to the normal range in the 1st week. These effects suggest biosynthetic inhibition of corticosteroids by spironolactone at the sites of 18-hydroxylation and/or 18-oxidation, because of the elevation of deoxycorticosterone and corticosterone concentrations (SAKAMOTO et al. 1990).

b) *Acne*

In the treatment of androgen-mediated skin disorders such as acne (SHAW 1991) and rosacea (AIZAWA and NIIMURA 1992), spironolactone provides a valuable therapeutic option. As far as its topical use is concerned, spironolactone seems to be highly effective and devoid of systemic effects. Local mild side effects were present in a small number of patients (MESSINA et al. 1990).

c) *Hirsutism*

CROSBY and RITTMASTER (1991) reported that low-dose spironolactone improved hirsutism in the majority of hirsute women, irrespective of age, severity or duration of hirsutism, menstrual status and serum hormone levels.

6. Other Disorders

There are some other diseases that can be treated with spironolactone, for example, mental disorders (GILLMAN and LICHTIGFELD 1986), premenstrual syndrome (BURNET et al. 1991) and Bartter's syndrome (CLEMENTSEN et al. 1989).

II. Dosage

Spironolactone is usually given in an initial dose of 100 mg daily. Subsequently the dose may be increased as required. Some patients may require doses of up to 400 mg daily. For children, a suggested initial dose of spironolactone is 3 mg/kg body weight in divided doses.

Canrenoate has been given in doses of 50–200 mg daily or on alternate days in one to three divided doses; up to 300 mg daily has been given.

In patients with liver dysfunction dosage adjustment is probably not necessary because the metabolism of spironolactone is unaltered (ABSHAGEN 1977). Furthermore, although the mean elimination half-life of canrenone, is longer in patients with chronic liver disease than in healthy subjects, there was no evidence of accumulation of canrenone in plasma (JACKSON et al. 1977).

In renal failure, spironolactone is usually avoided because of the risk of hyperkalemia and a further decline in renal function (BAILY 1978). However, BENNET et al. (1983) suggested that it could be administered to the patient whose glomerular filtration rate was over 10 ml/min, provided the dosage interval was adjusted.

III. Contraindications

The contraindications of spironolactone are: (a) significantly impaired renal function (serum creatinine concentrations >18 mg/l), (b) acute renal insufficiency and (c) anuria.

E. Side Effects and Toxicology

I. General Considerations

Spironolactone may give rise to headache and drowsiness, and gastrointestinal disturbances including cramps and diarrhea. Ataxia, mental confusion, deepening of the voice, menstrual irregularities, impotence and skin rashes have been reported as further side effects. Gynecomastia is not uncommon and in rare cases breast enlargement may persist. Transient increases in blood-urea-nitrogen concentrations may occur and mild acidosis has been reported.

Spironolactone has been demonstrated to cause tumors in rats. Since several side effects are sequelae of the renal or extrarenal actions of aldosterone antagonists, readers are referred to the above sections on these topics.

II. Main Side Effects

1. Hyponatremia, Hyperkalemia and Acid-Base Disturbances

A survey indicated that of 788 patients who received spironolactone, 164 developed side effects (Greenblatt and Koch-Weser 1973). These included hyperkalemia in 8.6%, dehydration in 3.4% and hyponatremia in 2.4%. These effects of aldosterone antagonists can be explained not only by their aldosterone antagonism but also by the secondary seven- to ten fold increase in circulating atrial natriuretic peptide, as evaluated in the conscious dog (Bie et al. 1990).

Hyperchloremia, which is usually observed together with hyperkalemia, may occur when aldosterone antagonists are used as the sole diuretic agent, resulting in dose-dependent metabolic acidosis.

2. Sexual Functions and Endocrine Disorders

The use of spironolactone in the treatment of hypertension has been limited by the occurrence of sexual side effects, mainly menstrual disturbances in women and gynecomastia and/or impotence in men. In a study of relatively young female patients, aged 28 to 50 years, who had received treatment with spironolactone commencing at 200 mg daily for 1–45 months, side effects were disturbance of menstruation (66%), breast tenderness (30%) and breast enlargement (27%) (Hughes and Cunliffe 1987).

Although several explanations have been provided, the most likely mechanism of these side effects is the lack of specificity of spironolactone as it binds, as already mentioned, not only to the renal aldosterone receptor, but also to the progesterone receptor and the dihydrotestosterone receptor (Miyake et al. 1978; Corvol et al. 1975; Bonne and Raynaud 1974).

In order to minimize this limitation of the use of an effective antialdosterone antihypertensive agent, De Gasparo et al. (1989) proposed two strategies: (a) a decrease in the daily dose of spironolactone, resulting in a dose-related decrease of gynecomastia and (b) an improvement in the receptor-binding specificity of spironolactone. Three $9\alpha,11\alpha$-epoxy derivatives have been characterized in vitro in rats and in rabbits, exhibiting a three- to tenfold decrease in the antiandrogenic and progestagenic effect, compared with spironolactone. In experimental animals, new aldosterone antagonists were characterized as less antiandrogenic agents than spironolactone (Nishino et al. 1988).

3. Carcinogenicity

There is a report of breast cancer in five patients who had taken spironolactone together with hydrochlorothiazide for prolonged periods (Loube and Quirk 1975). In addition, mutagenic metabolites unique to canrenoate were identified using rat liver, and shown to be directly acting mutagens in the mouse lymphoma assay (Cook et al. 1988).

4. Allergy

Eosinophilia and rash developed in two patients with alcoholic cirrhosis while taking spironolactone (Wathen et al. 1986).

5. Calcium Channel Antagonism

As stated above, spironolactone blocks slow calcium channels in plasma membrane as shown in patch-clamp experiments (Mironneau 1990).

F. Drug Interactions

I. Angiotensin-Converting Enzyme Inhibitors

Complete heart block was induced in a 72-year-old woman by hyperkalemia associated with treatment consisting of a combination of captopril and spironolactone (Lo and Cryer 1986). A similar, but fatal case with severe congestive heart failure was reported in an 85-year-old male patient taking furosemide, spironolactone and enalapril (Lakhani 1986). Hyperkalemia in a 55-year-old man taking spironolactone and enalapril persisted for over 48 h despite discontinuation of the drugs and treatment with ion-exchanger resin and intravenous glucose and insulin (Morton and Crook 1987).

From these cases with congestive heart failure, it might be suggested that potassium supplementation or potassium-sparing diuretics are not necessary when angiotensin-converting enzyme inhibitors are used together with loop or thiazide diuretics, and that blood urea and electrolyte concentrations must be known before treatment is started and has to be monitored during therapy (Lakhani 1986).

II. Ammonium Chloride

It has been reported that a 58-year-old woman receiving spironolactone 25 mg four times a day and potassium chloride 50 mmol daily developed acidosis about 20 days after starting ammonium chloride 4 g daily (Mashfold and Roberson 1972).

III. Aspirin

In healthy subjects acetylsalicylic acid has been reported to block the natriuresis induced by spironolactone in the presence of endogenous mineralocorticoids (Elliott 1962).

This finding was confirmed by giving a single oral dose of 600 mg aspirin together with spironolactone (20–100 mg) to ten healthy male physicians (Tweeddale and Ogilvie 1973) and six male volunteers (Ramsay et al. 1976). Since the reductions in urinary canrenone excretion correlated with its distal tubule effects, canrenone may be actively secreted at the proximal tubule, and this secretion is blocked by aspirin (or its conjugates) (RAMSAY et al. 1976; see also Chap. 6, this volume). Thus, careful consideration should be given to the choice of analgesics used during spironolactone therapy, and aspirin-containing products should be avoided. In low-renin hypertension, however, aspirin did not antagonize the antihypertensive effect of spironolactone (HOLLIFIELD 1976)

IV. Cyclosporin A

A potent immunosuppressive agent, cyclosporin A is widely used to prevent rejection in organ transplants by blocking the proliferation of T-helper cells, resulting in inhibition of the production of interleukin-2 and other lymphokines.

Spironolactone when used together with enalapril interrupts the cyclosporin-A-induced activation of the renin-angiotensin system, preventing peripheral vasoconstriction and more specifically the constriction of both glomerular arterioles and hepatic vessels. Hence, spironolactone diminishes cyclosporin A toxicity in both kidney and liver (IACONA et al. 1991).

V. Digitoxin

Concerning the effect of spironolactone on digitoxin half-life ($t_{1/2}$) in human subjects, controversial results have been reported. WIRTH et al. (1976) found that spironolactone decreased $t_{1/2}$ and that urinary elimination of unchanged digitoxin was delayed when spironolactone was given for at least 10 days to eight patients on oral maintenance digitoxin therapy. On the other hand, CARRUTHERS and DUJOVNE (1980) observed prolongation of digitoxin $t_{1/2}$ by subsequent administration of spironolactone to three healthy subjects. This controversy needs further investigation.

VI. Digoxin

Since both spironolactone and digoxin possess structural similarity and steroid-like nature, the former might influence the plasma concentration of the latter by interacting with the kinetics of digoxin in vivo. The uncharged steroid spironolactone and its metabolites might compete with the uncharged steroid digoxin, resulting in an inhibition of its renal clearance. Indeed, spironolactone when given to patients with congestive heart failure interacts with the elimination of digoxin. It has been reported that spironolactone reduces the renal clearance of digoxin (STEINESS 1974; WALDORFF et al. 1978; FENSTER et al. 1984; HEDMAN et al. 1992), thus causing a rise in the plasma concentration. Although reduction of nonrenal clearance of digoxin by spironolactone has also been postulated (FENSTER et al. 1984), a recent study in six male healthy subjects by HEDMAN et al. (1992) has shown no significant effect of spironolactone on the biliary clearance of digoxin. Nevertheless, these authors confirmed a significant reduction in renal clearance by an average of 13% (HEDMAN et al. 1992).

The mechanism of interactions requires further experimentation.

The similarities in chemical structure caused interference with the antibodies commonly used in radioimmunoassays to determine the plasma digoxin concentration (PHILLIPS 1973; THOMAS and MADDOX 1981; PALADINO 1984). To avoid this error, it is recommended either to use adequate antidigoxin-antibodies unaffected by spironolactone or its metabolites such as canrenone, 6α-thiomethylspironolactone and 7α-thimethylspironolactone, or to utilize a specific, combined HPLC/RIA technique (VETTICADEN et al. 1986).

VII. Fludrocortisone

RAMSAY et al. (1977) observed that in 12 healthy subjects receiving fludrocortisone, single doses of 100, 150 or 200 mg potassium canrenoate caused a paradoxical dose-related increase in urinary potassium excretion 12–16 h after administration. Similar results were obtained with spironolactone, though the dose-effect relationship was not clear. These observations may relate to the increased potassium excretion occasionally observed during the clinical use of aldosterone antagonists and may be caused by increased urine flow rate (see Chaps. 1, 8, this volume).

VIII. Mercurials

Although therapeutic use of mercurials has been largely abandoned, the use of mercurials in commercial and industrial products constitutes an important health risk. Using microorganisms, inorganic mercury can be converted to methyl mercury. These organic mercury compounds have different toxicodynamics and kinetics from the inorganic heavy metal. It has been reported that the toxic effects of mercury can be counteracted by spironolactone, spiroxazone or thiomesterone, all containing the S-CO-CH_3 group (SELYE 1970). KOUROUNAKIS et al. (1992) confirmed this action of spironolactone, and speculated on its molecular mechanism. The steroid may form a fairly stable, much less toxic complex with mercury, removing the heavy metal from its biological ligands and diverting it to the liver.

IX. Mitotane

Administration of mitotane up to 3 g daily to a 65-year-old patient with Cushing's syndrome who was also receiving spironolactone appeared to be ineffective and did not produce the side effects usually associated with mitotane (WORTSMAN and SOLAR 1977). Therefore, the two drugs cannot be used together in the management of Cushing's syndrome.

X. Analgesics

Licata and Bartter (1976) reported an interaction between spironolactone and propoxyphene in spironolactone-induced gynecomastia. In a 58-year-old patient, an allergic pleuritic reaction to the preparation of dextropropoxyphene, aspirin, caffeine and phenacetin was considered to trigger the spironolactone-induced gynecomastia. The pruritus and breast swelling disappeared when both the analgesic and spironolactone were withdrawn. The reintroduction of spironolactone produced no side effects, but the effects returned when the dextropropoxyphene was started again.

XI. Warfarin

Spironolactone has been associated with a reduction in warfarin activity and it has been suggested from studies in normal human subjects that this effect might be a consequence of the diuresis leading to an increase in the concentration of circulating clotting factors (O'Relly 1980).

References

Abshagen U (1977) Disposition kinetics of spironolactone in hepatic failure after single doses and prolonged treatment. Eur J Clin Pharmacol 11:169–176

Abshagen U, Rennekamp H, Luszpinski G (1976) Pharmacokinetics of spironolactone in man. Naunyn Schmiedebergs Arch Pharmacol 296:37–45

Agarwal MK, Kalimi M (1988) Paradoxical differences in the receptor binding of two new antimineralocorticoids. Biochem Biophys Acta 964:105–112

Agarwal MK, Kalimi M (1989) Analysis of the mineralocorticoid receptor in rat heart with the aid of two new spirolactone derivatives. Biochem Med Metabol Biol 41:36–45

Agarwal MK, Lazar G (1991) Antimineralocorticoids. Renal Physiol Biochem 14(6):217–223

Agarwal MK and Paillard J (1979) Paradoxical nature of mineralocorticoid receptor antagonism by progestins. Biochem Biophys Res Commun 89:77–84

Aizawa H, Niimura M (1992) Oral spironolactone therapy in male patients with rosacea. J Dermatol 19(5):293–297

Akamatsu H, Zouboulis CC, Orfanos CE (1993) Spironolactone directly inhibits proliferation of cultured human facial sebocytes and acts antagonistically to testosterone and 5-alpha-dihydrotestosterone in vitro. J Invest Dermatol 100(5): 660–662

Baily RR (1978) Diuretics in elderly. Br Med J 1:1618

Baulieu EE (1987) Steroid hormone antagonists at the receptor level: a role of the heat shock protein MW 90,000 (hsp 90). J Cell Biochem 35:161–174

Bennet WM, Aronoff GR, Morrison G, Golper TA, Pulliam J, Wolfson M, Singer I (1993) Drug prescribing in renal failure: dosing guidelines for adults. Am J Kidney Dis 3:155–193

Bie P, Wang BC, Leadley RJ Jr, Goetz KL (1990) Enhanced atrial peptide natriuresis during angiotensin and aldosterone blockade in dogs. Am J Physiol 258: R1101–R1107

Binder HJ, McGlone F, Sandle GI (1989) Effects of corticosteroid hormones on the electrophysiology of rat distal colon: implications for Na^+ and K^+ transport. J Physiol (Lond) 410:425–441

Bonne C, Raynaud JP (1974) Mode of spironolactone antiandrogenic action: inhibition of androstanolone binding to rat prostate androgen receptor. Mol Cell Endocrinol 2:59–67

Bonvalet JP (1987) Binding and action of aldosterone, dexamethasone, $1{,}25(OH)_2D_3$, and estradiol along the nephrone. J Steroid Biochem 27:953–961

Bonvalet JP, Blot-Chabaud M, Farman N (1991) Autoradiographic evidence of nuclear binding of spironolactone in rabbit cortical collecting tubule. Endocrinology 128:280–284

Brilla CG, Matsubara LS, Weber KT (1993) Antifibrotic effects of spironolactone in preventing myocardial fibrosis in systemic arterial hypertension. Am J Cardiol 71(3):12A–16A

Brown EA, Markandu ND, Sagnella GA, Ones BE, MacGregor GA (1984) Lack of effects of captopril on the sodium retention of nephrotic syndrome. Nephron 37:43–48

Burnet RB, Radden HS, Easterbrook EG, McKinnon RA (1991) Premenstrual syndrome and spironolactone. Aust N Z J Obstet Gynaecol 31(4):366–368

Carruthers SG, Dojovne CA (1980) Cholestyramine and spironolactone and their combination in digitoxin elimination. Clin Pharmacol Ther 27:184–187

Catania VA, Luquita MG, Carrillo MC, Mottino AD (1990) Sex differences in spironolactone induction of rat intestinal and hepatic p-nitrophenol UDP-glucuronyltransferase. Can J Physiol Pharmacol 68:1385–1387

Cella JA, Kagawa CM (1957) Steroidal lactones. J Am Chem Soc 79:4808–4809

Chang C, Kokontis J, Satchivi LA, Liao S, Tkeda H, Chang Y (1989) Molecular cloning of new human TR2 receptor: a class of steroid receptor with multiple ligand binding domains. Biochem Biophys Res Commun 165:735–741

Claire M, Rafestin-Oblin ME, Michaud A, Roth-Meyer C, Corvol P (1979) Mechanism of action of a new antialdosterone compound, prorenone. Endocrinology 104:1194–1200

Clement DL (1990) Peripheral action of spironolactone: plethysmographic studies. Am J Cardiol 65(23):12K–13K

Clementsen P, Hoegholm A, Hansen CL, Damkjaer M, Christensen P, Giese J (1989) Bartter's syndrome – treatment with potassium, spironolactone and ACE-inhibitor. J Intern Med 225(2):107–110

Cook CS, Hauswald CL, Schoenhard GL, Piper CE, Patel A, Radzialowski FM, Hribar JD, Aksamit W, Finnegan P, Bible RH, Oppermann JA (1988) Difference in metabolic profile of potassium canrenoate and spironolactone in the rat: mutagenic metabolites unique to potassium canrenoate. Arch Toxicol 61:201–212

Corvol P, Michaud A, Menard J, Freifield M, Mahoudeau J (1975) Antiandrogenic effect of spironolactones: mechanism of action. Endocrinology 97:52–58

Crosby PD, Rittmaster RS (1991) Predictors of clinical response in hirsute women treated with spironolactone. Fertil Steril 55:1076–1081

Dacquet C, Loirand G, Mironneau C, Mironneau J, Pacaud P (1987) Spironolactone inhibition of contraction and calcium channels in rat portal vein. Br J Pharmacol 92(3):535–544

Dahlstrom U, Karlsson E (1993) Captopril and spironolactone therapy for refractory congestive heart failure. Am J Cardiol 71(3):29A–33A

De Gasparo M, Whitebread SE, Preiswerk G, Jeunemaitre X, Corvol P, Menard J (1989) Antialdosterones: incidence and prevention of sexual side effects. J Steroid Biochem 32:223–227

De Kloet ER, Real JMHM (1987) Feedback action and tonic influence of corticosteroids on brain function: a concept arising from heterogeneity of brain receptor systems. Psychoneuroendocrinology 12:83–105

De Mendonca M, Grichois ML, Pernollet MG, Thorman B, Meyer P, Devynck MA, Garay R (1992) Hypotensive action of canrenone in a model of hypertension where ouabain-like factors are present. J Hypertens 3 [Suppl 3]:S73–S75

Doignon JL, Grognet JM, Thebault JJ, Caplain M, Capron MH, Pelletier B, Wehrlen M, Istin M (1990) Pharmacokinetics in healthy subjects of althiazide and spironolactone in a fixed combination for 2 doses. Therapie 45:475–481
Doucet A, Katz AI (1981) Mineralocorticoid receptors along the nephron: (3H)-aldosterone binding in rabbit tubule. Am J Physiol 241:F605–F611
Duax WL and Griffin JF (1989) Thyroid hormone receptor family gene regulation. In: Carlstedt-Duke J, Eriksson H, Gustafsson JA (eds) The steroids. Birkhäuser, Basel, pp 319–333
Dueymes JM (1990) Clinical update: spironolactone and altizide as monotherapy in systemic hypertension. Am J Cardiol 65(23):20K–23K
Elliott HC (1962) Reduced adrenocortical steroid excretion rates in man following aspirin administration. Metabolism 11:1015–1018
Ellison DH, Velazquez H, Wright FS (1989) Adaptation of the distal convoluted tubules of the rat. J Clin Invest 83:113–126
Farman N, Vandewalle A, Bonvalet JP (1982) Aldosterone binding in isolated tubules. II. An autoradiographic study of concentration dependency in rabbit nephron. Am J Physiol 242:F69–F77
Feller DR, Gerald MC (1971) Interactions of spironolactone with hepatic microsomal drug metabolizing enzyme systems. Biochem Pharmacol 20:1991–2000
Fenster PE, Hager WD, Goodman MM (1984) Digoxin-quinidine-spironolactone interaction. Clin Pharmacol Ther 36:70–73
Ferguson RK, Turek DM Rovner DR (1977) Spironolactone and hydrochlorothiazide in normal-renin and low renin essential hypertension. Clin Pharmacol Ther 21:62–69
Ganguly A (1992) Atrial natriuretic peptide-induced inhibition of aldosterone secretion: a quest for mediators. Am J Physiol 263(2, 1):E181–E194
Gardiner P, Schrode K, Quinlan D, Martin BK, Boreham DR, Rogers MS, Stubbs K, Smith M, Karim A (1989) Spironolactone metabolism: steady-state serum levels of the sulfur-containing metabolites. J Clin Pharmacol 29:342–347
Garg LC, Knepper MA, Burg MB (1981) Mineralocorticoid effects on Na-K-ATPase in individual nephron segments. Am J Physiol 240:F536–F544
Gerbes AL (1993) Medical treatment of ascites in cirrhosis. J Hepatol 17 [Suppl 2]:S4–S9
Gillman MA, Lichtigfeld FJ (1986) Synergism of spironolactone and lithium in mania. Br Med J 292:661–662
Gomez Sanchez EP, Fort CM, Gomez Sanchez CE (1990) Intracerebroventricular infusion of RU28318 blocks aldosterone-salt hypertension. Am J Physiol 258(3, 1):E482–484
Gomez Sanchez EP, Fort C, Thwaites D (1992) Central mineralocorticoid receptor antagonism blocks hypertension in Dahl S/JR rats. Am J Physiol 262(1, 1): E96–99
Greenblatt DG, Koch-Weser J (1973) Adverse reactions to spironolactone. J F Am Med Assoc 225(1):40–43
Grillo C, Vallee S, McEwen BS, De Nicola AF (1990) Properties and distribution of binding sites for the mineralocorticoid receptor antagonist [3H]ZK 91587 in brain. J Steroid Biochem 35(1):11–15
Gross JB, Imai M, Kokko JP (1975) A functional comparison of the cortical collecting tubule and the distal convoluted tubule. J Clin Invest 55:1284–1294
Hall PF (1985) Role of cytochromes P-450 in the biosynthesis of steroid hormones. Vitam Horm 42:315–368
Hedman A, Angelin B, Arvidsson A, Dahogvist R (1992) Digoxin interactions in man: spironolactone reduces renal but not biliary digoxin clearance. Eur J Clin Pharmacol 42:481–485
Herken H (1969) Aldosteron-Antagonisten. In: Herken H (Hrsg) Diuretica. Springer, Berlin Heidelberg New York (Handbuch der experimentellen Pharmakologie, Bd XXIV, S436–493)

Higashihara E, Carter NW, Pucacco L, Kokko JP (1984) Aldosterone effects on papillary collecting duct pH profile of the rat. Am J Physiol 246:F725–F731

Hollifield JW (1976) Failure of aspirin to antagonize the antihypertensive effect of spironolactone in low-renin hypertension. South Med J 69:1034–1036

Horisberger JD, Giebisch G (1987) Potassium-sparing diuretics. Renal Physiol 10: 198–220

Hughes BR, Cunliffe WJ (1987) Tolerance of spironolactone. Br J Dermatol 117 [Suppl]:32–38

Iacona A, Rossetti A, Filingeri V, Orsini D, Cresarini A, Cervelli V, Adorno D, Casciani CU (1991) Reduced nephrotoxicity and hepatoxicity in cyclosporin A therapy by enalapril and spironolactone in rats. Drugs Exp Clin Res 17:501–506

Jackson L, Branch R, Levine D, Ramsey L (1977) Elimination of canrenone in congestive heart failure and chronic liver disease. Eur J Clin Pharmacol 11: 177–179

Kalimi M, Opoku J, Agarwal M, Corley K (1990) Effects of antimineralocorticoid RU 26752 on steroid-induced hypertension in rats. Am J Physiol 258:E737–E739

Kamata S, Haga N, Mitsugi T, Kondo E, Nagata W, Nakamura M, Miyata K, Odaguchi K, Shimizu T, Kawabata T, Suzuki T, Ishibashi M, Yamada F (1985) Aldosterone antagonists. 1. Synthesis and biological activities of 11 beta, 18-epoxypregnane derivatives. J Med Chem 28:428–433

Kourounakis PN, Pouskoulelis GP, Rekka E (1992) Interaction of spironolactone with mercury. Drug Res 42:1025–1028

Lagrue G, Ansquer JC, Meyer Heine A (1990) Peripheral action of spironolactone: improvement in arterial elasticity. Am J Cardiol 65(23):9K–11K

Lakhani M (1986) Complete heart block induced by hyperkalaemia associated with treatment with a combination of captopril and spironolactone. Br Med J 293:271

Landau RL, Bergenstal DM, Lugibihl K, Kascht ME (1955) The metabolic effect of progesterone in man. J Endocrinol Metab 15:1194–1215

Licata AA, Bartter FC (1976) Spironolactone-induced gynaecomastia related to allergic reaction to "Darvon Compound". Lancet 2:905

Lo TCN, Cryer RJ (1986) Complete heart block induced by hyperkalaemia associated with treatment with a combination of captopril and spironolactone. Br Med J 292:1672

Lombes M, Farman N, Oblin ME, Baulieu EE, Bonvalet JP, Erlanger BF, Gasc JM (1990) Immunohistochemical localization of renal mineralocorticoid receptor by using an anti-idiopathic antibody that is an internal image of aldosterone. Cell Biol 87:1086–1088

Losert W, Bittler D, Buse M, Casals-Stenzel J, Haberey M, Laurent H, Nickisch K, Schillinger E, Wiechert R (1986) Mespirenone and other 15,16-methylene-17-spironolactones, a new type of steroidal aldosterone antagonists. Drug Res 36(2):1583–1600

Loube SD, Quirk RA (1975) Breast cancer associated with administration of spironolactone. Lancet I:1428–1429

Luzzani F, Glasser A (1984) Characterization of spironolactone binding sites distinct from aldosterone receptors in rat kidney homogenates. Biochem Pharmacol 33:2277–2281

Marver D, Stewart J, Funder JW, Feldman D, Edelman IS (1974) Renal aldosterone receptors: studies with (^{3}H)-aldosterone and the anti-mineralocorticoid (^{3}H)-spironolactone. Proc Natl Acad Sci USA 71:1431–1435

Mashford ML, Roberson MB (1972) Spironolactone and ammonium and potassium. Br Med J 4:298–299

Menard RJ, Guenthner TM, Kon H, Gillette JR (1979) Studies on the destruction of adrenal and testicular cytochrome P-450 by spironolactone. J Biol Chem 254: 1726–1733

Messina M, Manieri C, Musso MC, Pastorino R (1990) Oral and topical spironolactone therapies in skin androgenization. Panminerva Med 32(2):49–55
Mironneau J (1990) Calcium channel antagonist effects of spironolactone antagonist. Am J Cardiol 65:7K–8K
Mirshahi M, Pagano M, Razaghi A, Lazar G, Agarwal MK (1992) Immunophotochemical analysis of mineralocortin by polyclonal antibodies against the native receptor from rat kidney. Biochem Med Metabol Biol 47:133–144
Miyake A, Noma K, Nakao K, Morimoto Y, Yamamura Y (1978) Increase of serum oestrone and oestradiol following spironolactone administration in hypertensive men. Clin Endocrinol 9:523–533
Mori N, Yura K, Uozumi N, Sakai S (1991) Effect of aldosterone antagonist on the DC potential in the endolymphatic sac. Ann Otol Rhinol Laryngol 100(1): 72–75
Morton AR, Crook SA (1987) Hyperkalaemia and spironolactone. Lancet 1:1525
Murakami M, Suzuki H, Nakajima S, Nakamoto H, Kageyama Y, Saruta T (1989) Calcitonin gene-related peptide is an inhibitor of aldosterone secretion. Endocrinology 125(4):2227–2229
Nedelec L, Philibert D, Torelli V (1986) Recent development in the field of steroid antihormones. In: Lambert RW(ed) Proc 3rd SCI-RSC Medicinal Chemistry Symposium. Lond R Soc Chem [Suppl] 45:322–344
Nguyen TT, Lazure C, Babinski K, Chretien M, Ong H, De Lean A (1989) Aldosterone secretion inhibitory factor: a novel neuropeptide in bovine chromaffin cells. Endocrinology 124(3):1591–1593
Nishino Y, Schreder H, El Etreby MF (1988) Experimental studies on the endocrine side effects of new aldosterone antagonists. Drug Res 38:1880–1805
Omland E, Mathisen O (1991) Mechanism of ursodeoxycholic acid- and canrenoate-induced biliary bicarbonate secretion and the effect on glucose- and amino acid-induced cholestasis. Scand J Gastroenterol 26(5):513–522
Opoku J, Kalimi M, Agarwal M, Qureshi D (1991) Effect of a new mineralocorticoid antagonist mespirenone on aldosterone-induced hypertension. Am J Physiol 260:E269–E271
O'Reilly RA (1980) Spironolactone and warfarin interaction. Clin Pharmacol Ther 27:198–201
Overdiek HWPM, Hermens WAJJ, Merkus FWHM (1985) New insights into the pharmacokinetics of spironolactone. Clin Pharmacol Ther 38:469–474
Paladino A (1984) Influence of spironolactone on serum digoxin concentration. JAMA 251:470–471
Peterfalvi M, Torelli V, Fournex R, Rousseau G, Claire M, Michaud A, Corvol P (1980) Importance of the lactic ring in the activity of steroidal antialdosterones. Biochem Pharmacol 29:353–357
Phillips AP (1973) The improvement of specificity in radioimmunoassays. Clin Chim Acta 44:333–340
Pratt W (1990) Interaction of hsp90 with steroid receptors: organizing some diverse observations and presenting the newest concepts. Mol Cell Endocrinol 74: C69–C76
Rafestin-Oblin ME, Lombes M, Couette B, Baurieu EE (1992) Differences between aldosterone and its antagonists in binding kinetics and ligand-induced hsp90 release from mineralocorticoid receptor. J Steroid Biochem Mol Biol 41(3–8): 815–821
Ramsay LE, Harrison IR, Shelton JR, Vose CW (1976) Influence of acetylsalicylic acid on the renal handling of spironolactone metabolism in healthy subjects. Eur J Clin Pharmacol 10:43–48
Ramsay LE, Harrison IR, Shelton JR, Tidd MJ (1977) Paradoxical potassium excretion in response to aldosterone antagonists. Eur J Clin Pharmacol 11: 101–105

Ratka A, Sutanto W, Bloemers M, De Kloet ER (1989) On the role of brain mineralocorticoid (type I) and glucocorticoid (type II) receptors in neuroendocrine regulation. Neuroendocrinology 50(2):117–123

Rocco VK, Ware AJ (1986) Cirrhotic ascites: pathophysiology, diagnosis, and management. Ann Intern Med 105:573–585

Rossier BC, Clair M, Rafestin-Oblin ME, Geering K, Gaggeler HP, Corvol P (1983) Binding and antimineralocorticoid activities of spironolactones in toad bladder. Am J Physiol 244:C24–C31

Rourke KA, Bergstrom JM, Larson IW, Colby HD (1991) Mechanism of action of spironolactone on cortisol production by guinea pig adrenocortical cells. Mol Cell Endocrinol 81(1–3):127–134

Sadee W, Dagcioglu M, Schroeder R (1973) Pharmacokinetics of spironolactone, canrenone and canrenoate-K in humans. J Pharmacol Exp Ther 185:686–695

Sakamoto H, Ichikawa S, Sakamaki T, Nakamura T, Ono Z, Takayama Y, Murata K (1990) Time-related changes in plasma adrenal steroids during treatment with spironolactone in primary aldosteronism. Am J Hypertens 3:533–537

Sakauye C, Feldman D (1976) Agonist and antimineralocorticoid activities of spirolactones. Am J Physiol 231:93–97

Sale G, Luetscher JA Jr (1954) The effect of sodium-retaining corticoids, electrocortin, deoxycorticosterone and cortisone on renal function and excretion of sodium in adrenalectomized rats. Endocrinology 55:516–518

Schohn DC, Spiesse R, Wehrlen M, Pelletier B, Capron MH (1990) Aldactazine/captopril combination, safe and effective in mild to moderate systemic hypertension: report on a multicenter study of 967 patients. Am J Cardiol 65(23): 4K–6K

Schwartz GJ, Burg MB (1978) Mineralocorticoid effects on cation transport by cortical collecting tubules in vitro. Am J Physiol 235:F576–F585

Selye H (1970) Mercury poisoning: prevention by spironolactone. Science 169:775

Shapiro MD, Hasbargen J, Hensen J, Schrier RW (1990) Role of aldosterone in the sodium retention of patients with nephrotic syndrome. Am J Nephrol 10:44–48

Shaw JC (1991) Spironolactone in dermatologic therapy. J Am Acad Dermatol 24(2, 1):236–243

Sherry HJH, O'Donnell JP, Flowers L, Lacagnin LB, Colby HD (1986) Metabolism of spironolactone by adrenocortical and hepatic microsomes: relationship to cytochrome P-450 destruction. J Pharmacol Exp Ther 236:675–680

Solymoss B, Zsigmond G (1973) Effect of various steroids on the hepatic glucuronidation and biliary excretion of bilirubin. Can J Physiol Pharmacol 51:319–323

Stanton BA (1986) Regulation by adrenal corticosteroids of sodium and potassium transport in loop of Henle and distal tubule of rat kidney. J Clin Invest 78:1612–1620

Steiness E (1974) Renal tubular secretion of digoxin. Circulation 50:103–107

Stone D, Seldine D, Kokko J, Jacobson HR (1983) Mineralocorticoid modulation of rabbit medullary collecting duct acidification. J Clin Invest 72:77–83

Stripp B, Hamrick ME, Zampaglione NG, Gillette JR (1971) The effect of spironolactone on drug metabolism by hepatic microsomes. J Pharmacol Exp Ther 176:766–771

Thomas RW, Maddox RR (1981) The interaction of spironolactone and digoxin: a review and evaluation. Ther Drug Monit 3:117–120

Tweeddale MG, Ogilvie RI (1973) Antagonism of spironolactone-induced natriuresis by aspirin in man. N Engl J Med 289:198–200

Varin F, Tu TM, Benoit F, Villeneuve JP, Theoret Y (1992) High-performance liquid chromatographic determination of spironolactone and its metabolites in human biological fluids after solid-phase extraction. J Chromatogr 574:57–64

Vetticaden SJ, Barr WH, Beightol La (1986) Improved method for assaying digoxin in serum using high performance liquid chromatography-radioimmunoassay. J Chromatogr 383:186–193

Waldorff S, Andersen DJ, Heeboll-Nielsen N, Nielsen OG, Moltke E, Sorensen U, Steiness E (1978) Spironolactone-induced changes in digoxin kinetics. Clin Pharmacol Ther 24:162–167
Wathen CG, MacDonald T, Wise LA, Boyd SM (1986) Eosinophilia associated with spironolactone. Lancet 1:919–920
Weber KT, Villarreal D (1993) Aldosterone and antialdosterone therapy in congestive heart failure. Am J Cardiol 71(3):3A–11A
Weindel K, Lewicka S, Vecsei P (1991) Inhibitory effects of the novel antialdosterone compound mespirenone on adrenocortical steroidogenesis in vitro. Drug Res 41(9):946–949
Wirth KE, Frolich JC, Hollifield JW, Falkner FC, Sweetman BS, Oates JA (1976) Metabolism of digitoxin in man and its modification by spironolactone. Eur J Clin Pharmacol 9:345–354
Work J, Jamison RL (1987) Effect of adrenalectomy on transport in the rat medullary thick ascending limb. J Clin Invest 80:1160–1164
Wortsman J, Solar NG (1977) Mitotane. Spironolactone antagonism in Cushing's syndrome. JAMA 238:2527
Yamakawa M, Ezumi K, Shiro M, Nakai H, Kamata S, Matsuji T, Haga N (1987) Relationship of the molecular structure of aldosterone derivatives with their binding affinity for mineralocorticoid receptor. Mol Pharmacol 30:585–598
Zannad F (1993) Angiotensin-converting enzyme inhibitor and spironolactone combination therapy. New objectives in congestive heart failure treatment. Am J Cardiol 71(3):34A–39A

CHAPTER 10

Potassium-Retaining Diuretics: Amiloride

L.G. Palmer and T.R. Kleyman

A. Introduction

Amiloride was discovered in the late 1960s during an extensive screening process at the Merck Sharp and Dohme Research Laboratories. Starting with *N*-amidino-3-amino-6-bromopyrazinecarboxamide, over 300 compounds were tested for their ability to reverse the effects of mineralocorticoids in rats (Bicking et al. 1965). Of these amiloride was among the most potent in producing natriuresis without a concomitant kaliuresis. Since then over 1000 different analogs have been synthesized and studied (Kleyman and Cragoe 1988).

Although the initial screening assay was Na^+ excretion in adrenalectomized rats, it was soon discovered that amiloride was a potent inhibitor of transepithelial Na^+ transport in amphibian tissues such as skin and urinary bladder, which cold be studied in vitro (Bentley 1968; Ehrlich and Crabbé 1968; Salako and Smith 1970). According to the model of epithelial Na^+ transport of Koefoed-Johnsen and Ussing (1958), the two main steps in transport by such epithelia consist of passive Na^+ entry into the cells across the apical or mucosal membrane, and active extrusion of Na^+ (in exchange for K^+) across the basolateral membrane. Because amiloride acted rapidly and preferentially from the apical side of the epithelium, it was inferred that it inhibited the passive Na^+ permeability of Na^+ channels of the apical cell membrane.

Amiloride and its analogs have since become extremely useful tools in identifying and elucidating this and other transport mechanisms. Lindemann and Van Driessche (1977) used amiloride to produce fluctuations in Na^+ transport across the frog skin and were able to identify the apical transport process as an Na^+-specific channel or pore. Cuthbert and Shum (1976) used radiolabeled amiloride to estimate the number of binding sites (putative transporting sites) on epithelial cells. Benos and colleagues (1986) later used a photoactive analog of amiloride to identify specific apical membrane proteins which may form the channel. Other analogs of amiloride have been useful in studying other Na^+-transport processes, especially Na^+-H^+ exchange and Na^+-Ca^{2+} exchange. Thus the impact of this class of diuretics on laboratory research has been profound. These uses have been summarized in detail by Benos (1982) and Kleyman and Cragoe (1988).

As will be discussed in more detail below, the major target for amiloride in vivo is believed to be the epithelial Na^+ channel. These channels have been most extensively studied in amphibian tissues, particularly the frog skin and toad urinary bladder. In mammals they are important in controlling Na^+ reabsorption in the distal nephron, the colon, and the trachea. These are the most important sites of drug interaction in humans.

B. Structure-Function Relationships

The structure of amiloride is shown in Fig. 1. It consists of a substituted pyrazine ring with a carbonylguanidinium side chain. The ring and its substituents form a planar structure as can be seen in the space-filling view. A large number of analogs of this basic structure have been made and studied. Comprehensive descriptions of these analogs and their effects can be found in the literature (KLEYMAN and CRAGOE 1988). In this section we will selectively discuss several classes of analogs which may provide insight into the mechanism of action of the drug. For this purpose we will focus on changes in the guanidino side chain and on substituents at the 5- and 6-positions of the pyrazine ring.

Most of the data on structure-function relationships have been obtained using inhibition of Na^+ transport by amphibian epithelia (frog skin and toad

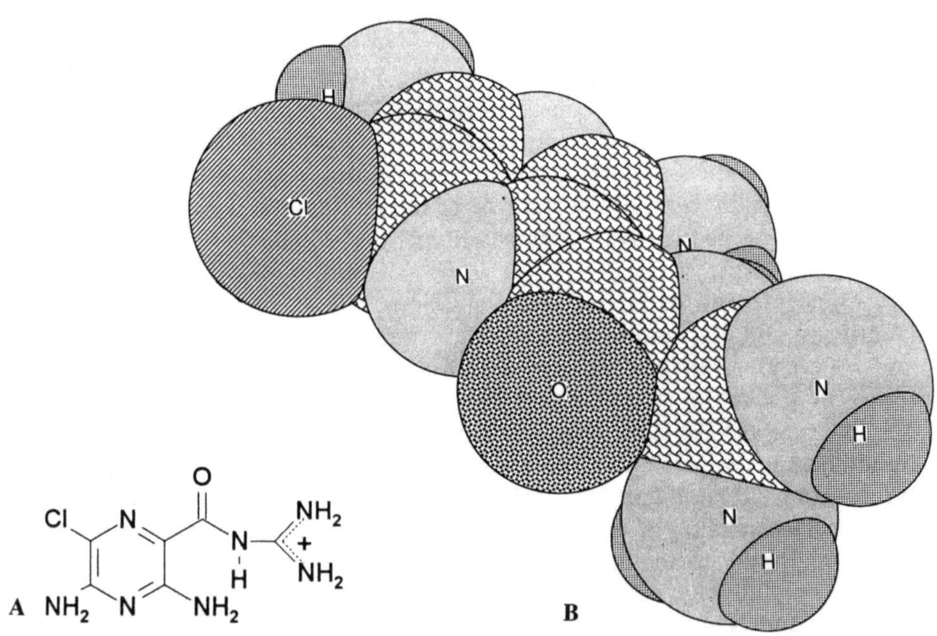

Fig. 1A,B. Structure of amiloride. **A** Chemical structure. **B** Space-filling model

bladder) in vitro as a measure of the diuretic potency. These methods permit much more precise and detailed examination of the effects of the drugs than do those involving intact animals. It should be emphasized, however, that it is believed that the epithelial Na^+ channel represents the main target of the drug in vivo. Furthermore, the functional properties of these Na^+ channels appear to be well preserved throughout the vertebrates (SMITH and BENOS 1991; PALMER 1992). Thus there is reason to believe that the information obtained using the amphibian systems will be relevant to the clinical effects of the drugs.

The most common assay is that of the short-circuit current (USSING and ZERAHN 1951). Here the Na^+ transport rate can be continuously monitored as an electrical signal, and the effects of the drugs at the apical suface of the epithelium can be ascertained rapidly and with high precision. In some studies fluctuation analysis of the short-circuit current has been used to measure the on- and off-rates for block by amiloride and its analogs (LI et al. 1985, 1987; HELMAN and BAXENDALE 1990). As will be discussed below, it is also possible to measure on- and off-rates using voltage-relaxation techniques in intact epithelia, or analysis of single-channel currents using patch-clamp methodology. However, limited information is available on amiloride analogs using the latter two approaches.

I. Guanidinium Substitutions

Guanidinium itself can block the epithelial Na^+ channel, albeit at relatively high concentrations. The concentration of guanidinium required for half-maximal block (K_i) in the toad bladder was about 90 mmol/l, whereas for amiloride in the same system the apparent K_i was around 100 nmol/l (PALMER 1985). Despite the differences in affinity of 6 orders of magnitude, the nature of the block by the two compounds appears to be similar in the sense that the voltage dependence was identical in both cases. As will be discussed below, this may imply a common site of action.

The charged guanidinium group also appears to be essential for high-affinity block of the channels by amiloride. The pK_a for amiloride is 8.67. At pH values where the guanidinium is uncharged the blocker becomes ineffective, and the potency correlates well with the calculated concentration of the positively charged species (BENOS et al. 1976; CUTHBERT 1976). Furthermore, compounds in which this group is replaced, for example with an OH group, are ineffective (KLEYMAN and CRAGOE 1988). Thus it is likely that the cationic guanidino moiety of the drug interacts specifically with the Na^+ channel, perhaps with a fixed negative charge on the channel protein.

On the other hand, at least one substituted pyrazine which lacks both the guanidinium and a positive charge can block the Na^+ channel, albeit with reduced affinity. The analog 6-chloro-3,5-diaminopyrazine-2-carboxamide (CDPC) has the guanidino group replaced by an amino group, and is therefore uncharged at physiological pH. The blocker has an apparent

K_i for the channel of about 100 μmol/l, at least 100-fold higher than that measured for amiloride (HELMAN and BAXENDALE 1990). It produces blocker-induced noise of the short-circuit current similar to that of amiloride, although its off rate is much higher than that for amiloride, accounting for its decreased affinity. Like amiloride its affinity is decreased in the presence of high Na^+ concentrations (PALMER 1991). Unlike amiloride, its effect is not voltage dependent. This difference would be expected from the absence of positive charge (PALMER 1991). The similarities in structure between the two compounds would suggest that they have a common site of interaction on the Na^+ channel, although competition studies have not been reported. It would therefore appear that both the guanidinium and pyrazine portions of amiloride interact with distinct parts of the pore. Either one of these interactions by itself may be sufficient to block the channel. However, both interactions are necessary to obtain a high-affinity block, i.e., one with an apparent K_i of less than 1 μmol/l.

Since the original description of amiloride, only one class of analogs has been identified which has a significantly higher potency for blocking the epithelial Na^+ channel. These compounds have a hydrophobic substitution on one of the amino groups on the guanidinium. They include benzamil, with a benzene substitution, and phenamil, with a phenol substitution, both of which are about ten times as potent as amiloride (KLEYMAN and CRAGOE 1988). These substituents presumably interact with a hydrophobic part of the Na^+ channel near the amiloride-binding site. The interaction does not appear to be an extremely specific one, as many different compounds having a substituted phenolic ring have been synthesized which appear to have an increased affinity for the channel (KLEYMAN and CRAGOE 1988). The major effect of these substitutions is to decrease the off rate for blocking the channel (LI et al. 1987). While these hydrophobic analogs appear to have a reduced affinity for the Na^+/H^+ exchanger, they all have an increased affinity for the Na^+/Ca^{2+} exchanger (KLEYMAN and CRAGOE 1988). Even so, the highest affinity of any amiloride analog for this exchanger is only about 10 μmol/l.

II. 6-Position Ring Substitutions

Amiloride has a chloro substitution on the 6-position of the pyrazine ring. This can be changed to a bromo substituent with little change in activity, but fluro and iodo substitutions as well as the unsubstituted compound have much reduced potencies (KLEYMAN and CRAGOE 1988). The interactions of compounds with frog skin were studied in detail using noise analysis (LI et al. 1985). Changes in affinity with H, F, I, Br and Cl at position 6 were strongly correlated with decreases in the off-rate constant for block, with the mean lifetime of the blocked state increasing from 6 ms to 255 ms. There was also a strong correlation with the decrease in pK_a of these compounds.

The findings were interpreted as indicating that an increase in partial negative charge at the 6-position may account for both the decreased pK_a and an increased stability of the blocked state. It was suggested that the 6-position of the ring may interact directly with the channel, perhaps at an electropositive residue.

III. 5-Position Ring Substitutions

The 5-position of the pyrazine ring requires an NH_2 substituent for optimal activity. Changing this group to H or Cl both reduced the on-rate and increased the off-rate constant for amiloride block (LI et al. 1985). The decreased on-rate constant is of particular interest since it indicates that the approach of the drug to its blocking site is not strictly diffusion limited. These substitutions, which do not increase the size of the molecule and are unlikely to produce a steric hindrance to binding, would not be expected to change the rate of diffusion up to the pore. The differences in on rate are consistent with the formation of a transitory pre-blocking drug-receptor complex in the blocking scheme. This will be discussed further below.

C. Pharmacodynamics

I. Sites of Action: Na^+ Transport

1. General

Amiloride is not specific for a particular organ or cell. It *is* specific for a particular ion transporter, namely the epithelial Na^+ channel. This is a channel which is expressed on the apical membrane of a variety of epithelial cells in different organs (Table 1). It serves the same general purpose – namely the selective transport of Na^+ into the cell down an electrochemical gradient – as do the Na^+ channels in nerve and muscle. It is functionally distinct from the excitable-cell Na^+ channels, however, in that it is not activated or inactivated by voltage and is not sensitive to the potent toxins tetrodotoxin and saxitoxin which block the action potential in neuronal cells. The two types of Na^+ channels also appear to be unrelated genetically (CANESSA et al. 1993).

The functional and biochemical properties of Na^+ channels have been reviewed recently (EATON and HAMILTON 1988; GARTY and BENOS 1988; SMITH and BENOS 1991; PALMER 1992). In brief, there are at least three types of epithelial Na^+ channel, all of which are blocked by amiloride. Indeed, sensitivity to amiloride has become the pharmacological hallmark of these channels. The three types can be distinguished according to their ion selectivity and their single-channel conductance measured using fluctuation

Table 1. Mammalian epithelia which are affected by amiloride

Tissue	Species	Dose/K_i (μmol/l)	Effects	References
Kidney				
DCT	Rat	10	J_{Na}, J_K, J_{Ca}	DUARTE et al. 1971; COSTANZO 1984
CCT	Rabbit	10/0.07	J_{Na}, J_K, V_t, R_a	O'NEIL and BOULPAEP 1979; KOEPPEN et al. 1983
OMCD	Rabbit	10	V_t, R_a	KOEPPEN 1986
IMCD	Rabbit	/0.07	V_{O2}	ZEIDEL et al. 1986
Urinary bladder	Rabbit	/0.3	I_{sc}, R_a	LEWIS and DIAMOND 1976
Intestine				
Ileum	Rat (salt-depleted)	50/1	I_{sc}	WILL et al. 1985b
Colon	Human	/1	I_{sc}	WILL et al. 1984
Respiratory tract				
Nasal epithelia	Human	100	J_{sc}, J_{Na}, R_a	BOUCHER et al. 1988
Trachea	Human	100	I_{sc}	WIDDICOMB et al. 1985
Bronchi	Human	100/0.7	J_{Na}, I_{sc}	KNOWLES et al. 1984
Salivary duct	Rat	10	J_{Na}, V_t	SCHNEYER 1970
Sweat duct	Human	100	J_{Na}	QUINTON 1981

V_t, decreases the (mucosa negative) transepithelial potential; I_{sc}, decreases the short-circuit current; J_{Na}, decreases the net absorptive flux of Na^+; J_K, decreases the net secretory flux of K^+; J_{Ca}, increases the net absorptive flux of Ca^{2+}; R_a, increases apical membrane resistance.

analysis and patch-clamp techniques (PALMER 1992). The major channel type has a low conductance (5 pS) and a very high selectivity (>100:1) for Na^+ over K^+. It has been observed in frog skin and toad bladder, and in mammalian cortical collecting tubule and colon. A second type, reported in toad kidney cells in culture, has intermediate conductance (9 pS) and selectivity (4:1). A third type has a large conductance (28 pS) and does not discriminate between Na^+ and K^+. It is observed in mammalian medullary collecting duct cells.

Candidate Na^+ channel proteins have been isolated using photoaffinity analogs of amiloride (SMITH and BENOS 1991). The best studied of these is a 700-kDa hetero-oligomer isolated from bovine kidney and from cultured toad kidney cells. It dissociates into at least six different peptides under reducing conditions. A single peptide of molecular weight 150 kDa binds amiloride. Antibodies against this protein indicate that it is localized predominantly on the apical membrane of epithelial cells known to express the Na^+ channel.

Recently, at least one component of the channel was cloned and sequenced from a rat colon cDNA library (CANESSA et al. 1993; LINGUEGLIA et al. 1993). The predicted amino acid sequence implies a core molecular weight of 78 kDa. Hydropathy analysis indicates that the protein could span the membrane from two to six times. RNA from the clone expressed an Na^+-selective, amiloride-sensitive conductance in *Xenopus* oocytes. However, the low degree of expression indicated that additional subunits may be necessary for full function of the channel.

The epithelial Na^+ channel is regulated by aldosterone in virtually every tissue in which it has been studied (GARTY 1986; ROSSIER and PALMER 1992). These include frog skin and toad urinary bladder and, in mammals, the cortical collecting tubule, colon, urinary bladder and the ducts of sweat and salivary glands. This has led to the suggestion that the biosynthesis of the channel may be under mineralocorticoid control. While this may indeed be the case, at least the short-term stimulatory effects of aldosterone on the channel appear not to involve the synthesis of new channel proteins (GARTY 1992). In any case, the correlation between the presence of Na^+ channels and the sensitivity of an epithelium to aldosterone explains the observations that amiloride is an effective antagonist of mineralocorticoids in humans and experimental animals.

2. Within the Kidney

Although amiloride will affect Na^+ channels, and therefore Na^+ reabsorption, in a variety of epithelia, the most important sites of action in terms of the diuretic effect of the drug are within the kidney. This is in part because the kidney handles much more salt than the other organs expressing the channels, and hence has a much larger capacity for salt excretion. In addition, amiloride is both filtered by the glomeruli and secreted into the urine by the proximal tubule. Thus it will be present in the tubular fluid at concentrations higher than those in plasma, especially in the distal nephron. Low doses of the drug will therefore preferentially affect Na^+ channels in the kidney because the drug concentration will be high in the distal tubular fluid. As will be discussed below, at clinically relevant doses of the drug, amiloride acts primarily on the distal nephron segments of the kidney.

a) *Proximal Tubule*

The proximal tubule is relatively insensitive to amiloride. Very high concentrations (1 mmol/l) were required to inhibit fluid reabsorption (Meng 1975). This inhibition may be accounted for by effects on Na^+/H^+ exchange and does not occur in a clinically relevant concentration range. The finding is consistent with the idea that most Na^+ reabsorption in the proximal tubule is accomplished by countertransport with hydrogen ion, which is blocked by amiloride only at higher concentrations, and by cotransport processes which are not affected by the drug. On the other hand, an apical membrane Na^+ channel has been observed in patch-clamp studies of the proximal straight tubule (Gögelein and Greger 1986). The channel was blocked by relatively high concentrations of amiloride on the *cytoplasmic* side of the membrane. Whether the channel is more sensitive to the drug on the extracellular surface has not been tested and whether this is a physiologically important route for Na^+ reabsorption in this segment, or a clinically important site of drug interaction, has also not been clearly established.

b) *Henle's Loop*

NaCl reabsorption in the thick ascending limb of Henle's loop is the major site of action of loop diuretics (see Chap. 7, this volume). No evidence for apical Na^+ channels or of an electrogenic Na^+ transport across the apical membrane has been presented. Indeed, even high concentrations (1 mmol/l) of amiloride did not change the apical membrane voltage or resistance (Greger 1985). The dominant ion conductance of the apical membrane is that to K^+ and is not affected by amiloride.

c) *Distal Convoluted Tubule*

Free-flow micropuncture studies showed that amiloride decreased distal Na^+ reabsorption in the rat distal convoluted tubule, although the effect was modest (Duarte et al. 1971). Notably, K^+ secretion was inhibited to an even greater extent (see below). Of particular clinical relevance was the finding that when amiloride was given in conjunction with furosemide, Na^+ excretion was augmented and the increase in K^+ excretion observed with this loop diuretic alone was suppressed (Hropot et al. 1985). In vivo microperfusion of the rat distal tubule confirmed the inhibition of Na^+ reabsorption in this segment by amiloride (Costanzo 1984). In that study it was found that amiloride-sensitive Na^+ transport is confined to the late portions of the distal convoluted tubule. A similar pattern of inhibition of K^+ secretion in the second half of the distal tubule was observed in micropuncture studies (DUARTE et al. 1971).

d) *Collecting Duct*

The action of amiloride has been demonstrated most directly in the cortical collecting tubule (CCT) and in the outer medullary collecting tubule (OMCT) (STONER et al. 1974; O'NEIL and BOULPAEP 1979; KOEPPEN et al. 1983; KOEPPEN 1986; SCHLATTER and SCHAFER 1987). In isolated pefused tubules, amiloride reduces net Na^+ reabsorption, the transepithelial electrical potential difference, and the conductance of the luminal membrane. In these respects the effects of amiloride appear to be similar to those observed in model amphibian epithelia such as frog skin and toad bladder (see below). In addition, patch-clamp experiments have directly documented the effects of amiloride on Na^+ channels in the CCT (PALMER and FRINDT 1986). In both the rabbit (SANSOM and O'NEIL 1985) and rat (TOMITA et al. 1985; REIF et al. 1986; FRINDT et al. 1990) CCT, the amiloride-sensitive Na^+ transport is enhanced by pre-treatment of the animals with mineralocorticoids or with a low-Na^+ diet to increase endogenous aldosterone secretion. Amiloride also inhibits transport in the inner medullary collecting duct (IMCD). This was demonstrated by inhibition of oxygen consumption in isolated IMCD cells (ZEIDEL et al. 1986). Direct effects of amiloride on Na^+ channels in cultured IMCD cells have also been directly demonstrated using patch-clamp techniques (LIGHT et al. 1990).

e) *Urinary Bladder*

The urinary bladder, at least in the rabbit, actively reabsorbs Na^+ through aldosterone-dependent, amiloride-sensitive Na^+ channels in the apical membrane (LEWIS and DIAMOND 1976). The rate of Na^+ reabsorption in the bladder per unit of surface area is low relative to that in the collecting duct, and the volume to surface area ratio is relatively large. It is not clear how important transport in this organ is to Na^+ homeostasis or whether inhibition of transport by amiloride is of quantiative relevance.

3. Other Epithelia

Na^+ channels which are blocked by amiloride are also found in other organ systems. In the gastrointestinal tract, amiloride sensitivity is commonly observed in the distal colon (FRIZZELL et al. 1976; ZEISKE et al. 1982; WILLS et al. 1984; WILL et al. 1985a). Under conditions of salt depletion, these channels are also expressed in the ileum (WILL et al. 1985a). In the respiratory tract, amiloride-sensitive Na^+ transport has been observed in nasal epithelia, the trachea, and bronchial epithelia (KNOWLES et al. 1984;

WIDDICOMB et al. 1985; BOUCHER et al. 1988). The reabsorptive sweat ducts (QUINTON 1981) and salivary glands (SCHNEYER 1970) also express amiloride-sensitive Na^+ channels.

II. Effects on Transport of K^+ and Other Ions

A strong impetus behind the development of amiloride and its major clinical uses is the prevention of kaliuresis. Indeed, amiloride has been shown to reduce urinary K^+ excretion in both experimental animals and in human subjects (GOMBES et al. 1966; BAER et al. 1967; BULL LARAGH 1968). As discussed above, the epithelial Na^+ channel appears to be the only important target for amiloride at submicromolar concentrations. The effects on renal tubular K^+ transport are therefore thought to be indirect and to involve changes in the driving forces for K^+ movement (Fig. 2).

According to the widely accepted model for ion transport in the CCT (GIEBISCH and FIELD 1989), Na^+ enters the cells across the apical membrane down an electrochemical concentration gradient and leaves the cells through an ATP-driven, K^+-coupled pump [$(Na^+ + K^+)$-ATPase] in the basolateral membrane. In this aspect the transport system is identical to that proposed by Koefoed-Johnsen and Ussing for the frog skin (KOEFOED-JOHNSEN and USSING 1958). In addition, however, the CCT also expresses a significant

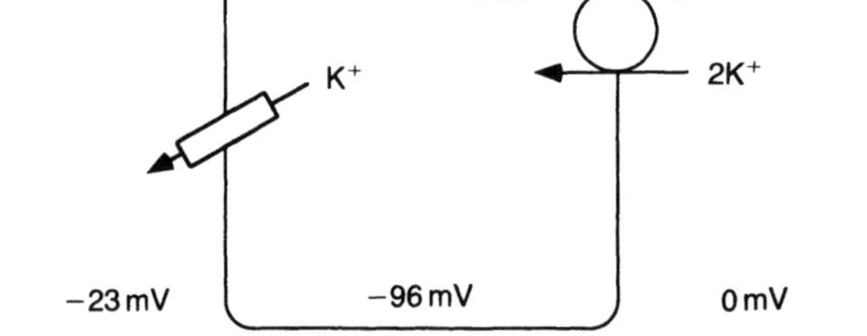

Fig. 2. Basic model for Na^+ and K^+ transport in the renal cortical collecting tubule. The cell is a control cell. The electrical potential difference across the luminal membrane (−73 mV) is smaller than the diffusional driving force for K^+ movement out of the cell. Thus K^+ is secreted into the lumen. In the presence of a maximal concentration of amiloride the cell hyperpolarizes such that the electrical potential difference across the luminal membrane (−91 mV) is approximately equal to the diffusional driving force for K^+ movement out of the cell. Thus K^+ secretion stops. (Voltages were taken from KOEPPEN et al. 1983)

apical membrane K^+ conductance which is mediated by K^+-selective ion channels distinct from those which transport Na^+ (WANG et al. 1992). The action of the $(Na^+ + K^+)$-pump will move K^+ into the cell across the basolateral membrane, and K^+ will be accumulated in the cell to levels above its electrochemical equilibrium. Movement of Na^+ into the cell through the apical channels will depolarize the luminal membrane, creating an even more favorable electrical driving force for K^+ movement from cell to lumen. In the absence of Na^+ transport, as occurs in the presence of amiloride, the apical membrane will hyperpolarize and the net driving force for K^+ secretion into the tubular fluid will diminish or perhaps vanish altogether. Thus K^+ secretion is expected to be strongly linked to Na^+ reabsorption in this segment, and indeed studies of rabbit (STOKES 1981) and rat (SCHAFER et al. 1990) CCTs confirm this dependence.

The kaliuresis produced by diuretics acting on the proximal tubule or the loop of Henle results in part from an increased delivery of *Na*$^+$ to the distal nephron (STANTON and GIEBISCH 1992). This in turn will increase Na^+ reabsorption in the DCT and CCT, with a concomitant increase in K^+ secretion. Thus blocking the Na^+ reabsorptive process in these segments with amiloride in combination with, for example, a loop diuretic, not only enhances the natriuresis produced by the first diuretic but will also reverse the increased K^+ excretion (HROPOT et al. 1985). These consideration may also explain why amiloride can be more effective as an antikaliuretic than as a natriuretic agent. In the distal convoluted tubule, a large portion of Na^+ reabsorption is mediated by an NaCl contransporter which is sensitive to thiazides but not to amiloride (see Velasquez Chap. 8, this volume). Because this transporter is electroneutral it will not affect the luminal membrane potential. Thus amiloride, acting on Na^+ channels, can have a disproportionately large effect on the membrane potential (and hence on K^+ secretion) relative to its effect on Na^+ reabsorption. In summary, inhibition of Na^+ channels can account very well for the observed effects of amiloride on both Na^+ and K^+ balance.

Changes in the electrical driving forces, either across the epithelium or across the apical membrane, probably also account for effects of amiloride on urinary acidification and divalent ion excretion. Amiloride impairs urinary acidification (GUIGNARD and PETERS 1970). Studies on the turtle bladder have demonstrated the strong dependence of the acidification mechanism in that tissue on the negative transepithelial potential (AL-AWQATI et al. 1977) and the reduced lumen-negative voltage probably accounts for the effects of the drug (ARRUDA et al. 1980). Amiloride also reduces urinary Mg^{2+} (DEVANE and RYAN 1983) and Ca^{2+} (COSTANZO and WEINER 1976) excretion. Using in vivo microperfusion techniques, COSTANZO (1984) showed that amiloride increased Ca^{2+} reabsorption in the late distal convoluted tubule, and that this effect was correlated with the inhibition of Na^+ transport by the drug. It is therefore likely that an amiloride-induced hyperpolarization of the luminal membrane increased the driving force for Ca^{2+} entry into the cell from the tubular fluid through an electrogenic transporter such as a channel.

III. Effects on Other Cellular Processes

Although the main pharmacological effects of amiloride can be accounted for by its interactions with the epithelial Na^+ channel, the drug also inhibits a number of other cellular processes, albeit at higher concentrations. A number of these amiloride-sensitive processes also involve the transmembrane transport of Na^+, both through cation exchangers and cation channels. This might reflect interactions of the drug with Na^+-binding sites on proteins. As discussed below, there is evidence in epithelial Na^+ channels that both amiloride and Na^+ interact with specific negatively charged sites in the mouth of the pore.

The most widely studied of the other amiloride-sensitive Na^+ transporters is the Na^+/H^+ exchanger (ARONSON 1985). In the apical membrane of the renal proximal tubule, this exchanger is involved in Na^+ and bicarbonate reabsorption (H^+ secretion). It is also expressed in the basolateral membrane of many epithelia and in most other cell types, where it plays a major role in regulating cell pH. The K_i for amiloride inhibition of Na^+/H^+ exchange varies from 10 to 100 μmol/l. However, some analogs of amiloride, particularly those such as ethyl isopropyl amiloride with substitutions at the 5-position of the ring, are much more potent against the Na^+/H^+ exchanger and less potent against the Na^+ channel (KLEYMAN and CRAGOE 1988). The Na^+/Ca^{2+} exchanger, which helps to regulate Ca^{2+} in the cytoplasm of many cell types, is also blocked by amiloride with a K_i in the range of 1 mmol/l (KACZOROWSKI et al. 1985). It has been possible to achieve 100- to 300-fold improvements in affinity with hydrophobic substitutions on both the side chain and 5-position on the pyrazine ring (KLEYMAN and CRAGOE 1988). The Na^+ pump, which is present in almost all cells, was inhibited by amiloride with half-maximal inhibition of around 1 mmol/l (SOLTOFF and MANDEL 1982).

Several other ion channels are also blocked by amiloride. There are two cases in which the sensitivity to amiloride is comparable to that of the classical epithelial Na^+ channel. Na^+-selective channels in the taste bud are blocked with a K_i of around 0.3 μmol/l (AVENET and LINDEMANN 1988). These channels, which can also be considered as epithelial channels, are thought to be involved in salt taste. Nonselective cation channels in thyroid cells inhibited by amiloride are completely blocked by 1 μmol/l amiloride (VERRIER et al. 1989). Other channels have a lower sensitivity to the drug. Olfactory cells contain odorant-stimulated channels which conduct Na^+ and K^+ equally well and are blocked by amiloride with a K_i of about 10 μmol/l (FRINGS and LINDEMANN 1988). Stretch-sensitive cation channels in the inner ear are inhibited with a K_i of about 50 μmol/l (JØRGENSEN and OHMORI 1988). T-type Ca^{2+} channels from the heart, which can also conduct Na^+ in the absence of extracellular Ca^{2+}, are blocked by amiloride with a K_i of around 30 μmol/l (TANG et al. 1988). A nonselective cation channel in endothelial cells is inhibited with a K_i of 10 μmol/l (VIGNE et al. 1989).

In addition to these Na^+-transporting proteins, there are least some amiloride-inhibitable processes which have no obvious connection to the binding or transport of Na^+. These include cAMP-dependent and tyrosine protein kinases, adenylate cyclase, kallikrein, and receptors for acetylcholine, catecholamines and atrial natriuretic factor. RNA synthesis, protein synthesis and oxidative phosphorylation can also be inhibited. For all of these effects, high concentrations of the blocker – usually 100 μmol/l or more – are required. A more complete description of these actions of amiloride is given by KLEYMAN and CRAGOE (1988).

IV. Interactions with the Epithelial Na^+ Channels

Since the earliest experiments on amphibian skin and urinary bladder many laboratories have found that amiloride has a rapid effect on Na^+ transport in vitro when added to the mucosal or apical solution, that this effect is readily reversible when the amiloride is removed from the solution, and that the drug has little or no effect, at least at concentrations of 10 μmol/l or less, from the serosal side. These observations indicate that amiloride affects the apical Na^+ transport system at a site outside the cell membrane.

1. Stoichiometry

In most tissues, the interactions of amiloride with the Na^+ transport system under any given conditions can be fairly well described by a simple 1:1 blocking reaction:

$$I = I_o/(1 + [\text{amiloride}]/K_i)$$

where I is the short-circuit current, I_o the current in the absence of amiloride, and K_i the apparent inhibition constant for amiloride (CUTHBERT 1974; CUTHBERT and FANELLI 1978). Typcal results are shown in Fig. 3. The amiloride-sensitive short-circuit current is generally accepted to represent the rate of Na^+ reabsorption through apical membrane channels. In addition, fluctuation analysis has indicated a linear relationship between reaction rate and amiloride concentration, consistent with this simple blocking scheme (LINDEMANN and VAN DRIESSCHE 1977; LI et al. 1982; LI and LINDEMANN 1983; ABRAMCHECK et al. 1985). In some cases, however, deviations from this behavior have been reported. BENOS (1982) found that the relationship between short-circuit current and amiloride concentration in the bullfrog skin was best described by raising the amiloride concentration to the 0.8 power (Hill coefficient of 0.82). This suggested negative cooperativity between amiloride-binding sites. An inhibition curve that was less steep than predicted for a simple blocking reaction was also noted by LEWIS and

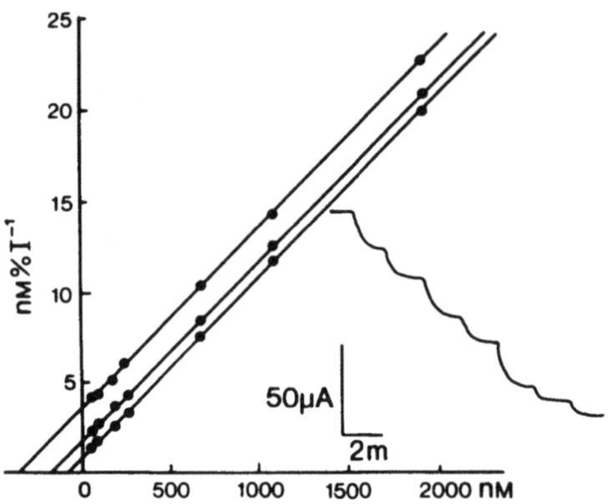

Fig. 3. Effects of amiloride on short-circuit current in frog skin. *Inset* shows a continuous trace of short-circuit current during increases in the amiloride concentration in the mucosal solution from 42 to 1915 nmol/l. Exposure to each concentration was 2 min. Reciprocal plots of concentration/percentage inhibition vs. concentration are shown for three different preparations. Values of K_i were obtained from the intercept of the straight lines with the abscissa and were 355, 180, and 86 nmol/l for the three cases. (Reproduced with permission from CUTHBERT and FANELLI 1978)

DIAMOND (1976) for the rabbit urinary bladder. The mechanisms for the apparent negative cooperativity may arise from the fact that the measurement (short-circuit current) involves the response of a whole tissue to the drug, rather than the amiloride receptor per se. As will be discussed below, the cells tend to adjust their transport properties in partial compensation for amiloride blockade.

2. Rate Constants

The rate constants for amiloride block have been estimated in three ways (Table 2). Fluctuation analysis of short-circuit current in the presence of amiloride reveals a Lorentzian power density spectrum (PDS) characteristic of blocker-induced noise (LINDEMANN and VAN DRIESSCHE 1977). This noise arises from spontaneous fluctuations in the number of *blocked* versus the number of *open* channels. Analysis of the corner frequency of the PDS as a function of amiloride concentration provides estimates of the on- and off-rates for block (Fig. 4). Typical values under standard experimental conditions (high Na^+ in the mucosal medium, room temperature) in a variety of epithelia are $k_{on} = 20\,sec^{-1}$ mmol/l and $k_{off} = 2\,s^{-1}$. A second method for estimating the rate constants makes use of the voltage dependence of the amiloride block (PALMER 1985; WARNCKE and LINDEMANN 1985). These methods measure relaxation behavior of the amiloride-sensitive current during either step or sinusoidal changes in membrane voltage. They are independent of the mechanisms underlying the voltage dependence. The techniques were applied to the toad urinary bladder and the results agreed well with estimates from fluctuation analysis. A third technique involves the

Table 2. Rate constants for amiloride block

Method/tissue	k_{on} (μmol/l s)	k_{off} (s^{-1})	K_i (μmol/l)	References
Noise analysis				
Toad bladder[a]	20	2	0.1	Li et al. 1982
Frog skin[a]	13	4	0.3	Li et al. 1987
Frog skin	23	3.5	0.2	Abramcheck et al. 1985
Frog colon	14	21	1.5	Krattenmacher et al. 1988
Frog lung	10	19	1.9	Fischer and Clauss 1990
Rabbit bladder[b]	46	13	0.3	Lewis et al. 1984
Rabbit colon[b]	68	8	0.4	Zeiske et al. 1982
Impedance				
Toad bladder	18	2.7	0.1	Warncke and Lindemann 1985
Frog skin	13	3.9	0.3	Warncke and Lindemann 1985
Toad bladder	22	3	0.1	Palmer 1985
Patch clamp				
Rat CCT	60	4	0.07	Palmer and Frindt 1986
A6 cells[c]	11	2.7	0.2	Eaton and Marunaka 1990
A6 cells[d]	34	30	0.9	Hamilton and Eaton 1985

[a] Basolateral membrane depolarized with K^+.
[b] 37°C.
[c] Highly selective channels.
[d] Poorly selective channels.

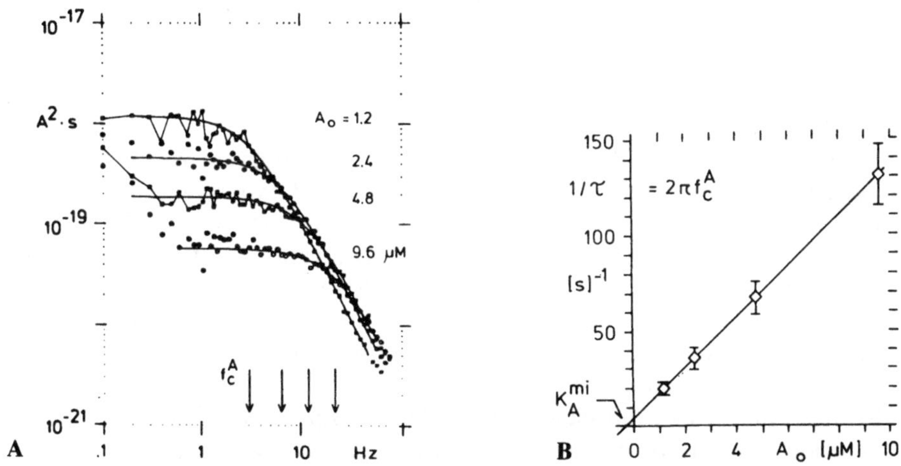

Fig. 4A,B. Amiloride-induced noise short-circuit current recordings from frog skin. In **A** the spectral density for shot noise is shown for four different amiloride concentrations, together with the best fit to a Lorentzian spectrum. In **B** the corner frequency (the point at which the noise amplitude is half-maximal) is plotted vs. the amiloride concentration. The on-rate constant is estimated from the slope of the relationship. The off-rate constant is estimated from the intercept with the abscissa. (Reproduced with permission from Li and Lindemann 1983)

measurement of currents through individual ion channels using the patch-clamp technique (Fig. 5). In the presence of a submaximal concentration of amiloride, the current through a single channel will fluctuate between the levels of the open and blocked states, revealing the mean lifetimes of these states (PALMER and FRINDT 1986; EATON and MARUNAKA 1990; LIGHT et al. 1990). Again, agreement with the other methods was reasonably good.

LI and LINDEMANN (1983) found that the *macroscopic* K_i for amiloride block of short-circuit current in frog skin was consistently higher than the *microscopic* block obtained from fluctuation analysis. They could interpret these results by assuming a closed state of the channel which was long lived compared to the open and amiloride-blocked states, and which had a relatively weak affinity for amiloride (Fig. 6a). A similar analysis has been carried out by HELMAN and BAXENDALE (1990), who found an increase in the number of blocked plus unblocked channels as the blocker concentration increased. This could also be explained if the channels could exist in a closed state which did not interact with amiloride, or reacted relatively weakly.

Patch-clamp results have confirmed the existence of long-lived closed states of the Na^+ channel. Using patch-clamp data from A6 cells EATON and MARUNAKA (1990) estimated that amiloride interacted much more slowly with the closed state of the channel. Since both the k_{on} and k_{off} were reduced to about the same extent, the affinities for the open and closed states were comparable. It is possible, however, that in other cells amiloride interacts more poorly with the closed channels, and that this might underlie the differences in macroscopic and microscopic K_i values, as well as the dependence of the (open + blocked) channel density on the amiloride concentration.

3. Competition with Na^+

The effects of mucosal Na^+ concentration on the apparent K_i for amiloride has been controversial. In a number of tissues and experimental conditions there was evidence for competitive interactions in that the K_i for block by amiloride increased with increasing Na^+ concentrations in the mucosal bath (CUTHBERT and SHUM 1976; SUDOU and HOSHI 1977; PALMER 1984). In studies of amiloride kinetics the on rate for amiloride block (k_{on}) was decreased at higher Na^+ concentrations, providing further evidence for competitive interactions between blocker and substrate (PALMER 1985; HOSHIKO and VAN DRIESSCHE 1986). The competitive effects of Na^+ on amiloride block have not always been observed, however (BENTLEY 1968; BENOS et al. 1979). For example, BENOS et al. (1979) found the inhibition to be competitive in some epithelia and noncompetitive in others. The reasons for the discrepant findings with respect to amiloride-Na^+ competition remain unclear.

The issue is complicated by the possibility that Na^+ and amiloride may interact in at least two ways. First, they might compete for a single binding

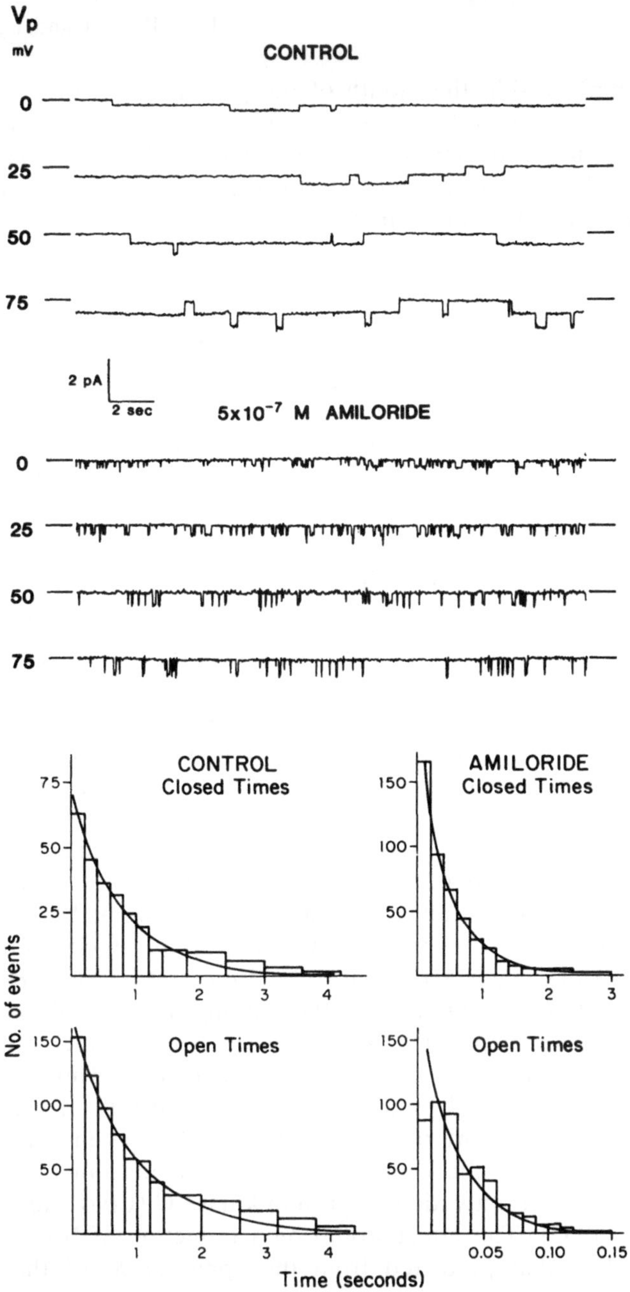

Fig. 5. Effect of amiloride on single-channel currents in the rat cortical collecting tubule. *Above*, currents in two cell-attached patches are illustrated. *Numbers to the left of the traces* indicate the applied potential across the patch, outside positive. Downward current deflections indicate channel openings, which are long lived in the absence of amiloride and short lived in the presence of amiloride. *Below* the open- and closed-time histograms are shown for the two cases. Both the open and closed times are reduced in the presence of amiloride. (Reproduced with permission from PALMER and FRINDT 1986)

site, for example within the mouth of the pore (see below). Second, Na^+ may promote a *closed* state of the channel, which as discussed above may interact differently with amiloride. This effect could be mediated directly through a regulatory Na^+-binding site on the outer surface of the channel, or indirectly through changes in the intracellular milieu. These two possibilities have been termed "self-inhibition" and "feedback inhibition" (LI and LINDEMANN 1983). If amiloride does not bind to or block the channel when it is in the closed state, this will give rise to competitive kinetics. If amiloride binds equally well to the open and closed states, this interaction will be noncompetitive. As mentioned above, patch-clamp recordings have confirmed the existence of the closed state. While there is no evidence for the self-inhibition phenomenon from single-channel records (PALMER and FRINDT 1988), feedback inhibition has been documented as described in the next section.

4. Feedback Response to Amiloride

While the response of an epithelium to a maximal dose of amiloride is rapid and stable, the response to a submaximal dose is more complex. In frog skin, when amiloride is applied at a dose sufficient to block 1/2 to 3/4 of the short-circuit current, the rapid fall was followed by a slower recovery toward the initial level (ABRAMCHECK et al. 1985; HELMAN and BAXENDALE 1990). Thus the apparent density of channels increases with the concentration of the blocker. This secondary response was termed "autoregulation" as it suggests that the epithelium is attempting to maintain a constant rate of Na^+ transport in spite of having a large percentage of channels blocked. Similar results were reported in studies of toad bladder (LI et al. 1982).

Analogous observations were made using the patch-clamp technique in cultured toad kidney cells and in rat CCT (LING and EATON 1989; FRINDT et al. 1993). Addition of amiloride to the bathing medium blocked channels outside the cell-attached patch but increased Na^+ channel activity in the patch which was protected from the drug by the pipette. Thus it appears that the channels are regulated by a feedback response system which tends to stabilize the Na^+ transport rate in the face of changes in Na^+ delivery to and/or Na^+ entry into the cells. The mechanism may involve changes in membrane voltage or activation of protein kinase C (LING and EATON 1989; FRINDT et al. 1993). The effect would be a reduction of Na^+ reabsorption that is less than that predicted from the apparent K_i of the drug. This phenomenon may account for the apparent negative cooperativity (Hill coefficient <1) of the effect of amiloride on short-circuit current discussed above.

5. Divalent Cation Requirements

CUTHBERT and WONG (1972) and BENOS et al. (1976) found that Ca^{2+} or a substitute divalent cation was required for amiloride inhibition of Na^+

transport in frog skin. They suggested that Ca^{2+} might be involved in the formation of the amiloride-channel complex. More recent data by DESMEDT et al. (1993) have challenged this conclusion. These authors reported that removal of divalent cations from the mucosal solution opened a second pathway for Na^+ transport which was nonselective among small cations and amiloride insensitive. They argued that current through the Na^+-selective pathway retained amiloride sensitivity even at very low divalent cation concentrations.

6. Model for Amiloride Block

CUTHBERT (1976) suggested that the block of Na^+ channels by amiloride might result from the competition of the positively charged guanidinium group and an Na^+ for a negatively charged binding site within the pore. The arguments for such a mechanism were based on the observations that (a) amiloride had to have a positive charge to effectively block the channels, (b) that the channel appeared to have an ionizable group with a pK_a of around 5 which needed to be deprotonated to support Na^+ transport and (c) that tetrodotoxin, which also contains a guanidinium group, has a similar blocking effect on voltage-gated Na^+ channels. The finding of competitive interactions between amiloride and Na^+ were consistent with this view, although as discussed above this competition has not been universally observed.

More direct evidence in favor of this idea came from studies of the voltage dependence of amiloride block. Making the cell voltage more negative increased the K_i for amiloride (PALMER 1984, 1985; WARNCKE and LINDEMANN 1985). Using impedance analysis or current relaxation after voltage steps, hyperpolarization was found to increase the k_{on} and decrease k_{off} for amiloride block. The findings were consistent with the idea tha the charged guanidinium portion of the amiloride molecule binds within the electric field of membrane, so that a negative cell potential pulls the blocker into its binding site. The most logical site at which this could occur is within the pore itself. Quantitative evaluation of the effect of voltage on K_i was consistent with a single positive charge sensing about 15%–20% of the transmembrane electric field at its binding site. If the portion of the molecule which fits within the channel contains a positive charge of less than +1 as a result of charge distribution along the molecule (WARNCKE and LINDEMANN 1985), then it would have to penetrate further into the field to account for the data.

Several additional pieces of evidence supported this model for amiloride block. First, the guanidinium ion itself was found to block Na^+ channels in the toad bladder (PALMER 1985). Although the affinity was very weak (K_i = 100 mmol/l) the voltage dependence for guanidinium block produced the same estimate of the fraction of the field sensed by the blocker at its site. This suggests that both guanidinium itself and the guanidinium side chain of amiloride bind to the same site within the pore. Furthermore, the competition between amiloride and either K^+ ions or guanidinium ions was voltage dependent (PALMER and ANDERSEN 1989). Hyperpolarization of the membrane increased the ability of these ions to displace amiloride from its binding site. This is predicted by the model in which all of these ions enter the pore from outside, preventing Na^+ from entering, but do not themselves go through the channel. Under these conditions the hyperpolarization will increase the occupancy of the pore by the competing ions. (The effect should be symmetric, such that amiloride will also compete more effectively with K^+ and guanidinium.) On the other hand, the permeant ions Na^+ and Li^+ also competed with amiloride. However, there was no discernible effect of voltage on ability of these ions to displace amiloride. These data could be explained by a model in which voltage increased the rates for both entry of Na^+ into the pore and exit of Na^+ from the pore to roughly the same extent. The occupancy of the pore by Na^+ was therefore a function of Na^+ concentration but was independent of voltage. Thus the site at which

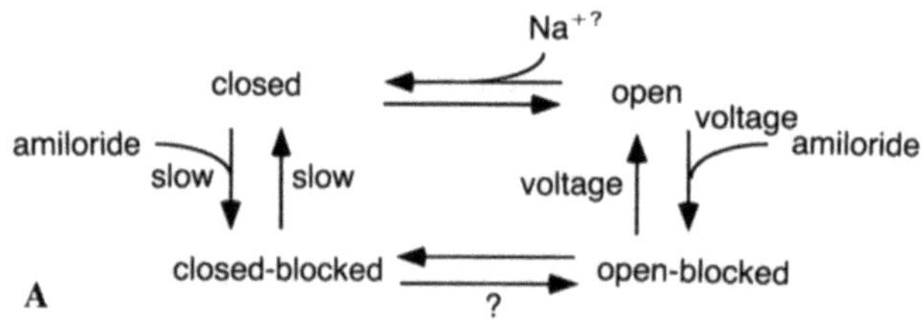

A

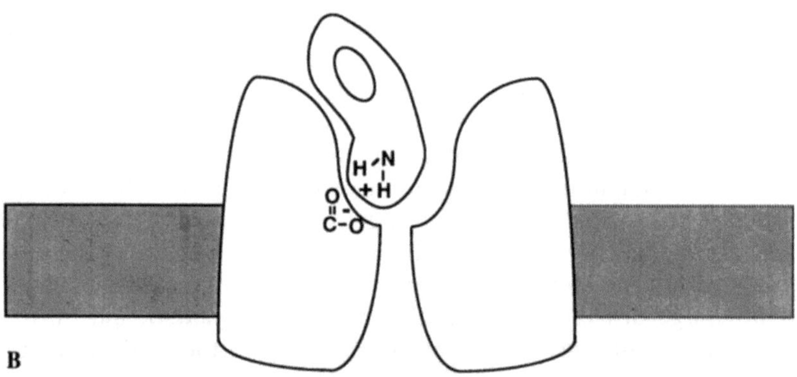

B

Fig. 6. Model for amiloride block of epithelial Na^+ channels. **A** Minimal kinetic scheme showing interactions of amiloride with the open and closed states of the channel. *Arrows marked with V* indicate that the rate constants are voltage dependent. The transition from open to closed states is facilitated by high concentrations of Na^+ either inside or outside the cell. The blocking reaction, at least for the open channels, may involve two separate reactions including the formation of an encounter complex (see text). **B** A structural model for amiloride block, showing that the guanidinium portion of the molecule may enter the pore

Na^+ competes with amiloride appears to be within the pore itself. These arguments have been reviewed in more detail (PALMER 1991).

LI et al. (1987) proposed a more elaborate version of the pore-plugging model for amiloride block to account for the dependence of on rate on the structure of the side chain. In their scheme, the molecule enters the pore under the influence of the membrane voltage and forms an "encounter complex" in which guanidinium side chain interacts with a carboxyl group in the mouth of the pore (Fig. 6b). The encounter complex has a short lifetime. The molecule will rapidly either leave the channel or form the relatively stable blocked state. The latter transition is not necessarily voltage dependent, but the dissociation of the blocker from the channel is inhibited by a negative cell potential.

D. Pharmacokinetics

Amiloride is administered in an oral preparation and is readily absorbed from the gastrointestinal tract. The drug is not altered by the liver and is recovered in the urine and stool unchanged (BAER et al. 1967; WEISS et al. 1969; VIDT 1981). Although amiloride is not metabolized, derivatives of amiloride, such as dimethylamiloride, are metabolized in the liver and converted to amiloride (BAER et al. 1967). Approximately 50% of an oral dose is excreted in the urine (WEISS et al. 1969; GRAYSON et al. 1971; SMITH and SMITH 1973; KNOWLES et al. 1992). Urinary excretion of amiloride occurs, in part, via glomerular filtration. The peak clearance of amiloride exceeds the clearance of inulin by four to fivefold, indicating that amiloride is likely secreted by the proximal tubule (WEISS et al. 1969; GRAYSON et al. 1971; SMITH and SMITH 1973). Several compounds, including creatinine and digoxin, compete with amiloride for tubular secretion (VIDT 1981; WALDORFF et al. 1981). Therefore, administration of amiloride may result in an elevation of serum creatinine levels, although the clearance of inulin, or glomerular filtration rate, remains unchanged. The remaining 50% of orally administered amiloride appears in the stool (WEISS et al. 1969; VIDT 1981; KNOWLES et al. 1992). Small amounts of amiloride can be found in tracheobronchial fluid, but at concentrations of less than 10% that of plasma (KNOWLES et al. 1992). Amiloride is not secreted into bile in dogs, suggesting that amiloride which appears in stool was not absorbed by the gastrointestinal tract (BAER et al. 1967). In man, amiloride was formerly assumed to be reabsorbed by about 50% of the given dose. However, more recent findings (SPAHN et al. 1987) have forced us to change our minds. In healthy controls, 50% of amiloride administered was recovered in the urine. In patients with liver disease, the amount excreted in the urine increased in proportion to their direct bilirubin to almost 100%. In conclusion, hepatobiliary clearance of amiloride is significant ($\approx$ 50%) and this route of elimination is impaired in liver disease. Thus, amiloride dosage and the patient's kidney function have to be cautiously considered in hepatic dysfunction. An increased half-life has been observed in patients with renal insufficiency and creatinine clearances of less than 20 ml/min (GEORGE 1980; VIDT 1981; SPAHN et al. 1987). It has not reported whether amiloride can be efficiently removed by dialysis.

A single oral dose of amiloride reaches peak serum levels 3–6 h after administration (Fig. 7). The serum half-life is 6–10 h (WEISS et al. 1969; GRAYSON et al. 1971; SMITH and SMITH 1973; KNOWLES et al. 1992). A half-life of 18 h has been found by others (SPAHN et al. 1987). Amiloride binds weakly to plasma proteins and can be detected in serum as long as 72–96 h following a single dose (WEISS et al. 1969; GRAYSON et al. 1971; KNOWLES et al. 1992). Approximately 25% of an administered dose is excreted within the first 10 h (WEISS et al. 1969; KNOWLES et al. 1992), and this percentage does not appear to vary with dosage. The volume of distribution of amiloride in dogs has been shown to be 270%–370% of estimated body volume, and in humans is approximately 5 l/kg body weight, suggesting extensive tissue uptake, although it is not concentrated within a specific organ (BAER et al. 1967; GRAYSON et al. 1971; SMITH and SMITH 1973; DOLLERY 1991).

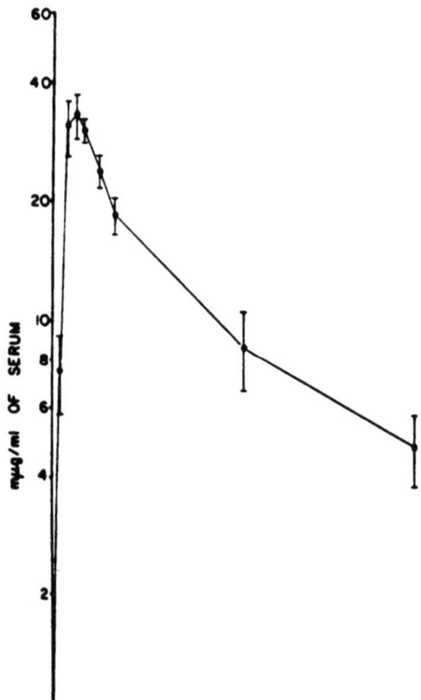

Fig. 7. Mean serum concentrations of amiloride hydrochloride in five volunteers following oral administration (n = 5, ±SEM). (Reproduced with permission from WEISS et al. 1969)

Urinary concentrations of amiloride following a single 20 mg dose reached levels of 4–30 mg/l and have an estimated concentration in the distal nephron of 3–20 μmol/l (GRAYSON et al. 1971; SMITH and SMITH 1973), more than sufficient to inhibit epithelial Na^+ channels. The peak effects of amiloride occur 6–10 h after ingestion and the effects last more than 10 h. Peak serum levels are approximately 0.1 μmol/l (GRAYSON et al. 1971; KNOWLES et al. 1992), suggesting that, despite proximal tubular secretion of amiloride, levels of amiloride in the proximal tubule are likely insufficient to inhibit Na^+-dependent transport processes, including Na^+/H^+ exchange, Na^+-glucose cotransport, Na^+-alanine cotransport, and Na^+-phosphate cotransport (KLEYMAN and CRAGOE 1988, 1990). In addition, serum levels of amiloride are insufficient to inhibit $(Na^+ + K^+)$-ATPase.

The pharmacokinetics of aerosolized amiloride have also been examined. The concentration of amiloride on the bronchial surface is greater than 10^{-6} mol/l for 3–5 h following administration of aerosolized amiloride ($\sim 10^{-2}$ mol/l in a volume of 3.3 ml), a concentration sufficient to inhibit Na^+ channels and depolarize tracheobronchial potential difference (KNOWLES et al. 1991, 1992). The half-life of amiloride at the bronchial surface is 35–40 min (KNOWLES et al. 1992). Approximately 20% of aerosolized amiloride is absorbed and recovered in urine and stool. The short-term administration of amiloride does not alter pulmonary function, including gas exchange, lung volume, or diffusing capacity. Long-term administration of amiloride in healthy volunteers does not alter pulmonary function, serum electrolytes, renal function, or blood pressure (KNOWLES et al. 1992).

E. Therapeutic Use

Diuretics are used in the treatment of hypertension and edematous states, including congestive heart failure and cirrhosis. Amiloride has been used alone as a diuretic under these circumstances, although more potent diuretics which affect different sites in the nephron, including thiazides and loop diuretics, are widely used in the treatment of these disorders. Both thiazide and loop diuretics induce kaliuresis.

Amiloride was initially developed as a diuretic which induces a mild natriuresis without an associated kaliuresis (BULL and LARAGH 1968). It is the K^+-sparing effects of amiloride which have been successfully exploitied clinically (KRISHNA et al. 1988). The coadministration of K^+-sparing diuretics, such as amiloride, with the kaliuretic thiazide or loop diuretics prevents hypokalemia and eliminates the need for oral K^+ supplementation (VIDT 1981; KRISHNA et al. 1988; DOLLERY 1991). Amiloride is usually administered in an oral dose of 5–10 mg/day with thiazide or loop diuretics. As hypokalemia may lead to further elevations in blood pressure in patients with hypertension, K^+-sparing diuretics may benefit patients who are receiving diuretics in the treatment of hypertension (KRISHNA 1990; KRISHNA and Kapoor 1991). The prevention of hypokalemia is also important in patients with congestive heart failure who receive digoxin.

Amiloride has Mg^{2+}-sparing properties in addition to K^+ sparing. The use of amiloride in combination with other diuretics, such as furosemide, prevents diuretic-induced Mg^{2+} wasting (Devane and Ryan 1983; Ryan 1986). Amiloride also has Ca^{2+}-sparing properties, decreasing renal Ca^{2+} excretion (Costanzo and Weiner 1976).

Li^+ also enters distal tubular epithelial cells through Na^+ channels and accumulates in the intracellular space. The administration of Li^+ in the treatment of manic/depressive disorder often results in a decrease in the cellular response to arginine vasopressin, leading to an increase in urinary free water excretion and hypernatremia (Boton et al. 1987). Administration of amiloride in these patients may prevent the intracellular accumulation of Li^+ and development of hypernatremia (Batlle et al. 1985; Boton et al. 1987).

Amiloride may also be used to prevent hypokalemia and the expansion of the extracellular fluid volume associated with elevated circulating levels of glucocorticoids or mineralocorticoids. Although it does not directly antagonize the binding of steroids to mineralocorticoid receptors, amiloride indirectly reverses the effects of aldosterone by blocking the Na^+ channels whose activity is controlled by mineralocorticoids. As discussed above, this will lead to inhibition of K^+ secretion as well as of Na^+ reabsorption, especially in the cortical collecting tubule.

In addition, amiloride may prove useful in the treatment of cystic fibrosis. Amiloride-sensitive Na^+ channels are present in airway epithelial cells (Cotton et al. 1987; Welsh 1990; Boucher et al. 1986, 1991; Disser et al. 1991). Mutations in a cAMP-regulated Cl^- channel, the cystic fibrosis transmembrane conductance regulator (CFTR), are responsible for the cystic fibrosis phenotype (Riordan et al. 1989; Anderson et al. 1991; Rich et a. 1991). The inability to activate airway epithelial Cl^- channels is one of the factors likely responsible for the thick, dehydrated secretions observed in cystic fibrosis (Welsh 1990). Increased Na^+ transport rates and abnormal regulation of Na^+ reabsorption by cAMP-mediated agonist pathways have also been observed in cystic fibrosis airway epithelium and in primary cultures of airway epithelial cells derived from patients with cystic fibrosis, and may also contribute to dehydration of airway fluids (Boucher et al. 1986, 1988, 1991; Cotton et al. 1987; Welsh 1990; Disser et al. 1991). Enhanced rates of Na^+ transport across cystic fibrosis airway epithelia are not simply a consequence of altered apical membrane Cl^- permeability; the mechanisms by which mutations in CFTR alters Na^+ channel function remain undefined.

Increased transepithelial Na^+ transport will lead to enhanced fluid *reabsorption* from the airway surfaces of patients with cystic fibrosis, exacerbating the decrease in fluid *secretion* which results from defected Cl^-

permeability. The increased Na^+ permeability could serve as an important factor contributing to the abnormal secretions observed in cystic fibrosis. Application of amiloride to the airway surface will counteract the increased Na^+ permeability. Clinical trials delivering amiloride to airway epithelia through inhalation of aerosolized drug are ongoing, and preliminary results do suggest that amiloride may slow the progressive deterioration of pulmonary function (KNOWLES et al. 1990, 1991, 1992). Improvements in airway surface liquid biorheology, and in mucociliary and cough clearance, were noted in patients with cystic fibrosis receiving aerosolized amiloride. A decrease in the rate of decline of forced vital capacity was also noted.

F. Side Effects and Toxicity

Amiloride inhibits renal K^+ secretion and the major side effect associated with administration of amiloride is hyperkalemia. Approximately 10% of patients receiving amiloride in the absence of kaliuretic diuretics will develop hyperkalemia, defined as serum K^+ concentration of greater than 5.5 mmol/l. The incidence of hyperkalemia decreases to 1%–2% when amiloride is administered with a thiazide diuretic (DOLLERY 1991; PHYSICIANS DESK REFERENCE 1993). The incidence of hyperkalemia may increase in patients receiving ß-adrenergic antagonists or converting enzyme inhibitors (JOHNSTON et al. 1992). In addition, the incidence of hyperkalemia is increased in patients with renal insufficiency, diabetes mellitus, metabolic or respiratory acidosis, and in the elderly (DOLLERY 1991; PHYSICIANS DESK REFERENCE 1993). The administration of amiloride may be associated with hyponatremia, elevations in serum creatinine and blood urea nitrogen (VIDT 1981; DOLLERY 1991; PHYSICIANS DESK REFERENCE 1993), and an increase in urinary pH and decrease in serum bicarbonate concentration as a result of decreased proton secretion in the distal nephron (BAER et al. 1967; GUIGNARD and PETERS 1970). The latter effect is presumably the result of a decreased transepithelial electrical potential difference in the distal nephron.

Amiloride has no known tumorogenic effect in laboratory animals and humans, and had no mutagenic activity in the Ames test. Amiloride has no known teratogenic effects in animals or in humans, although a decrease in rat pup growth and survival has been noted in animals receiving greater than five times the maximal daily dose for humans. Amiloride is excreted in rat milk and levels in milk exceed plasma levels. It is unknown whether amiloride is secreted in human milk. The LD_{50} of amiloride is 56 mg/kg in mice and between 36 and 85 mg/kg in rats (PHYSICIANS DESK REFERENCE 1993).

Aside from hyperkalemia, minor adverse side effects have been reported with the administration of amiloride. These include headache, weakness, dizziness, muscle cramps, nausea, anorexia, cough, dyspnea, abdominal pain, flatulence, impotence, and skin rash (DOLLERY 1991).

G. Drug Interactions

The concomitant administration of amiloride and K^+ may result in significant, potentially fatal, hyperkalemia and should be avoided. As mentioned above, the administration of amiloride with converting enzyme inhibitors or with β-adrenergic antagonists may increase the risk of hyperkalemia, and serum K^+ should be closely monitored. Administration of amiloride will decrease renal clearance of digoxin and lead to increases in serum digoxin levels (WALDORFF et al. 1981; IMPIVAARA and IISOLO 1985).

Although the administration of amiloride in patients receiving Li^+ may prevent hypernatremia, amiloride may reduce the renal clearance of Li^+ and result in elevated, potentially toxic, serum Li^+ levels (DOREVITCH and BARUCH 1986; ATHERTON et al. 1987). The administration of nonsteroidal antiinflammatory agents to patients receiving amiloride may reduce the effectiveness of the diuretic, and the dosage may need to be altered (DOLLERY 1991).

References

Abramcheck FJ, Van Driessche W, Helman SI (1985) Autoregulation of apical membrane Na^+ permeability of tight epithelia. Noise analysis with amiloride and CGS4270. J Gen Physiol 85:555–582

Al-Awqati Q, Mueller A, Steinmetz PR (1977) Transport of H^+ against electrochemical gradients in turtle urinary bladder. Am J Physiol 233:F502–F508

Anderson MR, Gregory RJ, Thompson S, Souza DW, Paul S, Mulligan RC, Smith AE, Welsh MJ (1991) Demonstration that CFTR is a chloride channel by alteration of its anion selectivity. Science 253:202–205

Aronson PS (1985) Kinetic properties of the plasma membrane Na^+-H^+ exchanger. Annu Rev Physiol 47:545–560

Arruda JA, Subbarayudu K, Dytko G, Mola R, Kurtzman NA (1980) Voltage-dependent distal acidification defect induced by amiloride. J Lab Clin Med 95:407–416

Atherton JC, Green R, Hughes S, McFall V, Sharples JA, Solomon LR, Wilson L (1987) Lithium clearance in man; effects of dietary salt intake, acute changes in extracellular fluid volume, amiloride and furosemide. Clin Sci 73:645–651

Avenet P, Lindemann B (1988) Amiloride-blockable sodium currents in isolated taste receptor cells. J Membr Biol 105:245–255

Baer JE, Jones CB, Spitzer SA, Russo HF (1967) The potassium-sparing and natriuretic activity of N-amidino-3,5-diamino-6-chloropyrazinecarboxamide hydrochloride dihydrate (amiloride hydrochloride). J Pharmacol Exp Ther 157: 472–485

Batlle DC, von Riotte AB, Gaviria M, Grupp M (1985) Amelioration of polyuria by amiloride in patients receiving long-term lithium therapy. N Engl J Med 312: 408–414

Benos DJ (1982) Amiloride: a molecular probe of sodium transport in tissues and cells. Am J Physiol 242:C131–C145

Benos DJ, Simon SA, Mandel LJ, Cala PM (1976) Effect of amiloride and some of its analogues on cation transport in isolated frog skin and thin lipid membranes. J Gen Physiol 68:43–63

Benos DJ, Mandel LJ, Balaban RS (1979) On the mechanism of the amiloride-sodium entry site interaction in anuran skin epithelia. J Gen Physiol 73:307–326

Benos DJ, Saccomani G, Brenner BM, Sariban-Sohraby S (1986) Purification and characterization of the amiloride-sensitive sodium channel from A6 cultured cells and bovine renal papilla. Proc Natl Acad Sci USA 83:8525–8529
Bentley PJ (1968) Amiloride: a potent inhibitor of sodium transport across the toad bladder. J Physiol (Lond) 195:317–330
Bicking JB, Mason JW, Woltersdorf OWJ, Jones JH, Kwong SF, Robb CM, Cragoe EJJ (1965) Pyrazine diuretics I. N-amidino-3-amino-6-halopyrazine-carboxamides. J Med Chem 8:638–642
Boton R, Gaviria M, Batlle DC (1987) Prevalence, pathogenesis, and treatment of renal dysfunction associated with chronic lithium therapy. Am J Kidney Dis 10:329–345
Boucher RC, Stutts MJ, Knowles MR, Cantley L, Gatzy JT (1986) Na^+ transport in cystic fibrosis respiratory epithelia. Abnormal basal rate and response to adenylate cyclase activation. J Clin Invest 78:1245–1252
Boucher RC, Cotton CU, Gatzy JT, Knowles MR, Yankaskas JR (1988) Evidence for reduced Cl^- and increased Na^+ permeability in cystic fibrosis human primary cell cultures. J Physiol (Lond) 405:77–103
Boucher RC, Chinet T, Willumsen N, Knowles MR, Stutts MJ (1991) Ion transport in normal and CF airway epithelia. Adv Exp Med Biol 290:105–115
Bull MB, Laragh JH (1968) Amiloride. A potassium-sparing diuretic. Circulation 37:45–53
Canessa CM, Horisberger J-D, Rossier BC (1993) Epithelial sodium channel related to proteins involved in neurodegeration. Nature 361:467–470
Costanzo LS (1984) Comparison of calcium and sodium transport in early and late rat distal tubules: effect of amiloride. Am J Physiol 246:F937–F945
Costanzo LS, Weiner IM (1976) Relationship between clearances of Ca^{2+} and Na^+: effect of distal diuretics and PTH. Am J Physiol 230:67–73
Cotton CU, Stutts MJ, Knowles MR, Gatzy JT, Boucher RC (1987) Abnormal apical cell membrane in cystic fibrosis respiratory epithelium. An in vitro electrophysiological analysis. J Clin Invest 79:80–85
Cuthbert AW (1974) Interactions of sodium channels in transporting epithelia: a two-state model. Mol Pharmacol 10:892–903
Cuthbert AW (1976) Importance of guanidinium groups for blocking sodium channels in epithelia. Mol Pharmacol 12:945–957
Cuthbert AW, Fanelli GM (1978) Effect of some pyrazine-carboxamides on sodium transport in frog skin. Br J Pharmacol 63:139–149
Cuthbert AW, Shum WK (1976) Induction of transporting sites in a sodium transporting epithelium. J Physiol (Lond) 260:223–235
Cuthbert AW, Wong PYD (1972) The role of calcium ions in the interaction of amiloride with membrane receptors. Mol Pharmacol 8:222–229
Desmedt L, Simaels J, Van Driessche W (1993) Ca^{2+}-blockade, poorly selective cation channels in the apical membrane of amphibian epithelia. UO_2^{2+} reveals two channel types. J Gen Physiol 101:85–102
Devane J, Ryan MP (1983) Dose-dependent reduction in renal magnesium clearance by amiloride during furosemide-induced diuresis in rats. Br J Pharmacol 80: 421–428
Disser J, Hazama A, Frömter E (1991) Some properties of sodium and chloride channels in respiratory epithelia of CF- and non-CF-patients. Adv Exp Med Biol 290:133–141
Dollery C (1991) Amiloride. Therapeutic drugs. Churchill Livingstone, New York, pp A79–A82
Dorevitch A, Baruch E (1986) Lithium toxicity induced by combined amiloride-HCl hydrochlorothiazide administration. Am J Psychol 143:257–258
Duarte CG, Chomety F, Giebisch G (1971) Effect of amiloride, ouabain, and furosemide on distal tubular function in the rat. Am J Physiol 221:632–639
Eaton DC, Hamilton KL (1988) The amiloride-blockable sodium channel of epithelial tissue. Ion Channels. Plenum, New York, pp 151–182

Eaton DC, Marunaka Y (1990) Ion channel fluctuations: "noise" and single-channel measurements. Curr Top Membr Transp 37:61–113
Ehrlich EN, Crabbé J (1968) The mechanism of action of amipramizide. Pflugers Arch 302:79–96
Fischer H, Clauss W (1990) Regulation of Na^+ channels in frog lung epithelium: a target for aldosterone action. Pflugers Arch 416:62–67
Frindt G, Sackin H, Palmer LG (1990) Whole-cell currents in rat cortical collecting tubule: low-Na^+ diet increases amiloride-sensitive conductance. Am J Physiol 258:F562–F567
Frindt G, Silver RB, Windhager EE, Palmer LG (1993) Feedback inhibition of Na^+ channels in rat CCT. II. Effects of inhibition of Na^+ entry. Am J Physiol 264:F565–F574
Frings S, Lindemann B (1988) Odorant response of isolated olfactory receptor cells is blocked by amiloride. J Membr Biol 105:157–172
Frizzell RA, Koch MJ, Schultz SG (1976) Ion transport by the rabbit colon: I. Active and passive components. J Membr Biol 27:297–316
Garty H (1986) Mechanisms of aldosterone action in tight epithelia (topical review). J Membr Biol 90:193–205
Garty H (1992) Regulation of Na^+ permeability by aldosterone. Semin Nephrol 12:24–29
Garty H, Benos DJ (1988) Characteristics and regulatory mechanisms of the amiloride-blockable Na^+ channel. Physiol Rev 68:309–373
George DV (1980) Amiloride handling in renal failure. Br J Clin Pharmacol 9:94–95
Giebisch G, Field MJ (1989) Mechanisms of segmental potassium reabsorption and secretion. The regulation of postassium balance. Raven, New York, pp 139–155
Gögelein H, Greger R (1986) Na^+-selective channels in the apical membrane of rabbit late proximal tubules (pars recta). Pflugers Arch 406:198–203
Gombes EA, Freis ED, Moghadam A (1966) Effects of MK-870 in normal subjects and hypertensive patients. N Engl J Med 275:1215–1220
Grayson MF, Smith AJ, Smith RN (1971) Absorption, distribution and elimination of ^{14}C-amiloride in normal human subjects. Br J Pharmacol 43:734P–744P
Greger R (1985) Ion transport mechanisms in thick ascending limb of Henle's loop of mammalian nephron. Physiol Rev 65:760–797
Guignard JP, Peters G (1970) Effects of triamterene and amiloride on urinary acidification and potassium excretion in the rat. Eur J Pharmacol 10:255–267
Hamilton KL, Eaton DC (1985) Single channel recordings from amiloride-sensitive epithelial soidium channel. Am J Physiol 249:C200–C207
Helman SI, Baxendale LM (1990) Blocker-related changes of channel density. Analysis of a three-state model for apical Na^+ channels of frog skin. J Gen Physiol 95:647–678
Hoshiko T, Van Driessche W (1986) Effect of sodium on amiloride- and triamterene-induced current fluctuations in isolated frog skin. J Gen Physiol 87:425–442
Hropot M, Fowler N, Karlmark B, Giebisch G (1985) Tubular action of diuretics: distal effects on electrolyte transport and acification. Kidney Int 28:477–489
Impivaara O, Iisolo E (1985) Serum digoxin concentrations in a representative digoxin-consuming adult population. Eur J Clin Pharmacol 27:627–632
Johnston RT, de Bono DP, Nyman CR (1992) Preventable sudden death in patients receiving angiotensin converting enzyme inhibitors and loop/postassium-sparing diuretic combinations. Int J Cardiol 34:213–215
Jørgensen F, Ohmori H (1988) Amiloride blocks the mechanoelectrical transduction channel of hair cells of the chick. J Physiol (Lond) 403:577–588
Kaczorowski GJ, Barrows GJ, Dethmers JK, Trumble MJ (1985) Inhibition of Na^+/Ca^{2+} exchange in pituitary plasma membrane vesicles by analogues of amiloride. Biochemistry 24:1394–1403
Kleyman TR, Cragoe EJ Jr (1988) Amiloride and its analogs as tools in the study of ion transport. J Membr Biol 105:1–21

Kleyman TR, Cragoe EJ Jr (1990) Cation transport probes: the amiloride series. Methods in enzymology. Cellular and subcellular transport: epithelial cell. Academic, Orlando, pp 739–755

Knowles M, Murray G, Shallal J, Askin F, Ranga V, Gatzy J, Boucher R (1984) Bioelectric properties and ion flow across excised human bronchi. J Appl Physiol 56:868–877

Knowles MR, Church NL, Waltner WE, Yankaskas JR, Gilligan P, King M, Edwards LJ, Helms RW, Boucher RC (1990) A pilot study of aerosolized amiloride for the treatment of lung disease in cystic fibrosis. N Engl J Med 322:1189–1194

Knowles MR, Church NL, Waltner WE, Yankaskas JR, Gilligan P, King M, Edwards LJ, Helms RW, Boucher RC (1991) Aerosolized amiloride as treatment of cystic fibrosis lung disease: a pilot study. Adv Exp Med Biol 290: 105–115

Knowles MR, Church NL, Waltner WE, Gatzy JT, Boucher RC (1992) Amiloride in cystic fibrosis: safety, pharmacokinetics and efficacy in the treatment of pulmonary disease. Amiloride and its analogs: unique cation transport inhibitors. VCH, New York, pp 301–316

Koefoed-Johnsen V, Ussing HH (1958) On the nature of the frog skin potential. Acta Physiol Scand 42:298–308

Koeppen BM (1986) Conductive properties of the rabbit outer medullary collecting duct: outer stripe. Am J Physiol 250:F70–F76

Koeppen BM, Biagi BA, Giebisch G (1983) Intracellular microelectrode characterization of the rabbit cortical collecting duct. Am J Physiol 244:F35–F47

Krattenmacher R, Fischer H, Van Driessche W, Clauss W (1988) Noise analysis of cAMP-stimulated Na^+ current in frog colon. Pflugers Arch 412:568–573

Krishna GG (1990) Effect of potassium intake on blood pressure. J Am Soc Nephrol 1:43–52

Krishna GG, Kapoor SC (1991) Potassium depletion exacerbates essential hypertension. Ann Intern Med 115:77–83

Krishna GG, Shulman MD, Narins RG (1988) Clinical use of potassium-sparing diuretics. Semin Nephrol 8:354–365

Lewis SA, Diamond JM (1976) Na^+ transport by rabbit urinary bladder, a tight epithelium. J Membr Biol 28:1–40

Lewis SA, Ifshin MS, Loo DD, Diamond J (1984) Studies of sodium channels in rabbit urinary bladder by noise analysis. J Membr Biol 80:135–151

Li JH-Y, Lindemann B (1983) Competitive blocking of epithelial sodium channels by organic cations: the relationship between macroscopic and microscopic rate constants. J Membr Biol 76:235–251

Li H-Y, Palmer LG, Edelman IS, Lindemann B (1982) The role of Na^+-channel density in the natriferic resonse of the toad urinary bladder to an antidiuretic hormone. J Membr Biol 64:77–89

Li JH-Y, Cragoe EJ Jr, Lindemann B (1985) Structure-activity relationship of amiloride analogs as blockers of epithelial Na^+ channels: I. Pyrazine-ring modifications. J Membr Biol 83:45–56

Li JH-Y, Cragoe EJ Jr, Lindemann B (1987) Structure-activity relationship of amiloride analogs as blockers of epithelial Na^+ channels. 2. Side-chain modifications. J Membr Biol 95:171–185

Light DB, Corbin JD, Stanton BA (1990) Amiloride-sensitive cation channel in apical membrane of inner medullary collecting duct. Am J Physiol 255: F278–F286

Lindemann B, Van Driessche W (1977) Sodium specific membrane channels of frog skin are pores: current fluctuations reveal high turnover. Science 195:292–294

Ling BN, Eaton DC (1989) Effects of luminal Na^+ on single Na^+ channels in A6 cells, a regulatory role for protein kinase C. Am J Physiol 256:F1094–F1103

Lingueglia E, Voilley N, Waldmann R, Lazdunski M, Barbry P (1993) Expression cloning of an epithelial amiloride-sensitive Na^+ channel. FEBS Lett 318:95–99
Meng F (1975) Comparison of the local effects of amiloride hydrochloride on the istonic fluid absorption in the distal and proximal convoluted tubule. Pflugers Arch 357:91–99
O'Neil RG, Boulpaep EL (1979) Effect of amiloride on the apical cell membrane cation channels of a sodium-absorbing, potassium-secreting renal epithelium. J Membr Biol 50:365–387
Palmer LG (1984) Voltage-dependent block by amiloride and other monovalent cations of apical Na^+ channels in the toad urinary bladder. J Membr Biol 80:153–165
Palmer LG (1985) Interactions of amiloride and other blocking cations with the apical Na^+ channel in the toad urinary bladder. J Membr Biol 87:191–199
Palmer LG (1991) The epithelial Na^+ channel: inferences about the nature of the conducting pore. Comments Mol Cell Biophysics 7:259–283
Palmer LG (1992) Epithelial Na^+ channels: function and diversity. Annu Rev Physiol 54:51–66
Palmer LG, Andersen OS (1989) Interactions of amiloride and small monovalent cations with the epithelial sodium channel. Inferences about the nature of the channel pore. Biophys J 55:779–787
Palmer LG, Frindt G (1986) Amiloride-sensitive Na^+ channels from the apical membrane of the rat cortical collecting tubule. Proc Natl Acad Sci USA 83: 2767–2770
Palmer LG, Frindt G (1988) Conductance and gating of epithelial Na^+ channels from rat cortical collecting tubules. Effects of luminal Na^+ and Li^+ J Gen Physiol 92:121–138
Physicians Desk Reference (1993) Medical Economic Company, Montvale, NJ
Quinton PM (1981) Effects of some ion transport inhibitors on secretion and reabsorption in intact and perfused single human sweat glands. Pflugers Arch 391:309–313
Reif MC, Troutman SL, Schafer JA (1986) Sodium transport by rat cortical collecting tubule. Effects of vasopressin and desoxycorticosterone. J Clin Invest 77: 1291–1298
Rich DP, Gregory RJ, Anderson MP, Manavalan P, Smith AE, Welsh MJ (1991) Effect of deleting the R domain on CFTR-generated chloride channels. Science 253:205–207
Riordan JR, Rommens JM, Kerem B, Alon N, Rozmahel R, Grzelczak Z, Zelenski J, Lok S, Plavsik N, Chao JL, Drumm ML, Iannuzzi MC, Collins FS, Tsui L-C (1989) Identification of the cystic fibrosis gene: cloning and characterization of complementary DNA. Science 245:1006–1072
Rossier BC, Palmer LG (1992) Mechanisms of aldosterone action on sodium and potassium transport. The kidney. Physiology and pathophysiology. Raven, New York, pp 1373–1409
Ryan MP (1986) Magnesium and potassium-sparing diuretics. Magnesium 5:282–292
Salako LA, Smith AJ (1970) Effects of amiloride on active sodium transport by the isolated frog skin: evidence concerning site of action. Br J Pharmacol 38:702–718
Sansom SC, O'Neil RG (1985) Mineralocorticoid regulation of apical cell membrane Na^+ and K^+ transport of the cortical collecting duct. Am J Physiol 248: F858–F868
Schafer JA, Troutman SL, Schlatter E (1990) Vasopressin and mineralocorticoid increase apical membrane driving force for K^+ secretion in rat CCD. Am J Physiol 258:F199–F210
Schlatter E, Schafer JA (1987) Electrophysiological studies in principal cells of rat cortical collecting tubules. Pflugers Arch 409:81–92
Schneyer LH (1970) Amiloride inhibition of ion transport in perfused excretory duct of rat submaxillary gland. Am J Physiol 219:1050–1055

Smith AJ, Smith RN (1973) Kinetics and bioavailability of two formulations of amiloride in man. Br J Pharmacol 48:646–649

Smith PR, Benos DJ (1991) Epithelial Na^+ channels. Annu Rev Physiol 53:509–530

Soltoff SP, Mandel LJ (1982) Amiloride directly inhibits the Na,K-ATPase activity of rabbit kidney proximal tubules. Science 220:957–959

Spahn H, Reuter K, Mutschler E, Gerok W, Knauf H (1987) Pharmacokinetics of amiloride in renal and hepatic disease. Eur J Clin Pharmacol 33:493–498

Stanton BA, Giebisch G (1992) Renal potassium transport. Handbook of physiology, section 8: renal physiology. Oxford University Press, New York, pp 813–874

Stokes JB (1981) Potassium secretion by cortical collecting tubule: relation to sodium reabsorption, luminal sodium concentration and transepithelial voltage. Am J Physiol 241:F395–F402

Stoner LC, Burg MB, Orloff J (1974) Ion transport in cortical collecting tubule: effect of amiloride. Am J Physiol 227:453–459

Sudou K, Hoshi T (1977) Mode of action of amiloride in toad urinary bladder. An electrophysiological study of drug action on sodium permeability of the mucosal border. J Membr Biol 32:115–132

Tang C-M, Presser F, Morad M (1988) Amiloride selectively blocks the low threshold (T) calcium channel. Science 240:213–215

Tomita K, Pisano JJ, Knepper MA (1985) Control of sodium and potassium transport in the cortical collecting tubule of the rat. Effects of bradykinin, vasopressin, and deoxycorticosterone. J Clin Invest 76:132–136

Ussing HH, Zerahn K (1951) Active transport of sodium as the source of electric current in the short-circuited isolated frog skin. Acta Physiol Scand 23: 110–127

Verrier B, Champigny G, Barbry P, Gerard C, Mauchamp J, Lazdunski M (1989) Identification and properties of a novel type of Na^+-permeable amiloride-sensitive channel in thyroid cells. Eur J Biochem 183:499–505

Vidt DG (1981) Mechanism of action, pharmacokinetics, adverse effects and therapeutic uses of amiloride hydrochloride, a new potassium-sparing diuretic. Pharmacotherapy 1:179–187

Vigne P, Champigny G, Marsault R, Barbry P, Frelin C, Lazdunski M (1989) A new type of amiloride-sensitive cationic channel in endothelial cells of brain microvessels. J Biol Chem 264:7663–7668

Waldorff S, Hanson PB, Kjoergard H, Buch J, Egebalb, Steiness E (1981) Amiloride-induced changes in digoxin dynamics and kinetics: abolition of digoxin-induced inotropism with amiloride. Clin Pharmacol Ther 30:172–176

Wang W, Sackin H, Giebisch G (1992) Renal potassium channels and their regulation. Annu Rev Physiol 54:81–96

Warncke J, Lindemann B (1985) Voltage dependence of Na^+ channel blockage by amiloride: relaxation effects in admittance spectra. J Membr Biol 86:255–265

Weiss P, Hersey RM, Dujovne CA, Bianchine JR (1969) The metabolism of amiloride hydrochloride in man. Clin Pharmacol Ther 10:401–406

Welsh MJ (1990) Abnormal regulation of ion channels in cystic fibrosis. FASEB J 4:2718–2725

Widdicomb JH, Coleman DL, Finkbeiner WE, Tuet IK (1985) Electrical properties of monolayers cultured from cells of human tracheal mucosa. J Appl Physiol 58:1729–1735

Will PC, Cortright RN, DeLisle RC, Douglas JG, Hopfer U (1985a) Regulation of amiloride-sensitive electrogenic sodium transport in the rat colon by steroid hormones. Am J Physiol 248:G124–G132

Will PC, Cortright RN, Groseclose RG, Hopfer U (1985b) Amiloride-sensitive salt and fluid absorption in small intestine of sodium-depleted rats. Am J Physiol 248:G133–G141

Wills NK, Alles WP, Sandle GI, Binder HJ (1984) Apical membrane properties and amiloride binding kinetics of the human descending colon. Am J Physiol 247: G749–G757

Zeidel ML, Seifter JL, Lear S, Brenner BM, Silva P (1986) Atrial peptides inhibit oxygen consumption in kidney medullary collecting duct cells. Am J Physiol 251:F379–F383

Zeiske W, Wills NK, Van Driessche W (1982) Na^+ channels and amiloride-induced noise in the mammalian colon. Biochim Biophys Acta 688:201–210

CHAPTER 11

Potassium-Retaining Diuretics: Triamterene

T. NETZER, F. ULLRICH, H. KNAUF, and E. MUTSCHLER

A. Chemical Structure and Properties

Triamterene belongs to the class of cyclic amidine diuretics along with amiloride. The core structure of triamterene (Fig. 3) is a pteridine ring which is substituted in the 2, 4, and 7 positions by amino groups (Fig. 7). In the 6 position a phenyl moiety is connected to the pteridine structure. Triamterene shows structural similiarities to folic acid and was originally synthesized by SPICKETT and TIMMIS as a folic acid antagonist (1954). Triamterene is a yellow, odorless, crystalline powder which is almost tasteless, but has a slightly bitter aftertaste. Although its pteridine ring system is substituted by three hydrophilic amino groups, its water solubility is lower than that of the unsubstituted pteridine. This effect is explained by the formation of intermolecular hydrogen bonds between the N atoms of the amino groups and the N atoms of the pteridine structure. The high melting point of triamterene (see Table 1) is also considered to be due to formation of intermolecular hydrogen bonds. Its characteristic fluorescence (λ_a = 369 nm; λ_e = 436 nm) allows its detection in body fluids following separation with thin-layer chromatography (GREBIAN et al. 1976).

B. Pharmacodynamics

I. Renal Effects

1. Structure-Activity Relationships of Pteridine Derivatives

Many structure-activity investigations have been performed using compounds with a pteridine ring system (WEINSTOCK et al. 1968; JONES and CRAGOE 1968; OSDENE 1961, 1966). However, the interpretation of these data is impaired for different reasons: First, most of these studies were performed in rats using oral administration of the compounds. Therefore lack of efficacy may also be related to reduced absorption and not only to a lack of pharmacodynamic effects. Second, different experimental conditions were used by different authors, thus impairing comparison of the obtained results. Third, in several studies only the effects on urine excretion were

Table 1. Physicochemical properties of triamterene

Molecular weight	253.3
Solubility	
Water	1:1000
Ethanol	1:3000
$CHCl_3$	1:4000
Ionization constants	
pK_{a1}	6.3
pK_{a2}	−1.2
Melting point	
From *n*-butanol	316°C
From DMF	327°C
Partition coefficient	
log P *n*-octanol/water	1.11

determined. Therefore these data cannot be interpreted with respect to antikaliuretic effects of the investigated compounds.

Four different groups of diuretically active pteridines are discussed:

1. 2,4-Diaminopteridines
2. 4,7-Diamino-6-carbamoylpteridines
3. 2,4,7-Triaminopteridines
4. 2,4,7-Triamino-6-phenylpteridines

a) 2,4-Diaminopteridines

The first pteridine which was found to be diuretically active was 2,4-diamino-6,7-dimethylpteridine (SKF 371; Fig. 1) (WEINSTOCK et al. 1968). The diuretic activity of this compound was only slightly impaired when one or both methyl groups were replaced by hydrogen. In contrast, longer alkyl chains or substitution with a phenyl moiety at position 6 and/or 7 almost eliminated diuretic activity. The same was true for the introduction of a 6-COOH or a 6-$CONH_2$ group at least in derivatives with unsubstituted position 7. Only the 6-COOMe-7-H derivative exerted diuretic effects. However, these effects could not be demonstrated in the Na^+-deficient rat. 6-Chloro substitution in 7-unsubstituted 2,4-diaminopteridines did not improve diuretic activity compared to SKF 371. Replacement of Cl by Br slightly reduced diuretic activity.

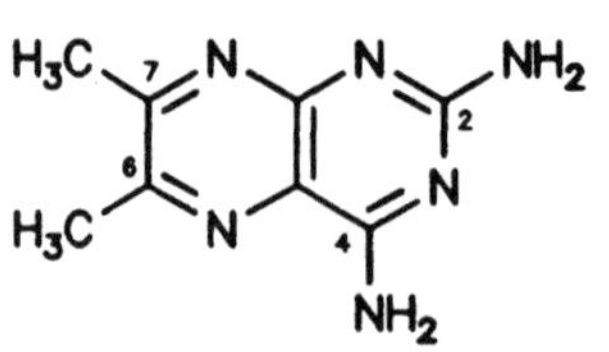

Fig. 1. Structural formula of SKF 371

b) 4,7-Diamino-6-carbamoylpteridines

In an early series of pteridine derivatives, 2-phenyl-4,7-diamino-6-carbamoylpteridine (SKF 6874; Fig. 2, No. 1) proved to be a promising diuretic compound (WEINSTOCK et al. 1968). It caused a more than 70% increase in urine volume at a dose of 15 mg/kg p.o. in saline-loaded rats and was also effective in Na^+-deficient rats.

Effects of Substitution of the Carbamoylhydrogens. WEINSTOCK et al. (1968) and OSDENE (1961, 1966) investigated derivatives of SKF 6874 which were substituted at the carbamoyl hydrogens. The effects of this substitution strongly depended on the substituent used. An increase in activity was seen with substituents containing basic amino nitrogen atoms at an optimal distance (2 carbon atoms) to the amide group. An example is WY-3654 (Fig. 2, No. 2). A 4 or 5 carbon atom bridge between the two nitrogens seemed to decrease diuretic activity. Substitution of both carbamoyl hydrogens generally decreased diuretic activity.

Effects of Substitution at Position 2. Diuretic activity was reduced or even eliminated upon substitution of a 2-phenyl nucleus by H or NH_2. However, the 2-NMe_2 compound was about as active as SKF 6874 (WEINSTOCK et al. 1968). In a series of derivatives which were substituted at the carbamoyl-hydrogens such as WY-5256 (Fig. 2, No. 3), OSDENE (1961, 1966) also showed the importance of a 2-aryl or 2-heterocyclic substituent. Compounds without such a substituent had considerably diminished diuretic activity.

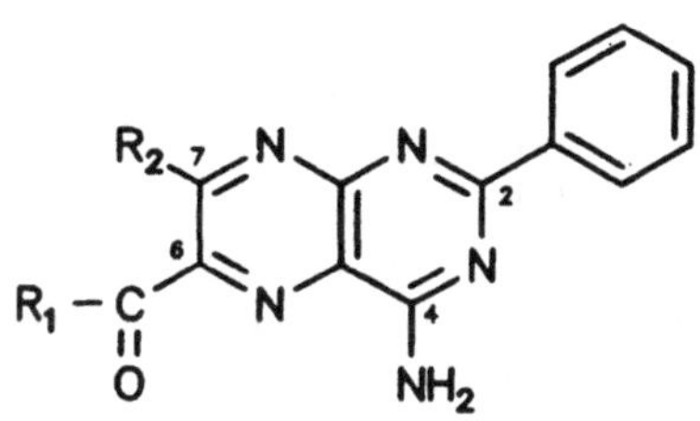

No.	R_1	R_2
1	NH_2	NH_2
2	NH — $(CH_2)_2$ — 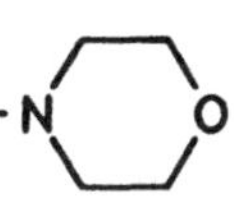	NH_2
3	$NH(CH_2)_2OCH_3$	$NH(CH_2)_2OCH_3$

Fig. 2. Structural formulas of SKF 6874 (*1*), WY-3654 (*2*), and WY-5256 (*3*)

c) *2,4,7-Triaminopteridines*

Effects of Substitution of Hydrogens at Amino Groups. The replacement of one hydrogen by an alkyl group reduced diuretic activity (WEINSTOCK et al. 1968). This tendency was generally more marked with longer alkyl groups. The substitution of both hydrogens at the amino group also resulted in a decrease in diuretic activity. The same was found for a substitution of an aliphatic amino group by a cyclic secondary amine such as piperidine. The position of the amino group which was modified was only of minor importance. In the chloramterene series the same effects were observed upon replacement of the hydrogens at amino groups (JONES and CRAGOE 1968).

Effects of Substitution at Position 6. WEINSTOCK et al. (1968) investigated the effect of different substitutents at position 6 of 2,4,7-triaminopteridine on diuretic activity. Their results show that a phenyl moiety which is directly connected to the pteridine system (= triamterene, SKF 8542; Fig. 3, No. 1) is superior to almost all other substituents. Nevertheless the 6-methyl and 6-ethyl analog were still fairly active. No diuretic activity was seen with the 6-COOH and the 6-$CONH_2$ compounds. Heterocyclic substitution generally decreased diuretic activity when compared to triamterene. Some compounds such as the 6-pyridyl or the 6-(4-thiazolyl) derivative were even found to be diuretically inactive. 2,4,7-Triamino-6-(2-furyl) pteridine (Fig. 3, No. 2), which is also known as furterene or furamterene, hardly exerted any diuretic effects in the investigations of WEINSTOCK et al. (1968). However, it has been developed as a diuretic by another company. Oral doses of 100 mg t.i.d. have been proven effective in patients with cirrhosis and heart failure (CRAGOE 1983). This illustrates fairly well the limitations of the results obtained in the screening tests of WEINSTOCK and coworkers.

JONES and CRAGOE (1968) reported that the 6-chloro analog of triamterene ("chloramterene"; Fig. 3, No. 3) was more potent than triamterene itself but less active than amiloride. More detailed investigations of 6-substituted 2,4,7-triaminopteridines have been reported by NETZER et al. (1992a, 1993a). In these studies the effects of compounds with side chains containing a tertiary amino group were investigated. The compounds were administered intravenously in conscious saline-loaded rats and renal Na^+, K^+, and Mg^{2+} excretion were determined.

In a first series of compounds without a phenyl nucleus in the side chain (Fig. 4, Nos. 1–3) the natriuretic effects were found to be weak. Even the antikaliuretic effects were attenuated in this series. However, a dose-dependent reduction of Mg^{2+} excretion was observed following administration of the compounds with a 2 or 3 carbon atom bridge between the pteridine ring and the basic amino nitrogen (Nos. 2, 3). None of the compounds impaired renal Ca^{2+} excretion. The highest selectivity for inhibition of Mg^{2+} excretion was demonstrated for 2,4,7-triamino-6-[2-(*N*,*N*-dimethylamino)ethyl]pteridine (RPH 3036) (Fig. 4, No. 2). This compound

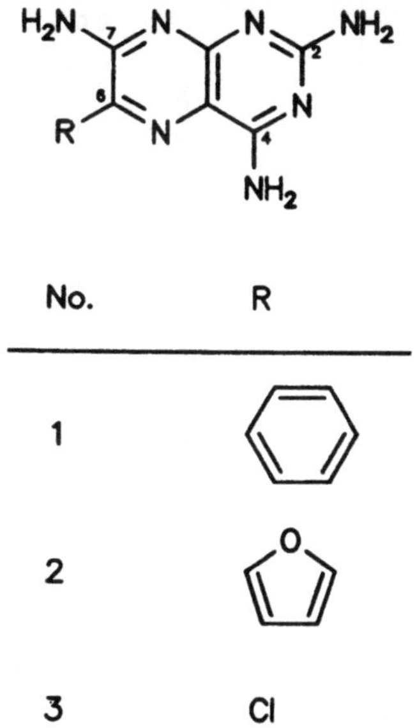

Fig. 3. Structural formulas of triamterene (SKF 8542, *1*), furamterene (F6613, *2*), and 2,4,7-triamino-6-chloropteridine (chloramterene, *3*)

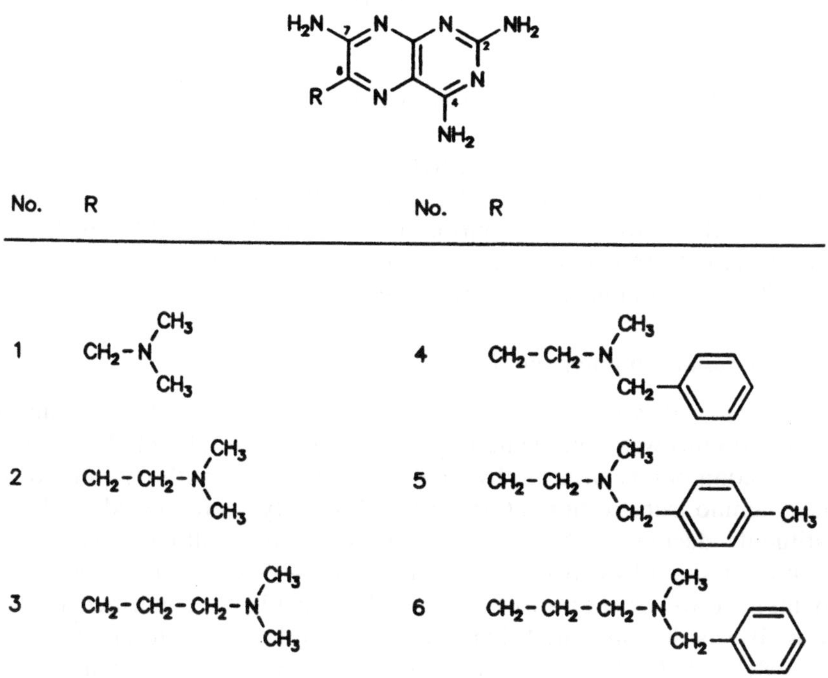

Fig. 4. Structural formulas of RPH 2997 (*1*), RPH 3036 (*2*), RPH 3038 (*3*), RPH 3028 (*4*), RPH 3040 (*5*), and RPH 3032 (*6*)

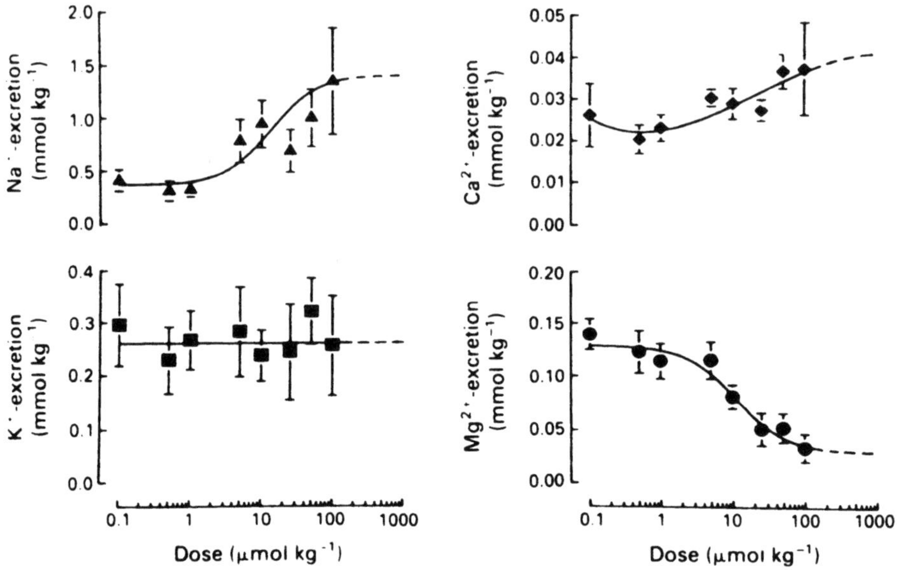

Fig. 5. Effects of different doses of RPH 3036 on urinary Na^+, K^+, Ca^{2+}, and Mg^{2+} excretion in the conscious saline-loaded rat. Urine was collected over a 2.5-h period. Each point is the mean of $n = 6$ rats/group; SEMs are shown by *vertical bars*. The ED_{50} values of natriuretic and antimagnesiuretic effects were 13.4 μmol/kg and 11.2 μmol/kg, respectively (NETZER et al. 1992a)

is an example of a pteridine derivative with a new electrolyte excretion "profile" (Fig. 5).

In a second series, one *N*-methyl group was replaced by an *N*-benzyl group (Fig. 4, Nos. 4–6). These derivatives offered more potent natriuretic effects and also showed – in contrast to RPH 3036 and RPH 3038 – antikaliuretic effects (NETZER et al. 1993a). The antimagnesiuretic properties were still more pronouced than their antikaliuretic effects.

d) 2,4,7-Triamino-6-phenylpteridines

Modifications at the Phenyl Ring. A first series of 2,4,7-triamino-6-phenylpteridines was investigated by WEINSTOCK et al. (1968). In this series different compounds with methyl, halogen, hydroxy, methoxy, amino, and *N*-acetylamino substitution at the phenyl moiety were tested. All these substituents decreased the diuretic activity, even resulting in compounds with no activity (4-hydroxy-, 4-amino-, and 4-*N*-acetylamino-phenyl substituents). However, not all of these findings could be confirmed by other groups. It was demonstrated that not only 4-methoxytriamterene (VOLLMER et al. 1981) but also 4-ethoxytriamterene (PRIEWER 1985) were diuretically as active as triamterene. In vitro investigations of 4-hydroxytriamterene (OH-TA) and its sulfuric acid ester (OH-TA ester), which are the phase I and phase II metabolites of triamterene, proved the effectiveness of 4-hydroxytriamterene derivatives in inhibiting Na^+ entry from lumen to cell (KNAUF et al. 1978). The in vivo diuretic and natriuretic activity of the OH-TA ester was demonstrated in conscious saline-loaded rats after intravenous administration (LEILICH et al. 1980). Despite the findings of WEINSTOCK et al. (1968), it therefore seemed possible to develop diuretically active ethers of 4-hydroxytriamterene.

Based on these findings, a variety of 4-hydroxytriamterene ethers have been synthesized and tested for their natriuretic, antikaliuretic, and antimagnesiuretic activity (PRIEWER et al. 1985a, 1986a,b; ULLRICH et al. 1991). The ED_{50} values (i.e., the potency) for the natriuretic effects were almost the same in all compounds tested (PRIEWER et al. 1987a). However, compounds with a basic side chain, such as 2,4,7-triamino-6-[4-(3-dimethylaminopropoxy)phenyl]pteridine (B3; Fig. 6, No. 1) proved to be much more potent antikaliuretics than triamterene itself. In contrast, the carboxylic acid 2,4,7-triamino-6-[4-(4-carboxybutoxy)phenyl]pteridine (S3; Fig. 6, No. 2) showed only minor antikaliuretic effects. The electrolyte excretion profile seen after administration of S3 was similar to that of the triamterene metabolite OH-TA ester. The neutral ethylamide of S3, 2,4,7-triamino-6-[4-(4-carboxybutoxy)phenyl]-pteridine ethylamide (A2; Fig. 6, No. 3), has about the same natriuretic and antikaliuretic potency as triamterene. These results, together with findings from studies with other pteridine derivatives, suggest a correlation between acidity of the side chain and antikaliuretic activity. However, recent studies do not fully support these considerations

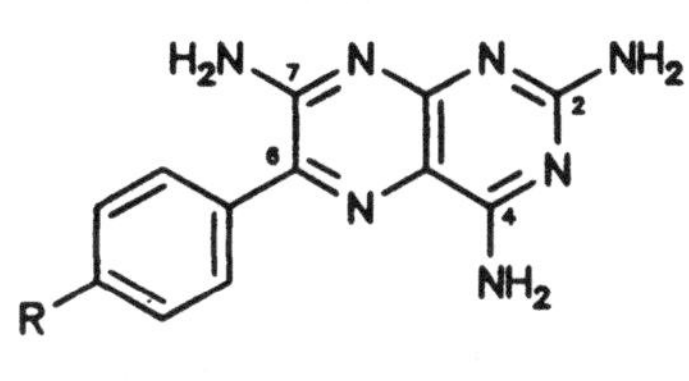

No.	R	No.	R
1	$O-(CH_2)_3-N(CH_3)_2$	4	$O-CH_2-CHOH-CH_2-N(CH_3)_2$
2	$O-(CH_2)_4-COOH$	5	$NH-SO_2-CH_3$
3	$O-(CH_2)_4-CONHC_2H_5$		

Fig. 6. Structural formulas of B3 (*1*), S3 (*2*), A2 (*3*), H3 (RPH 2823; *4*), and RPH 3048 (*5*)

(Netzer et al. 1992b). 2,4,7-Triamino-6-(4-methanesulfonamidophenyl) pteridine (RPH 3048; Fig. 6, No. 5) exerted marked antikaliuretic effects despite its acidic sulfonamide group. It was about as potent as the neutral amide A2. The assumption that pK_a value and antikaliuretic potency are correlated may therefore not be valid for all triamterene derivatives. However, the antikaliuretic properties of RPH 3048 do not completely contradict the above results: the sulfonamide group is bioisosteric to a carboxylic acid amide. In urine, at least 50% of RPH 3048 is not dissociated when a pK_a value of 8.2 is assumed for the sulfonamide moiety. Therefore the sulfonamide may act in a similar way to the carboxylic acid amide.

Interestingly, 4-*N*-acetylaminotriamterene, which is bioisosteric to RPH 3048, did not exert diuretic effects in the tests performed by Weinstock et al. (1968). This may be due to poor bioavailability after oral administration since RPH 3048 also did not show renal effects after oral dosing.

As already mentioned most of the triamterene derivatives discussed in the above paragraphs were also tested for their antimagnesiuretic effects (Mutschler et al. 1981; Priewer et al. 1985b, 1987b; Ullrich et al. 1991). All triamterene derivatives that exert acute antimagnesiuretic effects share some structural characteristics, the most important of which is a non-ionizable oxygen function (e.g., hydroxy, oxo, or carboxamide group) at a defined distance to the heterocyclic nucleus or as part of a flexible side chain (Priewer et al. 1991). Steric hindrance of that oxygen function leads to a decrease or loss of Mg^{2+}-retaining properties (Ullrich et al. 1991). In accordance with these requirements, 2,4,7-triamino-6-[4-(3-dimethylamino-2-hydroxypropoxy)phenyl]-pteridine (H3; Fig. 6, No. 4) showed antimagnesiuretic properties whereas B3 did not. Another triamterene derivative with pronounced antimagnesiuretic potency is A2 (Fig. 6, No. 3).

2. Triamterene

a) Effects on Electrolyte Excretion

The effects that have been observed after triamterene administration to rats (Mutschler et al. 1981) include a marked increase in renal Na^+ and Cl^- excretion and a substantial decrease in renal K^+ excretion. In addition, urinary output of protons was reduced, resulting in an increase in urine pH (Mohr et al. 1971). The excretion of the divalent cations Ca^{2+} and Mg^{2+} was influenced in two different ways: The excretion of Ca^{2+} was slightly increased whereas that of Mg^{2+} was reduced. However, the reduction of Mg^{2+} excretion was not observed until 6h after triamterene administration.

Similar effects of triamterene on urinary electrolyte excretion have been demonstrated in man (Knauf and Mutschler 1984). Na^+ and Cl^- excretion were increased by triamterene, but not to a similar extent as in rats. K^+ excretion was not reduced by triamterene but was kept constant. The constant K^+ excretion together with an increase in Na^+ excretion results in an increase in the Na^+/K^+ excretion ratio. This phenomenon is termed a *relative* K^+-sparing effect. In addition, a slight increase in Ca^{2+} excretion was found. Conflicting results have been obtained with respect to the acute effects of triamterene on Mg^{2+} excretion. In an early study by HÄNZE and SEYBERTH (1967), antimagenesiuretic effects were detected which were particularly impressive after more than 8 h. In contrast, in another study Mg^{2+} excretion was not changed by triamterene over a 24-h period (LOßNITZER and VÖLGER 1985). The discrepancies between these two studies might well be due to different doses. HÄNZE and SEYBERTH used a single oral dose of 300 mg triamterene whereas the other group only used 50 mg p.o. As has been demonstrated by KNAUF and MUTSCHLER (1984), an unambiguous antimagnesiuretic effect does not occur with triamterene doses <100 mg p.o. In contrast, 25 mg triamterene given orally produced 90% of the maximum natriuretic effect. K^+-sparing effects in man were evident in doses >15 mg.

b) Mechanism of Action

Triamterene is not an aldosterone antagonist as was initially speculated by WIEBELHAUS et al. (1961). This could be demonstrated by experiments in adrenalectomized animals where triamterene still proved to be effective (HERKEN and SENFT 1961; SCHAUMANN 1962). Nevertheless, triamterene acts – just as aldosterone antagonists – in the late distal tubule and collecting duct as shown by micropuncture studies (LACY et al. 1979). In that part of the nephron triamterene is thought to block luminal Na^+ channels, thereby inhibiting Na^+ reabsorption. Consequently, the transepithelial potential is decreased, thus reducing the driving force for K^+ secretion. As a result, renal Na^+ excretion is enhanced and K^+ excretion reduced. So far these considerations have not been proven in the nephron. Nevertheless, studies by KNAUF et al. (1976) in the rat salivary duct epithelium, which to a certain extent resembles that of the distal tubules of the kidney, support this concept. In addition, studies using amiloride were successful in substantiating in renal tissue the proposed concept of combined natriuretic and antikaliuretic effects (PALMER and FRINDT 1986).

However, this may only be one aspect of the mechanism of action of triamterene. The above-mentioned studies in rats with different triamterene derivatives (Sect. B.I.1.d) revealed that the natriuretic and antikaliuretic properties of these compounds may be separated. This does not fit in with the concept of just one transport system to be inhibited by triamterene. Therefore it has been speculated that one part of the effects of triamterene derivatives might be mediated by interaction with an Na^+/H^+ antiporter in the late distal tubule and collecting duct (KNAUF and MUTSCHLER 1984; GREGER 1986). According to this concept, the effects on Na^+ excretion should be mainly due to inhibition of the Na^+/H^+ antiport and the effects on K^+ should be caused by blockade of luminal Na^+ channels. Since the existence of an Na^+/H^+ antiport in the late distal tubule and collecting duct has not yet been proven and potent inhibitors of the Na^+/H^+ antiport such as amiloride analogs have no diuretic action (HENDRY and ELLORY 1988), this might not be the explanation for the observations made with triamterene derivatives. A concept which is currently under discussion is the blockade of apical Na^+ and K^+ channels by triamterene. The effects on K^+ channels may even be mediated by inhibition from the cytoplasmic side. However, so far there have been no experimental data to support this concept.

Another problem which has not yet been solved is the mechanism of action mediating the antimagnesiuretic effects of triamterene and other pteridine derivatives. Different findings in animal studies support the hypothesis that these effects are due to a direct renal action. The most important among these is the finding that pteridine diuretics do not reduce renal Ca^{2+} excretion (NETZER et al. 1992a). Thus, parathyroid hormone and calcitonin are probably not involved in these antimagnesiuretic effects. It has been hypothesized that the reduction of the transepithelial potential in the late distal tubule and collecting duct by blockage of luminal Na^+ channels might promote reabsorption of cations such as Mg^{2+}, thereby reducing their excretion (HEIDENREICH 1984; RYAN 1986). If this is true, any

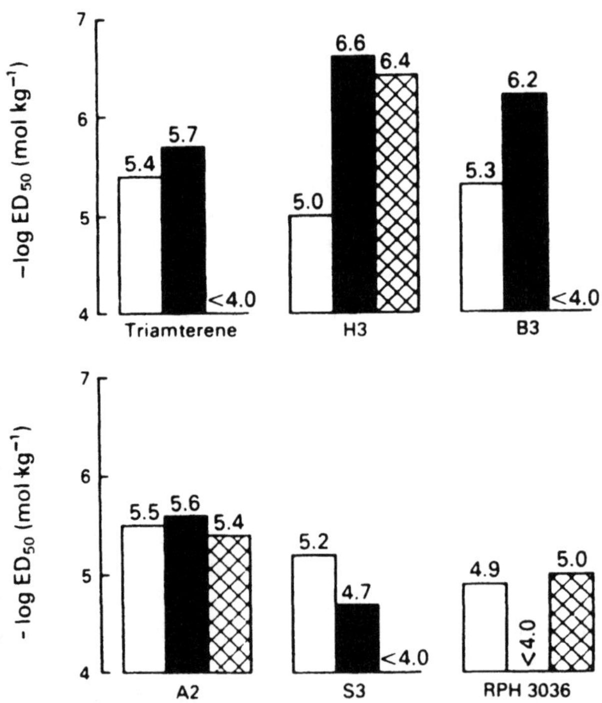

Fig. 7. Natriuretic (*open columns*), antikaliuretic (*solid columns*), and antimagnesiuretic (*cross-hatched columns*) properties of triamterene, H3, B3, S3, and RPH 3036 expressed as -log ED_{50} (mol/kg) values. (NETZER et al. 1992a)

antikaliuretic coumpound should also exert antimagnesiuretic effects. However, only certain triamterene derivatives have Mg^{2+}-retaining properties (Fig. 7). In addition, several antimagnesiuretic pteridine derivatives have been found which do not impair K^+ excretion at all (NETZER et al. 1992a). Therefore this hypothesis cannot account for the observed antimagnesiuretic effects of pteridine derivatives. The mode of action which is most consistent with the known experimental results is the selective inhibition of Mg^{2+} secretion in the late distal tubule and collecting duct. However, it is not unequivocally accepted that Mg^{2+} secretion occurs in that part of the nephron (MASSRY et al. 1969; ALFREDSON and WALSER 1970). Therefore further studies should be performed to examine this concept. In addition, more experimental work is needed to determine the molecular mechanism of this possible Mg^{2+} transport. It might well be that Mg^{2+} secretion is coupled to Na^+ reabsorption since RPH 3036 not only reduced renal Mg^{2+} excretion but also increased urinary Na^+ output. Inhibition of renal Na^+/K^+ ATPase by triamterene, which has been reported by OZEGOVIC and MILKOVIC (1982), seems to be an unspecific effect and is not considered to be related to the diuretic effects of triamterene (KNAUF et al. 1976).

II. Cardiac Effects

1. Structure-Activity Relationships

a) *In Vitro Studies*

In several studies in the 1970s triamterene was examined for cardiac effects. It was demonstrated that triamterene prolongs action potential duration and functional refractory period (LÜDERITZ et al. 1975), increases the toxic dose of digitalis (GÜTTLER et al. 1979), and increases myocardial force of contraction (GREEF and SCHUHMACHER 1979). These effects have been demonstrated in vitro and thus did not occur secondary to renal actions of triamterene. These results prompted structure-activity investigations with respect to antiarrhythmic properties of pteridine derivatives (BUSCH et al. 1991b; NETZER 1991). The determination of functional refractory period according to GOVIER (1965) was used as a screening test. In a first series only triamterene derivatives with a substituent at position 4 of the phenyl ring were investigated. In concentrations of 1–30 μmol/l lipophilic triamterene derivatives with neutral substituents (4-fluoro-, 4-chloro-, 4-methoxy-, 4-ethoxytriamterene) as well as 4-benzyltriamterene and its derivatives with basic substituents were more active than triamterene. 4-Hydroxytriamterene ethers with basic or acidic substituents such as S3 and H3 (for structural formulas see Fig. 6) and RPH 3048 (for structural formula see Fig. 6) showed no effects at all.

In a second series, 4,7-diamino-6-phenylpteridines were tested (NETZER 1991). The lack of the amino group at position 2 either slightly improved or slightly reduced but did not abolish activity. Furthermore, a group of 2,4,7-triaminopteridine derivatives (Fig. 4) with a tertiary amino group in the side chain were investigated for their effects on functional refractory period. Interestingly, the replacement of an *N*-methyl group by an *N*-benzyl group markedly increased activity. RPH 3040 (Fig. 4, No. 5) at 30 μmol/l was even more active than quinidine. The *N*-methyl compounds such as RPH 3036 and RPH 3038 were considerably less active than triamterene.

b) *In Vivo Studies*

To further evaluate the antiarrhythmic properties of pteridine derivatives, selected compounds were tested in the coronary artery ligated and reperfused rat (CAL-R) (BUSCH 1989; BUSCH et al. 1991a; NETZER 1991). Some compounds such as 4-benzyltriamterene could not be investigated following i.v. administration due to their poor solubility. At a dose of 30 μmol/kg, i.v. triamterene was found to offer protection against ventricular fibrillation in this model. The neutral triamterene derivatives, i.e., 4-fluoro-, 4-chloro-, 4-methoxy-, and 4-ethoxytriamterene, did not prove to be effective in the CAL-R model up to the highest tested dose (10 μmol/kg). Only 4-methoxytriamterene showed protection against ventricular fibrillation at a dose of 10 μmol/kg. However, most of the 4-benzyltriamterene derivatives with an amino group in the side chain exerted potent antiarrhythmic effects (BUSCH et al. 1991). At a dose of 30 μmol/kg they offered almost 100% protection against ventricular extrasystoles, ventricular tachycardia, and ventricular fibrillation. Following i.v. administration they were as potent as amiodarone (Fig. 8). As shown by NETZER (1991), the lack of an amino group at position 2 of the pteridine ring system reduced, but did not abolish, antiarrhythmic activity in the CAL-R model. The 2,4,7-triaminopteridine derivatives with an *N*-benzyl group in the side chain such as RPH 3040 (Fig. 4) in the CAL-R model were about as potent as the tested 4-benzyltriamterene compounds (NETZER 1991).

To determine the effects of the most interesting compounds after oral administration, the CAL-R model was evaluated for this purpose (NETZER 1991). In contrast to quinidine, none of the pteridine derivatives exerted evident antiarrhythmic effects after oral dosing. Drug plasma concentrations of the 4-benzyltriamterene derivatives were found to be low although heart tissue levels were high. As demonstrated in a pharmacokinetic study (NETZER 1991), this characteristic might be due to slow absorption from the GI tract, rapid distribution into tissue, but a slow redistribution into the central compartment. Since the plasma elimination constant was quite high, low plasma levels resulted.

c) *Electrophysiologic Studies*

Some of the investigated compounds have undergone electrophysiologic evaluation. 4-Benzyltriamterene (100 μmol/l) was shown to moderately

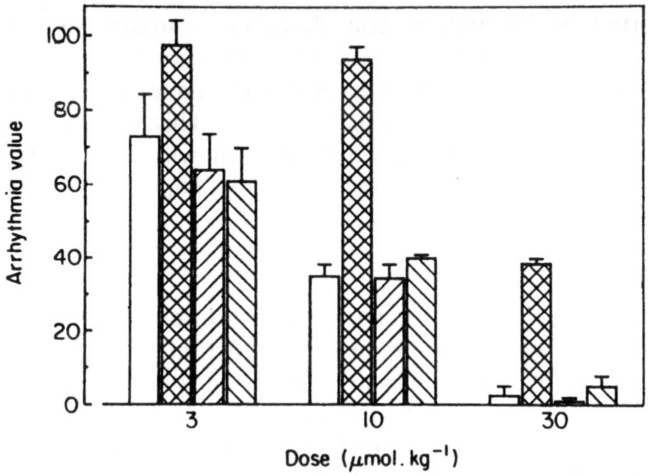

Fig. 8. Effects of amiodarone (□), triamterene (▩), and two benzyltriamterene derivatives [RPH 3007 (▨) and RPH 3008 (▧)] on reperfusion-induced arrhythmias in the coronary artery ligated and reperfused rat ($n = 4$; means ± SEM). (NETZER et al. 1992c)

prolong action potential duration (+26%) in guinea pig papillary muscle with only minor effects on V_{max} (−4%), which is indicative of a class-III antiarrhythmic mechanism of action (U. BORCHARD 1983, personal communication). BUSCH (1989) performed electrophysiologic studies with 4-(α-methylpiperazinobenzyl)triamterene using calf Purkinje and ventricular muscle fibers. In Purkinje fibers he found a moderate increase in action potential duration (30 μmol/l: +26%) together with a considerable decrease in V_{max} (30 μmol/l: −47%). In contrast, action potential duration in muscle fibers was markedly prolonged (30 μmol/l: +82%) with moderate effects on V_{max} (30 μmol/l: −26%). The combined effects on V_{max} and action potential duration point to a class-IA antiarrhythmic mechanism of action of this compound. In addition, a decrease of the action potential plateau was observed in both types of fibers. Therefore, 4-(α-methylpiperazinobenzyl)-triamterene obviously reduced the slow Ca^{2+} influx.

d) Interrelationship of Antiarrhythmic and Diuretic Actions of Pteridine Derivatives

All of the compounds that have been tested for antiarrhythmic effects have also been investigated for renal actions. Of these, pteridine derivatives were found which (a) have renal and cardiac effects (e.g., triamterene), which (b) only exert renal actions (e.g., RPH 3048; Fig. 6), and which (c) only develop cardiac effects (RPH 3041; NETZER et al. 1993b). This demonstrates that renal and cardiac actions of pteridine derivatives are mediated by different mechanisms. At least as far as Na^+ channels are concerned, this is also supported by theoretical considerations: cardiac Na^+ channels are voltage activated and can easily be blocked by tetrodotoxin; renal epithelial Na^+ channels are not voltage activated and cannot be blocked by tetrodotoxin (HORISBERGER and GIEBISCH 1987). This points to a different architecture of these channels, which in turn will require different compounds to achieve channel blockade (see also Chap. 10, this volume).

2. Triamterene

As already mentioned, triamterene was found to exert cardiac effects which are not secondary to alterations of renal function. First studies by SELLER et al. (1975a,b) in dogs demonstrated that amiloride but particularly triamterene diminished digitalis-induced K^+ loss from the heart and increased the dose of digitalis which induced toxic effects. Other groups confirmed in different species the finding that triamterene reduces digitalis toxicity and may increase the positive inotropic effects of cardiac glycosides (GREEF and KÖHLER 1975; GÜTTLER et al. 1986). These results were obtained by in vivo experiments and therefore it could not be excluded that these effects might be mediated by renal actions of triamterene. Nevertheless, in vitro studies by GÜTTLER et al. (1979) demonstrated that triamterene reduced the toxic effect of digitalis. In addition, Güttler et al. only found minor effects of triamterene on digitalis-induced K^+ loss from the heart. In contrast to SELLER et al. (1975a,b), they suggested that the antitoxic effects of triamterene on digitalis toxicity were due to its action potential prolonging activity (LÜDERITZ et al. 1975). This hypothesis was confirmed by the results of NAUMANN D'ALNONCOURT et al. (1976), who found that triamterene corrects digitalis-induced electrophysiological alterations such as a decrease of the action potential duration. The electrophysiological investigations of BORCHARD et al. (1981) demonstrated that the prolongation of the action potential by triamterene is not accompanied by a reduction of the maximum upstroke velocity up to a concentration of 100 μmol/l. Thus, a class III antiarrhythmic mechanism of action may be assumed for triamterene.

Besides the prolongation of its action potential and its antiarrhythmic effects, triamterene has been shown to have a positive inotropic action (GREEF and SCHUHMACHER 1979). It was demonstrated by SCHUHMACHER et al. (1983) that these effects are mainly due to inhibition of the cardiac phosphodiesterase (PDE). The IC_{50} values of triamterene and its phase II metabolite OH-TA ester are 127 μmol/l and 72 μmol/l, respectively. Since OH-TA ester is less lipophilic than triamterene, it may not penetrate the cells as well as triamterene. This may explain why OH-TA ester is – in contrast to its PDE inhibitory effect – about ten times less effective with respect to its inotropic actions.

Another effect which has been observed in cardiac cells with triamterene is the inhibition of the catecholamine-induced stimulation of the adenylate cyclase (AC) (BORCHARD et al. 1981). Detailed studies revealed that this was not due to binding of triamterene in the β-receptor but rather due to uncoupling of the signal transfer between the receptor and the catalytic site of the AC (NOACK and SCHUHMACHER 1983). The precise mechanism of this uncoupling is not known. However, the inhibition of AC may provide beneficial protection against catecholamine stimulation.

In man triamterene is rapidly metabolized to the OH-TA ester (see Sect. C). Since the cardiac effects of the OH-TA ester are less pronounced than those of triamterene, the clinical relevance of the effects mentioned above remains unclear.

III. Effects on Dihydrofolate Reductase

As already mentioned, triamterene was originally synthesized as a folic acid antagonist. However, in vitro considerable inhibition of dihydrofolate reductase (50%) has only been observed with triamterene concentrations as high as 5×10^{-6} mol/l (SCHALHORN and WILMANNS 1979). The metabolites of triamterene, *p*-hydroxytriamterene (OH-TA) and the sulfuric acid ester of *p*-hydroxytriamterene (OH-TA ester), needed even higher concentrations (7×10^{-5} mol/l) to cause the same degree of inhibition (SCHALHORN and WILMANNS 1979). Thus, at therapeutic concentrations (see below) of triamterene an effect on DNA synthesis is not likely. However, in patients with liver cirrhosis higher plasma concentrations of triamterene may occur, which have been reported to cause megaloblastic anemia and/or granulocytopenia (see below).

C. Pharmacokinetics

I. Metabolism in Man

In man triamterene is rapidly metabolized to the phase I metabolite *p*-hydroxytriamterene (OH-TA), which is subsequently conjugated with active sulfate to form the phase II metabolite *p*-hydroxytriamterene sulfuric acid ester (OH-TA ester) (LEHMANN 1965; see Chap. 5, this volume). Furthermore, some *N*-glucuronides of triamterene were described which are mainly excreted with bile (ANDRASH et al. 1971).

Protein binding was found to amount to 55% for triamterene and to 96% for its active metabolite OH-TA ester (MUTSCHLER et al. 1983a).

II. Pharmacokinetics in Healthy Volunteers

Intravenous Administration. After i.v. administration the plasma concentration of the phase II metabolite OH-TA ester is higher than that of triamterene 5 min after initiation of the infusion (GILFRICH et al. 1983; see Fig. 9). The phase I metabolite OH-TA can only be detected in trace

Table 2. Pharmacokinetic parameters of triamterene and OH-TA ester (oral administration)

Parameter	Triamterene	OH-TA ester
t_{max}	90 min	90 min
$t_{1/2}$	3 h	3 h
Cl_{ren}	0.22 l/min	0.17 l/min
Cl_{tot}	4.5 l/min	–
Bioavailability	50%	–
Plasma protein binding	55%	96%
V_d	13.4 l	–

Cl, clearance; V_d, volume of distribution.

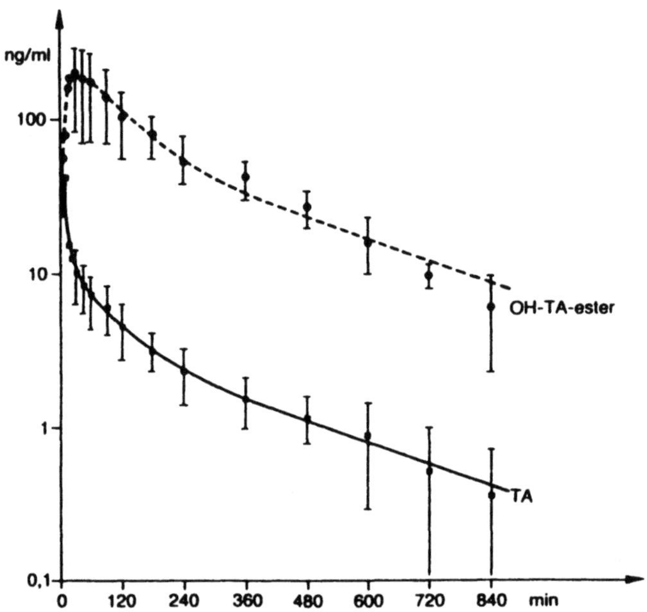

Fig. 9. Plasma levels of triamterene and OH-TA ester after i.v. administration of 10 mg triamterene during 10 min to healthy volunteers

amounts. The maximum plasma levels of the ester exceed by far those of the native compound. In agreement with the rapid biotransformation of triamterene, the total plasma clearance was found to be 4.5 l/min. The corresponding renal clearances (Cl_{ren}) were 0.22 l/min for triamterene and 0.17 l/min for the ester, respectively. Elimination half-lifes of both triamterene and its main metabolite were determined to be 3 h. The distribution volume of triamterene was calculated to be 13.4 l. In the urine

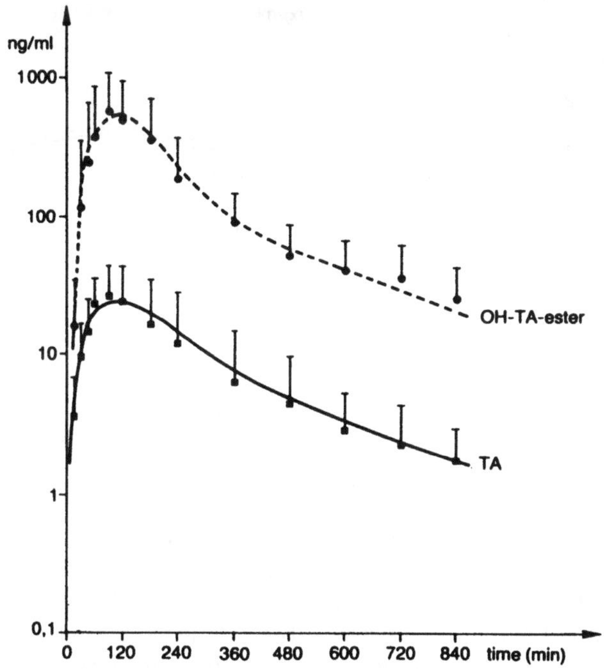

Fig. 10. Plasma concentrations of triamterene and OH-TA ester after oral administration of one 50-mg triamterene tablet

about 54% of the administered triamterene dose was recovered. Four percent was found to be unaltered triamterene and 50% was found to be OH-TA ester.

Oral Administration. At least 80% of a given triamterene dose is absorbed. However, due to a marked first-pass effect, its bioavailability is only about 50% (Mutschler et al. 1983a). After oral administration the plasma levels of OH-TA ester were found to be always higher than the corresponding values of triamterene (see Fig. 10). The ratio of the concentration of OH-TA ester vs. triamterene was nearly constant over the whole investigation period. Peak levels for both compounds were obtained 90 min after oral administration. The terminal half-life of triamterene and the OH-TA ester and their renal clearance did not significantly differ from the values found after i.v. treatment.

III. Pharmacokinetics in Patients with Liver Disease

In patients with liver cirrhosis the plasma concentrations of OH-TA ester are significantly lower than in healthy subjects, whereas the plasma levels of triamterene are increased resulting in almost equal levels of the native drug and the phase II metabolite (see Fig. 11). Furthermore, the elimination of

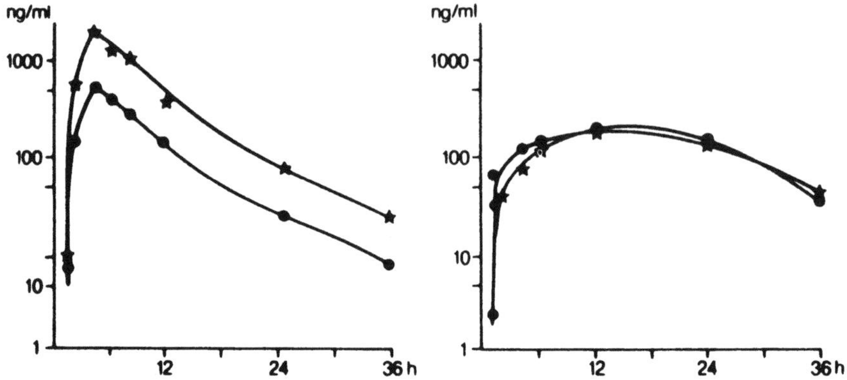

Fig. 11. Plasma concentrations of triamterene and OH-TA ester after oral administration of one 300-mg triamterene dose in a healthy volunteer (*left*) and in a patient with liver cirrhosis (*right*)

triamterene is markedly prolonged. In these patients the $t_{1/2}$ is about four times longer than in healthy volunteers whereas the elimination half-life of the OH-TA ester is not altered (MUTSCHLER et al. 1983b).

As mentioned above, in healthy subjects mainly the phase II metabolite is found in the urine. In contrast, patients with liver cirrhosis are excreting nearly the same amounts of triamterene and OH-TA ester. No OH-TA is found. This means that the phase I reaction, i.e., the hydroxylation of triamterene, is impaired in these patients. Due to the decrease in hydroxylation, accumulation of triamterene may occur in patients with liver cirrhosis if the normal dosage regimen is used. The ratio of OH-TA ester vs. triamterene in plasma reflects the impairment of liver function (ANTONIN et al. 1982).

Although hepatitis is not an indication for triamterene, this disease can be used as a good pharmacological model for studies on the biliary excretion of drugs. While in liver cirrhosis mainly the hydroxylation of triamterene is decreased, in patients with hepatitis the biliary excretion of triamterene and its metabolite(s) is strongly reduced. This means that those patients are not able to eliminate triamterene or its metabolites via the bile. Thus, the cumulative urinary excretion of triamterene and its metabolites was found to be increased from about 40%–50% of the given dose in healthy subjects to nearly 100% with a large amount of unaltered drug (MUTSCHLER et al. 1983a; see Fig. 12). This indicates that in patients with hepatitis the absorption of triamterene is not altered. However, almost the total amount of absorbed triamterene is excreted via the kidneys as native drug and phase II metabolite.

IV. Pharmacokinetics in Patients with Renal Disease

MUTSCHLER et al. (1983b) showed that the area under the curve (AUC) value of OH-TA ester was inversely related to the endogenous creatinine

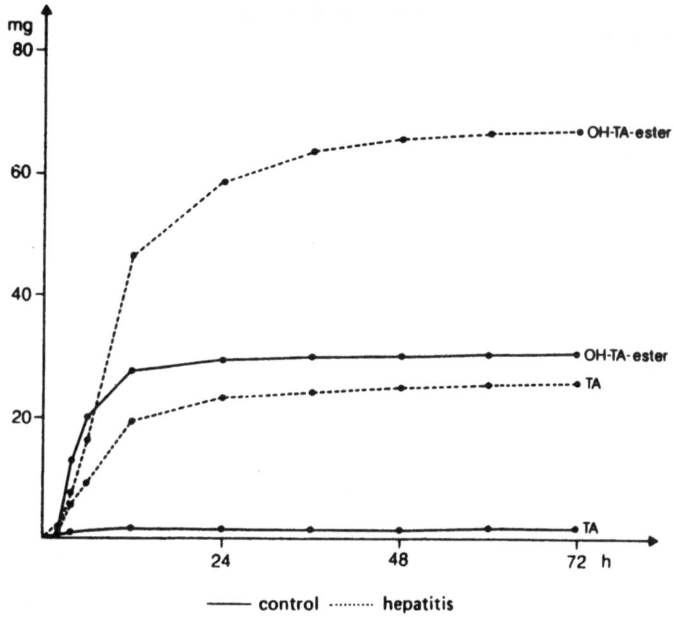

Fig. 12. Cumulative urinary excretion of triamterene and OH-TA ester after oral administration of 100 mg triamterene and 50 mg hydrochlorothiazide in a healthy control group and in patients with hepatitis

clearance (see Fig. 13). In contrast, the AUC value of triamterene is not – as expected – elevated in renal failure, due to the intact metabolic function of the liver. Interestingly, in patients with impaired kidney function the renal clearance of triamterene exceeds that of its metabolite. This can be explained by the lower binding of triamterene to plasma proteins (55%) than that of OH-TA ester (96%). It can be concluded that the phase II metabolite is almost completely excreted by tubular secretion, which is limited when kidney parenchyma is reduced. The extrarenal clearance of both triamterene and OH-TA ester remains constant and is independent of creatinine clearance. As a consequence, there is no compensatory extrarenal elimination which could prevent accumulation of OH-TA ester in renal insufficiency. Thus, the dose regimen of triamterene has to be adjusted (Bennett et al. 1987).

V. Pharmacokinetics in the Elderly

In a study of ten patients with a mean age of 74 years, it could be shown that t_{max} was not altered, indicating an unchanged absorption rate, but c_{max} values were about five times higher. T_{max} of OH-TA ester was increased to 3.5 h, indicating a reduced metabolic clearance of triamterene in older

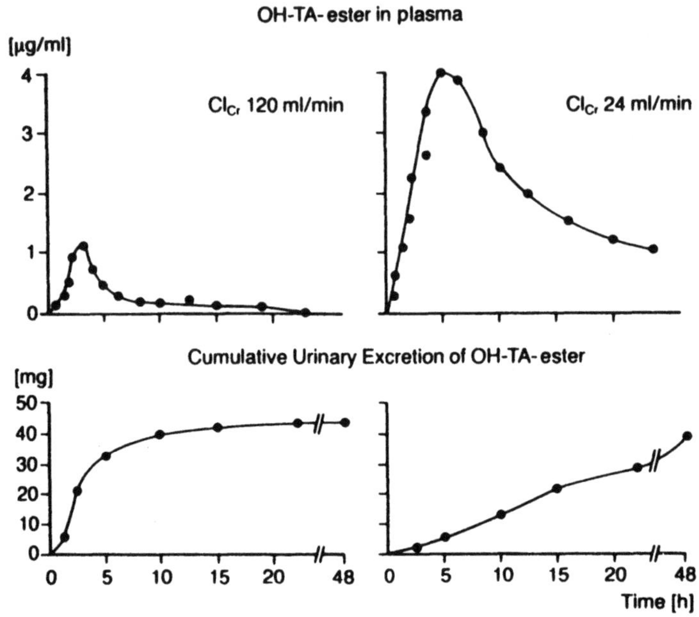

Fig. 13. *Top*, plasma levels of OH-TA ester following a single dose of 100 mg triamterene. *Left*, normal; *right*, reduced endogenous creatinine clearance (CL_{CR}, 24 ml/min). *Bottom*, cumulative urinary excretion of the active metabolite OH-TA ester in a healthy volunteer (*left*) and in a patient with reduced creatinine clearance (*right*)

patients. The AUC ratios of triamterene and OH-TA ester varied much more in the elderly than in younger patients.

D. Therapeutic Use

I. Indications

Due to its site of action in the late distal tubule and collecting duct of the kidney, triamterene has relatively weak diuretic properties. The clinical importance of triamterene is mainly based on its K^+-retaining effect. It is widely used in combination with more potent diuretics such as thiazides or loop diuretics to prevent the K^+ loss induced by these compounds. However, as a single drug, triamterene can be used in the treatment of edema associated with congestive heart failure, cirrhosis of the liver, nephrotic syndrome, steroid-induced edema, idiopathic edema, and edema due to secondary hyperaldosteronism. In combination with hydrochlorothiazide or furosemide, triamterene is also used for the treatment of hypertension and congestive heart failure.

II. Dosage

When used alone, the usual starting dose for the treatment of edema is 100 mg twice daily. For maintenance therapy, twice daily 50 mg is considered to be the optimum dose. In combination with hydrochlorothiazide (25 mg HCTZ/50 mg triamterene) the initial dose of triamterene for the treatment of hypertension or chronic heart failure is twice daily 50 mg. For controlling mild to moderate hypertension 50 mg triamterene and 25 mg HCTZ daily are effective. In mild to moderate renal failure (GFR <50 ml/min and >30 ml/min) the daily dose should not exceed 25 mg triamterene. In severe renal failure, the drug should not be used due to the danger of hyperkalemia.

A number of studies have been conducted with triamterene used in the treatment of patients with cirrhosis of the liver, but with contradictory results. Whereas some authors reported that dosages ranging from 200 to 600 mg/day caused no adverse events (GINSBERG et al. 1964; VESIN 1971), other reports indicated that the use of triamterene in patients with liver cirrhosis and ascites can be associated with renal insufficiency or megaloblastosis (CORCINO et al. 1970; RENOUX et al. 1976; BOSCH et al. 1973, 1974). Based on the fact that in patients with liver cirrhosis $t_{1/2}$ is prolonged to four times normal, a dose reduction was recommended by MUTSCHLER et al. (1983b). In geriatric patients a dose reduction is also suggested to avoid accumulation of both triamterene and OH-TA ester.

III. Side Effects

Endocrine/Metabolic Side Effects. As a result of its pharmacodynamic effects hyperkalemia may be seen with triamterene therapy in some patients. Due to increased renal bicarbonate excretion some patients may develop a metabolic acidosis. Upon initiation of triamterene administration serum urea and creatinine may increase, which is probably mediated by a decrease in GFR. However, this effect is reversible upon withdrawal of the drug. Further metabolic side effects include an increase in serum urate concentration and hyperthermia (SAFDI 1980). The latter is thought to be a hypersensivity reaction.

Central Nervous System Side Effects. Triamterene therapy may cause weakness, fatigue, and headache, symptoms which are frequently seen at the beginning of triamterene therapy and may be related to the diuretic effects of the drug. This may also be true for thirst and dry mouth, which have also been reported with triamterene therapy.

Gastrointestinal Side Effects. Known gastrointestinal side effects of triamterene are diarrhea, nausea, and vomiting. These adverse events are often avoided when triamterene is taken after meals. However, nausea and vomiting may also be due to electrolyte imbalances.

Renal Side Effects. It has been shown in different studies that triamterene may reduce GFR (LYNN et al. 1985; SICA and GEHR 1989). Concomitant use of nonsteroidal antiinflammatory agents in certain patients may precipitate transient renal failure, which is usually reversible when either triamterene or the nonsteroidal antiinflammatory drug is discontinued (SICA and GEHR 1989). In addition, triamterene may promote formation of urinary sediment (ETTINGER et al. 1979; PATEL 1981). The triamterene-induced casts may be responsible for an increased incidence of interstitial nephritis. However, interstitial nephritis has also been regarded as a drug-induced hypersensivity reaction (SICA and GEHR 1989). Following triamterene administration a change of urine color to blue-green with blue fluorescence has been reported (PAVLETICH and SAUSE 1982).

Hematologic Side Effects. Isolated cases of leukopenia, thrombocytopenia, and pancytopenia (RENOUX et al. 1976) have been reported. Patients with alcoholic liver cirrhosis and low serum levels of folic acid may develop a megaloblastic anemia (CORCINO et al. 1970). This has been associated with the antifolic acid effects of triamterene (see Sect. B.III).

Dermatologic Side Effects. Isolated cases of rash and photosensivity have been observed with triamterene treatment. These adverse events are regarded as hypersensivity reactions.

IV. Contraindications

Triamterene is contraindicated in patients with severe or progressive renal disease, hyperkalemia, severe hepatic disease, or hypersensivity to the drug. As mentioned above, the excretion of triamterene and OH-TA ester is delayed in patients with impaired renal function (KNAUF et al. 1983). Accumulation of the drug may lead to hyperkalemia. Furthermore, the use of triamterene in hyperkalemic states can also result in dangerous life-threatening hyperkalemia. Therefore, serum electrolyte concentrations should be monitored regularly during triamterene therapy.

Triamterene should be given with caution to patients with hyperuricemia or gout, or a history of nephrolithiasis (ETTINGER et al. 1979). Patients with depleted folic acid stores such as those with hepatic cirrhosis may be at increased risk of megaloblastosis.

V. Drug Interactions

The most common and important drug interactions of triamterene are related to its K^+-retaining activity. Concomitant use of triamterene and K^+ supplements or other K^+-conserving agents such as spironolactone, amiloride, and ACE inhibitors have been reported to result in severe hyperkalemia, which has been fatal in some cases (MASHFORD and ROBERTSON 1972; GREENBLATT and KOCH-WESER 1973).

Acute reversible renal failure was reported during combination therapy with triamterene and indomethacin (Favre et al. 1982). Indomethacin inhibits formation of prostaglandins which are assumed to protect the kidney against triamterene-mediated nephrotoxicity. The same interaction was observed with diclofenac (Todd and Sarkin 1988). Therefore, caution is generally required when using triamterene in combination with nonsteroidal antiinflammatory agents.

There is some evidence that concomitant amantadine and triamterene therapy (Wilson and Rajput 1983) may lead to a reduced plasma clearance of amantadine and an increased incidence of toxic reactions such as incoordination, agitation, and visual hallucinations.

Administration of triamterene and H_2-blockers such as cimetidine or ranitidine resulted in a significant decrease in renal clearance and hydroxylation of triamterene (Muirhead et al. 1988). Both H_2-blockers also reduced absorption of triamterene. However, since the natriuretic and antikaliuretic effects of triamterene were not altered, the clinical significance of this interaction appears to be questionable.

A useful drug interaction of triamterene is the basis of the combined therapeutic administration with more potent diuretic agents to prevent the K^+ loss caused by these compounds.

E. Toxicity

Toxicological studies have not demonstrated any specific problem of clinical relevance. Teratogenicity tests in rats and rabbits have shown no evidence of fetal abnormality or effect on live birth index and litter size. The oral LD_{50} values have been determined to be 285 mg/kg body weight in mice, >1280 mg/kg in rats, and 60 mg/kg in dogs: The corresponding values after i.v. administration were 41 mg/kg (mice), 40 mg/kg (rats), and 80–100 mg/kg (dogs) respectively.

In the case of overdosage electrolyte imbalance is expected to be the major concern. Disturbance of electrolyte balance may be associated with weakness, tiredness, state of confusion, paresthesia, increased neuromuscular excitability, a sharp, high T-wave due to hyperkalemia, arrhythmia, and a change in the blood status. Careful evaluation of the electrolyte homeostasis and fluid balance is mandatory. There is no specific antidote. Although triamterene is 55% protein bound, there might be some benefit to dialysis.

References

Alfredson KS, Walser M (1970) Is magnesium secreted by the rat renal tubule? Nephron 7:241–247

Andrash H, Fink T, Schmid E (1971) Über die Harnausscheidung des Antikaliuretikums Triamteren und seines phenolischen Metaboliten 2,4,7-Triamino-6-p-hydroxy-phenylpteridin bei Lebergesunden und Kranken mit Leberzirrhose. Z Gastroenterol 9:245–249

Antonin K-H, Antonin E, Moehrke W, Mutschler E, Voelger K-D (1982) Veraenderungen der Pharmakokinetik von Triamteren und Hydrochlorothiazid nach ein- und mehrmaliger Gabe an Patienten mit akuten und chronischen Lebererkrankungen. Therapiewoche 32:3905–3917

Bennett WM, Aronoff GR, Golper TA (1987) Drug prescribing in renal failure: dosing guidelines for adults. Am J Kidney Dis 3:155–193

Borchard U, Greef K, Hafner D (1981) The positive inotropic action of triamterene in isolated heart tissues. Its interaction with β-adrenergic agonists and electrophysiological investigations. Drug Res 31:1688–1693

Bosch J, Arroyo V, Rodes J et al. (1973) Renal insufficiency induced by diuretics in hepatic cirrhosis patients with ascites. Rev Clin Esp 131:375–382

Bosch J, Arroyo V, Rodes J et al. (1974) Spontaneous ascites compensation in liver cirrhosis. Rev Clin Esp 133:441–446

Busch AE (1989) Untersuchungen von Triamterenderivaten auf antiarrhythmische Eigenschaften. PhD thesis, Frankfurt/Main, Germany

Busch AE, Netzer T, Ullrich F, Mutschler E (1991a) Antiarrhythmic properties of benzyl-triamterene derivatives in the coronary artery ligated and reperfused rat. Drug Res 41:125–127

Busch AE, Ullrich F, Mutschler E (1991b) In vitro testing of triamterene derivatives for antiarrhythmic activity. Arch Pharm (Weinheim) 324:39–44

Corcino J, Waxman S, Herbert V (1970) Mechanism of triamterene-induced megaloblastosis. Ann Intern Med 73:419–421

Cragoe EJ (1983) Pteridines. In: Cragoe EJ (ed) Diuretics. Wiley, New York, pp 467–482

Ettinger B, Weil E, Mandel NS, Darling S (1979) Triamterene induced nephrolithiasis. Ann Intern Med 91:745–746

Favre L, Glasson P, Balloton MB (1982) Reversible acute renal failure from combined triamterene and indomethacin. Ann Intern Med 96:317–320

Gilfrich HJ, Kremer G, Moehrke W, Mutschler E, Voelger K-D (1983) Pharmacokinetics of triamterene after i.v. administration to man: determination of bioavailability. Eur J Clin Pharmacol 25:237–241

Ginsberg DJ, Saad A, Gabuzda GJ (1964) Metabolic studies with the diuretic triamterene in patients with cirrhosis and ascites. N Engl J Med 271:1229–1235

Govier WC (1965) The mechanism of the atrial refractory period change produced by ouabain. J Pharmacol Exp Ther 148:100–105

Grebian B, Geißler HE, Mutschler E (1976) Über die Bestimmung von Triamteren, Hydroxytriamteren und Hydroxytriamteren-schwefelsäureester in biologischem Material durch direkte Auswertung von Dünnschichtchromatogrammen. Drug Res 26:2125–2127

Greef K, Köhler E (1975) Animal experiments on the effect of triamterene and amiloride on heart and circulation and the toxicity of digoxin. Drug Res 25: 1766–1769.

Greef K, Schuhmacher P (1979) Comparison of the cardiac effects of triamterene and its phase-II metabolite p-hydroxytriamterene sulfuric acid ester. Drug Res 29:705–706

Greenblatt DJ, Koch-Weser J (1973) Adverse reactions to spironolactone: a report from the Boston Collaborative Drug Surveillance Program. Clin Pharmacol Ther 14:136

Greger R (1986) Diuretische Wirkungsmechanismen. In: Knauf H, Mutschler E (eds) Diuretika. Prinzipien der klinischen Anwendungen. Urban and Schwarzenberg, Munich, pp 3–17

Güttler K, Klaus W, Land E (1979) Antagonistic effect of triamterene to ouabain toxicity. Drug Res 29:623–628

Güttler K, Jenke J, Klaus W (1986) Experimental investigations on the antitoxic efficacy of triamterene in cardiac glycoside poisoning. Drug Res 36:684–688
Hänze S, Seyberth H (1967) Untersuchungen zur Wirkung der Diuretika Furosemid, Etacrynsäure und Triamteren auf die renale Magnesium- und Calciumausscheidung. Klin Wochenschr 45:313–314
Heidenreich O (1984) Mode of action of conventional and potassium-sparing diuretics – aspects with relevance to Mg-sparing effects. Magnesium 3:248–256
Hendry BM, Ellory JC (1988) Molecular sites for diuretic action. Trends Pharmacol Sci 9:416–421
Herken H, Senft G (1961) 2,4,7-Triamino-6-phenylpteridin als "Aldosteronantagonist". Klin Wochenschr 39:1205–1206
Horisberger JD, Giebisch G (1987) Potassium-sparing diuretics. Renal Physiol 10:198–220
Jones JH, Cragoe EJ (1968) Pyrazine diuretics. V. N-Amidino-3-aminopyrazinecarboxamidines and analogous 2,4-diaminopteridines. J Med Chem 11:322–325
Knauf H, Mutschler E (1984) Pharmakodynamik und -kinetik von Triamteren. In: Knauf H, Mutschler E (eds) 30 Jahre Triamteren. Wissenschafts-Verlag, Cologne, p 35
Knauf H, Wais U, Lübcke R, Albiez G (1976) On the mechanism of action of triamterene. Eur J Clin Invest 6:43–50
Knauf H, Mutschler E, Völger K-D, Wais U (1978) Pharmacological effects of phase-I and phase-II metabolites of triamterene. Drug Res 28:1417–1420
Knauf H, Moehrke W, Mutschler E (1983) Delayed elimination of triamterene and its active metabolite in chronic renal failure. Eur J Clin Pharmacol 24:453–456
Lacy FB, Dobyan DC, Jamison RL (1979) Effect of triamterene on the mammalian distal tubule in vivo. Renal Physiol 2:36–43
Lehmann K (1965) Trennung, Isolierung und Identifizierung von Stoffwechselprodukten des Triamterens. Drug Res 15:812–816
Leilich G, Knauf H, Mutschler E, Völger K-D (1980) Influence of triamterene and hydroxytriamterene sulfuric acid ester on diuresis and saluresis in rats after oral and intravenous application. Drug Res 30:949–953
Loßnitzer K, Völger K-D (1985) Die Wirkung von Diuretika auf den Magnesiumhaushalt. Therapiewoche 35:2201–2213
Lüderitz B, Naumann d'Alnoncourt C, Thomas E, Steinberg G (1975) Elektrophysiologische Untersuchungen über kardiale Wirkungen von Diuretika. Verh Dtsch Ges Kreisl Forsch 41:305–310
Lynn KL, Bailey RR, Swainson CP (1985) Renal failure with potassium-sparing diuretics. NZ Med J 98:629–633
Mashford ML, Robertson MB (1972) Spironolactone and potassium chloride. Br Med J 4:298–299
Massry SG, Coburn JW, Kleeman CR (1969) Renal handling of magnesium in the dog. Am J Physiol 216:1460–1467
Mohr C, Kersten L, Bräunlich H (1971) Einfluß von Triamteren auf den Ionenaustausch im distalen Nierentubulus bei Ratten unterschiedlichen Alters. Acta Biol Med Germ 26:361–369
Muirhead M, Bochner F, Somogyi A (1988) Pharmacokinetic drug interactions between triamterene and ranitidine in humans: alterations in renal and hepatic clearances and gastrointestinal absorption. J Pharmacol Exp Ther 244:734–739
Mutschler E, Vollmer G, Völger K-D (1981) Diurese und Salurese bei Ratten nach intravenöser Applikation von Pteridinderivaten. Magn Bull 3:46–50
Mutschler E, Gilfrich HJ, Knauf H, Moehrke W, Voelger K-D (1983a) Pharmocokinetics of triamterene. Clin Exp Hypertens [A] 5:249–269
Mutschler E, Gilfrich HJ, Knauf H, Moehrke W, Voelger K-D (1983b) Pharmakokinetik von Triamteren bei Probanden mit Leber- und Nierenfunktionsstoerungen. Klin Wochenschr 61:883–891

Naumann d'Alnoncourt C, Hornberger M, Lüderitz B (1976) Interaktionen von Triamteren und Herzglykosiden am Erregungsleitungssystem des Herzens. Verh Dtsch Ges Inn Med 82:1236–1239

Netzer T (1991) Untersuchungen zu renalen und kardialen Wirkungen von Pteridinderivaten. PhD thesis, Frankfurt/Main, Germany

Netzer T, Ullrich F, Priewer H, Majewski M, Mutschler E (1992a) Effects of a new pteridine derivative on urinary sodium, potassium and magnesium excretion in conscious saline-loaded rats. Br J Pharmacol 106:222–226

Netzer T, Ullrich F, Majewski M, Mutschler E (1992b) Synthesis, natriuretic, antikaliuretic and antimagnesiuretic properties of an acidic triamterene derivative. Drug Res 42:807–811

Netzer T, Knauf H, Mutschler E (1992c) Modulation of electrolyte excretion by potassium retaining diuretics. Eur Heart J 13 [Suppl G]:22–27

Netzer T, Ullrich F, Majewski M, Mutschler E (1993a) Antikaliuretic and antimagnesiuretic effects of new pteridine derivatives. In: Puschett JB (ed) Diuretics: chemistry, pharmacology and clinical applications. Elsevier, New York, pp 399–403

Netzer T, Knauf H, Mutschler E (1993b) Renal and cardiac effects of pteridine derivatives. In: Puschett JB (ed) Diuretics: chemistry, pharmacology and clinical applications. Elsevier, New York, pp 405–409

Noack E, Schuhmacher P (1983) The interaction of triamterene at the myocardial beta receptor site. J Mol Cell Cardiol 15:319–324

Osdene TS (1961) US Patent 2975180, March 14 1961

Osdene TS (1966) US Patent 3264295, August 2 1966

Ozegovic B, Milkovic S (1982) In vitro inhibition of rat kidney plasma membrane Na^+/K^+-ATPase activity by triamterene. Drug Res 32:1279–1281

Palmer LG, Frindt G (1986) Amiloride-sensitive Na channels from the apical membrane of the rat cortical collecting tubule. Proc Natl Acad Sci USA 83: 2767–2770

Patel KM (1981) Triamterene nephrolithiasis complicating dyazide therapy. J Urol 126:230

Pavletion KJ, Sause RB (1982) Counseling patients on use of OTC diagnostic products. Pharm Times 48:74–83

Priewer (1985) Struktur-Wirkungs-Untersuchungen an Ethern des 4-Hydroxytriamteren. PhD thesis, Frankfurt/Main, Germany

Priewer H, Kraft H, Mutschler E (1985a) Pharmacodynamics of acidic triamterene derivatives and their esters and amides. Drug Res 35:1544–1547

Priewer H, Kraft H, Mutschler E (1985b) Evidence for an acute antimagnesiuretic effect of triamterene derivatives. Pharm Res 2:90–93

Priewer H, Kraft H, Mutschler E (1986a) Pharmacodynamics of straight-chain and branched-chain acidic triamterene derivatives. Drug Res 36:213–215

Priewer H, Spahn H, Mutschler E (1986b) Pharmacodynamics of basic triamterene derivatives. Pharm Res 3:102–107

Priewer H, Wolf E, Kraft H, Knauf H, Mutschler E (1987a) Dissociation of the natriuretic and antikaliuretic properties of triamterene derivatives by dose-response experiments. Pharm Res 4:66–69

Priewer H, Kraft H, Bach N, Mutschler E (1987b) Triamterenderivate mit magnesiumsparenden Effekten. Magnes Bull 9:26–29

Priewer H, Ullrich F, Kleinsorge D (1991) Contribution to the mode of action of potassium and magnesium retaining triamterene derivatives. Magnes Bull 13: 94–99

Renoux M, Bernard JF, Amar M, Boivin P (1976) Acute pancytopenia and medullary megaloblastosis in a cirrhotic treated with triamterene. Nouv Pr Med 5:641–642

Ryan MP (1986) Magnesium and potassium-sparing diuretics. Magnesium 5:282–292

Safdi MA (1980) Fever secondary to triamterene therapy. N Engl J Med 303:701

Schalhorn A, Wilmanns W (1979) Inhibition of human leucocytes dihydrofolate reductase by triamterene and its metabolites. Drug Res 29:1409–1411
Schaumann W (1962) Zum Wirkungsmechanismus von 2,4,7-Triamino-6-phenylpteridin (Triamteren). Klin Wochenschr 40:756
Schuhmacher P, Greef K, Noack EA (1983) The effect of triamterene on myocardial phosphodiesterase and adenylate cyclase. Eur J Pharmacol 95:71–77
Seller RH, Banach S, Namey T, Neff M, Swartz C (1975a) Cardiac effects of diuretic drugs. Am Heart J 89:493–500
Seller RH, Greco J, Banach S, Seth R (1975b) Increasing the inotropic effect and toxic dose of digitalis by administration of antikaliuretic drugs – further evidence for a cardiac effect fo diuretic drugs. Am Heart J 90:56–67
Sica DA, Gehr TWB (1989) Triamterene and the kidney. Nephron 51:454–461
Spickett RGW, Timmis GW (1954) The synthesis of compounds with potential antifolic acid activity, part I. 7-Amino- and 7-hydroxy-pteridines. J Chem Soc 3:2887–2895
Todd PA, Sorkin EM (1988) Diclofenac: a reappraisal. Drugs 35:244–285
Ullrich F, Priewer H, Kleinsorge D, Mutschler E (1991) Investigations on the magnesium retaining properties of triamterene derivatives with an aromatic side chain. Magnes Bull 13:30–33
Vesin P (1971) Treatment of cirrhotic edema with triamterene, amiloride and spironolactone. New clinical and biological facts. Schweiz Med Wochenschr 101:398–401
Vollmer G, Mutschler E, Völger K-D (1981) On the diuretic and saluretic activity of p-hydroxytriamterene and methoxytriamterene. Drug Res 31:529–530
Weinstock J, Wilson JW, Wiebelhaus VD, Maass AR, Brennan FT, Sosnowski G (1968) Pteridines. XII. Structure-acitvity relationships of some pteridine diuretics. J Med Chem 11:573–579
Wiebelhaus VD, Weinstock J, Brennan FT, Sosnowski G, Larsen TJ (1961) A potent, non-steroidal orally active antagonist of aldosterone. Fed Proc 20:409
Wilson TW, Rajput AH (1983) Amantadine-Diazide(R) interaction. Can Med Assoc J 129:974–975

CHAPTER 12

Osmotic Diuretics: Mannitol

O.S. BETTER, I. RUBINSTEIN, and J. WINAVER

A. Introduction

Osmotic diuretics are freely filterable, low molecular weight substances that due to their limited reabsorption and small size create an osmotic force in the tubular fluid sufficient to retard the reabsorption of fluids and solutes (notably NaCl) along the nephron. Thus, osmotic diuresis results in urinary loss of water and sodium.

Mannitol, a six-carbon nonmetabolizable polyalcohol with a molecular weight of 182, is perhaps the oldest and most widely employed osmotic diuretic agent. Numerous studies in experimental animals and humans, under both physiologic and pathophysiologic conditions, have contributed significantly to our understanding of its renal and extrarenal actions. A compound with similar effects is sorbitol. Glucose may likewise act as an "endogenous" osmotic diuretic agent when present at sufficiently high concentrations in the glomerular filtrate, to exceed the reabsorption capacity of the proximal nephron. Other partially or poorly reabsorbable agents, such as urea and sulfates, were included in the past in the category of osmotic diuretics, because, when present in high concentrations in plasma, they may increase urine flow rate, in part, by a mechanism similar to that of mannitol (GENNARI and KASSIRER 1974).

Beyond its action as a diuretic agent, mannitol has been used by many renal physiologists, over the past 5 decades, as an important tool in the study of the mechanisms of salt and water transport at the tubular and cellular level. As a physiologic probe, mannitol has contributed significantly to our understanding of these transport processes. Furthermore, mannitol has important clinical and therapeutic applications, due to its actions in the kidney as well as at some extrarenal sites.

The introduction of mannitol for prophylaxis against postoperative acute renal failure (ARF) following cardiovascular surgery (BARRY et al. 1961; MYERS 1990) was associated with a dramatic decline in the occurrence of such ARF. In addition, our data suggest that mannitol is useful for the prevention of myoglobinuric ARF in man (BETTER and STEIN 1990). Our experience in the salvage of lives and limbs in victims of crush injury following mass disasters suggested that mannitol has beneficial extrarenal effects as well. Moreover, 20 years of experience with in vitro preservation of human cadaver organs suggest that beyond hypothermia the most important factor for successful transplantation is the presence of hyperoncotic activity in the perfusate (COLLINS and WINCOMB 1992).

Finally, in addition to being a hyperosmotic agent, mannitol has been shown also to be an effective scavenger of free hydroxyl radicals in many biologic systems (DELMAESTRO et al. 1980; MAGOVERN et al. 1984). Although most of the renal and extrarenal actions of mannitol have been attributed to its activity as a hyperosmotic agent, it is not unlikely that some of the beneficial effects of mannitol are due to the elimination of these highly cytotoxic free radicals.

The present chapter will summarize the current knowledge on the renal mechanisms of osmotic diuresis. Some extrarenal effects of mannitol, mostly based on our personal experience, will also be discussed.

B. Renal Effects

Intravenous administration of mannitol results in a brisk diuretic/natriuretic response. Depending on the dose administered, urine flow in man may reach 20%–30% of the filtered load of water, and up to 10%–15% of the filtered Na^+ may be excreted during the height of the diuresis (GENNARI and KASSIRER 1974). In the dog, even larger inhibition of water and salt reabsorption has been reported, with massive doses of mannitol injected (WESSON and ANSLOW 1948).

Concomitant with the diuresis and natriuresis produced by mannitol, the excretion of other ions such as Ca^{2+}, Mg^{2+}, phosphate and bicarbonate is also increased. Urinary concentration and dilution ability is severely impaired, and urine osmolality tends to approach isotonicity during the height of diuresis (GOLDBERG and RAMIREZ 1967). Many studies in the past have focused on the tubular effects of mannitol as the dominant mechanism by which the diuretic effect of the drug is achieved. However, in addition to its actions on salt and water reabsorption along the nephron, mannitol has important glomerular, systemic and renal hemodynamic effects, which, in turn, may influence and modulate its tubular actions. Furthermore, since injected mannitol is distributed and restricted primarily to the extracellular fluid compartment, the increase in osmotic pressure induced by mannitol results in water shifts from the intracellular compartment. Such redistribution of fluid may lead to extracellular volume expansion, as well as a decrease in plasma oncotic pressure, and blood viscosity and hematocrit values, which will further increase renal perfusion. Finally, the increase in extracellular fluid volume may activate endogenous natriuretic agents, and suppress antinatriuretic hormone systems, thereby contributing to the diuretic/natriuretic effect of mannitol. The major renal actions of mannitol are as follows:

1. Increase in cortical and medullary blood flow due to a decrease in RVR
2. Increase in GFR during renal hypoperfusion. Variable effects on GFR in kidneys with normal function
3. Prominent diuretic/natriuretic action due to inhibition of tubular reabsorption of water and salt
4. Increase in urinary excretion of K^+, Ca^{2+}, Mg^{2+}, phosphate and bicarbonate
5. Dissipation of medullary hypertonicity. Impairment of urinary concentration and dilution capacity
6. Increase in renal interstitial pressure and in intratubular pressure
7. Indirect actions on the kidney through: increase in ECF volume, dilution of plasma proteins and decrease in blood viscosity, release of prostaglandins and ANF and inhibition of renin angiotensin system

I. Renal Hemodynamic Actions

In most experimental settings, infusion of hypertonic mannitol increases renal blood flow (RBF), apparently by reducing renal vascular resistance (RVR) (LILIEN et al. 1963; BRAUN and LILIENFIELD 1963; STAHL 1965). The increase in total RBF is reflected by a concomitant increase in both cortical blood flow and medullary perfusion (VELASQUEZ et al. 1973; BUERKERT et al. 1981). Moreover, the effect of mannitol on RBF is observed under normotensive conditions (BRAUN and LILIENFIELD 1963), in animals made hypotensive by bleeding, and in the hypoperfused rat kidney, an experimental model of "prerenal failure" (MORRIS et al. 1972). The restoration of RBF and GFR following mannitol infuson in the hypoperfused kidney (SELKURT 1945; JOHNSTON et al. 1981; MORRIS et al. 1972) has served as the rationale for its use in the treatment and prophylaxis against imminent acute renal failure (see below).

On the other hand, massive doses of mannitol may cause a paradoxical increase in RVR, and decrease renal perfusion (LILIEN 1973). This phenomenon may be due to constriction of the efferent arteriole in response to the high concentration of mannitol. The mechanism by which mannitol exerts its effects on RBF is not thoroughly understood. In the "in situ" perfused dog kidney, GOLDBERG and LILIENFIELD (1965) demonstrated that systemic infusion of hypertonic mannitol resulted in a marked decrease in RVR. A similar decrease was produced by direct infusion of mannitol into the renal artery, and also by dextran and saline infusion, but not during infusion of whole blood. These decreases in vascular resistance correlated with diminished hematocrits of the perfusion fluids, suggesting that the decreased viscosity of blood following mannitol infusion could contribute to the increase in RBF (GOLDBERG and LILIENFIELD 1965). Earlier observations by LILIEN et al. (1963) have also suggested a similar linkage between the decrease in the hematocrit and RVR. However, assuming that blood at physiologic rates of flow behaves as a Newtonian fluid, the observed changes in hematocrit were not sufficient to explain the effects of mannitol on RVR. Therefore, additional mechanisms, unrelated to changes in blood viscosity, must be evoked as well. It is of interest that renal denervation and autonomic ganglioplegia did not influence the effects of hyperosmotic mannitol on RVR (GOLDBERG and LILIENFIELD 1965). This suggests that suppression of enhanced sympathetic activity is probably not involved in the mechanism of action of mannitol on vascular resistance.

It is likewise possible that mannitol may exert its renal vasodilatory action indirectly, i.e., by releasing other vasoactive agents. Thus, in the hypoperfused rat kidney, prior treatment with cyclooxygenase inhibitors markedly attenuated the renal vasodilatory effect of mannitol (JOHNSTON et al. 1981). During hypoperfusion, the intrarenal infusion of prostacyclin (PGI_2), but not of prostaglandin E_2 (PGE_2), resulted in an increase in RBF, and restored the renal vasodilatory response to mannitol. Interestingly, the renal vascular response to mannitol was not altered by pretreatment with an angiotensin-converting enzyme inhibitor, inhibitors of the kinin-kallikrein system or inhibitors of thromboxane synthesis. These findings suggest that in the ischemic hypoperfused rat kidney the vasodilatory response to mannitol administration is mediated largely by increased PGI_2 synthesis (JOHNSTON et al. 1981). An additional, indirect mechanism by which mannitol may exert its action on renal hemodynamics is by augmenting atrial natriuretic peptide (ANP) release (YAMASAKI et al. 1988). It has been recently demonstrated that mannitol infusion may increase plasma levels of this vasodilatory, natriuretic hormone, apparently due to the intravascular volume expansion induced by mannitol, or directly by the effect of hyperosmotic perfusion. It remains to be established if, and to what extent, ANP may mediate the natriuretic and vasodilatory response to mannitol.

The increased renal perfusion following mannitol is associated with a proportional increase in both cortical and medullary blood flow (STAHL 1965; VELASQUEZ et al. 1973; BUERKERT et al. 1981). The fact that medullary blood flow is significantly increased by hypertonic mannitol may contribute to the medullary "washout phenomenon," and is consistent with the marked dissipation of medullary hypertonicity observed during osmotic diuresis (MALVIN and WILDE 1959; GOLDBERG and RAMIREZ 1967). This, in turn, could contribute to the mechanism by which mannitol inhibits salt and water reabsorption in the loop of Henle (see below), and also explains, in part, the obliteration of urinary concentration and dilution capacities induced by osmotic diuresis.

II. Glomerular Filtration Rate

In contrast to the increase in RBF observed during mannitol diuresis, reports of the changes in GFR have conflicted. In the dog kidney, micropuncture studies have indicated a significant decrease in single nephron glomerular filtration rate (SNGFR) associated with a similar fall in whole kidney GFR (SEELY and DIRKS 1969).

In contrast, in the human and in the rat whole kidney, GFR is either unchanged or slightly increased during mannitol infusion (BLANTZ 1974; GOLDBERG et al. 1965; MALNIC et al. 1966). In addition to species variations, the change in GFR in response to mannitol depends largely on the baseline condition. Thus, in the hypoperfused kidney, including the dog kidney, infusion of mannitol tends to restore GFR towards normal levels (MORRIS et al. 1972). It appears, therefore, that several factors may influence the GFR response to mannitol administration.

The influence of mannitol on glomerular determinants in the rat kidney was evaluated by BLANTZ (1974) by micropuncture methodology. Mannitol had a consistent and profound effect upon the SNGFR in the hydropenic rat, increasing the filtration rate by 31%. This increment was due to both elevation in single nephron plasma flow and a decrease in afferent arterial oncotic pressure secondary to dilution of plasma protein. More than half of the alteration in SNGFR was due to a decrease in afferent oncotic pressure, suggesting that dilution of plasma protein has a major effect on SNGFR (BLANTZ 1974).

In a more recent study, BUERKERT and coworkers (1981) demonstrated that superficial and deep nephron filtration rates may be influenced differentially during mannitol infusion. In the latter study, whole kidney GFR remained unchanged during mannitol diuresis. However, SNGFR of superficial nephrons increased by 31% in response to mannitol infusion, whereas deep nephron SNGFR fell by approximately 40% (BUERKERT et al. 1981). The redistribution of GFR from juxtamedullary to superficial nephrons may be related to a decline in efferent resistance of deep nephrons secondary to the decrease in viscosity of the blood in the vasa recta as well as a decline in glomerular membrane permeability (BUERKERT et al. 1981).

The effects of mannitol on GFR under circumstances of reduced renal perfusion are of major interest, both in pathophysiologic terms and due to their clinical implications. Since its introduction by SELKURT (1945), mannitol has been widely used to prevent and treat experimental and clinical acute renal failure, because of its ability to maintain glomerular filtration during renal hypoperfusion. MORRIS et al. (1972) reported that in the hypoperfused rat kidney, an experimental model of prerenal failure, prior infusion of mannitol maintained glomerular filtration, which otherwise would stop when renal arterial pressure was reduced to 40 mmHg. Moreover, when given after hypoperfusion has been induced, mannitol will reestablish glomerular filtration which has already stopped (MORRIS et al. 1972). Under the same conditions, infusion of equal or greater volumes of isotonic saline did not maintain or reestablish GFR, suggesting that the effect of mannitol is probably independent of the influence of volume expansion per se, or the dilution of plasma protein. The authors proposed that mannitol exerted this effect by dilating the afferent arteriole, probably due to suppression of renin release and the intrarenal formation of angiotensin II (Vander and Miller 1964; Fojas and Schmid 1970).

A possible role for mannitol in maintaining GFR during renal ischemia, by reducing endothelial cell swelling, was originally suggested by Flores et al. (1972). These authors proposed that the failure of the blood flow to return to the kidney following transient ischemia, the so-called no reflow phenomenon, was due to swollen endothelial cells limiting the available vascular space. They further demonstrated that the "no reflow" and subsequent renal dysfunction were corrected by hypertonic mannitol, but not by equivalent expansion with isotonic saline or isotonic mannitol, indicating that the osmotic effects were primary (Flores et al. 1972). More recently, Mason et al. (1989) demonstrated by morphometric analysis that renal ischemia resulted in swelling of proximal tubular cells and thick ascending limb cells. This altered geometry led to depletion of the interstitial and vascular space in the cortex, resulting in vascular congestion, which, in turn, was responsible for the poor perfusion and impaired renal function. The injection of mannitol into the renal artery prior to ischemia reduced cell swelling and vascular congestion, and eliminated the occlusion of the thick ascending limb, thereby preventing the impairment in renal function (Mason et al. 1989).

The finding that mannitol might maintain glomerular filtration in the hypoperfused kidney is consistent with earlier observations indicating that mannitol prevented the decrease in GFR after hypotensive episodes (Braun and Lilenfield 1963; Peters and Brunner 1963). Similarly, in patients undergoing abdominal aortic aneurysmectomy (Barry et al. 1961) and open heart surgery (Etheredge et al. 1965), and in severely injured patients (Valdes et al. 1979), mannitol has been shown to increase GFR.

It is of interest that, despite the marked and impressive effects of mannitol on glomerular filtration under circumstances of low renal perfusion pressure, GFR remains unchanged or even decreased in intact animals and normal volunteers infused with mannitol (Braun and Lilienfield 1963; Stahl 1965; Williams et al. 1955; Windhager and Giebisch 1961). This finding may imply that the beneficial actions of mannitol observed during renal hypoperfusion are probably mediated by antagonizing systems that are activated when renal perfusion is severely compromised.

III. Tubular Salt and Water Reabsorption

Several studies using micropuncture and microcatheterization techniques have assessed the alterations in salt and water reabsorption in various nephron segments during osmotic diuresis. Based on these findings, it has become obvious that, in contrast to earlier studies (Windhager and Giebisch 1961; Malnic et al. 1966), which implicated the proximal nephron as the main site of action of mannitol, other segments contribute significantly to the altered reabsorption during osmotic diuresis. Of particular interest in that respect is the major inhibition in salt and water reabsorption in the loop of Henle (SEELY and DIRKS 1969; BUERKERT et al. 1981; WONG et al. 1979). Moreover, mannitol also inhibits salt and water reabsorption in the collecting duct (SONNENBERG 1978; BUERKERT et al. 1981), and has a more profound inhibitory effect on deep nephrons than superficial nephrons (BUERKERT et al. 1981). The following section will summarize the current views on the mechanisms mediating the tubular actions of mannitol.

1. Proximal Nephron

Mannitol has no direct inhibitory effect on Na^+ transport in the proximal nephron. Rather, mannitol indirectly influences Na^+ transport by virtue of its effect on water reabsorption. The obligatory presence of mannitol in the tubular fluid retards water reabsorption along this nephron segment. As a result, the concentration of Na^+ in the tubular fluid decreases, and the concentration gradient against which Na^+ must be transported increases (GENNARI and KASSIRER 1974). This, in turn, would diminish outward Na^+ transport, and at the same time increase the passive backflux of Na^+ into the tubule, leading consequently to a progressive reduction in net Na^+ reabsorption. Depending on the loads of mannitol, the proximal tubular fluid Na^+ concentration may fall to a level 30–40 mmol/l lower than that in peritubular fluid. At this point, a limiting gradient for Na^+ transport is reached, and net Na^+ reabsorption stops. Direct micropuncture measurements in the dog with filtrate mannitol of 60–70 mmol/l have shown that water reabsorption in the proximal nephron is reduced by approximately 10%, and proximal Na^+ reabsorption is reduced by 5% (SEELY and DIRKS 1969). In the rat, WINDHAGER and GIEBISCH (1961) demonstrated a greater inhibition of proximal reabsorption following mannitol infusion: 20% for water and 10% for Na^+. Similar observations were reported more recently by BUERKERT et al. (1981) and by WONG and coworkers (1979). Examination of the micropuncture data reported may lead to several important conclusions; first, delivery of Na^+ and water into the loop of Henle is increased following mannitol infusion. Second, the increase is proportionately greater for water than for Na^+ and, finally, the decrease in proximal Na^+ reabsorption is relatively small compared to the strikingly increased delivery of Na^+ and water to the distal nephron and in the final urine. The latter finding indicates that the reabsorption of Na^+ and water must be also markedly inhibited in the intervening segment, i.e., the loop of Henle.

2. Loop of Henle

SEELY and DIRKS (1963) were the first to indicate the importance of the loop of Henle as a predominant site of altered salt and water reabsorption during osmotic diuresis. As pointed out earlier, these authors demonstrated that, following mannitol infusion in the dog, fractional reabsorption of Na^+ and water decreased only mildly in the proximal tubule. The nonreabsorbable fraction, however, increased sharply at the distal nephron by 26% for Na^+ and 32% for water, indicating a large inhibitory effect of mannitol within the loop of Henle. The major influence of mannitol on salt and water reabsorption in the loop was later verified by other investigators (Wong et al. 1979; Buerkert et al. 1981).

Several mechanisms may account for the major alterations in salt and water handling in the loop during osmotic diuresis. Normally, approximately 10% of the filtered water and 15% of the filtered Na^+ are reabsorbed in the loop. However, water is reabsorbed mainly in the descending limb, as a result of osmotic extraction into the hypertonic medulla, whereas salt is reabsorbed predominantly in the ascending limb by mechanisms involving both active and passive components. It has been clearly demonstrated that medullary hypertonicity is abolished during osmotic diuresis (Goldberg and Ramirez 1967) (for mechanism see Chap. 2, this volume). Thus, the osmotic driving force for water extraction in the descending limb is diminished. Indeed, only 8% of the water entering the loop is reabsorbed during mannitol diuresis, as opposed to 50% reabsorption under control, nondiuretic conditions (Seely and Dirks 1969). The diminished water abstraction in the descending limb results in a decreased Na^+ concentration in the tubular fluid entering the thin ascending limb, which, in turn, may reduce Na^+ reabsorption in this nephron segment.

Thus, it is likely that the inhibition of Na^+ reabsorption in the loop of Henle is restricted mainly to the thin ascending limb (Gennari and Kassirer 1974). In contrast, salt reabsorption in the thick ascending limb occurs mainly by a secondary active mechanism mediated by an $Na^+2Cl^-K^+$ carrier (Greger 1985). Since salt reabsorption in this segment is load dependent, the increased delivery of salt and water during osmotic diuresis may actually enhance Na^+ reabsorption. This is supported by the observation that free water reabsorption is progressively increased in man during osmotic diuresis (Goldberg et al. 1965). Additionally, in the rat (Buerkert et al. 1981) but not in the dog (Seely and Dirks 1969), there is a significant reabsorption of salt and water between late proximal and early distal tubular sites of superficial nephrons.

In summary, the effects of mannitol diuresis on loop function involve a marked reduction in water and salt reabsorption in the descending and thin ascending limb, respectively, followed by incomplete recapture of the increased Na^+ load in the thick ascending limb (Gennari and Kassirer 1974). The magnitude of these effects may depend on the load of mannitol as well as on species differences.

Finally, mannitol has been shown to exert a more profound inhibitory effect on salt and water transport in deep nephrons than in superficial nephrons (Buerkert et al. 1981). Such a differential response is expected, if, indeed, the decreased medullary hypertonicity is a major factor mediating the altered loop reabsorption during osmotic diuresis.

3. Distal Tubule and Collecting Duct

Available data on salt and water transport in the final segments of the nephron during osmotic diuresis suggest that these portions, in particular the collecting duct, may contribute to the observed diuretic/natriuretic effect of mannitol. During osmotic diuresis the delivery of salt and water to the distal tubule and collecting duct is markedly increased. However, the final segments of the nephron fail to recapture the increased delivered loads of salt and water. Apparently, the high distal flow rate overwhelms the capacity for Na^+ reabsorption in the collecting duct (GENNARI and KASSIRER 1974).

The function of the collecting duct during osmotic diuresis was studied by the microcatheterization and micropuncture techniques (SONNENBERG 1978; BUERKERT et al. 1981). Both studies demonstrated that the reabsorption of delivered salt and water was diminished in the collecting duct during mannitol diuresis. Moreover, BUERKERT et al. (1981) compared the effects of saline diuresis with those of mannitol diuresis on the function of the terminal collecting duct. Both maneuvers increased salt and water delivery to the base of the medullary collecting duct. However, Na^+ and water reabsorption increased in this segment only during saline diuresis, not during mannitol diuresis. These findings are consistent with the microcatheterization data of SONNENBERG (1978), who demonstrated a complete inhibition of medullary collecting duct Na^+ transport with high doses of mannitol. The author suggested that the complete inhibition might reflect a direct action of mannitol on Na^+ transport mechanism, and not only its action as an osmotically active agent.

IV. Transport of Other Ions

Urinary excretion of other electrolytes such as K^+, the divalent ions Ca^{2+}, Mg^{2+} and phosphate, and substances such as urate and urea may be increased during osmotic diuresis. Such an effect may be secondary to the inhibition of water reabsorption, which blunts the increase in transtubular solute gradients favoring reabsorption (GENNARI and KASSIRER 1974). In addition, the increase in urine flow induced by osmotic diuresis may significantly limit the contact time required for completion of reabsorptive processes of certain ions.

Excretion of K^+ has been shown to increase during osmotic diuresis (MALNIC et al. 1966). This appears to be mainly due to increased distal delivery of Na^+ and high distal flow rate, both of which increase K^+ secretion in the collecting duct. With massive doses of mannitol, K^+ reabsorption may also be impaired in the proximal nephron and loop of Henle (MALNIC et al. 1966).

The effects of osmotic diuresis on renal excretion of the divalent cations Ca^{2+} and Mg^{2+} have been studied by several investigators (WESSON 1962;

DUARTE and WATSON 1967; BETTER et al. 1966; WONG et al. 1979). Early clearance studies demonstrated an increased excretion of both ions, which correlated with the augmented natriuresis (WESSON 1962). DUARTE and WATSON (1967) were the first to apply micropuncture methodology to localize the action of mannitol on renal Ca^{2+} transport. Mannitol caused an inhibition of Ca^{2+} and Na^{+} reabsorption in the proximal tubule of rats, which could completely account for the increase in urinary Ca^{2+} excretion. In a later study, WONG and coworkers (1979) found that in the dog kidney mannitol infusion inhibited Ca^{2+} and Mg^{2+} transport mainly in the loop of Henle, which is similar to its effect on Na^{+} reabsorption in this species. Ca^{2+} reabsorption in the proximal tubule was minimally affected, whereas proximal Mg^{2+} reabsorption was "paradoxically" increased during mannitol diuresis (WONG et al. 1979). The mechanism of the differential effect of mannitol on proximal reabsorption of Ca^{2+} and Mg^{2+} is not clear.

The effect of osmotic diuresis on renal phosphate handling depends largely on the experimental settings. In animals with an intact parathyroid gland, mannitol infusion either does not affect phosphate excretion (WESSON 1962) or it causes phosphaturia only with massive doses (MAESAKA et al. 1976). The latter effect was attributed to dilution of serum ionized Ca^{2+} with secondary release of parathyroid hormone. However, in the parathyroidectomized dog, WONG et al. (1979) demonstrated a phosphaturic response to high doses of mannitol, but not during low-dose mannitol diuresis. This parathyroid-independent phosphaturia was due to inhibition of phosphate reabsorption in the proximal tubule and in the loop of Henle (WONG et al. 1979). Finally STINEBAUGH et al. (1971) have demonstrated in dogs that infusion of hypertonic mannitol increased urinary bicarbonate excretion. This effect was observed also in acetazolamide-treated dogs, suggesting that mannitol diminished bicarbonate reabsorption by a carbonic anhydrase independent mechanism.

With the exception of K^{+} depletion, which has been noted during prolonged osmotic diuresis, such as produced by chronic hyperglycemia, alterations in plasma concentrations of other ions are of rare occurrence.

V. Urinary Concentration and Dilution

During osmotic diuresis the ability of the kidney to concentrate and dilute the urine is markedly impaired. When mannitol is infused in the hydropenic state, urine osmolality decreases curvilinearly with increasing urine flow rate and osmolal clearance (C_{osm}). Concomitantly, the calculated tubular reabsorption of solute free water ($T^{c}_{H_2O}$) rises at a low rate of solute excretion, levels off at moderate rates and often decreases at very high excretory loads (GOLDBERG et al. 1965; GOLDBERG 1973). In the dog, but not in man or the rat, a hypotonic urine may occasionally be produced, even in the presence of a supramaximal amount of antidiuretic hormone (WESSON and ANSLOW 1948). This, however, occurs only during massive diuresis, when the hypotonic distal tubular fluid is delivered unchanged into the final urine (GENNARI and KASSIRER 1974). During maximal water diuresis, urine osmolality rises toward isotonicity as urine flow, solute excretion and free water clearance C_{H_2O} rise (WESSON and ANSLOW 1952). Thus, regardless of the initial water hydration state, urine osmolality will approach isotonicity with increasing degrees of osmotic diuresis (GENNARI and KASSIRER 1974).

A major factor contributing to the observed alterations in urinary concentration and dilution during osmotic diuresis is, undoubtedly, the dissipation of medullary hypertonicity (MALVIN and WILDE 1959; GOLDBERG and RAMIREZ 1967). This phenomenon may be due to a combination of several mechanisms: an increase in medullary blood flow resulting in "medullary washout" of solutes; decreased Na^+Cl^- reabsorption mainly in the thin ascending loop of Henle, but also in the collecting duct; and, finally, inhibition of medullary urea recycling. The relative contribution of each mechanism to the final decrease in medullary osmolality remains to be determined. Direct tissue analysis of solute contents in medullary sections reveals a major decrease in papillary urea concentration, and a smaller change in papillary sodium content (GOLDBERG and RAMIREZ 1967). This finding, however, does not indicate a dominant contribution of impaired urea recycling, since other mechanisms, in particular increased blood flow to the papilla, may also account for the observed phenomenon.

The ability of the kidney to dilute the urine maximally is likewise impaired during mannitol diuresis (GOLDBERG 1973; WESSON and ANSLOW 1952). This may reflect the obligatory presence of the osmotic agent in the urine, and the inability of the distal nephron and collecting duct to further dilute the delivered tubular fluid, even when ADH secretion is maximally suppressed. The alterations in urinary concentration and dilution outlined above may have important clinical implications. Because urine osmolality tends to approach isotonicity, regardless of whether the patient is hydropenic or water loaded, it cannot be used as a reliable index of the hydration state. Moreover, the concentrating defect may augment urinary water losses, resulting in severe water depletion and hyperosmolarity.

VI. Miscellaneous Effects

In addition to the direct renal tubular, glomerular and hemodynamic effects of mannitol, this agent may influence renal function indirectly by affecting other endocrine or local systems. As pointed out in earlier sections, the effects of hypertonic mannitol on the release of ANF (YAMASAKI et al. 1988) and vasodilatory prostaglandins (JOHNSTON et al. 1981) on the one hand, combined with the suppression of the renin angiotensin system (VANDER and MILLER 1964), may be important factors in determining the diuretic/natriuretic actions of mannitol.

Moreover, infusion of hypertonic mannitol may significantly affect local pressures within the kidney, which may secondarily influence tubular reabsorption of salt and water. Thus, direct measurements of renal interstitial pressure by chronically implanted capsules in dog kidney have clearly demonstrated a marked increase during mannitol infusion (OTT et al. 1971). Evidently, the tubular lumen is not compressed by the elevated interstitial pressure, since intratubular hydrostatic pressure is also markedly increased during mannitol infusion (BRENNER et al. 1972). In fact, direct recordings of hydrostatic pressures have demonstrated a widening of the pressure gradients between the intratubular pressure in the proximal nephron and the adjacent peritubular capillaries (BRENNER et al. 1972). The increased luminal hydrostatic pressure following mannitol infusion may assist in keeping the patency of the tubular lumen, and prevent its occlusion during ischemic and toxic insults to the kidney.

C. Beneficial Extrarenal Effects of Hypertonic Mannitol

Due to its hyperosmotic properties, mannitol may affect other systems in addition to the kidney. Some of these extrarenal effects, e.g., the actions on the central nervous system, the muscles and the cardiovascular system, are of major clinical importance. The following section will briefly summarize a few aspects of these extrarenal effects, some of which are based on our personal experience.

By redistributing fluid from edematous organs and tissues, mannitol may improve their function while, at the same time, expand the depleted intravascular space in human casualties (BETTER and STEIN 1990). Studies in our laboratory have shown that mannitol may enhance the recovery from experimental compartment syndrome in the dog (BETTER et al. 1991). Acute compartment syndrome is a devastating complication of rhabdomyolysis caused by muscle tamponade, secondary to increased intracompartmental pressure. In dogs with experimental compartment syndrome, induced by injection of autologous plasma into the anterolateral compartment of the hind limb, systemic infusion of hypertonic mannitol caused a larger decrement in intracompartmental pressure than in animals treated with saline only (Fig. 1). The decompressive effect of mannitol was presumably due to its osmotic activity, which tended to augment fluid removal from the interstitium of the injured muscle. Similarly, more recently, we have shown that hypertonic mannitol enhances the recovery of injured muscle in rats with experimental crush injury induced by applying mechanical pressure on the hind limb (RUBINSTEIN et al. 1993). Hypertonic mannitol caused a significant increase in twitch amplitude following direct electrical stimulation of the crushed muscle (Fig. 2). Moreover, the effects of mannitol on the recovery from crush syndrome were augmented by hyperbaric oxygen treatment. Others have shown that hyperbaric oxygen in useful in the management of experimental compartment syndrome in the dog (STRAUSS et al. 1986). Taken together, these studies in experimental crush syndrome suggest that the combination of intravenous hypertonic mannitol and hyperbaric oxygen

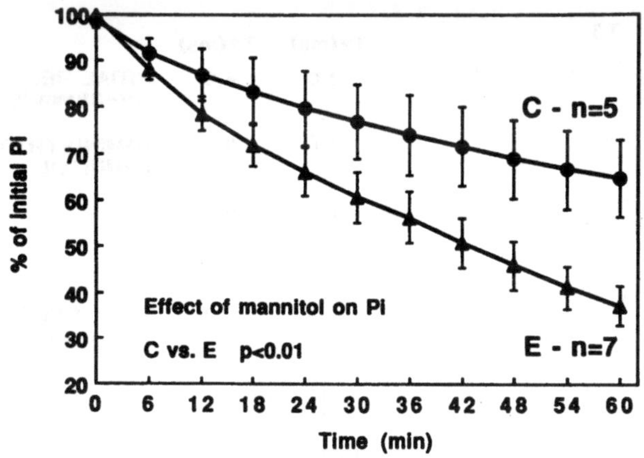

Fig. 1. Alterations in hind limb intracompartmental pressure (*Pi*) vs. time in dogs with experimental acute compartment syndrome. *E*, dogs treated with intravenous hypertonic mannitol; *C*, dogs treated with saline only. (From Better et al. 1991)

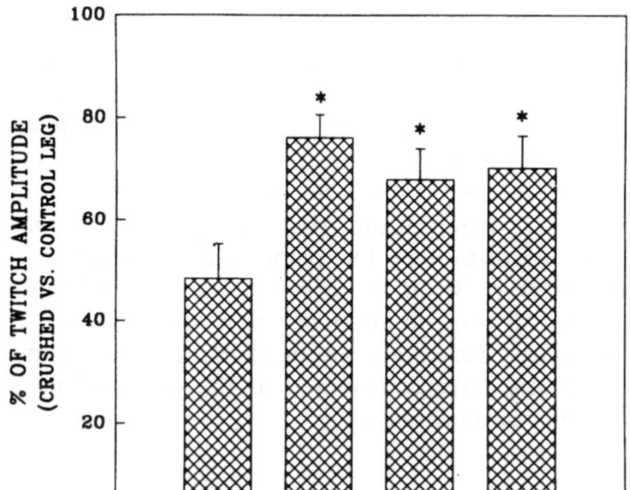

Fig. 2. Effects of intravenous hypertonic mannitol (*M*), hyperbaric oxygen (*HBO*) or a combination of both on the response of injured muscle to direct electrical stimulation in rats with experimental crush syndrome. The amplitude of the injured muscle is expressed as a percentage of the amplitude of its control counterpart limb

may, in the future, be an important adjunct in the treatment of some forms of crush injury in humans.

Finally, mannitol has been used to relieve cerebral edema and to treat increased intracranial pressure in man for several decades. Intravenous

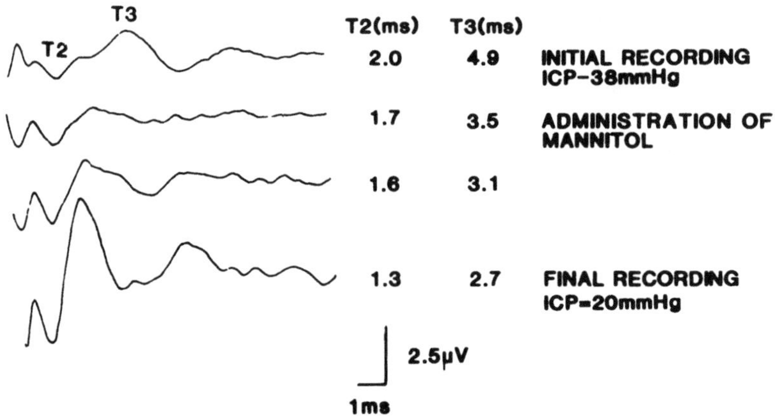

Fig. 3. Temporal alterations in recorded trigeminal evoked potentials following i.v. hypertonic mannitol infusion in patients with brain trauma and increased intracranial pressure. See text for further details

hypertonic mannitol induces an impressive reduction of intracranial pressure and is also known to cause a definite increase in cerebral blood flow (JOHNSTON et al. 1972; MILLER 1973). An attempt was made in our hospital to correlate the decompressive effect of mannitol with central nervous system function in human victims of head trauma (M. FEINSOD, unpublished observations). Thus, within 30 min the decompressive action of mannitol was followed by correction of the pathologic prolongation of trigeminal evoked potentials (Fig. 3). It is likely that this normalization of evoked potentials was due to amelioration of brain edema and the decrease in intracranial pressure. However, an incidental improvement in cerebral circulation could also have contributed to the observed normalization of measured brain electrical activity.

D. Effects on the Cardiovascular System

The influence of mannitol in expanding extracellular and intravascular volume and improving circulatory function was alluded to in previous sections. The beneficial effect of mannitol on the circulation is due to its ability to increase venous return and cardiac preload, thus stabilizing arterial pressure and organ perfusion. However, in addition, mannitol has a direct positive inotropic effect on the heart as shown in the isolated working heart model (BEN-HAIM et al. 1992a,b). Such an effect of hypertonic solution in improving myocardial contractility may be due to the increase in intracellular calcium concentration in the myocytes exposed to hyperosmolarity (BEN-HAIM et al. 1992b).

Finally, mannitol has been shown to exert favorable effects on myocardial edema and infarct size in experimental models of myocardial infarction and reperfusion. Thus, intracoronary perfusion with hyperosmotic mannitol significantly reduced the early myocardial edema and infarct size following coronary occlusion and reperfusion (GARCIA-DORADO et al. 1992). In another study, mannitol-treated hearts demonstrated improved postischemic left ventricular function, greater coronary blood flow and improved structural preservation compared with hearts reperfused with isoosmotic solutions (LUCAS et al. 1980). It was postulated that early treatment with mannitol during postischemic reflow prevented cell swelling, thereby maintaining cell volume and enhancing myocardial viability. However, other mechanisms, not related directly to the osmotic properties of mannitol, e.g., its action as a potent hydroxyl free radical, may also be involved in protecting the ischemic myocardium (MAGOVERN et al. 1984).

E. Clinical Use

I. Clinical Applications

Mannitol is used in the prophylaxis and treatment of acute renal failure, for the reduction of pressure and volume of the cerebrospinal fluid and in ophthalmology for short-term reduction of intraocular pressure and vitreous volume. Some authors consider mannitol to be the most effective drug for reducing intracranial pressure associated with Reye's syndrome (REYNOLDS 1990). It is also indicated as an urinary bladder irrigant during transurethral prostate resection or bladder surgery to prevent hemolysis. In hemodialysis, mannitol infusion can be helpful in correcting cramps and hypotension associated with ultrafiltration. Cerebral edema and ophthalmoplegia due to diabetic ketoacidoses may be reversed by prompt therapy with mannitol (FRANKLIN et al. 1982). Furthermore, mannitol can be used to enhance the excretion of drugs in drug intoxication (e.g., with barbiturates, bromide, aspirin) (WARREN and BLANTZ 1981). The combination of mannitol and normal saline infusion has been used safely and effectively to treat the cerebral edema and hyponatremia associated with water intoxications in infants and children (NUTMAN and HILL 1992).

II. Pharmacokinetics

After oral administration, approximately 20% of mannitol is absorbed (REILLY 1976). Following intravenous administration mannitol is rapidly distributed primarily in the extracellular space (GRABIE et al. 1981; GILMAN et al. 1985) and excreted unchanged in the urine, undergoing no metabolism (REYNOLDS 1990). Onset of diuresis or decrease in intracranial pressure occurs within 15–30 min after i.v. administration (MADSEN et al. 1970;

REILLY 1976). The drug does not cross the blood-brain barrier, its volume of distribution is 0.47 l/kg (CLOYD et al. 1986), and its half-life 70–100 min (REYNOLDS 1990).

III. Dosage

To treat oliguria 50–100 g mannitol may be used intravenously as a 15%–20% solution over 24 h. To reduce intracranial or intraocular pressure the usual dose is 1.5–2 g/kg intravenously over 30–60 min also as a 20% solution. When used to enhance the excretion of toxins the total dose should not exceed 200 g. In pediatric patients (17 years of age and younger) the maximum recommended dose is 2 g/kg (ANON 1984).

IV. Precautions

If patients with renal insufficiency are treated with mannitol a test dose should always be given. During mannitol therapy urinary output should be carefully monitored and fluid and electrolyte balance maintained. In addition, it is necessary to filter solutions of 15% and 20% mannitol.

V. Adverse Reactions

If mannitol is administered rapidly in high doses, this may cause excessive expansion of the extracellular space, leading to congestive heart failure and/or pulmonary edema. In addition, rapid administration of 20% mannitol has been reported to cause hypotension in patients undergoing craniotomy (DOMAINGUE and NYE 1985). BORGES and coworkers described mannitol intoxications (central nervous system depression, confusion, lethargy) in eight patients with renal failure who received about 300 g mannitol (BORGES et al. 1982). Furthermore, hyponatremia and hyperkalemia were seen after high-dose intravenous mannitol administration (BERRY and PETERSON 1981; MANNINEN et al. 1987). Several patients with a history of chronic renal insufficiency developed acute renal failure receiving doses between 200 and 400 g mannitol (WEAVER and SICA 1987).

VI. Contraindications

Mannitol is contraindicated in patients with anuria or renal failure and should also not be used in patients with severe pulmonary congestion, severe congestive heart failure, severe dehydration or acute intracranial bleeding (except during craniotomy). Hypertonic mannitol solutions should not be mixed with whole blood, because agglutination of red blood cells can occur due to the high osmotic pressure.

Acknowledgments. The authors are grateful to Ms. Ruby Snyder and Deborah Shapiro for their secretarial assistance in the preparation of the manuscript.

References

Anon (1984) Guidelines for administration of intravenous medications to pediatric patients, 2nd edn. American Society of Hospital Pharmacists. Bethesda, MD

Barry KC, Cohen A, Knochel JP, Whelan TJ, Beisel WR, Vargas CA, LeBlanc PC (1961) Mannitol infusion II. The prevention of acute functional renal failure during resection of an aneurysm of the abdominal aorta. N Engl J Med 264: 967–971

Ben-Haim SA, Edoute Y, Hayam G, Better OS (1992a) Sodium modulates inotropic response to hyperosmolarity in isolated working rat heart. Am J Physiol 263 (Heart Circ Physiol 32):H1154–H1160

Ben-Haim SA, Hayam G, Edoute Y, Better OS (1992b) Effect of hypertonicity on contractility of isolated working rat left ventricle. Cardiovasc Res 26:379–382

Berry AJ, Peterson ML (1981) Hyponatremia after mannitol administration in the presence of renal failure. Anesth Analg 60:165–167

Better OS, Stein JH (1990) Early management of shock and prophylaxis of acute renal failure in traumatic rhabdomyolysis. N Engl J Med 322:825–829

Better OS Jr, Gonick HC, Chapman LC, Varady PD, Kleeman CR (1966) Effect of urea-saline diuresis in renal clearance of calcium, magnesium, and inorganic phosphate in man. Proc Soc Exp Biol Med 121:592–596

Better CS, Zinman C, Reis DN, Har-Shai Y, Rubinstein I, Abassi Z (1991) Hypertonic mannitol ameliorates intracompartmental tamponade in model compartment syndrome in the dog. Nephron 58:344–346

Blantz RC (1974) Effect of mannitol on glomerular ultrafiltration in the hydropenic rat. J Clin Invest 54:1135–1143

Borges HF, Hocks J, Kjellstrand CM (1982) Mannitol intoxication in patients with renal failure. Arch Intern Med 142:63–66

Braun WE, Lilienfield LS (1963) Renal hemodynamic effects of hypertonic mannitol infusions. Proc Soc Exp Biol Med 114:1–6

Brenner BM, Troy JL, Daugharty TM (1972) Pressures in cortical structures of the rat kidney. Am J Physiol 222:246–251

Buerkert J, Martin D, Prasad J, Trigg D (1981) Role of deep nephrons and the terminal collecting duct in a mannitol-induced diuresis. Am J Physiol 240 (Renal Fluid Electrolyte Physiol 9):F411–F422

Cloyd JC, Snyder BD, Cleeremans B, Bundlie SR, Blomquist CH, Lakatua TJ (1986) Mannitol pharmacokinetics and serum osmolality in dogs and humans. J Pharmacol Exp Ther 236:301–306

Collins GM, Wicomb WN (1992) New organ preservation solutions. Kidney Int 42(38):S197–S202

Delmaestro RF, Thaw HH, Bjork J, Planker M, Arfors KE (1980) Free radicals as mediators of tissue injury. Acta Physiol Scand 492 [Suppl I]:43–57

Domainque CM, Nye DH (1985) Hypotensive effect of mannitol administered rapidly. Anaesth Interns Care 13:134–136

Duarte CG, Watson JF (1967) Calcium reabsorption in proximal tubule of the dog nephron. Am J Physiol 212:1355–1360

Etheredge EE, Levitin HL, Nakamura K, Glenn WWL (1965) Effect of mannitol on renal function during open-heart surgery. Ann Surg 161:53–62

Flores J, DiBona DR, Beck CH, Leaf A (1972) The role of cell swelling in ischemic renal damage and the protective effect of hypertonic solute. J Clin Invest 51:118–126

Fojas E, Schmid HE (1970) Renin release, renal autoregulation and sodium excretion in the dog. Am J Physiol 219:464–468

Franklin B, Lui J, Ginsberg-Fellner F (1982) Cerebral edema and ophthalmoplegia reversed by mannitol in a new case of insulin-dependent diabetes millitus. Pediatrics 69:87–90

Garcia-Dorado D, Theroux P, Munoz R, Alonso J, Elizaga J, Fernandez-Aviles F, Botas J, Solares J, Soriano J, Duran JM (1992) Favorable effects of hyperosmotic reperfusion on myocardial edema and infarct size. Am J Physiol 262:H17–H22

Gennari FJ, Kassirer JP (1974) Osmotic diuresis. N Engl J Med 291:714–720
Gilman AG, Goodman LS, Rall TW, Murad F (eds) (1985) Goodman and Gilman's the pharmacological basis of therapeutics, 7th edn. Macmillan, New York
Goldberg M (1973) The renal physiology of diuretics. In: Orloff J, Berliner RW (eds) Handbook of physiology, renal physiology. American Physiological Society, Washington DC, pp 1003–1031
Goldberg AH, Lilienfield LS (1965) Effects of hypertonic mannitol on renal vascular resistance. Proc Soc Exp Biol Med 119:635–642
Goldberg M, Ramirez MA (1967) Effects of saline and mannitol diuresis on the renal concentrating mechanism in dogs: alterations in renal tissue solutes and water. Clin Sci 32:475–493
Goldberg M, McCurdy D, Ramirez M (1965) Differences between saline and mannitol diuresis in the hydropenic man. J Clin Invest 44:182–192
Grabie MT, Gipstein RM, Adams DA, Hepner W (1981) Contraindications for mannitol in aphakic glaucoma. Am J Ophthalmol 91:265–267
Greger R (1985) Ion transport mechanism in thick ascending limb of Henle's loop of mammalian nephrons. Physiol Rev 65:760–797
Johnston IH, Paterson A, Harper AM, Jennet WB (1972) The effect of mannitol on ICP and CBF. In: Brock M, Dietz H (eds) Intracranial pressure. Springer, Berlin, Heidelberg, New York, pp 176–180
Johnston PA, Bernard DB, Perrin NS, Levinsky NG (1981) Prostaglandins mediate the vasodilatory effect of mannitol in the hypoperfused rat kidney. J Clin Invest 68:127–133
Lilien OM (1973) The paradoxical reaction of renal vasculature to mannitol. Invest Urol 10:346–353
Lillien OM, Jones SG, Mueller CB (1963) The mechanism of mannitol diuresis. Surg Gynecol Obstet 117:221–228
Lucas SK, Gardner TJ, Flaherty JT, Bulkley BH, Elmer EB, Gott VL (1980) Beneficial effects of mannitol administration during reperfusion after ischemic arrest. Circulation 62 [Suppl I]:34–41
Madsen PO, Knuth OE, Wagenknecht LV, et al. (1970) Induction of diuresis following transurethral resection of the prostate. J Urol 104:735–738
Maesaka JK, Bergen ML, Bornia ME, Abramson RB, Levitt MF (1976) Effect of mannitol on phosphate transport in intact and acutely thyroparathyroidectomized rats. J Lab Clin Med 87:680–691
Magovern GJ, Jr, Bolling SF, Casale AS, Bulkley BH, Gardner TJ (1984) The mechanism of mannitol in reducing ischemic injury: hyperosmolarity or hydroxyl scavenger? Circulation 70 [Suppl I]:91–95
Manninen PH, Lam AM, Gelb AW, Brown SC (1987) The effect of high-dose mannitol on serum and urine electrolytes and osmolality in neurosurgical patients. Can J Anaesth 34:442–446
Malnic G, Klose R, Griebish G (1966) Micropuncture study of the distal tubular potassium and sodium transport in the rat nephron. Am J Physiol 211:529–547
Malvin RL, Wilde WS (1959) Washout of renal countercurrent Na gradient by osmotic diuresis. Am J Physiol 197:177–180
Mason J, Joeris B, Welsch J, Kriz W (1989) Vascular congestion in ischemic renal failure: the role of cell swelling. Miner Electrolyte Metab 15:114–124
Miller JD (1973) The effects of hyperbaric oxygen at 2 and 3 atmospheres absolute and intravenous mannitol on experimentally increased intracranial pressure. Eur Neurol 10:1–11
Morris CR, Alexander EA, Bruns FJ, Levinsky NG (1972) Restoration and maintenance of glomerular filtration by mannitol during hypoperfusion of the kidney. J Clin Invest 51:1555–1564
Myers BD (1990) Nature of postischaemic renal injury following aortic or cardiac surgery. In: Bihari D, Nield G (eds) Acute renal failure in the intensive therapy unit. Springer Verlag, London, pp 167–180

Nutman J, Hill JH (1992) Treatment of water intoxication with mannitol (letter). Am J Dis Child 146:1130–1131
Ott CE, Navar LG, Guyton AC (1971) Pressures in static and dynamic states from capsules implanted in the kidney. Am J Physiol 221:394–400
Peters G, Brunner H (1963) Mannitol diuresis in hemorrhagic hypotension. Am J Physiol 204:555–562
Reilly MJ (ed) (1976) Pocket consultant: diuretics. Am Soc Hosp Pharm, Washington DC
Reynolds JEF (ed) (1990) Martindale: The Extra Pharmacopoeia (CD-ROM Version). Micromedex, Denver
Rubinstein I, Kerem D, Melamed Y, Better OS (1993) Hyperbaric O_2 (HBO) & 20% I.V. mannitol (M) enhance recovery from muscular crush injury. In: Reinertsen RE, Brubakk AO, Bolstad G (eds) Proceedings of the 19th Annual Meeting of the European Underwater Biomedical Society on Diving and Hyperbaric Medicine, Trondheim, Norway, 17–20 Aug 1993, pp 137–141
Seely JF, Dirks JH (1969) Micropuncture study of hypertonic mannitol diuresis in the proximal and distal tubule of the dog kidney. J Clin Invest 48:2330–2340
Selkurt EE (1945) Changes in renal clearance following complete ischemia of kidney. Am J Physiol 144:395–403
Sonnenberg H (1978) Effects of furosemide, acetazolamide, and mannitol on medullary collecting-duct function in the rat kidney. Pflugers Arch 373:113–123
Stahl WM (1965) Effect of mannitol on the kidney. Changes in intrarenal hemodynamics. N Engl J Med 272:381–386
Stinebaugh BJ, Bartow SA, Eknoyan G, Martinez-Maldonado M (1971) Renal handling of bicarbonate: effect of mannitol diuresis. Am J Physiol 220:1271–1274
Strauss MB, Hargens AR, Gershuni DH, Hart GB, Akeson WH (1986) Delayed use of hyperbaric oxygen for treatment of a model anterior of compartment syndrome. J Orthop Res 4:108–111
Valdes ME, Landau SE, Shah DM, Newell JC, Scovill WA, Stratton H, Rhodes GR, Powers Jr SR (1979) Increased glomerular filtration rate following mannitol administration in man. J Surg Res 26:473–477
Vander AJ, Miller R (1964) Control of renin secretion in the anesthetized dog. Am J Physiol 207:537–546
Velasquez MT, Notargiacomo AV, Cohn JN (1973) Comparative effects of saline and mannitol on renal cortical blood flow and volume in the dog. Am J Physiol 224:322–327
Warren SE, Blantz RC (1981) Mannitol. Arch Intern Med 141:493–497
Weaver A, Sica D (1987) Mannitol-induced acute renal failure. Nephron 45:233–235
Wesson LG (1962) Magnesium, calcium and phosphate excretion during osmotic diuresis in the dog. J Lab Clin Med 60:422–432
Wesson LG Jr, Anslow WP Jr (1948) Excretion of sodium and water during osmotic diuresis in the dog. Am J Physiol 153:465–474
Wesson LG Jr, Anslow WP (1952) Effect of osmotic and mercurial diuresis on simultaneous water diuresis. Am J Physiol 170:255–269
Williams TF, Hollander W Jr, Strauss MB, Rossmeisl EC, McLean R (1955) Mechanism of increased renal sodium excretion following mannitol infusion in man. J Clin Invest 34:595–601
Windhager EE, Giebisch G (1961) Micropuncture study of renal tubular transfer of sodium chloride in the rat. Am J Physiol 200:581–590
Wong NLM, Quamme GA, Sutton RAL (1979) Effects of mannitol on water and electrolyte transport in the dog kidney. J Lab Clin Med 94:683–692
Yamasaki Y, Nishiuchi T, Kojima A, Saito H, Saito S (1988) Effects of an oral water load and intravenous administration of isotonic glucose, hypertonic saline, mannitol and furosemide on the release of atrial natriuretic peptide in men. Acta Endocrinol (Copenh) 119:269–278

CHAPTER 13

Clinical Uses of Diuretics

J.B. PUSCHETT

A. Introduction

Diuretics remain among the most frequently prescribed therapeutic agents. Their use, over time, has been found to be generally safe and effective, although, as with all other drugs, side effects and adverse reactions do occur. Nevertheless, the risk/benefit ratio is weighted heavily in favor of their beneficial effects, and they have been successfully employed in the treatment of both edematous and nonedematous states. Regardless of the clinical category of the diuretics, or the site(s) within the nephron at which they act, all such drugs share in common the capacity to interfere with Na^+ reabsorption. Because this feature of their actions results in the excretion of tubular fluid associated with the unreabsorbed ions, these agents are also diuretic. Furthermore, all diuretics impair ionic reabsorption by acting at the luminal surface of the renal tubular cell. Thus, they must gain access to the tubular stream in order to be effective. This phenomenon represents one of the limiting features of their activities in various disease states as outlined below.

The basic principles of diuretic action have already been discussed explicitly in previous chapters in this volume. Nevertheless, from the specific perspective of clinical application of diuretics in man, a concise and simplified summary of the renal handling of electrolyte transport and the mechanisms of diuretic action will be provided in the following section.

B. Physiological Basis of Diuretic Action and Clinical Implications of Physiological Principles

An understanding of the sites and mechanisms of the major reabsorptive sites of the principal ions of the glomerular filtrate and of tubular water is important to the rational and judicious use of diuretics in the treatment of various disease states (PUSCHETT 1985). Provided in Fig. 1 is a diagrammatic summary of these nephron loci. The observations to be cited and the characterizations of the nephron transport sites are based largely upon a compilation of the data obtained from studies of the human kidney utilizing the clearance technique as well as both free-flow micropuncture and in vivo microperfusion data, derived primarily from experiments performed in the rat and dog. It is clear that, especially when dealing with the late distal

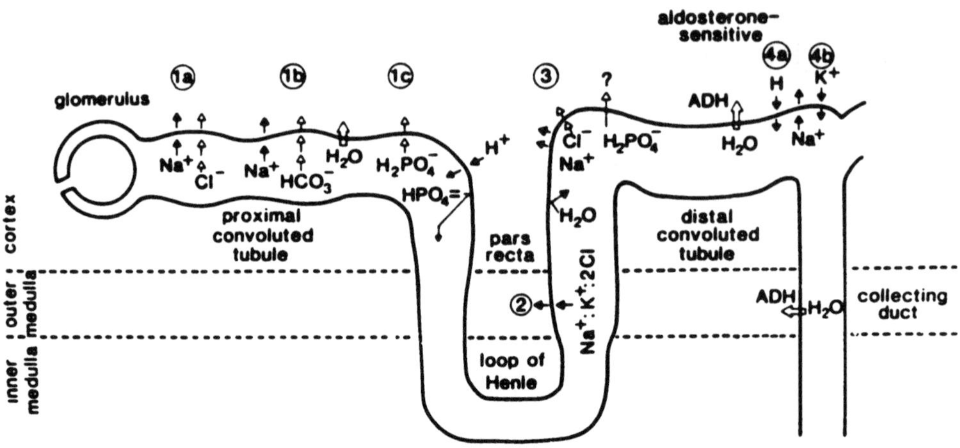

Fig. 1. Delineation of the major electrolyte and fluid transport sites within the nephron. See text for details

nephron (beyond the medullary ascending limb of the loop of Henle), the details of transport differ in the various mammalian species that have been studied. This is especially true when one attempts to compare in vivo studies in the rat with in vitro microperfusion investigations performed in the rabbit, with respect to the identification of nephron segments which may possess differing functions. For example, the connecting tubule of the rabbit has no counterpart in the rat. Nevertheless, good agreement has generally been found to occur between observations obtained with indirect measures of overall nephron functioning in man and direct determinations performed in most mammalian species.

Both glomerular and tubular functioning play important roles in determining the response of the kidney to diuretic administration. With regard to the former, renal hemodynamic factors affect the natriuretic and diuretic actions of these agents in two general ways. First, the level of Na^+ excretion is, of course, a function of the amount of Na^+ filtered. Filtered Na^+ load (FNa^+) is substantially dependent upon glomerular filtration rate (GFR), since $FNa^+ = GFR \times PNa^+$ (plasma sodium concentration). Thus, reductions in GFR will reduce FNa^+ and result in decrements in the amount of Na^+ reaching the nephron system, where diuretics act. Reduced levels of renal function therefore compromise the ability of diuretics to act and may lead to diuretic resistance and/or the need to increment the drug dose (see below). Second, some diuretics (e.g., the thiazide group and the carbonic anhydrase inhibitors) tend to reduce GFR as part of their actions (PUSCHETT and WINAVER 1992). Accordingly, increments in absolute Na^+ excretion secondary to the inhibitory effects of these compounds on tubular Na^+ reabsorption would have to exceed any hemodynamic action of the drug to reduce FNa^+, in order for the drug to be an effective natriuretic and diuretic agent.

Table 1. Primary and secondary sites of activity of diuretics within the nephron and their absolute and relative natriuretic potencies (adapted from PUSCHETT and WINAVER 1992)

Drug	Primary site of action[a]	Secondary site (s) of action[a]	Natriuretic potency	Maximum percentage of filtered Na^+ excreted in response to the drug (%)
Acetazolamide	Proximal tubule (1b, 1a, 1c)	Collecting duct	Mildly potent	3–5
Benzolamide	Proximal tubule (1b, 1a, 1c)		Mildly potent	3–5
Ethacrynic acid	Ascending limb of the loop of Henle (2)		Very potent	20–25
Furosemide	Ascending limb of the loop of Henle (2)	Proximal tubule (1b)	Very potent	20–25
Bumetanide	Ascending limb of the loop of Henle (2)	Proximal tubule(1a, 1b)	Very potent	20–25
Torasemide	Ascending limb of the loop of Henle (2)		Very potent	20–25
Piretanide	Ascending limb of the loop of Henle (2)	Proximal tubule(1a, 1b)	Very potent	20–25
Tripamide	Ascending limb of the loop of Henle (2)	?Proximal tubule (1)[b]	ID[c]	ID[c]
Azosemide	Ascending limb of the loop of Henle (2)	?Proximal tubule (1)[b]	Very potent	20–25
Muzolimine	Ascending limb of the loop of Henle (2)	?Proximal tubule (1)[b]	Very potent	15–25
Indacrinone	Ascending limb of the loop of Henle (2)		ID[c]	ID[c]
Mannitol	Ascending limb of the loop of Henle (2)	Proximal tubule (1)	Very potent	20–30
Mercurials	Ascending limb of the loop of Henle (2)		Very potent	15–20
Chlorothiazide	Early distal convoluted tubule (3)	Proximal tubule (1b, 1c)	Moderately potent	8–10
Metolazone	Early distal convoluted tubule (3)	Proximal tubule (1a, 1c)	Moderately potent	8–10
Indapamide	Early distal convoluted tubule (3)	Proximal tubule (1)	Moderately potent	8–10
Xipamide	Early distal convoluted tubule (3)		ID[c]	ID[c]
Amiloride	Late distal convoluted tubule and collecting duct (4b)		Mildly potent	2–3
Spironolactone	Late distal convoluted tubule and collecting duct (4a)		Mildly potent	2–3
Triamterene	Late distal convoluted tubule and collecting duct (4b)		Mildly potent	2–3

[a] Numbers in parentheses refer to the nephro site(s) depicted in Fig. 1.
[b] Preliminary observations; additional investigation will be required to firmly establish secondary site(s) of action.
[c] ID, insufficient data available to classify as regards potency.

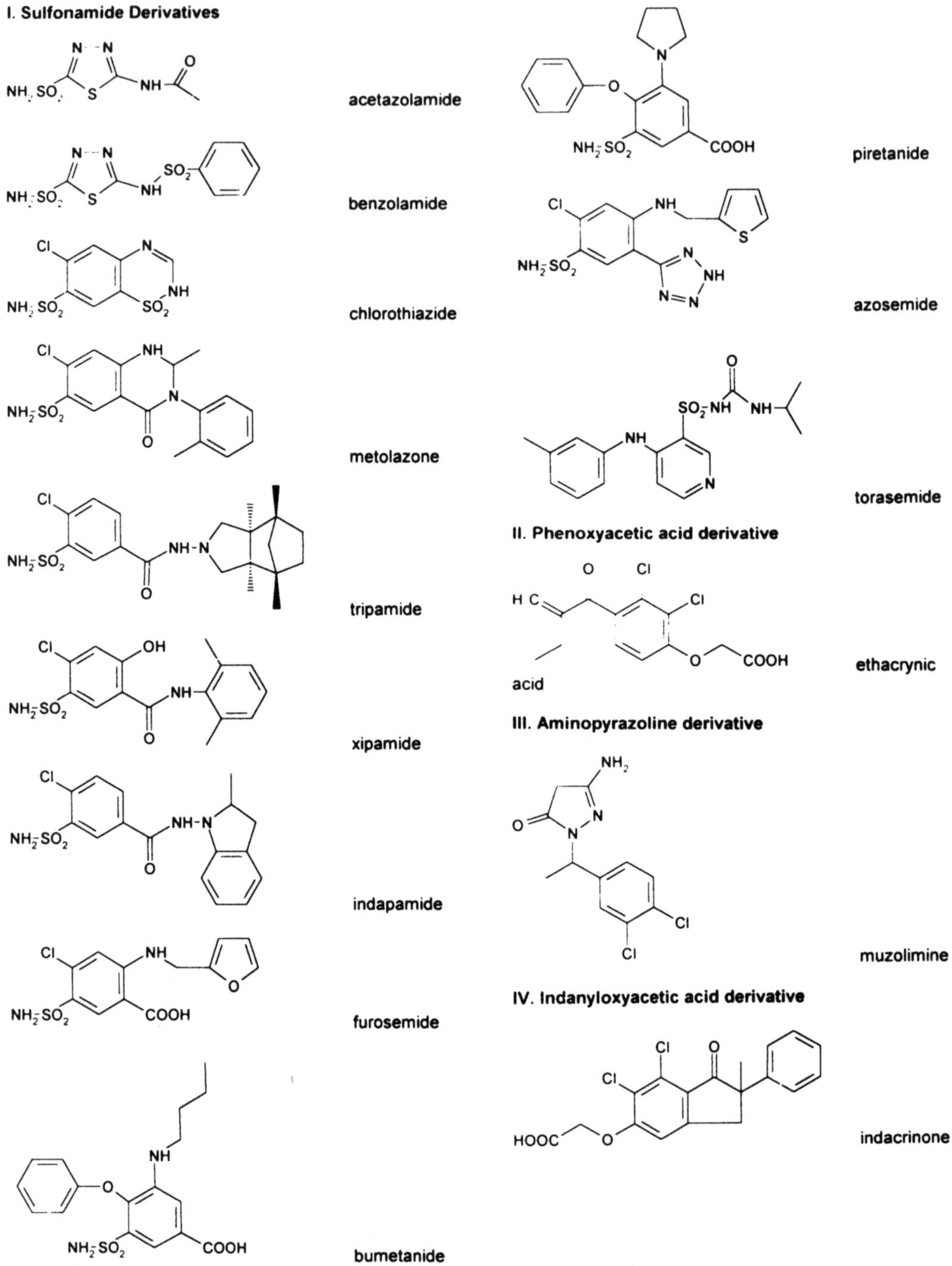

Fig. 2. Chemical structures of the currently available diuretic agents categorized by chemical class. [Modified from Puschett and Winaver (1992) with permission of the editors]

V. Mercurial

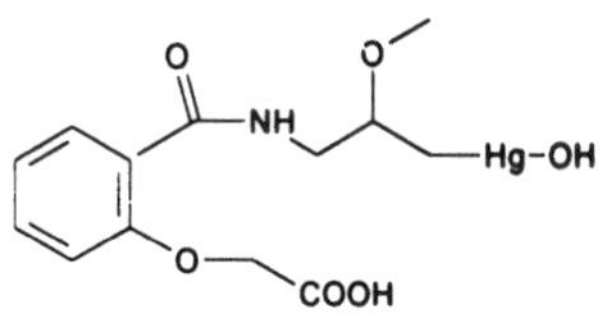

mersalyl

VI. Aldosterone antagonist

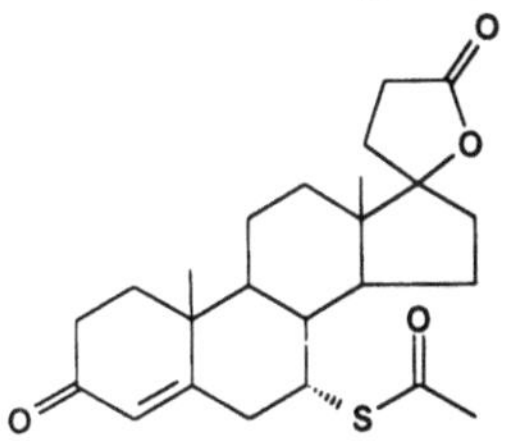

spirolactone

VII. Pteridine derivative

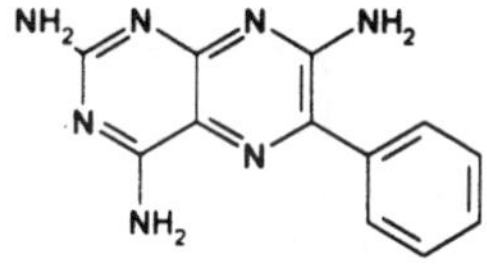

triamterene

VIII. Pyrazinoylguanidine derivative

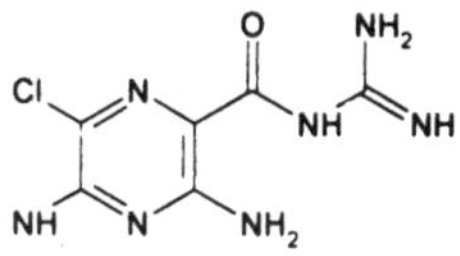

amiloride

Fig. 2. *Continued*

I. Proximal Tubule

Provided in Fig. 1 is a diagram depicting the major sites of electrolyte transport throughout the nephron, and in Table 1 are listed the primary and secondary site(s) of activity of the diuretics Fig. 2 along with estimates of their natriuretic potencies. The physiological principles governing transport within the nephron have important clinical implications. The majority (estimated at 60%–70%) of filtered Na^+ is reabsorbed in the proximal nephron (Cortney et al. 1965; Lassiter et al. 1961). Those drugs which have their primary sites of activity in this nephron segment are the carbonic anhydrase inhibitors acetazolamide and benzolamide (Table 1, Fig. 2). The carbonic anhydrase inhibitors impair Na^+ reabsorption by interfering with the process by which bicarbonate is "regenerated" or "reclaimed" from the tubular contents (Fig. 3) (Koeppen et al. 1989) and therefore act at the site labelled 1b in the nephron diagram (Fig. 1). These drugs impair the "reabsorptive" process for bicarbonate at several steps as shown in Fig. 3, the ultimate result of which is a reduction in the amount of hydrogen ion available for secretion. Acetazolamide is the archetypical carbonic anhydrase inhibitor (Fig. 2). Its ability to inhibit the enzyme is related to its chemical structural characteristics, which include an unsubstituted sulfonamido ($SO_2NH_2^-$) grouping (Maren et al. 1954). This structure is a necessary but not always a sufficient requirement for the conferral of carbonic anhydrase inhibitory activity (Puschett and Winaver 1992). Its administration to animals or man routinely leads to the elaboration of a urine containing large amounts of bicarbonate (Maren 1974; Puschett and Goldberg 1969). As demonstrated in Fig. 3, the bicarbonate "reabsorption" process actually involves the transformation of filtered bicarbonate through the evanescent

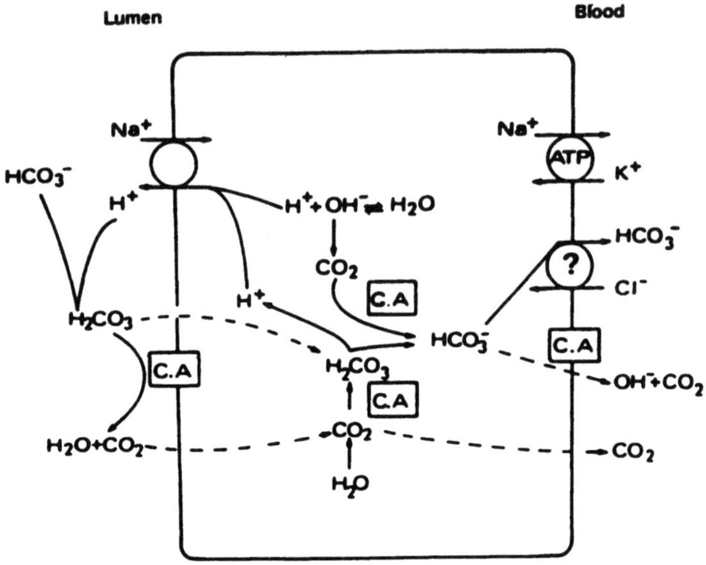

Fig. 3. Details of the transport processes which result in the reclamation of bicarbonate from the proximal tubular contents

Table 2. Comparison of in vitro carbonic anhydrase inhibitory activity of diuretic agents (structures see Fig. 2) with that of acetazolamide [data reprinted from PUSCHETT and GOLDBERG (1968) with permission of the editors]

Diuretic	Concentration required for 50% inhibition of carbonic anhydrase (*l*)
Acetazolamide	1.81×10^{-8}
Chlorothiazide	1.29×10^{-7}
Furosemide	9.88×10^{-7}
Ethacrynic acid	5.81×10^{-3}

formation of carbonic acid to water and CO_2 gas. This series of reactions occurs in the tubular lumen because of the presence of the enzyme in the proximal tubular luminal brush border microvilli. The CO_2 gas diffuses easily into the cell because of its small size, its noncharged nature and because it is a gas. Its reconversion to bicarbonate is likewise catalyzed by carbonic anhydrase, allowing for its extrusion into the peritubular capillary blood, thus completing the "reabsorption" cycle (Fig. 3). Agents which have similar chemical structures to those of acetazolamide and benzolamide (Fig. 2) are candidates for carbonic anhydrase inhibitory activity. Some of these agents (e.g., chlorothiazide and furosemide) do possess inhibitory capability when compared in vitro with the activity of acetazolamide (Table

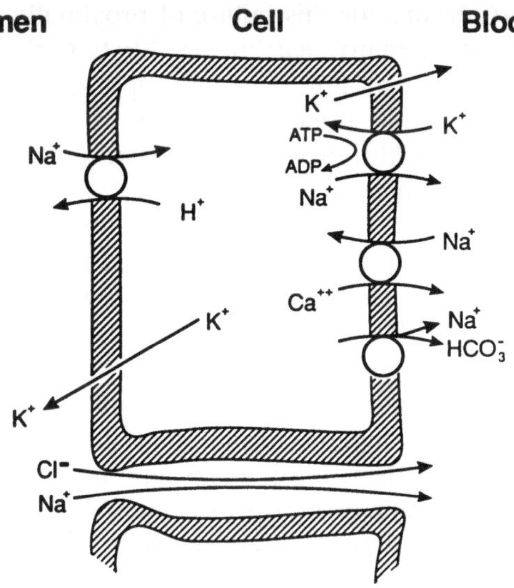

Fig. 4. Major transport pathways of an "idealized" proximal tubular cell. Not shown is the transfer of Na^+ from the tubular lumen associated with the reabsorption of glucose and amino acids, prominent constituents of the proximal glomerular filtrate. [Reproduced with permission of the editors from PUSCHETT and O'DONOVAN (in press)]

2; PUSCHETT and GOLDBERG 1968) while others do not (PUSCHETT and WINAVER 1992).

The transport sites for the proximal nephron (both the convoluted and straight portions) presented in Fig. 1 depict Na^+ as being reabsorbed associated with three of the major anions of the glomerular filtrate: chloride (site 1a), bicarbonate (includes the activity of the Na^+/H^+ antiporter as described above, site 1b), and phosphate (site 1c). Not depicted but of signal importance in the accomplishment of proximal reabsorption is the entry also of Na^+ into the renal tubular cell by means of electrogenic cotransport with amino acids and glucose. Water flow, which is isosmotic, is passive, following the osmotic gradient generated by the transfer of the ions and other solutes. As described elsewhere in this volume (see Chap. 2), the force for Na^+ reabsorption is generated by the presence in the basolateral membrane of the ubiquitous enzyme $(Na^+ + K^+)$-ATPase (Fig. 4), which reduces cellular Na^+ concentration such that an electrochemical gradient exists from tubular contents (reflecting the extracellular fluid concentration) to renal tubular cell. Because the proximal nephron accounts for the majority of Na^+ reabsorption in the nephron, it might be reasoned that inhibition of proximal transport would produce a major natriuresis. However, that this is not the case is pointed up by the fact that the carbonic anhydrase inhibitors are only mildly potent, usually causing the excretion of 5% or less of filtered Na^+ when given to man in therapeutic dosage (Table 1). There are principally two reasons for this failure of proximally active agents, when given alone, to cause a major natriuresis First, sodium ions rejected for reabsorption at the level of the proximal nephron resulting from diuretic administration have ample opportunities to be reabsorbed later in the nephron (see Fig. 1). Additionally, only a portion of total proximal reabsorption is interfered with by a given agent.

Mannitol, an osmotic diuretic, is also inhibitory of transport in the proximal tubule, although its primary site of action is the loop of Henle (Table 1) (SEELY and DIRKS 1969; DIRKS et al. 1966; BUERKERT et al. 1981; PUSCHETT and WINAVER 1992). This complex sugar is diuretic and natriuretic largely because it is poorly reabsorbed from the tubular lumen and is slowly metabolized, so that it acts as a nonreabsorbable solute (cf. Chap. 5, this volume). Therefore, it obligates intratubular fluid (and Na^+) causing an increased excretion of both with attendant anions. For a fuller discussion of osmotic diuresis, the reader is referred to Chap. 12).

Several other agents have secondary effects in the proximal nephron (cf. Table 1). In the cases of chlorothiazide and furosemide (Fig. 2), this (additional) proximal activity is related to the capacity of these agents to inhibit carbonic anhydrase, although less strongly than is the case for acetazolamide (Table 2). Although a member of the sulfonamide group, bumetanide (Fig. 2), which evidences proximal activity in the human subject (BOURKE et al. 1973; JAYAKUMAR and PUSCHETT 1977; KARLANDER et al. 1973; LANT 1975; PUSCHETT 1981), as well as in the experimental animal (GUTSCHE et al. 1983; HIGASHIO et al. 1978; PUSCHETT et al. 1978) and in vitro (FORBUSH and PALFREY 1983), does not appear to inhibit proximal reabsorption by inhibiting carbonic anhydrase (OLSEN 1977; PUSCHETT et al. 1978) but decreases bicarbonate reabsorption by some other (as yet unknown) mechanism (PUSCHETT et al. 1978). Piretanide (Fig. 2), another sulfonamide derivative, has also been determined to have proximal tubular activity as well based upon clearance studies in man (MARONE and REUBI 1980; POZET et al. 1980; TEREDESAI and PUSCHETT 1979) as well as micropuncture investigations in the dog (WINAVER et al. 1980). In the latter studies, proximal Na^+ bicarbonate transport was reduced but Na^+ phosphate reabsorption in that nephron segment was unaffected by the drug. Thus, both bumetanide and piretanide uncoupled phosphate and bicarbonate transport at the proximal level. It had previously been presumed, based upon clearance studies, that agents and/or maneuvers that influence the transport of one of these ions also altered the reabsorption of the other in the same direction. The mechanism by which piretanide reduces proximal transport is not clear but could include carbonic anhydrase inhibitory activity based upon its chemical structure and recently completed studies of its in vitro effects (HROPOT, unpublished observations).

Three of the drugs that are natriuretic primarily because they inhibit Na^+ transport in the early distal convoluted tubule (site 3, Fig. 1) also have proximal tubular activity (PUSCHETT and WINAVER 1992). Chlorothiazide (STEINMULLER and PUSCHETT 1972; HEINEMANN et al. 1959; EARLEY et al.

1961; SUKI et al. 1965; KUNAU et al. 1975; FERNANDEZ and PUSCHETT 1973), metolazone and indapamide (Fig. 2), all of which are sulfonamide derivatives, have been demonstrated to have proximal actions. However, their proximal effects do not add to their natriuretic potency when the drugs are given alone, since ions rejected at the level of the proximal nephron are reabsorbed in the loop of Henle. Chlorothiazide is thought to act proximally by virtue of its ability to inhibit carbonic anhydrase (Table 2) (PITTS et al. 1958; PUSCHETT and GOLDBERG 1968). The proximal action of metolazone has also been carefully studied (STEINMULLER and PUSCHETT 1972; SUKI et al. 1972; FERNANDEZ and PUSCHETT 1973). Despite its sulfonamide structure (Fig. 2), metolazone has not been demonstrated to possess carbonic anhydrase inhibitory activity either in vivo (STEINMULLER and PUSCHETT 1972) or in vitro (BELAIR et al. 1969). Therefore, the mechanism of its proximal effect is unknown (PUSCHETT et al. 1973). Both indirect observations and direct measurements have verified that indapamide has secondary proximal tubular effects of modest degree (BURKE et al. 1983). The basis for this activity has not been studied.

II. Loop of Henle

Located in the outer medullary or thick portion of the ascending limb of the loop of Henle is a transport site at which another 20%–25% of filtered Na^+ is reabsorbed. This locus, labelled site 2 in Fig. 1, transports Na^+, K^+ and 2 chloride ions in a secondary active electroneutral process (Fig. 5)

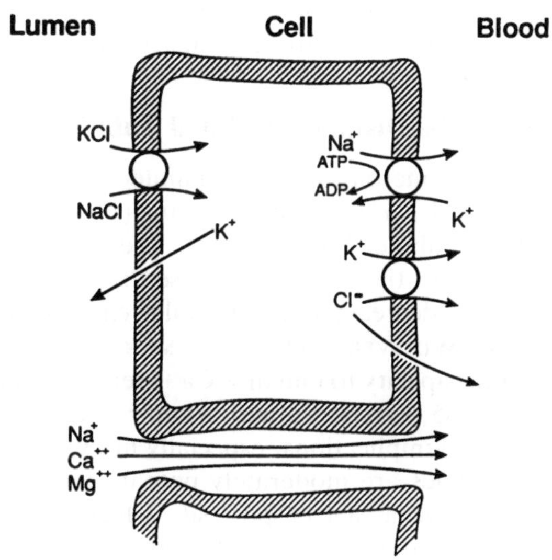

Fig. 5. Transport pathways in a loop of Henle cell. The major luminal transport site transfers 1 Na^+, 1 K^+ and 2 Cl^- in an electroneutral, secondary active process. [Reproduced with permission from PUSCHETT and O'DONOVAN (in press)]

(GREGER 1981, 1984) (see also Chap. 7 for extensive discussion of this transport site). As in other renal tubular cells, the gradient between intracellular and extracellular Na^+ is the result of the action of basolaterally located (Na^++K^+)-ATPase. There is also a basolateral K^++Cl^- transporter. Furthermore, conductance pathways exist for K^+ luminally and for chloride basally and there is also paracellular ionic transport (Fig. 5). Those agents which have their primary site of action in the loop of Henle are listed in Table 1. Furosemide, the first loop of Henle diuretic to be developed, is thought to act by virtue of its ability to bind to the chloride-binding site of the cotransporter (HANNAFIN et al. 1983). Bumetanide and piretanide are also thought to share this cellular mechanism of action, but there is some evidence for alternative actions of ethacrynic acid and muzolimine (PUSCHETT and WINAVER 1992; SCHLATTER et al. 1983). Furosemide, bumetanide, piretanide, tripamide, torasemide and azosemide are all sulfonamide derivatives, whereas ethacrynic acid, muzolimine and indacrinone are not (Fig. 2). As mentioned previously, mannitol acts largely as a nonreabsorbable solute.

The mercurial diuretics have largely disappeared from clinical use for principally two reasons: first they must be given parenterally and second, with the development of furosemide and the other drugs already discussed, equipotent, orally effective agents have supplanted the mercurials in the treatment of moderate to severe edema.

The loop of Henle agents are currently the most potent natriuretic drugs available (Table 1). Their very potent activity results from two factors: (a) they act at a site in the nephron at which large amounts of Na^+ are transported (20%–25%) and (b) Na^+ ions rejected for transport in this nephron segment have limited opportunities for reabsorption at more distal sites in the tubular system, which are less capacious (sites 3, 4, Fig. 1).

III. Early Portion of the Distal Convoluted Tubule

The thiazide diuretics, metolazone, indapamide and xipamide (Fig. 2) inhibit the electroneutral Na^+Cl^- cotransporter (Fig. 6) in the cells of the early distal convoluted tubule (Fig. 1, site 3). The mechanism of this action is thought to be related to the binding of these drugs to a thiazide-sensitive transporter leading to a decrease in intracellular chloride activity (FANESTIL 1993; PUSCHETT and WINAVER 1992; FRIEDMAN and GESEK 1993). The thiazides also have the capacity to enhance Ca^{2+} reabsorption in this nephron segment (COSTANZO 1985; COSTANZO and WEINER 1974). This feature of the drug's action has clinical implications, especially in hypercalciuric conditions (see below). The thiazides are moderately potent diuretic agents (Table 1) because they are effective at a transport site which is responsible for the reabsorption of approximately 5%–8% of filtered Na^+. In addition, failure of Na^+ ions to be rejected at this nephron locus (Fig. 1, site 3) virtually assures their appearance in the urine, since the ability of more distal sites (site 4, Fig. 1) to reabsorb them is extremely limited. The thiazides and their analogues are discussed in extenso in Chap. 8 of this volume.

IV. Late Portion of the Distal Convoluted Tubule and the Collecting Duct

Several complex transport functions which eventually lead to the secretion of K^+ and H^+ in exchange for the reabsorption of Na^+ are defined by the activities of a group of specialized cells located in the most distal reaches of the nephron (Fig. 1, site 4). In these nephron segments, the distal portion of the distal convoluted tubule and the collecting duct, regulation of the transport processes can be divided into that portion which is under control of the mineralocorticoid system (Fig. 1, site 4a) and that which functions even in the absence of these steroids (site 4b). In fact, three cell types appear to be involved, as depicted in Fig. 7. K^+ transport in these nephron segments is largely a function of the principal cell, the major collecting duct cellular type, which receives Na^+ through an apically located Na^+ channel and extrudes K^+ through both luminal and basolateral conductancting. As a result resulting channels; Na^+/K^+ exchange occurs. The intercalated cells secrete hydrogen ion into the tubular lumen as the result of the activity of a proton-translocating ATPase located in the luminal cell membrane. Bicarbonate ions, the product of the action of cellular carbonic anhydrase on CO_2 and water, exits the cell basolaterally as part of the action of a bicarbonate/chloride antiporter (intercalated cell type A. Fig. 7). Another type of intercalated cell (type B, Fig. 6) is the site of luminal bicarbonate secretion.

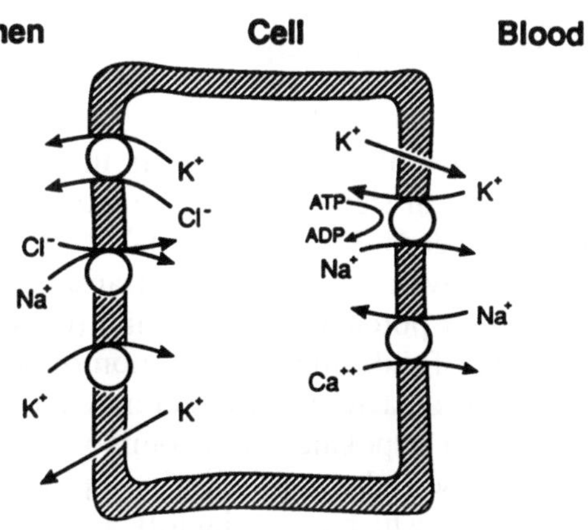

Fig. 6. Diagrammatic representation of the transport pathways in the early distal convoluted tubule cell. The thiazide diuretics and metolazone as well as indapamide and xipamide inhibit the electroneutral NaCl cotransporter located in the apical membrane. [Reproduced with permission of the editors from PUSCHETT and O'DONOVAN (in press)]

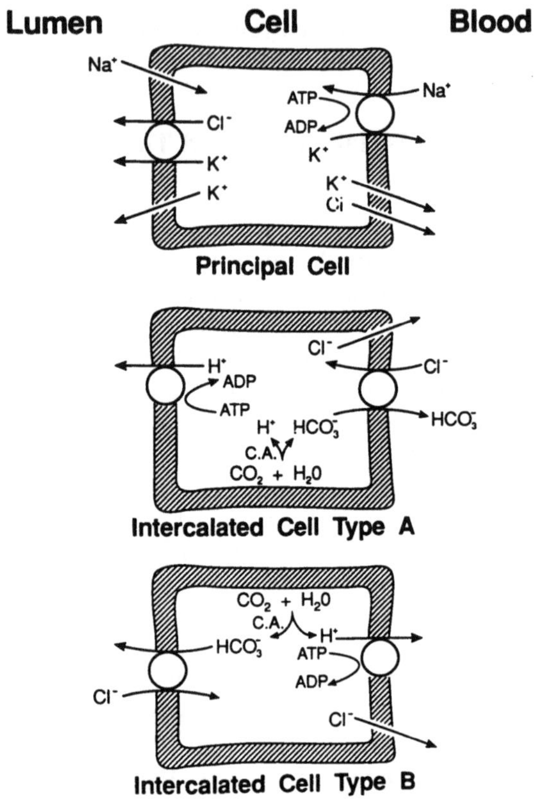

Fig. 7. Transport pathways in those renal tubular cells located in the late portion of the distal convoluted tubule and in the collecting duct which are responsible for the exchange of Na^+ for H^+ and K^+. The principal cells are largely responsible for K^+ exchange and the intercalated cells for H^+ secretion. [Reproduced with permission from PUSCHETT and O'DONOVAN (in press)]

Although the activities of the transporters in these nephron segments have profound physiological significance, the reabsorption of Na^+ at these sites probably never exceeds 3% of filtered Na^+ load. Accordingly, the diuretics which are effective in impairing the transport of Na^+ at site 4 are given not because of their potency, which is mild (cf. Table 1), but because they have the capability of reducing the excretion of K^+. This characteristic of these diuretics, spironolactone, triamterene and amiloride (Fig. 2), lessens the possibility that hypokalemia, a common complication of diuretic therapy, will occur, or else reduces its severity. Spironolactone is a steroid-related compound (Fig. 2) which acts primarily as a competitive inhibitor of aldosterone, although it can also inhibit adrenal steroid synthesis (CORVOL et al. 1981; LIDDLE 1957). Amiloride, a derivative of pyrazinoylguanidine (Fig. 2), interferes with Na^+ reabsorption at this site either by blocking the apical Na^+ channel or by causing the channel to close (BURG and SARIBAN-

SOHRABY 1984; O'NEIL and BOULPAEP 1979; BENOS 1982; KOEPPEN et al. 1983). The mechanisms of action of triamterene have been less well studied, but the latter agent appears to act by mechanisms similar to those described for amiloride. Extensive discussions of the effects of these drugs are provided in Chaps. 9–11 of this volume.

The reabsorption of filtered K^+ is virtually complete by the time the tubular stream reaches the end of the ascending limb of the loop of Henle (WRIGHT and GIEBISCH 1992). Accordingly, K^+ which appears in the final urine is a consequence of the secretory activities of the cells in the late portions of the distal convoluted tubule and the collecting duct as just described. Thus, incremental K^+ excretion can occur from two sources: (a) the inhibition of Na^+ transport at sites proximal to site 4 (Fig. 1) with the presentation of additional Na^+ ions for exchange or (b) the impairment of K^+ reabsorption at the level of the proximal tubule, the loop of Henle, or the early distal convoluted tubule (sites 1, 2, and 3, Fig. 1). Thus, in consideration of the concept that increased distal delivery of Na^+ will result in enhanced exchange for K^+ (Fig. 8), it is clear that in general the more potent the natriuretic agent, the greater the potential for increased kaliuresis. Furthermore, the production of volume contraction which often follows the administration of diuretics leads to the activation of the renin-angiotensin-aldosterone system. Since, as described earlier, the distal exchange mechanism is under at least partial control of the mineralocorticoid system, the

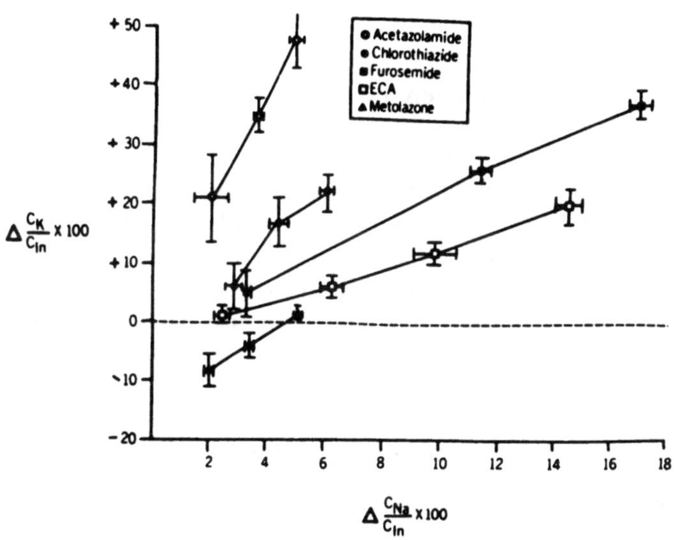

Fig. 8. Alterations in the percentage of filtered K^+ excreted into the urine ($\Delta C_K/C_{In} \times 100$) as a function of distal Na^+ delivery as expressed by increments in the percentage of filtered Na^+ excreted ($\Delta C_{Na}/C_{In} \times 100$) for five diuretic agents. In a general way and regardless of the drug employed, K^+ excretion increased as Na^+ delivery to the distal nephron was incremented. [Reproduced from PUSCHETT and RASTEGAR (1974) with permission of the editors]

development of secondary hyperaldosteronism abets this process. If unrecognized and untreated, persistent kaliuresis will invariably lead to the development of hypokalemia (see also Sect. E below).

C. Diuretics in the Treatment of Edematous States and Disorders Associated with Abnormalities of Renal Function

I. General Principles

Regardless of the pathogenesis of the disease process, the development of edema in systemic disease has as its common pathway the retention of Na^+ and water in a clinical situation in which, in fact, a *decrease* in Na^+ reabsorption would be a more appropriate response on the part of the kidney. This misinterpretation of body volume status is related to alterations in sensory mechanisms monitoring blood flow that incorrectly signal the kidneys that body volume is threatened. There follows activation of the renin-angiotensin-aldosterone system and stimulation of other humoral and neural influences, all of which contribute to edema formation. Accordingly, in each disease instance, the physician must first attempt to uncover the pathogenetic mechanisms and endeavor to correct them. Important adjuncts to this primary activity are the imposition of dietary Na^+ restriction and the use of diuretics. The latter two maneuvers assume positions of primacy in those clinical circumstances in which the underlying pathophysiology cannot be remedied.

Dietary therapy is absolutely essential in edema states. The placement of patients on diuretics while allowing *ad libitum* dietary Na^+ intake is counterproductive and will almost universally lead to failure in the therapy of the expanded extracellular fluid volume. On the other hand, dietary therapy that is too strict will invite noncompliance. Na^+-restricted diets ranging from 250 mg to 4 g Na^+/day are listed in Table 3. Diets of 1 g or less are unachievable except in hospitalized patients. Furthermore, diets of 500 mg or less are so unpalatable that their use cannot be recommended except under exceptional circumstances or for just a few days. If the physician

Table 3. Na^+-restricted diets (reproduced from GREENBERG and PUSCHETT 1991)

Na^+	Content	Description
4 g	175 mmol	No added salt; suitable for outpatients
2 g	87 mmol	Mild restriction; suitable for outpatients; excludes obviously salty foods
1 g	43 mmol	Moderate restriction; rarely achievable by outpatients
500 mg	22 mmol	Severe restriction
250 mg	11 mmol	Rigid restriction; unpalatable

explains the need for and importance of Na^+ restriction, it is often possible to obtain the cooperation of the patient and his/her family in adhering to a 2-g Na^+ diet as an outpatient. Often, however, patients can only achieve a reduction in Na^+ intake from the 300 mmol/day (approximately 7 g) intake that typifies the average individual to 3–4 g Na^+/day. A familiar and recurrent circumstance is the response of an edematous patient to hospitalization and the imposition of Na^+ restriction, whose medications are unchanged from his/her outpatient prescription.

Bed rest is likewise an important issue in the therapy of edema. It was recognized several years ago that simply putting patients to bed and restricting their Na^+ intake was effective in treating patients with conditions such as ascites from cirrhosis of the liver (Pecikyan et al. 1967). As patients assume a recumbent position, fluid redistributes so that central blood volume increases and the perfusion of the renal and splanchnic circulations improves (Greenberg and Puschett 1991). In patients whose edema is the result of congestive heart failure, rest in bed reduces oxygen consumption and the demand on the heart.

Successful diuresis depends upon several factors. First and foremost, there must be enough residual renal function such that the drug may gain access to the tubular lumen and so that inhibition of Na^+ transport will cause the excretion of sufficient amounts of Na^+ such that negative Na^+ balance may be established. Furthermore, as discussed previously, the impairment of Na^+ reabsorption related to the action of the diuretic will be ineffective if the Na^+ ions rejected at the site(s) of the diuretic's activity are reabsorbed elsewhere in the nephron.

Diuretic resistance will be discussed in detail below. However, an important principle which embraces the issue of "apparent" resistance to diuretics deserves mention at this juncture.

Many patients with edematous disorders seem to behave as if their renal tubular systems are avid for Na^+, their edema notwithstanding. This appears to result from a misinterpretation by the kidney of the status of overall body volume and Na^+ as mentioned above. The latter circumstance in turn is probably related to the fact that receptors located most likely in the renal vascular bed and in the nephron itself sense inadequate arterial filling. Thus, in disorders such as severe congestive heart failure and advanced liver disease, the kidneys behave as if proximal tubular reabsorption has been dramatically enhanced. Under those circumstances, inhibition of Na^+ transport at the loop of Henle or at more distal sites will be unsuccessful in interfering with the reabsorption of the majority of filtered Na^+. Accordingly, combination diuretic therapy will be required (see Sect. C.X, below).

II. Congestive Heart Failure

As is the case in the therapy of all disease processes, the employment of diuretics in the treatment of congestive heart failure (CHF) must be tailored to the severity of the disease as well as to the urgency with which it must be treated. General guidelines for the therapy of mild, moderate, and severe disease and for the special circumstance represented by acute pulmonary edema are provided in Table 4. The dosage ranges, times of onset and durations of action of commonly employed diuretic agents are listed in Table 5. The primary and secondary sites and mechanisms of action of selected diuretic agents are presented in Table 6 and the New York Heart Association (NYHA) classification and the myocardial oxygen consumption data for patients with varying degrees of CHF are provided in Table 7. Patients described in Table 4 as having mild CHF correspond to class I NYHA. Those in the moderate range can be considered to be represented in classes II and III, and those designated class IV represent severe disease. Although the approach to the therapy of CHF has changed somewhat over time, diuretics remain a mainstay of its medical management (Hilleman and Mohiuddin 1991; Pitt 1992; Greenberg and Puschett 1991). Evidence has accrued, however, that the angiotensin-converting enzyme (ACE) inhibitors have a place in the therapy of not only severe (the Consensus Trial Study Group 1987) but also mild and moderate degrees of CHF (the Solvd Investigators 1990, 1991). Digoxin remains appropriate therapy in patients with systolic dysfunction (the Captopril-Digoxin Multicenter Research Group 1988), and pre-load reduction with nitrates or other agents continues to be useful (Cohn et al. 1986).

Na^+ restriction is mandatory. In inpatients, 1 g/24 h is appropriate with gradual liberalization to 2 g as the patient improves and approaches discharge (Table 3). Diuretic dosage adjustment may be required as the salt intake is altered. In the mildest forms of CHF (class I, NYHA, Table 7), the patients is often asymptomatic; and his functional abnormality usually is restricted to right heart failure with the sole manifestation being an elevated central venous pressure. Renal hemodynamics ordinarily are undisturbed. Peripheral edema, representing the sequestration of salt and water in the interstitial space, especially in the extremities, may or may not be present. In these patients, therapy with one of the thiazide agents metolazone or indapamide should be undertaken (Puschett 1977). If the patient is compliant with his or her Na^+ restriction, daily therapy may not be necessary. Instead, dosing every other day (especially with the longer-acting thiazides such as chlorthalidone and metolazone) or 3–4 days/week may be sufficient. The latter regimen will also tend to minimize side effects and is usually begun with a dose of hydrochlorothiazide of 50 mg (or its equivalent) (Table 5). The dose can be doubled if necessary.

The thiazide will not be effective in patients with renal insufficiency whose levels of glomerular filtration rate (GFR) are below 30–40 ml/min (Reubi and Cottier 1961; Frazier and Yager 1973). In the latter instance, metolazone or one of the loop of Henle drugs will be necessary. Metolazone, in addition to its ability to be effective in patients with GFR reductions, also has a prolonged action. Therapy may be started with 2.5–5.0 mg/day, which

Table 4. Therapeutic guidelines in heart failure [adapted from GREENBERG and PUSCHETT (1991) with permission of the editors]

Disease chronicity	Disease severity	Clinical findings	Renal blood flow	Glomerular filtration rate	Therapeutic measures
Chronic	Mild	None, or mild peripheral edema with modest exertional dyspnea	Normal or slightly decreased	Normal	Na^+ restriction, thiazide or metolazone
Chronic	Moderate	Moderate peripheral edema; vascular congestion and/or pleural effusion	Moderately reduced	Moderately reduced	Digitalis (in patients with systolic dysfunction), loop of Henle drugs, may add antikaluretic agent; angiotensin-converting enzyme inhibitor as unloading agent once patient is stabilized
Chronic	Severe	Marked peripheral edema, dyspnea at rest (may include paroxysmal nocturnal dyspnea), orthopnea, marked pulmonary vascular congestion	Markedly reduced	Markedly reduced	Digitalis (in patients with systolic dysfunction). Combination diuretic therapy: loop of Henle drugs plus metolazone; angiotensin-converting enzyme inhibitors; nitrate may be helpful for preload reduction
Acute	Severe	Pulmonary edema	Markedly reduced	Markedly reduced	Intravenous loop of Henle drugs, follow with oral therapy once improvement occurs. Depending upon degree of improvement, may require combination diuretic drug therapy ("sequential nephron blockade"). Digitalis, angiotensin-converting enzyme inhibitors, nitrates may all be needed, especially acutely

Table 5. Diuretic dosages (Stmelurs See Fig. 2), time of onset and duration of action [adapted from PITTS (1985) with permission of the editors]

Drug class	Diuretic	Usual dosage range (mg)	Dosage frequency per day	Time of onset after oral dose	Time of onset after i.v. dose	Duration of action, oral (i.v.)	Comments
Carbonic anhydrase inhibitors	Acetazolamide	250–1000	1–2×	1–2 h	30–60 min	8–12 h	Renal effect is reduced by presence of metabolic acidosis; drug action can lead to metabolic acidosis, thereby limiting its effectiveness. Can lower GFR
Drugs acting primarily in the early distal convoluted tubule (see Fig. 1)	Hydrochloro-thiazide	25–100	1–2×	2 h	–	6–12 h	Thiazides are ineffective effective in moderate renal insufficiency. Lowest dosages are recommended in the therapy of mild to moderate hypertension; begin with 25 mg/day (12.5 mg in the elderly)
	Chorthalidone	25–50	1×	2 h	–	48–72 h	Long-acting thiazides (e.g., chlorthalidone, indapamide) cause hypokalemia more predictably than short-acting congeners except when given in low dosage
	Metolazone	2.5–20	1×	1 h	–	12–24 h	While similar to the thiazides, differs from them in being effective even in advanced renal insufficiency. When used in combination with loop of Henle drugs in patients with resistant edema, provides a very potent diuretic effect
	Indapamide	2.5–5.0	1×	1–2 h	–	34–36 h	
Loop of Henle blockers[a]	Furosemide	20–320	1–2×	0.5 h	5 min	6–8 h (2–3 h)	Dosage of agents in this group is best determined by titration on an individual basis. Because of their

						natriuretic potency (see Table 1), these drugs are markedly kaliuretic. They remain effective in advanced renal insufficiency. Frequent, high-dosage administration intravenously (especially in patients who are simultaneously receiving aminoglycosides) predisposes patients to vestibular and/or auditory toxicity
Ethacrynic acid	50–250	1–2×	0.5 h	5 min	6–8 h (2–3 h)	When given intravenously, ethacrynic acid may cause a "burning" sensation or poresthesias. This usually responds to a slower rate of administration
Bumetanide	0.5–20.0	1–2×[b]	0.5 h	5 min	4–6 h (2–3 h)	Occasionally effective in patients who have demonstrated resistance to furosemide and ethacrynic acid (and vice versa). Higher dosages reserved for patients with advanced renal insufficiency or renal failure
Torasemide	2.5–20.0	1×	10 min	<1 h	6–12 h	Torasemide has been effective in patients with renal failure in doses up to 100 mg i.v. and 200 mg p.o. once daily
Piretanide	3.0–6.0	1–2×	1 h	10–30 min	4–6 h (3 h)	Dosages of piretanide of up to 60 mg/day have been utilized in patients with renal failure
Indacrinone						Uricosuric by virtue of its ability to impair urate reabsorption in the proximal tubule
Mannitol	50–200	Variable	No oral preparetion	Immediate		Constant infusion of 50–200 g over 24 h usually results in a urine flow rate of 30–50 ml/h. For acute glaucoma, cerebral edema, and preparation for ocular or

Table 5. *Continued*

Drug class	Diuretic	Usual dosage range (mg)	Dosage frequency per day	Time of onset after oral dose	Time of onset after i.v. dose	Duration of action, oral (i.v.)	Comments
							neurosurgery, the usual dosage is 1.5–2 g/kg over 0.5–1 h. Generally a 50–100 g slow i.v. bolus is used for prophylaxis of acute renal failure at the time of the potentially harmful event (surgery, prolonged hypotension, IVP, etc.)
Potassium-retaining drugs	Spironolactone	25–400	1–4×	48–72 h	–	3–4 days after cessation of therapy	Effect of spironolactone depends upon the presence of aldosterone. The drug has a prolonged time of onset, and its effect is long-lasting even after its discontinuation
	Amiloride	5–20	1×	2 h	–	24 h	Because of its Ca^{2+}-retaining effect, is often used in patients with renal stone disease. Besides being potassium retaining, also causes conservation of Mg^{2+}
	Triamterene	50–300	1–2×	2–4h	–	7–9 h	Triamterene causes K^+ retention irrespective of the patient's adrenal mineralocorticoid status. May produce azotemia and hyperkalemia. Simultaneous administration of nonsteroidal antiinflammatory agents should be avoided

[a] Equipotent dosages of these drugs are: furosemide, 40 mg; bumetanide, 1 mg; ethacrynic acid, 50 mg; torasemide, 10–20 mg; piretanide, 12 mg. The relative potency of bumetanide may be less in renal insufficiency.
[b] In special circumstances, especially when given intravenously and for short periods, the drug may be administered up to 3–4×/day with careful monitoring.

Table 6. Sites and mechanisms of action of selected diuretic agents[a]

Drug	Primary site of action	Secondary site(s) of action	Comments
Chlorothiazide	Early distal convoluted tubule (site 3)	Proximal nephron (sites 1b, 1c)	Primary action is related to inhibition of NaCl cotransporter. Secondary effect related to interference with $NaHCO_3$ "reabsorption" associated with carbonic anhydrase inhibitory activity
Metolazone	Early distal convoluted tubule (site 2)	Proximal nephron (sites 1a, 1c)	Primary action same as that for chlorothiazide. Secondary action ot related to carbonic anhydrase inhibitory activity, but due to interference with Na phosphate cotransport
Furosemide	Ascending limb of the loop of Henle (site 2)	Proximal nephron (site 1b)	Primary action related to an ability to interfere with activity of the Na^+ $2Cl^-$ K^+ cotransporter in the luminal membrane of the loop of Henle cell. Secondary activity in the proximal nephron is due to (modest) carbonic anhydrase inhibitory effect (see text)
Bumetanide	Ascending limb of the loop of Henle (site 2)	Proximal nephron (sites 1a, 1c)	Primary action same as that for furosemide. Secondary effect similar to that for metolazone
Torasemide	Ascending limb of the loop of Henle (site 2)		Primary action same as furosemide
Ethacrynic acid	Ascending limb of the loop of Henle (site 2)		Action similar to that for furosemide and bumetanide. Not a sulfonamide derivative, so not a candidate for carbonic anhydrase inhibition
Spironolactone	Late distal convoluted tubule and collecting duct (site 4a)		Competitive inhibitor of aldosterone
Amiloride	Late distal convoluted tubule and collecting duct (site 4b)		Blocks Na^+ channel or renders it closed
Triamterene	Late distal convoluted tubule and collecting duct (site 4b)		Actions most likely similar to those of amiloride

[a] Site references relate to nephron loci depicted in Fig. 1.

Table 7. Classification of degree of disability in congestive heart failure (reproduced with permission from HILLEMAN and MOHIUDDIN 1991)

Symptoms	NYHA Class	MVO_2 (ml/kg min)
Asymptomatic with ordinary physical activity	I	≥21
Symptomatic with ordinary physical activity	II	16–20
Symptomatic with less than ordinary physical activity	III	10–15
Symptomatic at rest or unable to perform activity	IV	≤9

NYHA, New York Heart Association classifiction; MVO_2, myocardial oxygen consumption.

need be given only once daily or every other day. The dose in patients with severe renal insufficiency should probably not exceed 20 mg/day (Table 5) (DARGIE et al. 1972; CRASWELL et al. 1973; PATON and KANE 1977).

As CHF worsens, effective arterial filling is further compromised so that renal blood flow (RBF) falls. Whether or not this will lead to a reduction in GFR depends upon the degree of RBF decline since, if not too severe, the kidneys' ability to perform GFR autoregulation will not be exceeded. Once it is, GFR will be reduced. In addition, as mentioned previously, sensors located in the kidney will respond to the decrease in effective arterial filling by enhancing renal tubular Na^+ reabsorption. This phenomenon, mediated by the renal nerves as well as humoral mechanisms (including but not limited to the renin-angiotensin-aldosterone axis) functions independently of the altered renal hemodynamics (SKORECKI and BRENNER 1982). As a result of these changes, the entire renal tubular system behaves as if it is avid for Na^+, leading to the development of peripheral edema as well as pulmonary vascular congestion and/or pleural effusion. Digoxin administration is indicated in most of these patients, who should also receive a loop of Henle diuretic. The dosage of digoxin depends upon the level of renal function. The reader is referred to the review by BENNETT et al. (1983) for specific recommendations which can be summarized as follows: usual digitalizing dose (GFR >50% of normal), 1.0–1.5 mg; maintenance dose, 0.25 mg/day. For patients with GFR values of 10–15 ml/min, reduce dosage to 25%–75% of standard dose. Patients with GFR values of <10 ml/minute, reduce dosage to 10%–25% of the usual dose. Dialysis patients should receive 0.125 mg every other day or every 3rd day (GREENBERG and PUSCHETT 1991).

As far as loop diuretics are concerned, one may begin with 40 mg furosemide or its equivalent p.o., incrementing the amount of drug administered until an effective dose is reached. When that dosage has been established, it may be necessary to repeat the dose more than once daily to achieve adequate diuresis. Alternatively, combination diuretic therapy (sequential nephron blockade; see Sect. C.X, below) may prove necessary. Many patients with moderate to severe edematous disorders such as CHF behave as if they have an intense stimulus for enhanced proximal reabsorption. In these patients, especially if GFR is moderately or markedly compromised, diuresis with a single agent will often be difficult. Data from Brater and his colleagues suggest that beyond a dose of 120 mg of intravenous furosemide (or 240 mg by mouth), 3–6 mg bumetanide, or 10 mg of torasemide, the likelihood for the recruitment of additional Na^+ excretion is minimal (BRATER 1993). Accordingly, since distal delivery of Na^+ is reduced, the author and his colleagues have favored the addition of a drug which acts proximally in conjunction with a loop of Henle diuretic in an attempt to deliver Na^+ ions from the proximal nephron past the blockaded loop of Henle. Good results have been obtained, in our hands, with a combination of metolazone and one of the loop of Henle agents. Metolazone has its primary effect at the nephron transport site located in the early distal convoluted tubule (Fig. 1, site 3). However, it also possesses proximal tubular activity as discussed earlier in this chapter (Table 6). When given alone, the proximally rejected ions are reabsorbed in the loop of Henle (FERNANDEZ and PUSCHETT 1973; PUSCHETT et al. 1973). Therefore, when combined with, for example, furosemide, sequential nephron blockade of sites 1, 2, and 3 is obtained; and a substantial natriuresis generally results. In addition, it is possible in some clinical circumstances to add a site 4 (Fig. 1) agent such as spironolactone, triamterene or amiloride both to complete the impairment of Na^+ transport along the nephron as well as to attempt to reduce the substantial kaliuresis that can be expected. These patients not only have enhanced Na^+ transport at this site in the nephron but usually also evidence secondary hyperaldosteronism stimulated by the reduced effective arterial filling described earlier (HENSEN et al. 1991; WEBER and VILLARREAL 1993; VAN VLIET et al. 1993). Because metolazone is not available in an intravenous preparation and to arrange it so that the absorption of the drug occurs at approximately the same time as the furosemide is administered, it is given 1 h before the loop of Henle blooker is administered. A suggested regimen would be 5–10 mg of metolazone p.o. followed in 1 h by 40–80 mg intravenous furosemide, 1–2 mg of bumetanide or 50–100 mg of ethacrynic acid intravenously. The thiazides may also be used in combination with the loop of Henle drugs but are less utilitarian when used in patients with reductions in GFR and/or metabolic acidosis. An intravenous preparation of chlorothiazide is available. Once a diuresis has been established and the patient's edema has improved, it is often possible to give both drugs orally (e.g., 2.5–5.0 mg of metolazone or 50–100 mg of hydrochlorothiazide with 40–120 mg of furosemide or 1–3 mg of bumetanide in the morning). The effective dose of the loop of Henle drug and hydrochlorothiazide may be repeated later in the day, as necessary, whereas metolazone is long-acting.

In patients with severe CHF whose RBF and GFR are usually both markedly reduced, intense Na^+ avidity is the result of activation of the neurohumorally mediated stimuli for tubular Na^+ reabsorption added to substantial reductions in filtered Na^+ load. The latter condition is a consequence of the reduction in GFR. Furthermore, since RBF usually falls proportionately more than GFR, the filtration fraction rises, which favors Na^+ reabsorption at the proximal tubular level. In fact, although it has been difficult to duplicate these circumstances in the experimental animal (SCHNEIDER et al. 1971; STUMPE et al. 1974), patients with severe CHF do behave as if they have substantially incremented proximal tubular Na^+ reabsorption (PUSCHETT 1977; BELL et al. 1964). In patients with this severe degree of physiological derangement, it is unusual for a single diuretic to be effective, even in large dosage. Massive doses of the loop of Henle drugs are to be avoided for three reasons: (a) as described above, additional natriuresis is rarely seen when the recommended maximum doses are exceeded (BRATER 1993; PUSCHETT 1977, 1985), (b) escalation of the dose seems only to result in an increased incidence of side effects, especially when the intravenous route is employed, and (c) when combined with other agents, the effective dose of the loop of Henle drug can often be adjusted substantially downward from the amounts utilized when it is administered alone.

When diuresis has been successfully induced in patients with moderately advanced to severe CHF, either renal function will improve as cardiac output is restored or the patient will simply become volume depleted. In the former circumstance, the blood urea nitrogen (BUN) and serum creatinine will improve; and the disproportionate increase in BUN to creatinine ("prerenal azotemia") will decline. In the latter situation, these values will rise; and the BUN will rise proportionately more than the serum creatinine. This phenomenon is the result of slow flow along the nephron with greater opportunities for urea to be reabsorbed from the collecting duct than is the case for creatinine. In some patients, the CHF will improve but only at the cost of renal functional compromise as reflected in elevated levels of these two substances. In these cases, the physician will have to determine how much renal impairment can be tolerated in order to attain a satisfactory level of cardiac compensation. It should be remembered that ACE inhibition can cause renal dysfunction especially in the setting of volume contraction caused by diuretics so that their use may have to be discontinued or the dosage reduced for varying periods of time (OSTER and MATERSON 1992; SCHWARTZ et al. 1991). In a minority of patients, dialysis may be required because of intractable edema related to severely compromised cardiac function or because aggressive therapy of CHF precipitates end-stage renal disease (GREENBERG and PUSCHETT 1991). Finally, continuous infusion of furosemide has been suggested as an alternative to bolus administration in CHF (LAHAV et al. 1992) as well as in renal insufficiency (RUDY et al. 1991).

Pulmonary edema represents a special circumstance from both pathophysiologic and therapeutic standpoints because of the acuity of its development and its potential for resulting in an unfavorable outcome. The therapy of this illness, which used to depend upon such maneuvers as rotating tourniquets and even venesection to reduce central volume and venous pressure, has been revolutionized by the availability of the rapidly acting, intravenously administered loop of Henle drugs (Table 5). In addition to their effects on vascular volume, these agents may importantly affect venous capacitance acutely, resulting in improvement even before their diuretic actions become manifest (DIKSHIT et al. 1973). Despite the fact that both RBF and GFR are usually severely reduced (Table 4), patients may respond to the intravenous administration of a loop of Henle drug (for example, 40–120 mg furosemide or 1–3 mg of bumetanide). In patients with pre-existing renal disease, combination diuretic therapy as described above is often necessary. Treatment also should be begun promptly with ACE inhibitors and pre-load reducers as well as digoxin if there are no contraindications to their use. For those patients in desparate condition who do not respond adequately to the agents described, hemofiltration with or without dialysis may be helpful (GERHARDT et al. 1979; GOLPER 1985).

Data for pharmacodynamic relationships of four loop of Henle diuretics are presented in Fig. 9, which provide potency rank order for the drugs on a weight basis (BRATER 1991). Regardless of the drug utilized, the slopes and shapes of the curves are all similar and indicate that Na^+ excretion rate approaches a maximum value as dosage (and, therefore, diuretic delivery into the urine) is increased. In CHF, both pharmacokinetic and pharmacodynamic alterations in loop diuretics occur, changing diuretic excretion rate (BRATER et al. 1982, 1984). By increasing the dosages of these drugs, it is possible to effect a diuresis by increasing tubular fluid delivery of the agent, but maximal diuresis is seldom achieved (BRATER et al. 1980).

III. Nephrotic Syndrome

The nephrotic syndrome is a pathophysiologic process that results in the excretion of 3–3.5 g or more of protein in the urine per day. As the disease progresses, the patient goes into negative protein balance; and the full-blown syndrome (proteinuria, hypoalbuminemia, edema and hyperlipidemia) develops. It should be recognized, however, that prior to the development of all of the hallmarks of this illness, nephrotic range proteinuria and edema may occur. The edema formation is related to at least two major factors: (a) lowered oncotic pressure in the vascular compartment, allowing transudation of fluid from the venous circulation into the interstitial space to occur and (b) reduced effective renal arterial filling leading to the stimulation of enhanced tubular reabsorption of Na^+ by mechanisms similar to those described for CHF (see Sect. C.II above). Attempts to localize this enhanced reabsorption to the proximal or distal portions of the nephron have not proven to be definitive (GRAUSZ et al. 1973; KURODA et al. 1979; FIRTH and LEDINGHAM 1991; SHAPIRO et al. 1990), but nephrotic patients behave as if the entire nephron is involved as is so often the case in CHF. In some cases, a third factor may be superimposed related to the development of renal insufficiency by the same renal disease process that has resulted in the

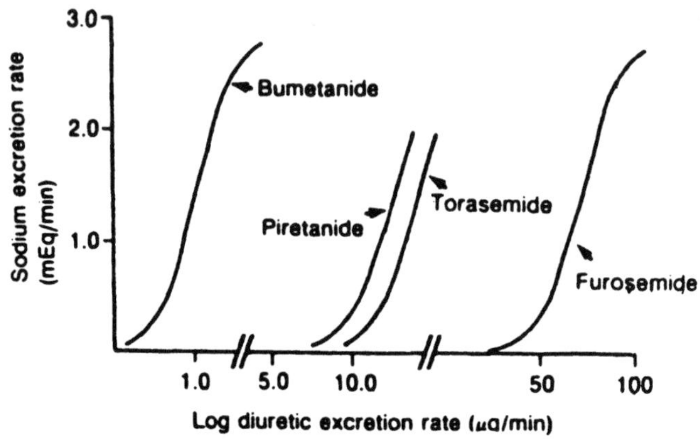

Fig. 9. The pharmacodynamic profile of four loop of Henle diuretic agents (structures see Fig. 2) expressed as the relationship between the drug's urinary excretion rate (*abscissa*) and the absolute rate of Na^+ excretion induced by the drug (*ordinate*). [Reproduced from BRATER (1991) with permission]

nephrotic syndrome. Pathologically, this process represents abnormalities in the glomerular basement membrane which are often the result of an immunologic process that involves the deposition of antigen-antibody complexes in this structure. Thus, as is the case in CHF, renal insufficiency can complicate the nephrotic syndrome leading to a reduction in filtered Na^+ load and therefore reducing Na^+ excretion as a phenomenon independent of the tubular reabsorptive derangement just described.

Definitive therapy of this illness should be directed toward an investigation of the pathogenetic mechanisms (which will often include a renal biopsy) followed by attempts to reverse the underlying disease process. Less specific therapy includes Na^+ restriction and diuretic administration. While the thiazides have been utilized effectively in mild forms of this illness, the loop of Henle drugs have achieved prominence in the treatment of the nephrotic syndrome (GREENBERG and PUSCHETT 1991). As little as 40 mg of furosemide, 1 mg of bumetanide, 6 mg of piretanide or 2.5 mg of torasemide may be effective in this syndrome. However, some nephrotic patients behave as if they are resistant to the actions of the loop of Henle drugs (KELLER et al. 1982; SMITH et al. 1985; KIRCHNER et al. 1990). This resistance has been ascribed, at least in part, to the binding of the drug to urinary albumin, thus rendering it unavailable for interaction with the renal tubular co-transporter (KIRCHNER et al. 1990, 1991). As a consequence, the dosage of the drug must often be incremented in this disorder. Hyperaldosteronism also appears to play a role in the pathogenesis of the edema formation of at least some patients with the nephrotic syndrome. Therefore, spironolactone may add to the therapeutic effect of the loop of Henle agents in this disorder (SHAPIRO et al. 1990) as may other site 4 agents (Fig. 1, Table 1). Indeed, some patients behave as if they are resistant to diuretic action. These individuals, combination therapy is often necessary as described above for

CHF (see also Sect. C.X, below). In some patients, this may take the form of combining intravenous drugs that impair transport in the loop of Henle with oral metolazone (for details, see Sect. C.II, above) or the blockade of proximal reabsorption with a thiazide or acetazolamide plus a loop of Henle agent. In the latter case, one could use chlorothiazide, 500 mg, or acetazolamide, 500 mg, plus furosemide, 80–120 mg, with the drugs all given intravenously. The development of reduced renal perfusion due to over-aggressive diuretic therapy will lead to renal insufficiency and prerenal azotemia as described for CHF (above). In some cases, this may accelerate the progression of the patient's underlying renal disease process toward the development of end-stage renal disease.

Some patients with resistant edema secondary to the nephrotic syndrome have been successfully treated with hyperoncotic albumin. The beneficial effects of this form of therapy when they occur have been considered to result from an elevation in plasma oncotic pressure resulting in the mobilization of sequestered salt and water from the interstitial space into the vascular compartment. However, it is clear that the albumin has several effects and that this mechanism of action of the agent has been over-emphasized in the past (FIRTH and LEDINGHAM 1991; GREENBERG and PUSCHETT 1991). Albumin can have profound effects on glomerular and tubular function. Its administration can elevate GFR, can reduce proximal tubular reabsorption because of its volume expansion effect, and can also *enhance* proximal tubular reabsorption because of its effect to elevate proximal tubular capillary oncotic pressure. Accordingly, whether or not albumin will contribute to a natriuresis and diuresis depends upon which of these effects of this agent predominate. Perhaps the most effective manner in which to employ the albumin is to combine its use with intravenous loop of Henle diuretic administration in the hope that its proximal tubular transport inhibitory effect will exceed its influence on postglomerular capillary reabsorption. If the latter effect should predominate, an *antinatriuresis* could conceivably occur (GRAUSZ et al. 1973). In any case, the effect of intravenous albumin is rather transitory (DAVISON et al. 1974). An example of combined therapy with a loop of Henle agent is the intravenous administration of 12.5–25 g salt-poor albumin during which 120–240 mg of furosemide is given, also intravenously.

IV. Liver Disease

Progressive liver disease (usually the result of cirrhosis) frequently presents with ascites with or without the presence of significant peripheral edema. Ascites represents obstruction to venous outflow and results from an inflammatory process in the liver which ultimately leads to anatomic disorganization and fibrosis of the portal triads, thus interfering with normal venous and lymphatic drainage. Hypoalbuminemia may also play a role in edema formation by virtue of a decrease in oncotic pressure with resultant fluid transudation from the vascular compartment. However, as in the instances of CHF and the nephrotic syndrome, the final common pathway of edema formation is the response by the kidney to what it perceives to be inadequate arterial filling resulting in Na^+ retention (SKOREKI and BRENNER 1982). This occurs despite the fact that in many such patients the plasma volume is either normal or even actually expanded (GREENBERG and PUSCHETT 1991; LIEBERMAN et al. 1970; TRISTANI and COHN 1967).

Patients with this disorder often have a modest impairment in renal function adding to the pathogenesis of the edema and/or ascites formation. In patients with more advanced disease, oliguria may develop. The latter circumstance, called the hepatorenal syndrome, is associated with rising levels of BUN and serum creatinine. It may respond initially to volume expansion and later to diuretics but is essentially unresponsive to therapy unless the underlying liver disease improves or liver transplantation is performed. In some cases, hepatorenal syndrome may result from the injudicious use of diuretics, but an attempt must be made to distinguish this disorder from acute renal injury related to drug toxicity, hypotension, infection, etc.

Diuretic administration to patients with liver disease must always be preceded by careful thought given the potential for baleful consequences of over-aggressive therapy. These include the precipitation of hepatic coma, the development of the hepatorenal syndrome or of acute functional renal insufficiency brought about by the supervention of hypovolemia and hypokalemia. In many cases, the physician should be willing to allow the presence of modest degrees of edema if the patient is not uncomfortable and if therapy is aimed largely at achieving a cosmetic result. Frequently, bed rest and Na^+ restriction will be efficacious in these patients especially if diuretics have not previously been administered. In more advanced cases, a satisfactory natriuresis and diuresis may be achieved with Na^+ restriction plus spironolactone. The latter should be given in doses of at least 100 mg/day. Furthermore, because the onset of action of this drug is delayed (Table 5), adjustments in dosage should not be made more often than every 2–3 days. In those patients who do not respond to this approach, a loop of Henle drug or a metozalone may either be added or substituted for the spironolactone. The advantage of the use of the latter agent is that it often assists in the prevention of hypokalemia. K^+ depletion elevates renal ammonia production and can either worsen or precipitate hepatic encephalopathy. The recommendation that a diuresis of no more than 1–2 kg/day be the target in these patients results from the observation that cirrhotic patients can mobilize only approximately 900 ml of ascitic fluid/day (SHEAR et al. 1970), although when one also mobilizes edema fluid, larger amounts of weight loss can be safely achieved (POCKHOS and REYNOLDS 1986) as long as prerenal azotemia does not supervene (GREENBERG and PUSCHETT 1991). Albumin infusion has sometimes been utilized as adjunct therapy in this disorder. As described above for nephrotic syndrome, this maneuver may have no effect;

but, even if effective, its action is always transitory and its use should therefore be reserved for truly resistant cases.

As discussed in a recent review (GREENBERG and PUSCHETT 1991), use of ascites reinfusion and or the placement of a peritoneovenous (LeVeen) shunt has shown inconsistent results and has been associated with a number of complications. The latter include venous thrombosis, disseminated intravascular coagulation and variceal bleeding (BLENDIS et al. 1979). Therefore, although the shunt procedure has its enthusiastic proponents (WAPNICK et al. 1978), the author and his colleagues do not recommend its employment based upon experience with it thus far (LINAS et al. 1986). Therapeutic paracentesis combined with albumin infusion has been successfully employed in patients with tense ascites (ARROYO et al. 1992).

Brater and his colleagues have examined the pharmacokinetics and pharmacodynamics of the loop of Henle in patients with cirrhosis (BRATER 1992; FREDRICK et al. 1991). They found that in patients with hepatic disease whose renal function was still normal the drug was subject to altered pharmacodynamics but that pharmacokinetics were normal (Fig. 10). These observations suggest that the drug dosage will need to be incremented in such patients or that combination diuretic therapy will be required as described earlier.

V. Idiopathic Edema

A syndrome characterized by the occurrence of cyclical edema not related to the menstrual cycle in women in the child-bearing age group has been termed idiopathic edema (FELDMAN et al. 1978). The disease process tends to wax and wane, often occurs in the setting of emotional lability and usually subsides as the patient ages, but is unassociated with menopause. These

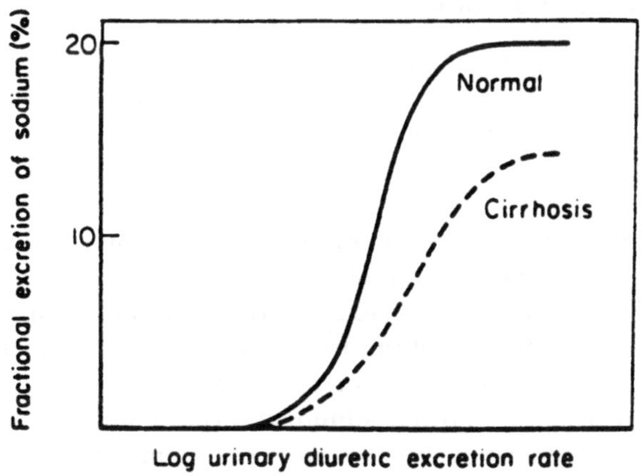

Fig. 10. Diagammatic representation of the change in the relationship between Na^+ excretion and the urinary excretion of loop diuretics in patients with cirrhosis. [Reproduced with permission from BRATER (1992)]

patients show a remarkable relationship between the upright posture and their edema. In fact, the latter feature of this illness has led to one of the therapeutic strategies employed. Often if patients can be induced to remain in bed for 48–72 h, diuresis occurs and the edema subsides (Feldman et al. 1978; Streeten et al. 1973). Patients with this disorder demonstrate an intradiem weight gain of greater than 1 kg and excrete most of their fluid nocturnally. This can be documented by having the patient weigh herself upon arising and 12 h later or by the collection of the urine in two daily aliquots (from 7:00 a.m. to 7:00 p.m. and from 7:00 p.m. to 7:00 a.m.). These patients have a tendency to abuse diuretics and usually live in great fear that the process of weight gain and edema formation will be resistant to therapy and will be progressive. Treatment should be directed toward reassurance, postural therapy and careful, intermittent use of diuretics, should they prove necessary. When diuretics are employed, mild agents such as the thiazides should be used and only resistant patients should be given more potent agents. In general, daily dosing should be avoided. For example, 25–50 mg of hydrochlorothiazide or 2.5–5.0 mg of metolazone every other day or 4 days/week should be attempted. In some patients, a K^+-sparing agent given in conjunction with a thiazide or metozalone may be helpful when therapy with the latter agent alone has been unsuccessful. Under no circumstances should the patient be given a large number of diuretic tablets at any one time since overuse is common, often resulting in tachyphylaxis and diuretic complications. Perhaps the most important aspect of the therapy of this disorder, therefore, is for the physician to gain the confidence of the patient so that injudicious diuretic self-medication can be avoided. Ordinarily, such patients are quite compliant with reduced Na^+ intakes.

VI. Premenstrual Syndrome

This disorder differs from idiopathic edema principally in the fact that there is a distinct and well-characterized relationship between edema formation and the menstrual cycle. Generally, symptoms and signs develop from 5 to 10 days prior to the onset of menses. They include irritability and emotional lability which are often severe including clinical depression as well as headache and peripheral edema. Patients often complain also of swollen and tender breasts, increased abdominal girth and ankle edema. Weight gain approximates 1–3 kg in general, and all of the signs and symptoms described begin to resolve once menstruation occurs. The pathogenetic mechanisms of this disorder are unknown (Daly et al. 1983). Symptomatic therapy consisting of Na^+ restriction and the administration of small doses of mildly to moderately potent diuretics usually suffice to ameliorate or obviate the patients' symptoms and signs. Hydrochlorothiazide in a dose of 25 mg (rarely, 50 mg) or metolazone, 2.5 mg/day, taken for several days prior to menstruation is usually effective. Spironolactone has also been used with good results. Occasionally, ataractic agents and progesterone have been required.

VII. Acute Glomerulonephritis

Patients who develop this inflammatory disease of the kidney often present with edema. Usually, this is mild and is frequently described as "puffiness" of the periorbital region or of the hands, noted by the patient in the morning, which tends to subside during the day. On occasion, substantial peripheral edema may develop but is rare unless the patient also develops nephrotic syndrome. The Na^+ and water retention seen in this disease process is thought to result from reduced GFR levels as well as from enhanced renal tubular reabsorption of Na^+ (GLASSOCK 1980). Whether the latter phenomenon is a result of the decreased GFR or can occur independently is unclear. In general, these patients are volume expanded. Accordingly, hypertension may develop as well. In patients with marked diminutions of GFR, loop of Henle drugs will no doubt be necessary. These agents, in addition to resolving the edema, are frequently effective in treating the hypertension, making the administration of other antihypertensive agents usually unnecessary. Only in those patients who develop oliguric acute renal failure as part of their glomerulonephritis picture will the diuretics prove unsuccessful. In the latter patients, dialysis will be necessary, but the prognosis for the ultimate recovery of renal function is good, especially in children.

VIII. Acute Renal Failure

There is no evidence that the administration of diuretics results in a beneficial effect on GFR in patients with an established renal parenchymal lesion (acute tubular necrosis) (RODRIGUEZ-ITURBE 1984; EPSTEIN et al. 1975; KLEINECHT et al. 1976; BROWN et al. 1981). Indeed, the overwhelming preponderance of the data support the view that the outcome of renal function and the morbidity and mortality of the disease process which led to the acute renal failure are not impacted by diuretic administration. However, in the first 24–48 h of oliguria, a state of so-called functional renal failure may exist which may respond to either volume expansion in the case of a volume contracted patient or to diuretics in the euvolemic or hypervolemic patient. Obviously, the response of the patient will depend importantly upon the etiology of the acute renal failure. Regardless of the influence of diuretic administration on the GFR, however, it seems clear that converting the patient from oliguria to nonoliguria will make his (her) management easier. Accordingly, in the volume replete patient, an attempt to do so with diuretics seems reasonable. Nonoliguric patients have fewer problems with volume overload and hyperkalemia. The therapeutic strategy should be similar to that discussed above for the nephrotic syndrome. Once diagnostic efforts have been made and any corrective measures taken to reverse the underlying pathophysiologic process, an assessment of the patient's fluid status should be made. In patients who are not hypovolemic, therapy should first be attempted with loop of Henle drugs. Given the marked renal insufficiency, little benefit will be obtained in this group of patients with small doses. Accordingly, 80–100 mg of furosemide or equivalent dosages of bumetanide, furosemide, piretanide, or ethacrynic acid should be administered by the intravenous route. If no response is obtained in 3–4 h, the dose should be doubled. If at this point, there has been no response, combination diuretic therapy should be attempted. The details of this approach have been outlined above for CHF and the nephrotic syndrome. Thiazide diuretics, even in combination, are generally ineffective in this clinical circumstance because of the marked reduction in GFR. However, metozalone still may be useful.

It has become commonplace in the past several years to attempt to reverse oliguria in the patient with acute renal failure who is in the intensive care unit either by infusing low "renal" doses of dopamine or by combining the latter agent with diuretics. There is, however, no evidence that dopamine infusion alone improves GFR in the setting of acute renal failure (PARKER et al. 1981; LINDNER 1983; GRAZIANI et al. 1984). Further, as pointed out by SZERLIP (1991), those studies which have suggested a beneficial effect on the course of acute renal failure consequent to the addition of diuretic administration to the dopamine infusion have been uncontrolled, often anecdotal reports (HENDERSON et al. 1980; PARKER et al. 1981; LINDNER 1983; GRAZIANI et al. 1984).

In those patients who appear to be in the stage of functional renal failure, mannitol may be helpful (LUKE et al. 1970). Exactly why this agent is sometimes effective is unclear. However, it has been postulated that the intratubular obligation of fluid by this essentially nonreabsorbable solute may result in the relief of intratubular obstruction caused by casts, pigment or tubular debris resulting from (proximal) tubular cell necrosis. Mannitol should be used with caution or not at all in hypervolemic patients so that the development of pulmonary edema and/or symptomatic hyponatremia is/are not precipitated.

K^+-sparing agents are contraindicated in these patients because they already have difficulty disposing of K^+ and tend to be hyperkalemic. Since they also usually have metabolic acidosis, acetazolamide most likely will not be effective. Once the kidney begins to recover and especially in the initial stages of a spontaneous diuresis, the diuretics may be helpful in reducing the edema formation in a more rapid fashion than would be the case if one simply waited for renal function to improve fully. There appears to be no disadvantage to this approach as long as volume depletion and hypokalemia are not allowed to supervene.

IX. Chronic Renal Failure

In patients with reduced renal function, the major limitations of diuretic effectiveness are represented by (a) the difficulties associated with delivering the diuretic into the tubular fluid in sufficient concentration to impair Na^+

reabsorption and (b) the fact that progressively smaller amounts of Na^+ are presented for tubular reabsorption because of the reduced levels of GFR. Thus, although fractional Na^+ excretion is increased in patients with chronic renal failure, more likely reflecting the "hyperfiltration per nephron" originally postulated by BRICKER (1972), nevertheless, absolute Na^+ excretion is clearly reduced as a result of the diminished Na^+ load reflecting the GFR decrement.

Because GFR levels are usually reduced below 30 ml/min in this group of patients, the thiazides are ineffective. Accordingly, use of the loop of Henle drugs represents the treatment of choice. Patients with renal insufficiency have also been determined to have reduced renal clearances of the loop diuretics too. In addition, the drugs compete for secretion with retained organic acids, thus further limiting their entry into the tubular lumen (BRATER 1990, 1992). Accordinly, the dose must be incremented substantially, especially in patients with severe renal failure so that sufficient drug reaches the tubular lumen (VOELKER et al. 1987). Surprisingly, despite the fact that fractional Na^+ excretion is already rather high in these patients, the diuretics are able to increase Na^+ excretion further in most cases (BRATER 1990, 1992). Furosemide, bumetanide, piretanide and torasemide have all proven useful in this clinical circumstance (VOELKER et al. 1987; SCHULZ et al. 1990; MARONE 1993; LUFT et al. 1993). Indeed, an increase in urine flow rate and Na^+ excretion may even be possible to achieve in patients on dialysis who have some residual renal function, although the price to be paid is the administration of very high doses, e.g., up to 1000 mg/day of furosemide (VAN OLDEN et al. 1993), 100 mg of torasemide (STOLEAR et al. 1990), and 192 mg of piretanide (MARONE 1993).

In patients with chronic, severe renal insufficiency, single doses of 120–160 mg of furosemide i.v. or 240–320 mg orally appear to approximate the maximally effective doses of the drug. Whereas in normal subjects the equivalency of bumetanide to furosemide is 1 to 40, in renal insufficiency it is closer to 1 to 20, probably related to differences in nonrenal clearances of the drug (VOELKER et al. 1987; BRATER et al. 1986). Although the maximal dosage of torasemide and piretanide have not been determined as yet, it has been estimated that for piretanide it is approximately 36–48 mg i.v. and for torasemide, up to 100 mg i.v. and 200 mg by mouth (Table 5).

X. Resistant Edema

Resistance to the actions of diuretics may occur for a number of reasons. First, of course, one must rule out the possibility that the situation represents apparent rather than true resistance (BRATER 1993; PUSCHETT and O'DONOVAN, to be published; ELLISON 1991). Causes of the latter include failure of the patient to adhere to a Na^+-restricted diet, lack of compliance with the diuretic regimen and the administration of short-acting diuretics at intervals which are too far apart. The latter is a common problem especially

Table 8. Requirements for diuretic effectiveness [reproduced from PUSCHETT and O'DONOVAN (in press) with permission of the editors]

1. Glomerular filtration that is not severely compromised
2. Delivery of the diuretic to its site of action within the kidney
3. Delivery of Na^+ to the nephron segment at which the diuretic acts
4. Responsiveness of the tubular site to the diuretic
5. Absence of major reabsorption of rejected Na^+ at nephron sites distal to that of the diuretic's action

with respect to the loop diuretics which act quickly but whose effects are usually gone in from 3 to 4 h (Table 5). Consequently, if their dosage is not repeated during the day, "rebound" Na^+ retention will supervene so that a net natriuresis and diuresis for the 24-h period will not occur (WILCOX et al. 1983; BRATER 1993; LOON et al. 1989). In Table 8, the requirements for diuretic effectiveness are provided. Diuretics act from the luminal surface of the renal tubular cell. Therefore, they must not only gain access to the blood stream but also must be secreted into the tubular lumen in sufficient quantity to inhibit Na^+ transport mechanisms. Obviously, they must also have sufficient substrate upon which to act, hence the requirement for GFR and for delivery of enough Na^+ to the site at which the diuretic acts so that Na^+ reabsorption impairment may occur. As a part of this process, the transporter must be responsive to the diuretic. Finally, once sodium ions have been rejected for reabsorption at a given nephron locus, they must be able to escape transport from the tubular contents at some more distal nephron site(s).

The causes of true diuretic resistance are provided in Table 9. Perusal of this list reveals that patients behave as if they are resistant because one or more of the necessary conditions for diuretic effectiveness (Table 8) has (have) not been met. Some patients with advanced congestive heart failure or with anasarca behave as if they have reduced or delayed absorption of the diuretic from the gastrointestinal tract. In patients with renal failure or in those with circulatory compromise related to inadequate arterial filling (e.g., CHF or advanced liver disease), there is reduced secretion of the diuretic into the tubular lumen. This phenomenon has also been reported in a group of renal transplant patients who were found to respond less well to furosemide than other patients (SMITH et al. 1981). As described previously in patients with nephrotic range proteinuria, there is evidence for binding of the drug to tubular albumin, resulting in a reduced amount of unbound diuretic that is available for interaction with the transport site.

Certain drugs interfere with the actions of the diuretics. While it is unclear now the nonsteroidal anti-inflammatory drugs (NSAIDs) impair the natriuretic effectiveness of furosemide, this phenomenon does not appear to result from a competition with NSAIDs for tubular secretion (CHENNAVASIN et al. 1980). Captopril, but not other ACE inhibitors, has been reported to

Table 9. Mechanisms of diuretic resistance [adapted from PUSCHETT and O'DONOVAN (in press) with permission of the editors]

1. Failure of the diuretic to reach the tubular site of action of the diuretic
a) Reduced or delayed gastrointestinal absorption
b) Reduced tubular secretion
c) Decreased availability within the tubular lumen
2. Interference with the diuretic by other drugs
3. Tubular adaptation to prolonged diuretic usage
4. Failured of the diuretic to impair Na^+ reabsorption at a major site of reabsorption

interfere with the diuretic action of furosemide but not hydrochlorothiazide (TOUSSAINT et al. 1989). Of course, any antihypertensive agent that impairs renal blood flow and GFR by causing hypotension and/or by reducing the portion of cardiac output perfusing the kidneys will adversely affect the activity of diuretics. Interference with the tubular secretion of amiloride and triamterene is the most likely cause for impairment of their actions by cimetidine (MUIRHEAD et al. 1986; SOMOGYI et al. 1989).

The phenomenon of tubular adaptation to prolonged diuretic administration has now been documented both in the experimental animal (KAISSLING et al. 1985; KAISSLING and STANTON 1988; STANTON and KAISSLING 1988; WALTER and SHIRLEY 1986; ELLISON et al. 1989) and in man (LOON et al. 1989; ALMESHARI et al. 1993). In the experimental animal, when either furosemide or hydrochlorothiazide is given over prolonged periods, the cells of the distal convoluted tubule (Fig. 1, site 3) hypertrophy and can be demonstrated to increase their capacity to transport Na^+. The latter ion appears to serve as a growth factor for these cells (KAISSLING and STANTON 1988; STANTON and KAISSLING 1988; ELLISON et al. 1989). In man, a similar phenomenon is thought to occur. Evidence in favor of this view is provided by the results of experiments which demonstrate that the diuretic resistance encountered when furosemide has been administered chronically can be overcome by interfering with Na^+ reabsorption in the early distal convoluted tubule (ELLISON 1993; LOON et al. 1989).

The technique of sequential nephron blockade has been advocated by the author and his colleagues for those patients manifesting intense renal tubular Na^+ avidity (PUSCHETT 1985; GREENBERG and PUSCHETT 1991). This therapeutic technique is most likely to be successful when the pathophysiology of the patient's disease process includes markedly enhanced proximal tubular reabsorption. Certain clinical conditions, which have already been described above, have this pathogenetic mechanism in common. These include severe chronic congestive heart failure, advanced stages of the nephrotic syndrome and cirrhosis with the formation of ascites with or without peripheral edema. In each of these situations, the kidney is stimulated to increment proximal reabsorption by hormonal and neural influences which sense decreased effective arterial filling despite the fact that the venous circuit is overloaded. Because these patients behave as if the delivery of Na^+ and water to more distal nephron sites beyond the proximal tubule is diminished, in many cases marbedly so, diuretic agents which act in the loop of Henle (site 2, Fig. 1) or beyond will not be effective despite their delivery to the appropriate Na^+ transporter. Accordingly, rather than incrementing the dose of the loop of Henle agent, we recommend the addition of another drug which has proximal activity. As described previously, this will have the effect of attacking the source of the problem, that is, enhanced proximal reabsorption. Because the loop of Henle, a major reabsorption locus, is also blocked, the proximally rejected ions will escape reabsorption in the loop. The author and his colleagues have successfully combined metolazone with a loop of Henle agent for this purpose. When given alone, metolazone owes its natriuretic capability to its capacity to interfere with Na^+/Cl^- reabsorption in the early distal convoluted tubule (site 3, Fig. 1). However, when it is administered with a loop of Henle agent, its proximal activity becomes manifest because proximally rejected ions are delivered past the transporter at the level of the ascending limb of the loop of Henle (FERNANDEZ and PUSCHETT 1973; PUSCHETT et al. 1973). An example of sequential nephron blockade is presented in the studies depicted in Fig. 11. In these experiments, the addition of metolazone to steady-state ethacrynic acid diuresis resulted in an augmentation of Na^+ excretion (PUSCHETT and RASTEGAR 1974). Accordingly, sequential blockade of sites 1 through 3 is accomplished. Acetazolamide or a thiazide may also be employed for this purpose. It is also possible to add a site 4 agent (Table 1) as well, both to complete the nephron blockade as well as to attempt to reduce the expected kaliuresis.

When dealing with a patient who appears to be resistant to diuretic therapy, one must first assure that a true state of resistance does exist. Once

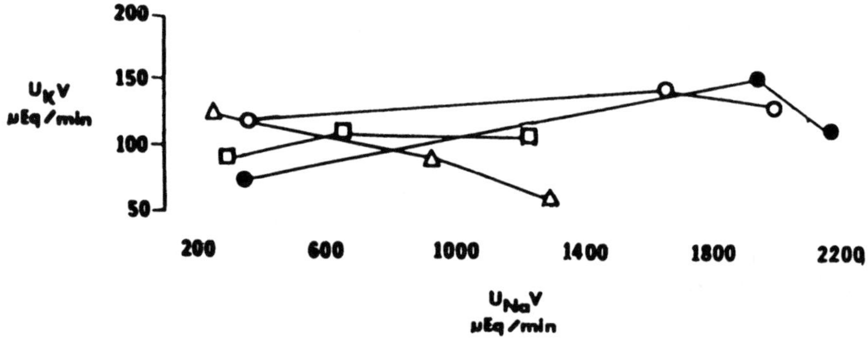

Fig. 11. The effects on absolute Na^+ excretion ($U_{Na}\dot{V}$) of the addition of metolazone (*symbols on the right*) to subjects undergoing steady-state ethacrynic acid diuresis (*middle set of symbols*) superimposed upon water diuresis (*symbols on the left*). These observations, obtained in four subjects, are plotted vs. the simultaneously determined absolute K^+ excretion ($U_K\dot{V}$). [Reproduced from PUSCHETT and RASTEQAR (1974) with permission of the editors]

that situation is confirmed, the approach to the patient includes the following steps: (1) Verify that a Na^+-restricted diet has been prescribed and is being followed. (2) Examine whether or not efforts to determine and correct the underlying disorder have been expended. (3) If the patient has not responded to large doses of loop of Henle drugs given by mouth, attempt the intravenous route. In these cases, reduce the dose given intravenously from that which was employed orally since many patients who have not responded to oral agents will develop a major natriuresis and diuresis from much smaller amounts of the agent when given directly into the circulation. (4) In patients with normal renal function, begin with 40–80 mg of intravenous furosemide or its equivalent (Table 5). As described above, the dose will have to be substantially higher in patients with moderate to severe renal insufficiency. (5) If the patient does not respond within 3–4 h, increment the dose (e.g., double the dose). One should probably not exceed 240 mg of furosemide (or its equivalent) as a single dose. In some circumstances, a continuous infusion may be more effective than a bolus injection (LAHAV et al. 1982; RUDY et al. 1991). (6) If the results are still unsatisfactory, add a second agent. For example, metolazone, 5–10 mg p.o. (there is no commercially available i.v. preparation) followed in 1 h by 80–120 mg of intravenous furosemide (or its equivalent). Administration of the loop of Henle agent may be repeated later in the day or may be incremented if necessary. Metolazone is long-acting so that only a single daily dose is required (Table 5). (7) Add spironolactone, amiloride or triamterene (Table 5). The latter drugs should never be used in patients with renal insufficiency because of the possibility that hyperkalemia will develop and should be used with caution in patients with acidemia. (8) In special circumstances, albumin infusion may prove helpful. The rationale and basis for this therapeutic maneuver was discussed extensively earlier (see Sect. C.III, above). When given in combination with loop of Henle drgus, albumin infusion may result in a diuresis unachievable with the diuretic alone.

D. Diuretics in the Treatment of Nonedematous Disorders

I. Hypertension

Diuretics have been utilized as “first-line” agents in the therapy of hypertension for over 30 years. They have proven to be safe and effective. Over the past several years, concerns about their use as the primary agent in hypertension have centered about possible adverse effects of these drugs on mortality from coronary heart disease in the hypertensive patient population. The risks and benefits of their employment in hypertension will therefore be discussed below.

Blood pressure represents the product of blood flow and vascular resistance ($P = Q \times R$). Flow is a function of cardiac output in the human body, which in turn is dependent upon venous return to the heart. The latter

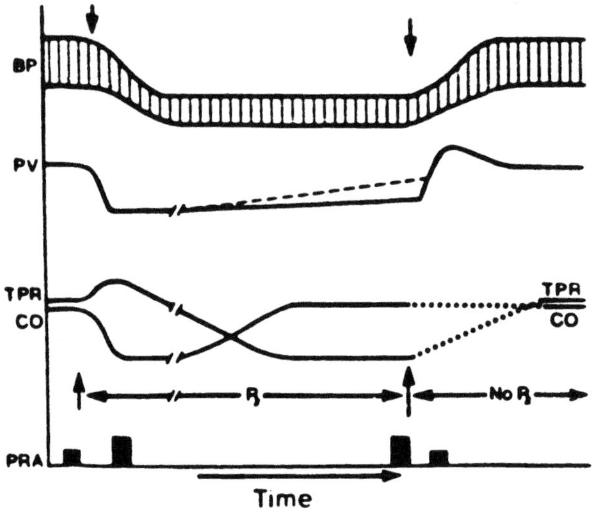

Fig. 12. Diagrammatic representation of the alterations in blood pressure (*BP*), plasma volume (*PV*), total peripheral resistance (*TPR*) and cardiac output (*CO*) during administration of a thiazide for the treatment of hypertension. The drug was administered *at the arrow on the left* and discontinued *at the arrow on the right*. [Reproduced with permission from MOSER (1987)]

is influenced substantially by extracellular fluid (ECF) volume. Vascular resistance is largely a function of the vessels in the renal circulation. Blood pressure may therefore be decreased either by reducing ECF volume or by reducing vascular resistance or both. Studies carried out many years ago have demonstrated that diuretics reduce blood pressure initially by causing ECF volume contraction, but that in the chronic situation they do so by reducing peripheral resistance (GIFFORD et al. 1961; VILLARREAL et al. 1962; TARAZI et al. 1970). These findings are summarized in Fig. 12. The initial reduction in blood pressure is related to the acute-subacute decline induced in plasma volume. This observation has an important clinical implication. That is, the administration of diuretics for hypertension *must* be accompanied by Na^+ restriction. Failure to do so can obviate the antihypertensive effects of this group of agents (WILSON and FREIS 1959). Within weeks to months following the initiation of diuretic therapy and although blood pressure remains reduced, plasma volume tends to return to or toward baseline values (dashed line in Fig. 12). However, vascular resistance falls. Cardiac output, which shows an initial increase, declines over time so that it eventually shows no change or is mildly reduced (VILLARREAL et al. 1962; TARAZI et al. 1970).

In the early and middle 1980s, principally two observations resulted in a re-examination of the safety and effectiveness of the diuretics in the therapy of hypertension. Data provided from the National Center for Health

Statistics study indicated a divergence in the mortality rate from stroke versus that from coronary heart disease (CHD). The age-adjusted reduction in stroke mortality was 50%, whereas that from CHD was found to be 30% (FEINLEIB 1984). The implication was that, for some reason possibly related to the major form of therapy for hypertension (i.e., diuretics), the improvement in CHD mortality had lagged behind that which should have been expected based on the stroke mortality data. Coupled with these observations were those provided by a randomized primary prevention trial performed in almost 13000 hypertensive men, assigned either to a special intervention program (SI) or referred to their usual health care provider (UC). In this study, the Multiple Risk Factor Intervention Trial (MRFIT), there was no statistically significant difference in mortality from CHD between the SI and UC groups (MULTIPLE RISK FACTOR INTERVENTION TRIAL RESEARCH GROUP 1982). In a study performed in Norway of 785 hypertensive men, no difference in mortality could be found between the treated and untreated groups (HELGELAND 1980). With these data serving as a stimulus, three possibilities for this putative failure of antihypertensive therapy (principally involving diuretics) to result in improved CHD-related adverse events were identified/suspected: (a) diuretics cause hypokalemia (and hypomagnesemia) which predispose to the development of arrhythmias and, potentially, to sudden death; (b) diuretics increase serum lipids, thus resulting in the exacerbation of a risk factor for the development of CHD; and (c) diuretics, as opposed to some other antihypertensives, do not result in the regression of left ventricular hypertrophy even when they normalize blood pressure, failing to obviate an additional cardiac risk factor. The arguments on both sides of this issue have been reviewed recently (O'DONOVAN et al. 1992; PUSCHETT, in press) and are summarized in Table 10. The incidence of hypokalemia has been reduced with the employment of lower dosages of the thiazides than were utilized in the past (PUSCHETT, in press, JNC-V 1993), in addition to which the initial spate of articles suggesting a relationship between diuretic therapy and the incidence of arrhythmias has been followed by several reports indicating no increase in incidence. Furthermore, hypokalemia is easily treated either with K^+ supplementation or the addition of potassium-sparing agents (Tables 5, 6). Hyperlipidemia may occur, but is ordinarily rather modest and usually does not persist. Furthermore, this potential side effect can be managed with appropriate dietary therapy. While the regression of left ventricular hypertrophy (LVH) would appear, *a priori*, to be desirable, it is currently unclear whether or not LVH represents an independent risk factor for cardiac mortality. Nevertheless, recent reports have appeared indicating that diuretics are, in fact, effective in reducing left ventricular size (for reviews of these issues, see PUSCHETT 1994; O'DONOVAN et al. 1992; MOSER 1993; WEINBERGER 1988). Recently, the Committee on High Blood Pressure of the American Heart Association has issued their fifth commission report including recommendations for the therapy of hypertension (JOINT NATIONAL COMMITTEE ON DETECTION, EVALUATION AND

Table 10. Evidence for and against baleful effects of diuretics on coronary heart disease (CHD) [adapted from O'DONOVAN et al. (1992) with permission of the editors]

Side effects		Mitigating factors
Hypokalemia	Diuretics cause hypokalemia (and, possibly, hypomagnesemia), which predisposes to the development of cardiac arrhythmias leading to sudden death	Hypokalemia may not occur (especially in patients treated with low-dose thiazides) and can easily be treated. Initial reports of an increased incidence of arrhythmias in diuretic-treated patients have not been corroborated by more recent studies
Hyperlipidemia	Diuretics increase serum lipid levels providing an increased risk for the development of CHD	When increases in serum cholesterol do occur, they average 7%, can be controlled with diet and do not appear to persist
Left ventricular hypertrophy (LVH)	Diuretic therapy of hypertension does not result in the regression of LVH despite the normalization of blood pressure. LVH is a risk factor for mortality from hypertensive heart disease	Recent studies suggest that a reduction in left ventricular size *does* occur with diuretic therapy. There is controversy currently as to whether LVH is an independent risk factor for cardiac mortality

TREATMENT OF HIGH BLOOD PRESSURE 1993). An important aspect of their report was the provision of the latest data for the reduction in mortality from stroke and from CHD as of 1989, demonstrating that these two rates of decline now approximate each other (Fig. 13). This group suggested that diuretics and β-blockers continue to be utilized as first-line therapy for hypertension. Specific recommendations developed by the author and his colleagues are listed in Table 11.

Diuretics are especially useful in those groups of patients who appear to have primarily a form of "volume expansion" essential hypertension: African-Americans, the obese and the elderly. As discussed earlier, based

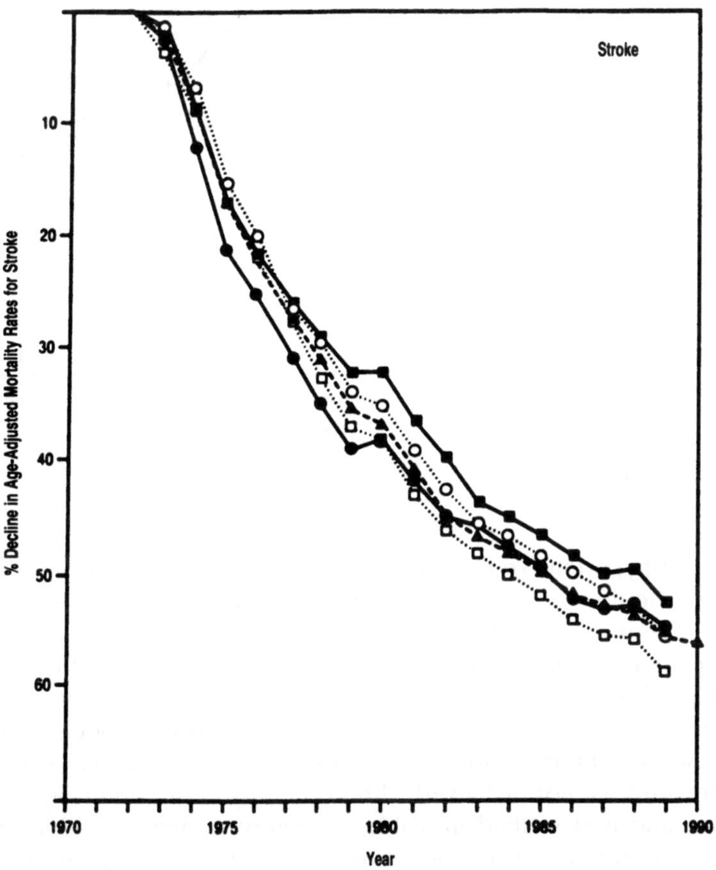

Fig. 13. Decline in age-adjusted mortality for stroke (*diagram on the left*) and for coronary heart disease (*CHD*, *on the right*) by race and sex since 1972. *Triangles* represent the total population. *Solid squares* provide data for black men, *Solid circles* for black women, *open squares* for white men, and *open circles* for white women. [Reproduced from the Fifth Report of the JOINT NATIONAL COMMITTEE ON DETECTION, EVALUATION AND TREATMENT OF HIGH BLOOD PRESSURE (1993) with permission of the editors]

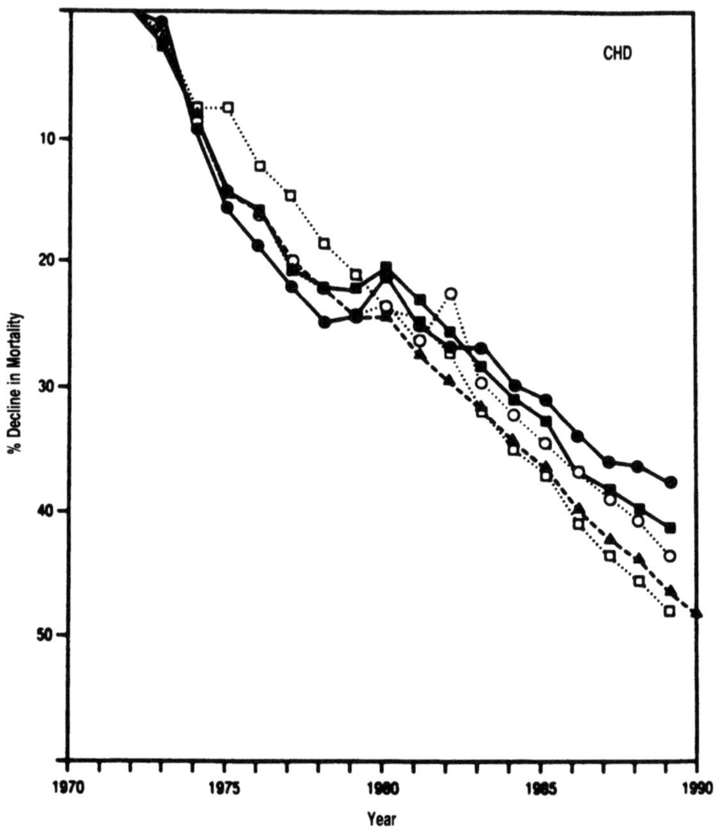

Fig. 13. *(Continued)*

upon the antihypertensive mechanisms of action of the diuretics, dietary salt restriction should be imposed except in patients who have renal "salt wasting." The use of low doses of diuretics is recommended. This will mean that the patient will most likely not evidence a reduction in blood pressure for at least 3–4 weeks once therapy and Na^+ restriction are instituted. Rather than exceeding a dose of 50 mg of hydrochlorothiazide (25 mg in the elderly), we recommend adding a second agent (see Table 11) if target blood pressure is not achieved. Monotherapy with other agents is an alternative and is described in Table 11. Diuretics may be added to these regimens. Careful monitoring or serum chemistry values is recommended as outlined.

Table 11. Recommendations for antihypertensive therapy with diuretics

1. The diuretics continue to be useful as monotherapy in mild to moderate hypertension and are preferred in certain demographic groups (African Americans, the obese and the elderly).
2. Patients should be instructed to simultaneously restrict their Na^+ intakes (except in the case of renal "Na^+ wasters").
3. Therapy should be initiated with 25 mg of hydrochlorothiazide or its equivalent (but 12.5 mg in the elderly), and dosage adjustment should not be attempted (except under urgent circumstances) for 3–4 weeks. Compliance with diet and drug dose should be verified. A maximum dose of 50 mg of hydrochlorothiazide (or its equivalent) is recommended (but 25 mg in the elderly).
4. Diuretics complement the actions of other antihypertensives. Therefore, if target blood pressure is not achieved, a β-blocker, Ca^{2+} channel antagonist, ACE inhibitor or an agent from another class of drugs may be added.
5. In demographic groups other than those cited in (1) (above), therapy with a β-blocker, Ca^{2+} channel antagonist or another agent represents a reasonable alternative. This may be dictated by the clinical circumstance (e.g., ACE inhibitor in a patient with CHF or a β-blocker in a patient with a recent myocardial infarction etc). Diuretics may be added to these agents, if necessary.
6. Careful monitoring of metabolic parameters is indicated.
   a) Check serum K^+ before treatment and at the first visit after initiation of therapy (3–4 weeks), again in 2 months, then every 3–6 months and whenever necessary. Serum K^+ concentrations of 3.3 mmol/l (or lower) should be treated either with KCl supplementation or with the addition of a K^+-sparing agent. However, in patients taking digoxin or in whom recent myocardial infarction has occurred, levels of 3.5 mmol/l (or below) should be treated.
   b) Determinations of fasting glucose, lipid levels, BUN, creatinine, electrolytes, and Mg^{2+} should be undertaken at the first visit (3–4 weeks) and then every 6 months if stable. More frequent assessments will be required if abnormalities occur. Abnormalities of serum lipids should be treated with appropriate dietary restriction.

II. Toxemia of Pregnancy

The term toxemia of pregnancy includes two clinical syndromes, preeclampsia and eclampsia. The former involves the development of hypertension, edema and proteinuria in patients who have reached the 20th week of gestation. The term eclampsia is employed when, in addition, convulsions occur. The pathogenesis of these disorders is unknown. For many years, these conditions were treated with the administration of thiazide diuretics (Menzies 1964; Finnerty and Bepko 1966). However, more recently it has been determined that although these patients have edema, plasma volume may be reduced compared to normal pregnant women. Therefore, to protect the circulation of the maternal-fetal unit, the use of salt restriction and diuretics has largely been abandoned except in the case of the development of CHF. These conditions respond to delivery of the fetus. Hydralazine has been proposed as the agent of choice in the therapy of the hypertension of the toxemias (Wonley 1986) because of its lack of effect on uterine blood flow.

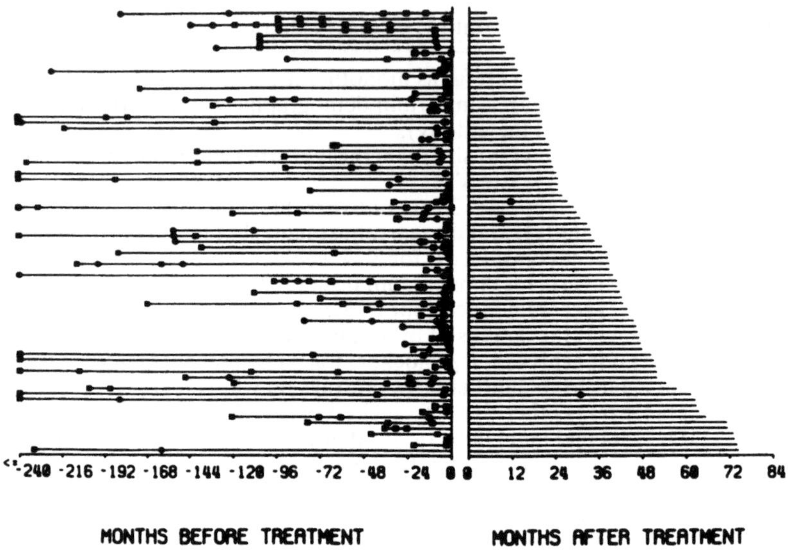

Fig. 14. Renal stone episodes (*represented as symbols along the time lines*) in patients with idiopathic hypercalciuria before and after the institution of chlorthalidone therapy. Each line represents and individual patient. [Reproduced from Coe (1977) with permission]

III. Hypercalcemia

Therapy of hypercalcemia with loop of Henle diuretics takes advantage of the fact that the loop is a major site of Ca^{2+} reabsorption (Fig. 5). Ca^{2+} and Na^+ are handled similarly by the nephron. Therefore, approximately 20%–25% of filtered Ca^{2+} load is reabsorbed in the ascending limb of the loop of Henle (Puschett 1989). In the study initially describing the utility of this methodology, Suki and his coworkers combined intravenous furosemide administration with saline infusion to prevent the patients from becoming volume depleted (Suki et al. 1970). Fractional Ca^{2+} excretion rates of up to 10% (from baseline values of 1%–2%) may be achieved and serum Ca^{2+} may fall by 1–1.5 mmol/l. The thiazides are not utilized for this purpose because they cause an enhancement of Ca^{2+} reabsorption. Furthermore, administration of the thiazides has occasionally been associated with the development of hypercalcemia. As with all other disorders, the source of the hypercalcemia should be sought. In those cases in which the pathogenetic mechanisms cannot be reversed (e.g., malignancy), therapy other than diuretics and/or saline infusion must be instituted (diphosphonates, gallium nitrate, corticosteroids, mithramycin, etc.).

IV. Renal Stone Disease

The thiazide diuretics have been employed for many years for the therapy of those renal stone formers who are identified as having idiopathic hypercalciuria with beneficial effects (YENDT and COHANIM 1987; AGUS 1984; COE 1977). The thiazides have been determined to be effective in this disorder at least in part because they reduce Ca^{2+} excretion (YENDT and COHANIM 1978). They do so by *enhancing* Ca^{2+} reabsorption in the early distal convoluted tubule (site 3, Fig. 1), while at the same time inhibiting Na^+ reabsorption at that site (COSTANZO and WINDHAGER 1978; COSTANZO 1984, 1985). Other factors may contribute to the reduction in stone formation often seen associated with thiazide therapy (Fig. 14). These include the increase in urinary volume, an increase in pyrophosphate excretion and the possibility that the drug itself may retard the formation, growth and aggregation of crystal nuclei (AGUS 1984). The actions of amiloride are synergistic with those of the thiazides in reducing Ca^{2+} excretion (LEPPLA et al. 1983). This is related to the fact that amiloride also promotes Ca^{2+} reabsorption but in a different nephron segment than do the thiazides. Data from microperfusion experiments performed by Costanzo indicate that while the thiazides are effective in the early distal convoluted tubule, amiloride acts in the later portion of that nephron segment (COSTANZO 1984, 1985). Accordingly, amiloride has often been used effectively in combination with the thiazides (and, on occasion, alone) in the therapy of this disorder (LEPPLA et al. 1983; MASCHIO et al. 1981). Interestingly, triamterene, another of the K^+-sparing agents (Table 5), must be avoided in this illness because of its tendency to cause stone formation (ETTINGER et al. 1979; PATEL 1981).

V. Diabetes Insipidus

This disorder, characterized by polyuria (which is occasionally massive) and polydipsia, results either from the absence of antidiuretic hormone (ADH) or because the renal tubular epithelium of the collecting duct is either resistant to or unresponsive to the action of the hormone ("nephrogenic diabetes insipidus"). The central nervous system form of this disorder is ordinarily treated with injections of vasopressin tannate in oil or the nasal insufflation of DDAVP (desamino-8-D-arginine vasopressin), a synthetic analog of the parent compound. Because the response to these agents in central diabetes insipidus is ordinarily so satisfactory, diuretics are rarely used in this disorder (although they are effective). However, the thiazides have been employed with success in the nephrogenic form of the disease (CRAWFORD and KENNEDY 1959). Apparently, their ability to lower urinary volume has to do with the development of volume contraction. Evidence for this thesis is of two types: (a) simultaneous restriction of dietary Na^+ enhances the drug's effect on urine flow rate and (b) high salt intake obviates the drug's effectiveness. Micropuncture studies performed in the Brattleboro rat (an animal with hereditary central diabetes insipidus) indicate that the drug enhances proximal tubule Na^+ reabsorption (WALTER et al. 1979) probably because of a volume stimulus, and chronic thiazide administration has been shown to have a distal nephron effect as well (SHIRLEY et al. 1982). Other diuretics, therefore, are also effective assuming that Na^+ restriction is imposed so that volume depletion occurs (SUTTON and DRANCE 1986).

VI. Hyperkalemia

The rationale for the therapy of hyperkalemia with diuretics is based upon the renal physiological principles outlined in Sect. B, above. First, K^+ reabsorption occurs throughout the proximal nephron and in the loop of Henle, such that filtered K^+ has been virtually completely transported from the luminal contents by the time the tubular fluid enters the distal convoluted tubule (WRIGHT and GIEBISCH 1992). Therefore, diuretics which interfere with Na^+ transport in these nephron segments will also impair K^+ reabsorption leading to its downstream delivery where capture of these ions by more distal transport sites is much more limited. Second, under normal circumstances, most of the K^+ that appears in the urine is that which has been added by secretory mechanisms located in the late portions of the distal convoluted tubule and in the collecting duct. Accordingly, drugs that interfere with Na^+ reabsorption proximal to site 4 (Fig. 1) will present additional Na^+ ions for exchange to the principal cells of the late distal nephron (Fig. 8). This exchange mechanism is regulated, in part, by the mineralocorticoid system (site 4a, Fig. 1). Patients who are treated with diuretics often demonstrate that the activity of the renin-angiotensin-aldosterone axis has been stimulated. Accordingly, two factors lead to the excretion of increased amounts of K^+ in the urine: an incremental presentation of Na^+ to the exchange sites for transfer and an enhancement of the activity of site 4 by increased amounts of circulating aldosterone. As shown in Fig. 8, kaliuretic tendency is a function of the natriuretic potency of diuretics.

As is the case in all disorders, a search should be made for the source of the hyperkalemia. This most often occurs in a clinical setting of chronic renal insufficiency with the imposition of a drug or maneuver that further hampers the disposal of K^+. Examples include the administration of K^+-sparing drugs, an ACE inhibitor or the nonsteroidal agents. Some patients on restricted Na^+ intake utilize large amounts of salt substitutes, many of which contain K^+ as the major cation.

E. Diuretic Side Effects and Adverse Reactions

I. Volume Contraction

Overaggressive use of diuretics will result in an excessive loss of extracellular fluid with the result that mild to major clinical symptoms and signs will occur, dependent upon the degree of volume loss. This may vary from orthostatic hypotension and tachycardia to vascular collapse. Not only is the absolute amount of ECF that has been lost important, but the rapidity of the decrement is a significant factor as well. Thus, patients who lose, for example, 1 kg/day for several days will fare better than someone whose body volume decreases by 3–4 liters in a 24-h period. Although the latter may be appropriate in some clinical situations, such as pulmonary edema and gross volume overload, as the patient approaches euvolemia, this sort of loss will be much more poorly tolerated. One of the hallmarks of this side effect of diuretics is the development of "pre-renal azotemia." In this circumstance, either the BUN is elevated and the serum creatinine is unchanged, or both are elevated but the BUN rises proportionately more than the creatinine. This phenomenon is explainable on the basis that slow flow along the nephron secondary to volume contraction has differing effects on the reabsorption of urea from that on creatinine. Given increased contact time in, for example, the collecting duct, urea will be more readily reabsorbed than creatinine because it is a smaller molecule and is uncharged.

Patient symptomatology may include fatigue, malaise, weakness, postural dizziness or lightheadedness, as well as thirst and a dry mouth. Poor skin turgor and decreased perspiration may be present (SILVER 1985). All of the diuretics can cause this complication, but oviously the more potent drugs have a greater potential for doing so than do those that are only mildly natriuretic (Table 1). Correction of this problem involves discontinuation of the diuretic; liberalization of Na^+ intake; and in cases of severe volume depletion, may include saline adminstration.

II. Hyponatremia

Patients treated with diuretics may develop hyponatremia because of the interference by the drug with the ability of the patient to generate solute-free water. Diuretics may do so in one or more of several ways: (1) directly, by impairing the process of urinary dilution at those sites in the nephron where this takes place. The major urinary dilution site is in the loop of Henle (site 2, Fig. 1) and a secondary such locus is located in the early distal convoluted tubule (site 3, Fig. 1). (2) Diuretics may compromise solute-free water generation by diminishing the delivery of salt and water to the diluting sites. This can occur either because they reduce GFR either by an effect on renal blood flow or because of severe volume contraction or by causing an enhancement of proximal tubular reabsorption. The latter is a result of volume contraction and resultant activation of neural and humoral stimuli favoring Na^+ retention as described earlier in this chapter (see above). Additionally, diuretics can cause an increase in the elaboration of ADH, thus favoring free-water reabsorption from the collecting duct rather than its excretion. They do so by initiating a volume stimulus for ADH secretion.

Hyponatremia of mild or modest degrees will respond to discontinuation of the diuretic and fluid restriction. In patients with central nervous system symptoms and signs, however, intravenous hypertonic saline infusion will be required.

III. Hypokalemia

The causes of diuretic-induced kaliuresis which underlie the development of hypokalemia have been discussed extensively in the preceding paragraphs and will not be repeated here. If the decrement in serum K^+ is modest, the diuretic may be continued if clinically necessary, and the K^+ deficit replaced. Hypokalemia should be anticipated when a therapeutic regimen causing an aggressive diuresis has been pursued. Patients may be relatively asymptomatic or may complain of fatigue and lethargy. Marked degrees of hypokalemia may result in such severe degrees of muscular weakness that paraplegia or quadriplegia and/or respiratory difficulty may occur. Electrocardiographic abnormalities include flattened T waves, the development of U waves, S-T segment depression and, on occasion, the occurrence of arrhythmias. Patients receiving digitalis are especially susceptible to the latter complication.

It has been estimated that a fall in serum K^+ from 4.0 to 3.0 mmol/l represents anywhere from a 100- to a 400-mmol total body deficit (JONES et al. 1982; SCRIBNER and BURNELL 1956). Further reductions in serum K^+ are not linearly related to the additional decrements in total body K^+. K^+ repletion can usually be performed by mouth unless the decrement is large and the patient is demonstrating major abnormalities as noted above. Because hypokalemia is usually associated with the simultaneous development of metabolic alkalosis, it should be treated with the chloride salt. In those situations in which K^+ cannot be given orally, it may be given intravenously at a rate that should generally not exceed 10 mmol/h. However, in life-threatening circumstances, up to 40 mmol/h may be given for short periods of time while the patient is simultaneously undergoing ECG monitoring (SILVER 1985). Replacement of the K^+ deficit with alkaline salts (bicarbonate, acetate, gluconate) will fail to correct and may perpetuate an accompanying metabolic alkalosis which, in turn, increases K^+ loss. Furthermore, K^+ repletion will be more difficult.

Alternatively, K^+-sparing agents may be combined with other diuretics. This has been an especially popular method for preventing or treating hypokalemia resulting from diuretic therapy of hypertension.

IV. Hypomagnesemia

A number of reports have appeared in which increased excretion of Mg^{2+} by diuretics has been documented. These have largely been acute studies in which the loop of Henle agents induced a brisk Mg^{2+} diuresis (SUKI et al. 1970; DUARTE 1968; EKNOYAN et al. 1970). This phenomenon would be expected given the fact that Mg^{2+} is largely reabsorbed in the loop of Henle (QUAMME and DIRKS 1989). The thiazides have a much more modest effect (WONG et al. 1982; PARFITT 1969; REYES and LEARY 1983). Whether the chronic use of diuretics results in a *clinically significant* decline in the Mg^{2+} concentration is a matter of controversy. This issue is complicated by the fact that serum Mg^{2+} concentrations levels are not a completely reliable indicator of total body (or tissue levels of) Mg^{2+} (ALFREY et al. 1974). While persuasive arguments have been presented on each side of this issue (for a fuller treatment, see PUSCHETT, in press) no resolution has thus far evolved. Accordingly, the author believes that, especially in the setting of myocardial disease, the possibility of Mg^{2+} depletion from diuretics should be entertained. Furthermore, therapy should be considered which does not increase the serum Mg^{2+} concentration above the normal range. When Mg^{2+} repletion is necessary, it can usually be performed orally. Magnesium oxide, 300 mg two to four times per day, is often employed with the limiting feature of the higher dosage being related to the development of diarrhea. The K^+-sparing diuretics spironolactone and amiloride also cause Mg^{2+} retention (but see Chap. 11, this volume). For more emergent situations, $MgSO_4$, 50% solution, may be utilized intramuscularly or intravenously in a dose of 2–4 ml given every 4–6 h as needed.

V. Acid-Base Disorders

1. Metabolic Alkalosis

Metabolic alkalosis is the most common acid-base disturbance resulting from the use of the thiazides and the loop of Henle agents. It is generated by a process which includes the loss of fluid containing disproportionately more chloride than bicarbonate (as compared with the composition of the ECF) as well the loss of hydrogen ions. The latter transpires because increased amounts of Na^+ are presented for exchange in the late portions of the distal nephron for *both* H^+ and K^+. The latter phenomenon is intensified by the development of secondary hyperaldosteronism in diuresed patients. The metabolic alkalosis is then sustained by the concomitant volume contraction causing the kidney to reabsorb Na^+ more avidly, associated with the available anions. Since the serum bicarbonate concentration has become elevated, filtered bicarbonate ions are readily available for reabsorption, thus helping to maintain the alkalosis (SILVER 1985). Treatment of this disorder should include measures to provide additional chloride so that the kidney can excrete bicarbonate, and/or correction of the volume contraction. In those patients in whom the diuretics cannot be discontinued for clinical reasons, KCl administration or the addition of K^+-sparing diuretics may suffice to remedy the situation, since K^+ depletion usually accompanies the alkalosis. Persistent, severe alkalemia (pH > 7.55) may be associated with the development of cardiac arrhythmias, but generally this condition is well-tolerated. Tetany and hyperactive reflexes may result from the effects of the alkaline pH on extracellular ionized Ca^{2+}, and patients may complain of muscle cramps and weakness.

2. Metabolic Acidosis

This acid-base disturbance can occur as the result of the use of the carbonic anhydrase inhibitors or with the administration of the K^+-sparing drugs. In the former circumstance, $Na^+HCO_3^-$ "reabsorption" is impaired, leading to a bicarbonaturia. Over time, therefore, the serum bicarbonate falls. In the latter situation, Na^+ exchange for *both* H^+ as well as K^+ ion is impaired. This is ordinarily not a problem unless the patient already has some limitation in the ability to excrete acid (e.g., renal insufficiency).

VI. Hyperglycemia

There have been suggestions since the late 1950s that diuretics, especially the thiazides, might be involved in the development of glucose intolerance (WILKINS 1959; GOLDNER et al. 1960; FINNERTY 1959; BRECKENRIDGE et al. 1967; LEWIS et al. 1976). Most of the earlier reports were flawed by virtue of the existence of one or more of the following problems: (a) they were uncontrolled, often anecdotal observations; (b) rather large doses of the thiazides were employed; (c) the studies did not report the ages of the subjects; and/or (d) they were performed in the aged population, in which group the incidence of glucose intolerance can be expected to increase (PUSCHETT 1993). In a prospective study performed at the Hammersmith Hospital in London, mean fasting blood sugar and 2-h glucose tolerance test values in 34 nondiabetic hypertensives treated with oral diuretics began to become abnormal at 6 years after initiation of therapy and became frankly elevated at 14 years (MURPHY et al. 1982). However, there was no control group. Interestingly, there was no correlation between elevation of the blood sugar and serum K^+ concentration. Two well-controlled, randomized studies have examined this issue. The European Working Party on Hypertension in the Elderly followed 119 elderly hypertensives for from 1 to 3 years after random assignment either to placebo or to 25–50 mg of hydrochlorothiazide and 50–100 mg of triamterene daily. Approximately half of the treated group also received 0.5–2.0 g of α-methyldopa/day. In the placebo group, no changes in fasting or 1- or 2-h glucose concentrations occurred. In patients treated with diuretics, fasting blood sugar levels rose significantly at 2 and 3 years (by 96 and 138 mg/l, respectively). There was a correlation between blood sugar elevation and serum K^+ fall at 2 years in the diuretictreated patients (AMERY et al. 1978). In the other study worthy of note, BERGLUND and ANDERSSON (1981) compared the effects on fasting blood sugar of 2.5–5.0 mg of bendroflumethiazide to 80–160 mg of propranolol/day administered randomly to two groups of previously untreated nondiabetic, middle-aged male hypertensives. Over a 6-year period, they found no differences in either fasting blood sugar or the 1-h oral glucose test between the two groups. The mean serum K^+ level of 4.0 ± 0.4 mmol/l in the thiazide-treated group was lower than that of the propanol group (4.4 ± 0.4, $p < 0.0001$), but total body K^+ levels did not differ. BENGTSSON et al. (1984) also found no difference between diuretics and β-blockers in an evaluation of more than 1000 women, with respect to the risk of developing glucose intolerance, when the effects of diuretics were compared to those of β-blockers. However, they did note an increased risk when treated groups were compared to a nontreated cohort. The conclusion one draws from a review of the available data, therefore, is that the thiazides *do* seem to cause glucose intolerance in susceptible populations of patients (e.g., the elderly) and that the risk of this complication developing is heightened by the use of large doses of the thiazides and by prolonged administration of these drugs.

The mechanism of this effect of the drugs remains unclear. Several items of information suggest that hypokalemia plays a role. First, a number of clinical conditions in which hypokalemia is a cardinal feature may include glucose intolerance (Bartter's syndrome, ectopic ACTH production, primary hyperaldosteronism and others). GORDEN (1972) studied the relationship of hypokalemia induced by the administration of 9α-fluohydrocortisone and chlorothiazide in five young, healthy volunteers. Glucose intolerance developed as hypokalemia was induced and was corrected by K^+ repletion. In a study performed on normal volunteers by HELDERMAN et al. (1983), glucose intolerance could be prevented if thiazide-induced hypokalemia was obviated by the simultaneous administration of KCl. In their evaluation of the mechanisms of the induction of abnormal glucose metabolism by the K^+ depletion, HELDERMAN et al. (1983) concluded that it was the result of decreased sensitivity of the pancreatic β cell to glucose. Other studies have reported impaired insulin secretion as well. Indeed, it has been suggested recently that hyperinsulinism and glucose intolerance are common accompaniments of essential hypertension (SOWERS 1990). Thus, SWISLOCKI et al. (1989) have provided evidence that untreated patients with hypertension demonstrate hyperinsulinism, hyperglycemia and insulin resistance compared to normotensives. Unfortunately, this study involved only a small number of patients and some K^+ concentrations were not studied.

In summary, the observations available indicate that glucose intolerance may develop in diuretic-treated patients but is not inevitable. Additionally, while K^+ depletion may be related to the development of glucose intolerance, it appears that abnormalities in glucose metabolism related to diuretic administration may occur independently of this disturbance in electrolyte balance. The mechanism appears to be both an insensitivity of the pancreatic β cell to hyperglycemia as well as tissue resistance to insulin.

Finally, it has been reported that in hypertensive diabetics treatment of the hypertension with diuretics is associated with excess mortality (Warram et al. 1991). The observations provided in this retrospective study have been criticized by Parving (1992) on several bases including the facts that neither the drugs utilized in the study nor the drug dosages were given, that the indications for diuretic use and the presence or absence of complicating conditions (CHD, CHF, etc.) were not provided and that preexisting conditions were not enumerated so that other risks for mortality in these patient groups cannot be evaluated.

VII. Hyperlipidemia

The elevation of serum cholesterol by diuretics has been discussed previously in Sect. D.I on the use of diuretics in the treatment of hypertension. The thiazides and related agents do appear to raise serum cholesterol from 7% to 13% with a mean increase over baseline values of 6%–7% when studied over periods ranging from 4 to 52 weeks following the institution of therapy (Moser 1989). When the results of longer-term studies are examined, however, ranging from 1 to 6 years, there appears to be a lack of persistent and consistent elevation of this value or else the increment is rather modest (Multiple Risk Factor Intervention Trial Resarch Group 1982; Greenberg et al. 1984; Heyden et al. 1985). Whether or not, therfore, short-term elevations in serum cholesterol and/or adverse effects on serum triglycerides and lower-density lipoproteins related to diuretic administration represent a significant increase in CHD mortality cannot be determined at present. It seems likely, however, that adoption of the lower-dose diuretic regimen advocated by JNC-V is likely to have a moderating effect on this complication.

VIII. Hyperuricemia

This complication of diuretics has recently been reviewed (Steele 1993). Hyperuricemia results from diuretic administration for principally two reasons. First, it occurs because of volume depletion which tends to elevate the level of uric acid in the blood; and, second, it results from the effects of volume depletion on the secretion of urate by the proximal tubule. Diuretic-induced volume depletion enhances urate reabsorption as much as in the case for Na^+ (Steele 1969; Steele and Oppenheimer 1969). The action of volume depletion to reduce urate secretion appears to be mediated by activation of the renin-angiotensin system resulting either in renal vasoconstriction and/or redistribution of renal blood flow (Ferris and Gorden 1968; Steele 1993). Additionally, certain of the diuretics may compete for secretion with urate along a shared proximal tubular pathway.

Although hyperuricemia occurs frequently with diuretic therapy, it seldom leads to acute gouty arthritis. However, in susceptible individuals, the likelihood of this complication is increased.

IX. Ototoxicity

The development of both vestibular and auditory abnormalities has been observed with diuretic therapy. These complications have largely been restricted to the administration of loop of Henle agents, especially furosemide and ethacrynic acid. They appear to occur most often when these agents are given in large dosages, intravenously, and in those patients who are also receiving other nephrotoxic and/or ototoxic agents (e.g., aminoglycosides). Preexistent renal insufficiency also seems to be a predisposing factor (SILVER 1985). Damage to the cochlear hair cells leading to ataxia may precede the development of hearing loss (PUSCHETT 1994). Fortunately, these complications are often (but not always) reversible when the offending agents are discontinued.

X. Nephrotoxicity

The thiazide diuretics and furosemide have been reported to cause acute interstitial nephritis (BENNETT 1986). which is thought to represent a hypersensitivity reaction. The presence of fever, eosinophilia and skin rash as well urinary eosinophils has not been constant. Accordingly, the diagnosis has been made only by biopsy in some instances. Acute renal insufficiency commonly accompanies this disease process, and chronic interstitial disease does develop. The disorder ordinarily reverses once the offending agent has been withdrawn. Corticosteroid therapy has also been associated with recovery.

XI. Hyperkalemia

The handling of K^+ by the kidney has been discussed extensively in the preceding paragraphs. Hyperkalemia from diuretic administration occurs only with the use of the K^+-sparing agents. In general, this complication can be expected to develop only in those patients who already have difficulty excreting K^+. This includes patients with preexistent renal insufficiency, patients who are volume contracted and individuals who are also receiving K^+ supplementation simultaneously. Accordingly, the K^+-sparing drugs should be used with great caution or not at all in patients with renal insufficiency and should probably never be used in the latter clinical setting in combination with K^+ supplementation. Excessive K^+ intake in the diet should also be searched for. Patients on Na^+-restricted diets who are using salt substitutes may be using preparations that contain large amounts of K^+. Unless urgent therapy is necessary, hyperkalemia will reverse once the offending agent is discontinued and/or modifications are made in the patient's diet. If symptoms and signs of K^+ toxicity are present (marked lethargy, electrocardiographic changes, etc.), emergency treatment with K^+ exchange resin, alkalinization, glucose and insulin, Ca^{2+} infusion and even dialysis may be required.

References

Agus ZS (1984) The role of diuretics in the therapy of renal stone disease. In: Puschett JB, Greenberg A (eds) Diuretics: chemistry, pharmacology and clinical applications. Elsevier, New York, pp 189–193

Alfrey AC, Miller NL, Butkus D (1974) Evaluation of body magnesium stores. J Lab Clin Med 84:153–162

Almeshari K, Ahlstrom NG, Capraro FE, Wilcox CS (1993) A volume-independent component to postdiuretic sodium retention in humans. J Am Soc Nephrol 3:1878–1883

Amery A, Bulpitt C, de Schaepdryver A, Fagard R, Hellemans J, Mutsers A, Berthaux P, Deruyttere M, Dollery C, Forette F, Lund-Johansen P, Tuomilehto J (1978) Glucose intolerance during diuretic therapy. Results of trial by the European working party on hypertension in the elderly. Lancet 1:681–683

Arroyo V, Ginès P, Planas R (1992) Treatment of ascites in cirrhosis. Diuretics, peritoneovenous shunt, and large-volume paracentesis. Gastroenterol Clin North Am 21:237–256

Belair FK, Van Denburg B, Borrelli A, Lawlor R, Panasevich R, Yelnosky J (1969) Pharmacology of SR 720-22. Arch Int Pharmacodyn Ther 177:71–87

Bell NH, Schedel MP, Bartter FC (1964) An explanation for abnormal water retention and hypo-osmolality in congestive heart failure. Am J Med 36:351–360

Bengtsson C, Blohmé G, Lapidus L, Lindquist O, Lundgren H, Nyström E, Petersen K, Sigurdsson JA (1984) Do antihypertensive drugs precipitate diabetes? Br Med J 289:1495–1497

Bennett WM (1986) Diuretic toxicity and drug interactions. In: Dirks JH, Sutton RAL (eds) Diuretics: physiology, pharmacology and clinical use. Saunders, Philadelphia, pp 362–372

Bennett W, Aronoff GR, Morrison G, Golper TA, Pulliam J, Wolfson M, Singer I (1983) Drug prescribing in renal failure: dosing guidelines for adults. Am J Kidney Dis 3:155–193

Benos DJ (1982) Amiloride: a molecular probe of sodium transport in tissues and cells. Am J Physiol 242 (Cell Physiol 11):C131–C145

Berglund G, Andersson O (1981) Beta-blockers or diuretics in hypertension? A six year follow-up of blood pressure and metabolic side effects. Lancet 1:744–747

Blendis LM, Grieg PD, Langer B, Baigrie RS, Ruse J, Taylor BR (1979) The renal and hemodynamic effects of the peritoneo-venous shunt for intractable hepatic ascites. Gastroenterology 77:250–257

Bourke E, Asbury MJA, O'Sullivan S, Gatenby PBB (1973) The sites of action of bumetanide in man. Eur J Pharmacol 23:283–289

Brater DC (1990) Diuretics. In: Williams RL, Brater DC, Mordenti J (eds) Rational therapeutics: a clinical pharmacologic guide for the health professional. Dekker, New York, pp 269–315

Brater DC (1991) Clinical pharmacology of loop diuretics. Drugs 41 [Suppl 3]:14–22

Brater DC (1992) Clinical pharmacology of loop diuretics in health and disease. Europ Heart J 13 [Suppl G]:10–14

Brater DC (1993) Resistance to diuretics: mechanisms and clinical implications. In: Advances in nephrology, vol 22. Mosby, St Louis, pp 349–369

Brater DC, Chennavasin P, Seiwell R (1980) Furosemide in patients with heart failure. Shift of the dose-response relationship. Clin Pharmacol Ther 28:182–186

Brater DC, Seiwell R, Anderson S, Burdette A, Dehmer G, Chennavasin P (1982) Absorption and disposition of furosemide in congestive heart failure. Kidney Int 22:171–176
Brater DC, Day B, Burdette A, Anderson S (1984) Bumetanide and furosemide in heart failure. Kidney Int 26:183–189
Brater DC, Anderson SA, Brown-Cartwright D (1986) Response to furosemide in chronic renal insufficiency: rationale for limited doses. Clin Pharmacol Ther 40:134–139
Breckenridge A, Dollery CT, Welborn TA, Fraser R (1967) Glucose tolerance in hypertensive patients on long-term diuretic therapy. Lancet 1:61–64
Bricker NS (1972) On the pathogenesis of the uremic state. An exposition of the "trade-off" hypothesis. N Engl J Med 286:1093–1099
Bricker NS, Fine LG, Kaplan NA, Epstein M, Bourgoignie JJ, Licht A (1978) "Magnification phenomenon" in chronic renal disease. N Engl J Med 299: 1287–1293
Brown CB, Ogg CS, Cameron JS (1981) High dose furosemide in acute renal failure: a controlled trial. Clin Nephrol 15:90–96
Buerkert J, Martin D, Prasad J, Trigg D (1981) Role of deep nephrons and the terminal collecting duct in a mannitol-induced diuresis. Am J Physiol 240: F411–F422
Burg M, Sariban-Sohraby S (1984) Control of amiloride-sensitive sodium channels in cultured renal epithelial cells. In: Puschett JB, Greenberg A (eds) Diuretics: chemistry, pharmacology and clinical applications. Elsevier, Amsterdam, pp 329–334
Burke TJ, Nobles EM, Wolf PE, Erickson AL (1983) Effect of indapamide on volume-dependent hypertension, renal haemodynamics, solute excretion and proximal nephron fractional reabsorption in the dog. Curr Med Res Opin 8 [Suppl 3]:25–37
Captopril-Digoxin Multicenter Research Group (1988) Comparative effects of therapy with captopril and digoxin in patients with mild to moderate heart failure. JAMA 259:539–544
Chennavasin P, Seiwell R, Brater DC (1980) Pharmacokinetic-dynamic analysis of the indomethacin-furosemide interaction in man. J Pharmacol Exp Ther 215: 77–81
Coe FL (1977) Treated and untreated recurrent calcium nephrolithiasis in patients with idiopathic hypercalciuria, hyperuricosuria, or no metabolic disorder. Ann Intern Med 87:404–440
Cohn JN, Archibald DG, Ziesche S, Franciosa JA, Harston WE, Tristani FE, Dunkman WB, Jacobs W, Francis GS, Flohr KH, Goldman S, Cobb FR, Shah PM, Saunders R, Fletcher RD, Loeb HS, Hughes VC, Baker B (1986) Effect of vasodilator therapy on mortality in chronic congestive heart failure: results of a Veterans Administration cooperative study. N Engl J Med 314:1547–1552
Consensus Trial Study Group (1987) Effects of enalapril on mortality in severe congestive heart failure: results of cooperative North Scandinavian enalapril survival study. N Engl J Med 316:1429–1435
Cortney MA, Mylle M, Lassiter WE, Gottschalk CW (1965) Renal tubular transport of water, solute, and PAH in rats loaded with isotonic saline. Am J Physiol 209:1199–1205
Corvol P, Claire M, Oblin ME, Geering K, Rossier B (1981) Mechanism of the antimineralocorticoid effects of spirolactones. Kidney Int 20:1–6
Costanzo LS (1984) Comparison of calcium and sodium transport in early and late rat distal tubules: effect of amiloride. Am J Physiol 247:F937–F945
Costanzo LS (1985) Localization of diuretic action in microperfused rat distal tubules: Ca and Na transport. Am J Physiol 248 (Renal Fluid Electrolyte Physiol 17): F527–F535
Costanzo LS, Weiner IM (1974) On the hypocalciuric action of chlorothiazide. J Clin Invest 54:628–637

Costanzo LS, Windhager EE (1978) Calcium and sodium transport by the distal convoluted tubule of the rat. Am J Physiol 235:F492–F506
Craswell PW, Ezzat E, Kopstein J, Varghese Z, Moorhead JF (1973) Use of metolazone, a new diuretic, in patients with renal disease. Nephron 12:63–73
Crawford JD, Kennedy GC (1959) Chlorothiazide in diabetes insipidus. Nature 183:891–892
Daly MJ, Winn H, Wilson JR (1983) Sexual responses of women, dysmenorrhea and premenstual tension. In: Willson JR, Garrington ER, Ledger WJ (eds) Obstetrics and gynecology. Mosby, St Louis, pp 96–106
Dargie HJ, Allison MEM, Kennedy AC, Gray MJB (1972) High dosage metolazone in chronic renal failure. Br Med J 4:196–198
Davison AM, Lambie AT, Virth AH, Cash JD (1974) Salt-poor albumin in the management of the nephrotic syndrome. Br Med J 1:481–484
Dikshit K, Vyden JK, Forester JS, Chatterjie K, Prakesh R, Sevan HJC (1973) Renal and extra-renal hemodynamic effects of furosemide in congestive heart failure after acute myocardial infarction. N Engl J Med 288:1087–1090
Dirks JH, Cirksena WJ, Berliner RW (1966) Micropuncture study of the effect of various diuretics on sodium reabsorption by the proximal tubule of the dog. J Clin Invest 45:1875–1885
Duarte CG (1968) Effects of ethacrynic acid and furosemide on urinary calcium, phosphate and magnesium. Metabolism 17:867–876
Earley LE, Kahn M, Orloff J (1961) The effects of infusions of chlorothiazide on urinary dilution and concentration in the dog. J Clin Invest 40:857–866
Eknoyan G, Suki WN, Martinez-Maldonado M (1970) Effect of diuretics on urinary excretion of phosphate, calcium, and magnesium in thyroparathyroidectomized dogs. J Lab Clin Med 76:257–266
Ellison DH (1991) The physiologic basis of diuretic synergism: its role in treating diuretic resistance. Ann Intern Med 114:886–894
Ellison DH (1993) Epithelial cell hypertrophy: a physiological cause of diuretic resistance. In: Puschett JB, Greenberg A (eds) Diuretics IV: chemistry, pharmacology and clinical applications. Elsevier, Amsterdam, pp 427–434
Ellison DH, Velázquez H, Wright FS (1989) Adaptation of the distal convoluted tubule of the rat: structural and functional effects of dietary salt intake and chronic diuretic infusion. J Clin Invest 83:113–126
Epstein M, Schneider NS, Befeler B (1975) Effect of intrarenal hemodynamics in acute renal failure. Am J Med 58:510–515
Ettinger B, Weil E, Mandel NS, Darling S (1979) Triamterene-induced nephrolithiasis. Ann Intern Med 91:745
Fanestil DD (1993) Regulation of the renal pharmacological receptor for thiazide-type diuretics. In: Puschett JB, Greenberg A (eds) Diuretics IV: chemistry, pharmacology and clinical applications. Elsevier, Amsterdam, pp 303–308
Feinleib M (1984) The magnitude and nature of the decrease in coronary heart disease mortality rate. Am J Cardiol 54:2C–6C
Feldman HA, Jayakumar S, Puschett JB (1978) Idiopathic edema: a review of etiologic concepts and management. Cardiovasc Med 3:475–488
Fernandez PC, Puschett JB (1973) Proximal tubular actions of metolazone and chlorothiazide. Am J Physiol 225:954–961
Ferris TF, Gorden P (1968) Effect of angiotension and norepinephrine upon urate clearance in man. Am J Med 44:359–365
Finnerty FA (1959) Discussion. In: Moyer JF (ed) Hypertension. Saunders, Philadelphia, p 653
Finnerty FA Jr, Bepko F Jr (1966) The real value of thiazides in the pregnant juvenile. Clin Res 14:108
Firth JD, Ledingham JGG (1991) Renal sodium retention in the nephrotic syndrome. Aust NZ J Med 21:893–901
Forbush B III, Palfrey HC (1983) [^{3}H]bumetanide binding to membranes isolated from dog kidney outer medulla. J Biol Chem 258:11787–11792

Frazier HS, Yager H (1973) The clinical use of diuretics. N Engl J Med 288:246–249
Fredrick MJ, Pound DC, Hall SD, Brater DC (1991) Furosemide absorption in patients with cirrhosis. Clin Pharmacol Ther 49:241–247
Friedman PA, Gesek FA (1993) Mechanism of action of thiazide diuretics on sodium and calcium transport by distal convoluted tubules. In: Puschett JB, Greenberg A (eds) Diuretics IV: chemistry, pharmacology and clinical applications. Elsevier, Amsterdam, pp 319–376
Gerbes AL (1993) Medical treatment of ascites in cirrhosis. J Hepatol 17 [Suppl 2]:S4–S9
Gerhardt RE, Abdulla AM, Mach SJ, Hudson JB (1979) Isolated ultrafiltration in the treatment of fluid overload in cardiogenic shock. Arch Intern Med 139:358–359
Gifford RW, Mattox VR, Orvis AL, Jones DA, Rosevear JW (1961) Effect of thiazide diuretics on plasma volume, body electrolytes and the excretion of aldosterone in hypertension. Circulation 24:1197–1205
Glassock RJ (1980) Sodium homeostasis in acute glomernephritis and the nephrotic syndrome. Contrib Nephrol 23:181–203
Goldner MG, Zarowitz H, Akgun S (1960) Hyperglycemia and glycosuria due to thiazide derivatives administered in diabetes mellitus. N Engl J Med 262:403–405
Golper TA (1985) Continuous ateriovenous hemofiltration in acute renal failure. Am J Kidney Dis 6:373–396
Gorden P (1972) Glucose intolerance with hypokalemia: failure of short-term potassium depletion in normal subjects to reproduce the glucose and insulin abnormalities of clinical hypokalemia. Diabetes 22:544–551
Grausz H, Lieberman R, Earley LE (1973) Effect of plasma albumin on sodium reabsorption in patients with nephrotic syndrome. Kidney Int 1:47–54
Graziani G, Cantaluppi A, Casati S, et al. (1984) Dopamine and furosemide in oliguric acute renal failure. Nephron 37:39–42
Greenberg A, Puschett JB (1991) Treatment of edematous states. In: Suki WN, Massry SG (eds) Therapy of renal diseases and related disorders, 2nd edn. Kluwer, Boston, pp 27–43
Greenberg G, Brennan PJ, Midel JE (1984) Effects of diuretic and beta-blocker therapy in the Medical Research Council trial. Am J Med 85 [Suppl 2A]:45–51
Greger R (1981) Chloride reabsorption in the rabbit cortical thick ascending limb of the loop of Henle. Pflugers Arch 390:38–43
Greger R (1984) The Na^+, K^+, $2Cl^-$ system in the diluting segment of rabbit kidney. Fed Proc 43:2473–2476
Gutsche HU, Muller-Ott K, Brunkhorst R, Niedermayer W (1983) Dose-related effects of furosemide, bumetanide, and piretanide on the thick ascending limb function in the rat. Can J Physiol Pharmacol 61:159–165
Hannafin J, Kinne-Saffran E, Friedman D, Kinne R (1983) Presence of a sodium-potassium chloride cotransport system in the rectal gland of Squalus acanthias. J Membr Biol 75:73–83
Heinemann HO, Demartini FE, Laragh JH (1959) The effect of chlorothiazide on renal excretion of electrolytes and free water. Am J Med 26:853–861
Helderman JH, Elahi D, Andersen DK, Raizes GS, Tobin JD, Shocken D, Andres R (1983) Prevention of the glucose intolerance of thiazide diuretics by maintenance of body potassium. Diabetes 32:106–111
Helgeland A (1980) Treatment of mild hypertension: a five year controlled drug trial. The Oslo Study. Am J Med 69:725–732
Henderson IS, Beattie TJ, Kennedy AC (1980) Dopamine hydrochloride in oliguric states. Lancet 2:827–829
Hensen J, Abraham WT, Dürr JA, Schrier RW (1991) Aldosterone in congestive heart failure: analysis of determinants and role in sodium retention. Am J Nephrol 11:441–446

Heyden S, Borhani NO, Tyroler HA, Schneider KA, Langford HG, Hames CG, Hutchinson R, Oberman A (1985) The relationship of weight change to changes in blood pressure, serum uric acid, cholesterol and glucose in the treatment of hypertension. J Chronic Dis 38:281–288

Higashio T, Abe Y, Yamamoto K (1978) Renal effects of bumetanide. J Pharmacol Exp Ther 207:212–220

Hilleman DE, Mohiuddin SM (1991) Changing strategies in the management of chronic congestive heart failure. DICP Ann Pharmacother 25:1349–1354

Jayakumar S, Puschett JB (1977) Study of the sites and mechanisms of action of bumetanide in man. J Pharmacol Exp Ther 201:251–258

Joint National Committee on Detection, Evaluation and Treatment of High Blood Pressure (1993) The fifth report of the Joint National Committee on Detection, Evaluation and Treatment of High Blood Pressure (JNC V). Arch Intern Med 153:154–183

Jones JW, Sebastian A, Hulter HN, Schambelan M, Sutton JM, Biglieri EG (1982) Systemic and renal acid-base effects of chronic dietary potassium depletion in humans. Kidney Int 21:402–410

Kaissling B, Stanton BA (1988) Adaptation of distal tubule and collecting duct to increased sodium delivery. I. Ultrastructure. Am J Physiol 255:F1256–F1268

Kaissling B, Bachmann S, Kriz W (1985) Structural adaptation of the distal convoluted tubule to prolonged furosemide treatment. Am J Physiol 248:F374–F381

Karlander S-G, Henning R, Lundvall O (1973) Renal effects of bumetanide, a new saluretic agent. Eur J Clin Pharmacol 6:220–233

Keller E, Hoppe-Seyler G, Schollmeyer P (1982) Disposition and diuretic effect of furosemide in nephrotic syndrome. Clin Pharmacol Ther 32:442–449

Kirchner KA, Voelker JR, Brater DC (1990) Intratubular albumin blunts the response to furosemide – a mechanism for diuretic resistance in the nephrotic syndrome. J Pharmacol Exp Ther 252:1097–1101

Kirchner KA, Voelker JR, Brater DC (1991) Binding inhibitors restore furosemide potency in tubule fluid containing albumin. Kidney Int 40:418–424

Kleinecht D, Ganeval D, Gonzalez-Duque LA, Fermanian J (1976) Furosemide in acute oliguric renal failure, a controlled trial. Nephron 17:51–58

Koeppen BM, Biagi BA, Giebisch GH (1983) Intracellular microelectrode characterization of the rabbit cortical collecting duct. Am J Physiol 244 (Renal Fluid Electrolyte Physiol 13):F35–F47

Koeppen B, Giebisch GH, Malnic G (1989) Mechanism and regulation of tubular acidification. In: Seldin DW, Giebisch GH (eds) The kidney: physiology and pathophysiology. Raven, New York, pp 1441–1525

Kunau RT Jr, Weller DR, Webb HL (1975) Classification of the site of action of chlorothiazide in the rat nephron. J Clin Invest 56:401–407

Kuroda S, Aynedjian HS, Bank N (1979) A micropuncture study of renal sodium retention in nephrotic syndrome in rats: evidence for increased resistance to tubular fluid flow. Kidney Int 16:561–571

Lahav M, Regev A, Ra'anani P, Theodor E (1992) Intermittent administration of furosemide vs continuous infusion preceded by a loading dose for congestive heart failure. Chest 102:725–731

Lant AF (1975) Effects of bumetanide on cation and anion transport. Postgrad Med J 51:35–42

Lassiter WE, Gottschalk CW, Mylle M (1961) Micropuncture study of net transtubulin nondiuretic mammalian kidney. Am J Physiol 200:1139–1146

Leppla D, Browne R, Hill K, Pak CYC (1983) Effect of amiloride with or without hydrochlorothiazide on urinary calcium and saturation of calcium salts. J Clin Endocrinol Metab 57:920–924

Lewis PJ, Kohner EM, Petrie A, Dollery CT (1976) Deterioration of glucose tolerance in hypertensive patients on prolonged diuretic treatment. Lancet 1:564–566

Liddle GW (1957) Sodium diuresis induced by steroidal antagonists of aldosterone. Science 126:1016–1018
Lieberman FL, Denison EK, Reynolds TB (1970) The relationship of plasma volume, portal hypertension, ascites and renal sodium retention in cirrhosis: the overflow theory of ascites formation. Ann NY Acad Sci 170:202–206
Linas SL, Schaefer JW, Moore EE, Good JT, Gransiracusa R (1986) Peritoneovenous shunt in the management of the hepatorenal syndrome. Kidney Int 30:736–740
Lindner A (1983) Synergism of dopamine and furosemide in diuretic-resistant oliguric acute renal failure. Nephron 33:121–126
Loon NR, Wilcox CS, Unwin RJ (1989) Mechanism of impaired natriuretic response to furosemide during prolonged therapy. Kidney Int 36:682–689
Luft FC, Brier ME, Klein K, Mann JF (1993) Pharmacodynamics and pharmacokinetics of piretanide and furosemide in patients with chronic renal insufficiency. In: Puschett JB, Greenberg A (eds) Diuretics IV: chemistry, pharmacology and clinical applications. Elsevier, Amsterdam, pp 121–125
Luke RG, Briggs JD, Allison MEM, Kennedy AC (1970) Factors determining response to mannitol in acute renal failure. Am J Med Sci 259:168
Maren TH (1974) Chemistry of the renal reabsorption of bicarbonate. Can J Physiol Pharmacol 52:1041–1050
Maren TH, Mayer E, Wadsworth BC (1954) Carbonic anhydrase inhibition. I. The pharmacology of Diamox – acetylamino-1,3,4,-thiadiazole-5-sulfonamide. Bull Johns Hopkins Hosp 95:199–255
Marone C (1993) Piretanide in the treatment of renal and hepatic oedema and nephrotic syndrome. In: Knauf H, Mutscher E (eds) The diuretic agents, vol 2, Piretanide. Marais, Carnforth, UK, pp 71–80
Marone C, Reubi FC (1980) Effects of a new diuretic piretanide compared with furosemide on renal diluting and concentrating mechanisms in patients with the nephrotic syndrome. Eur J Clin Pharmacol 17:165–171
Maschio G, Tessitore N, D'Angelo A, Fabris A, Pagano F, Tasca A, Graziani G, Aroldi A, Surian M, Colussi G, Mandressi A, Trinchieri A, Roccò F, Ponticelli C, Minetti L (1981) Prevention of calcium nephrolithiasis with low-dose thiazide, amiloride and allopurinol. Am J Med 71:623–626
Menzies DN (1964) Controlled trial of chlorothiazide in treatment of early preeclampsia. Br Med J 1:739–742
Moser M (1987) Diuretics in the management of hypertension. Med Clin North Am 71:935
Moser M (1989) Lipid abnormalities and diuretics. Am Fam Physician 40:213–220
Moser M (1993) Diuretics should continue to be recommended as initial therapy in the treatment of hypertension. In: Puschett JB, Greenberg A (eds) Diuretics IV: chemistry, pharmacology and clinical applications. Elsevier, Amsterdam, pp 465–476
Muirhead MR, Somogyi AA, Rolan PE, Bochner F (1986) Effect of cimetidine on renal and hepatic drug elimination: studies with triamterene. Clin Pharmacol Ther 40:400–407
Multiple Risk Factor Intervention Trial Research Group [MRFIT] (1982) Multiple risk factor intervention trial. Risk factor changes and mortality results. JAMA 248:1465–1477
Murphy MB, Lewis PJ, Kohner E, Schumer B, Dollery CT (1982) Glucose intolerance in hypertensive patients treated with diuretics; a fourteen-year follow-up. Lancet 2:1293–1295
O'Donovan, Muhammedi M, Puschett JB (1992) Diuretics in the therapy of hypertension: current status. Am J Med Sci 304:312–318
Olsen UB (1977) The pharmacology of bumetanide. Acta Pharmacol Toxicol 41:1–29
O'Neil RG, Boulpaep EL (1979) Effect of amiloride on the apical cell membrane cation channels of a sodium-absorbing, potassium-secreting renal epithelium. J Membr Biol 50:365–387

Oster JR, Materson BJ (1992) Renal and electrolyte complications of congestive heart failure and effects of therapy with angiotensin-converting enzyme inhibitors. Arch Intern Med 152:704–710
Parfitt AM (1969) The acute effects of mersalyl, chlorothiazide and mannitol on the renal excretion of calcium and other ions in man. Clin Sci 36:267–282
Parker S, Carlon GC, Isaacs M, Howland WS, Kahn RC (1981) Dopamine administration in oliguria and oliguric renal failure. Crit Care Med 9:630–632
Parving H-H (1992) Excess mortality associated with diuretic therapy in diabetes mellitus. Arch Intern Med 152:1093–1094
Patel KM (1981) Triamterene nephrolithiasis complicating dyazide therapy. J Urol 126:230
Paton RR, Kane RE (1977) Long-term diuretic therapy with metolazone of renal failure and nephrotic syndrome. J Clin Pharmacol 17:243–251
Pecikyan R, Kanzaki G, Berger EY (1967) Electrolyte excretion during the spontaneous recovery from the ascitic phase of cirrhosis of the liver. Am J Med 42(3):359–367
Pitt B (1992) Congestive heart failure: new therapeutic strategies. Clin Cardiol 15 [Suppl I]: I-2–I-4
Pitts RF, Kruck F, Lozano R, Taylor DW, Heidenreich OPA, Kessler RH (1958) Studies on the mechanism of diuretic action of chlorothiazide. J Pharmacol Exp Ther 123:89–97
Pitts TO (1985) Relative potency of diuretic drugs; onset and duration of action. In: Puschett JB, Greenberg A (eds) The diuretic manual. Elsevier, New York, pp 153–163
Pockros PJ, Reynolds TB (1986) Rapid diuresis in patients with ascites from chronic liver disease: the importance of peripheral edema. Gastroenterology 90:1827–1833
Pozet N, Hadj-Aissa A, Pellet M, Traeger J (1980) Activity of a new high efficiency diuretic in man: piretanide (HOE 118). Br J Clin Pharmacol 9:577–583
Puschett JB (1977) Physiologic basis for the use of new and older diuretics in congestive heart failure. Cardiovasc Med 2:119–134
Puschett JB (1981) Renal effects of bumetanide. J Clin Pharmacol 21:575–580
Puschett JB (1985) Edema formation and the physiological basis for diuretic usage. In: Puschett JB, Greenberg A (eds) The diuretic manual. Elsevier, New York, pp 3–15
Puschett JB (1989) Renal handling of calcium. In: Massry SG, Glassock NJ (eds) Textbook of nephrology, 2nd edn, vol I. Williams and Wilkins, Baltimore, pp 293–299
Puschett JB (1993) Diuretics in the therapy of hypertension in the diabetic patient. In: Puschett JB, Greenberg A (eds) Diuretics IV: chemistry, pharmacology and clinical applications. Elsevier, Amsterdam, pp 239–247
Puschett JB (1994) Diuretics in hypertension. In: Singh BN, Dzau VJ, Vanhoutte P, Woosley RL (eds) Cardiovascular pharmacology and therapeutics. Churchill Livingstone, New York, pp 885–908
Puschett JB, Goldberg M (1968) The acute effects of furosemide on acid and electrolyte excretion in man. J Lab Clin Med 71:666–677
Puschett JB, Goldberg M (1969) The relationship between the renal handling of phosphate and bicarbonate in man. J Lab Clin Med 73:956–969
Puschett JB, O'Donovan R (in press) Renal actions and uses of diuretics. In: Massry SG, Glassock RJ (eds) Textbook of nephrology, 3rd edn. Williams and Wilkins, Baltimore
Puschett JB, Rastegar A (1974) Comparative study of the effects of metolazone and other diuretics on potassium excretion. Clin Pharm Ther 15:397–405
Puschett JB, Winaver J (1992) Effects of diuretics on renal function. In: Giebisch GH, Windhager E (eds) Renal physiology, section 8, Handbook of physiology. Oxford, New York, pp 2335–2406

Puschett JB, Steinmuller SR, Rastegar A, Fernandez P (1973) Metolazone: mechanism and sites of action. In: Lant AF, Wilson GM (eds) Modern diuretic therapy in the treatment of cardiovascular and renal disease. Excerpta Medica, Amsterdam, pp 168–178

Puschett JB, Sylk D, Teredesai PR (1978) Uncoupling of proximal sodium bicarbonate from sodium phosphate transport of bumetanide. Am J Physiol 235 (Renal Fluid Electrolyte Physiol 4):F403–F408

Quamme GA, Dirks JH (1989) Renal handling of magnesium. Massry SG, Glassock RJ (eds) Textbook of nephrology, 2nd edn. Williams and Wilkins, Baltimore, p 315

Reubi FC, Cottier PT (1961) Effect of reduced glomerular filtration rate on responsiveness to chlorothiazide and mercurial diuretics. Circulation 23:200–210

Reyes AJ, Leary WP (1983) Magnesium deficiency provoked by diuretics. S Afr Med J 63:410–412

Rodriguez-Iturbé B (1984) The use of diuretics in acute renal failure. In: Puschett JB, Greenberg A (eds) Diuretics: chemistry, pharmacology and clinical applications. Elsevier, New York, pp 461–469

Rudy DW, Voelker JR, Greene PK, Esparza FA, Brater DC (1991) Loop diuretics for chronic renal insufficiency: a continuous infusion is more effective than bolus theapy. Ann Intern Med 115:360–366

Schlatter E, Greger R, Weidtke C (1983) Effect of "high ceiling" diuretics on active salt transport in the cortical thick ascending limb of Henle's loop of rabbit kidney. Pflugers Arch 396:210–217

Schneider EG, Dresser TP, Lynch RE, Knox FG (1971) Sodium reabsorption by proximal tubule of dogs with experimental heart failure. Am J Physiol 220:952–957

Schulz W, Dörfler A, Strickle L, Achhammer I (1990) Double-blind clinical trial investigating the efficacy and long-term tolerance of torasemide 200 mg p.o. compared with furosemide 500 mg p.o. and placebo p.o. in patients with chronic renal failure on hemodialysis, a multicentre study. Fischer, Stuttgart, pp 249–257 (Progress in pharmacology and clinical pharmacology, vol 8/1)

Schwartz D, Averbuch M, Pines A, Kornowski R, Levo Y (1991) Renal toxocity of enalapril in very elderly patients with progressive, severe congestive heart failure. Chest 100:1558–1561

Scribner BH, Burnell JM (1956) Interpretation of the serum potassium concentration. Metabolism 5:468–479

Seely JF, Dirks JH (1969) Micropuncture study of hypertonic mannitol diuresis in the proximal and distal tubule of the dog kidney. J Clin Invest 48:2330–2340

Shapiro MD, Hasbargen J, Hensen J, Schrier RW (1990) Role of aldosterone in the sodium retention of patients with nephrotic syndrome. Am J Nephrol 10:44–48

Shear L, Ching S, Gabuzda GJ (1970) Compartmentalization of ascites and edema in patients with hepatic cirrhosis. N Engl J Med 282:1391–1396

Shirley DG, Walter SJ, Laycock JF (1978) The role of sodium depletion in hydrochlorothiazide-induced antidiuresis in Brattleboro rats with diabetes insipidus. Clin Sci Mol Med 54:209–215

Silver MR (1985) Complications of diuretic drugs. In: Puschett JB, Greenberg A (eds) The diuretic manual. Elsevier, New York, pp 107–114

Skorecki KL, Brenner BM (1982) Body fluid homeostasis in congestive heart failure and cirrhosis with ascites. Am J Med 72:323–338

Smith DE, Gambertoglio JG, Vincenti F, Benet LZ (1981) Furosemide kinetics and dynamics after kidney transplant. Clin Pharmacol Ther 30:105–113

Smith DE, Hyneck ML, Berardi RR, Port FK (1985) Urinary protein binding, kinetics and dynamics of furosemide in nephrotic patients. J Pharm Sci 74:603–607

SOLVD Investigators (1990) Studies of left ventricular dysfunction (SOLVD) – Rationale, design and methods: two trials that evaluate the effect of enalapril in patients with reduced ejection fraction. Am J Cardiol 66:315–322

SOLVD Investigators (1990) Studies of left ventricular dysfunction (SOLVD) – Rationale, design and methods: two trials that evaluate the effect of enalapril in patients with reduced ejection fraction. Am J Cardiol 66:315–322
SOLVD Investigators (1991) Effect of enalapril on survival in patients with reduced left ventricular ejection fractions and congestive heart failure. N Engl J Med 325:293–302
Somogyi AA, Hovens CM, Muirhead MR, Bochner F (1989) Renal tubular secretion of amiloride and its inhibition by cimetidine in humans and in an animal model. Drug Metab Disp 17:190–196
Sowers JR (1990) Relationship between hypertension and subtle and overt abnormalities of carbohydrate metabolism. J Am Soc Nephrol 1:S39–S47
Stanton BA, Kaissling B (1988) Adaptation of distal tubule and collecting duct to increased Na^+ delivery. II. Na^+ and K^+ transport. Am J Physiol 255:F1269–F1275
Steele TH (1969) Evidence for altered renal urate reabsorption during changes in volume of the extracellular fluid. J Lab Clin Med 74:288–299
Steele TH (1993) Importance and pathogenesis of diuretic-induced hyperuricemia. In: Puschett JB, Greenberg A (eds) Diuretics IV: chemistry, pharmacology and clinical applications. Elsevier, Amsterdam, pp 231–238
Steele TH, Oppenheimer S (1969) Factors affecting urate excretion following diuretic administration in man. Am J Med 47:564–574
Steinmuller SR, Puschett JB (1972) Effects of metolazone in man: comparison with chlorothiazide. Kidney Int 1:169–181
Stolear IC, Achhammer I, Georges B (1990) Efficacy of torasemide in the treatment of patients with high-grade renal failure on dialysis. In: Krück F, Mutschler E, Knauf H (eds) Progress in pharmacology and clinical pharmacology, vol 8/1. Fischer, Stuttgart, pp 259–267
Streeten DHP, Dalakos TG, Souma M, Fellerman H, Clift GV, Schletter FE, Stevenson CT, Speller PJ (1973) Studies of the pathogenesis of idiopathic oedema. The roles of postural changes in plasma volume, plasma renin activity, aldosterone secretion rate and glomerular filtration rate in the retention of sodium and water. Clin Sci Mol Med 45:347–373
Stumpe KO, Reinelt B, Ressel C, Klein H, Krück F (1974) Urinary sodium excretion and proximal tubule reabsorption in rats with high-output heart failure. Nephron 12:261–274
Suki W, Rector FC Jr, Seldin DW (1965) The site of action of furosemide and other sulfonamide diuretics in the dog. J Clin Invest 44:1458–1469
Suki WN, Yium JJ, Von Minden M, Saller-Hebert C, Eknoyan G, Martinez-Maldonado M (1970) Acute treatment of hypercalcemia with furosemide. N Engl J Med 283:836
Suki WN, Dawoud F, Eknoyan G, Martinez-Maldonado M (1972) Effects of metolazone on renal function in normal man. J Pharmacol Exp Ther 180: 6–12
Sutton RAL, Drance SM (1986) Other uses of diuretics. In: Dirks JH, Sutton RAL (eds) Diuretics: physiology, pharmacology, and clinical use. Saunders, Philadelphia, pp 273–283
Swislocki ALM, Hoffman BB, Reaven GM (1989) Insulin resistance, glucose intolerance and hyperinsulinemia in patients with hypertension. Am J Hypertens 2:419–423
Szerlip HM (1991) Renal-dose dopamine: fact and fiction. Ann Intern Med 115(2): 153–154
Tarazi RC, Dustan HP, Frohlich ED (1970) Long-term thiazide therapy in essential hypertension. Circulation 41:709–717
Teredesai P, Puschett JB (1979) Acute effects of piretanide in normal subjects. Clin Pharmacol Ther 25:331–339
Toussaint C, Masselink A, Gentges A, Wambach G, Bönner G (1989) Interference of different ACE-inhibitors with the diuretic action of furosemide and hydrochlorothiazide. Klin Wochemschr 67:1138–1146

Tran JM, Farrell MA, Fanestil DD (1990) Effect of ions on binding of the thiazide-type diuretic metolazone to kidney membrane. Am J Physiol 258 (Renal Fluid Electrolyte Physiol 27):F908–F915

Tristani FE, Cohn JN (1967) Systemic and renal hemodynamics in oliguric hepatic failure: effect of volume expansion. J Clin Invest 46:1894–1906

Van Olden RW, Van Meyel JJM, Gurlay PGG (1993) Acute and long-term efficacy of high dose furosemide in hemodialysis patients with residual renal function. In: Puschett JB, Greenberg A (eds) Diuretics IV: chemistry, pharmacology and clinical applications. Elsevier, Amsterdam, pp 117–120

Van Vliet AA, Donker AJM, Nauta JJP, Verheugt FWA (1993) Spironolactone in congestive heart failure refractory to high-dose loop diuretic and low-dose angiotensin-converting enzyme inhibitor. Am J Cardiol 71:21A–28A

Villarreal H, Exaire JE, Rivollo A, Soni J (1962) Effects of chlorothiazide on systemic hemodynamics in essential hypertension. Circulation 26:405–408

Voelker JR, Cartwright-Brown D, Anderson S, Leinfelder J, Sica DA, Kokko JP, Brater DC (1987) Comparison of loop diuretics in patients with chronic renal insufficiency. Kidney Int 32:572–578

Walter SJ, Shirley DG (1986) The effect of chronic hydrochlorothiazide admimistration on renal function in the rat. Clin Sci 70:379–387

Walter SJ, Laycock F, Shirley DG (1979) A micropuncture study of proximal tubular function after acute hydrochlorothiazide administration to Battleboro rats with diabetes insipidus. Clin Sci 57:427–434

Wapnick S, Grosberg S, Kinney M, Azzara V, Le Veen HH (1978) Renal failure in ascites secondary to hepatic, renal and pancreatic disease. Arch Surg 113:581–585

Warram JH, Laffel LMB, Valsania P, Christlieb AR, Krolewski AS (1991) Excess mortality associated with diuretic therapy in diabetes mellitus. Arch Intern Med 151:1350–1356

Weber KT, Villarreal D (1993) Aldosterone and antialdosterone therapy in congestive heart failure. Am J Cardiol 71:3A–11A

Weinberger MH (1988) Diuretics and their side effects. Hypertension 2:S16–S20

Wilcox CS, Mitch WE, Kelly RA, Skorecki K, Meyer TW, Friedman PA, Souney PF (1983) Response of the kidney to furosemide. I. Effects of salt intake and renal compensation. J Lab Clin Med 102:450–458

Wilkins RW (1959) New drugs for the treatment of hypertension. Ann Int Med 50:1–10

Wilson IM, Freis ED (1959) Relationship between plasma and extracellular fluid volume depletion and the antihypertensive effect of chlorothiazide. Circulation 20:1028

Winaver J, Sylk DB, Teredesai PR, Robertson JS, Puschett JB (1980) Dissociative effects of piretanide on proximal tubular PO_4 and HCO_3 transport. Am J Physiol 238 (Renal Fluid Electrolyte Physiol 7):F60–F68

Wong NLM, Quamme GA, Sutton RAL, Dirks JH (1979) Effects of mannitol on water and electrolyte transport in the dog kidney. J Lab Clin Med 94:683–692

Wonley RJ (1986) Pregnancy-induced hypertension. In: Danforth N, Scott JR, Di Saia PJ, Hammond CB, Spellacy WN (eds) Obstetrics and gynecology. Lippincott, Philadelphia, pp 446–466

Wright FS, Giebisch GH (1992) Regulation of potassium excretion. In: Seldin DW, Giebisch G (eds) The kidney: physiology and pathophysiology, 2nd edn. Raven, New York, pp 2209–2248

Yendt ER, Cohanim M (1978) Prevention of calcium stones with thiazides. Kidney Int 13:397–409

Subject Index

Springer-Verlag and the Environment

We at Springer-Verlag firmly believe that an international science publisher has a special obligation to the environment, and our corporate policies consistently reflect this conviction.

We also expect our business partners – paper mills, printers, packaging manufacturers, etc. – to commit themselves to using environmentally friendly materials and production processes.

The paper in this book is made from low- or no-chlorine pulp and is acid free, in conformance with international standards for paper permanency.

in Hamburg, Germany

MIX
Papier aus verantwortungsvollen Quellen
Paper from responsible sources
FSC® C105338

If you have any concerns about our products,
you can contact us on
ProductSafety@springernature.com

In case Publisher is established outside the EU,
the EU authorized representative is:
Springer Nature Customer Service Center GmbH
Europaplatz 3, 69115 Heidelberg, Germany

Printed by Libri Plureos GmbH
in Hamburg, Germany